Spherical Cellulose

Regeneration, Applications, and Technological Challenges

Edited by

Mohd Shabbir and Annu

CRC Press
Taylor & Francis Group
Boca Raton London New York

CRC Press is an imprint of the
Taylor & Francis Group, an **informa** business

First edition published 2027
by CRC Press
2385 NW Executive Center Drive, Suite 320, Boca Raton FL 33431

and by CRC Press
4 Park Square, Milton Park, Abingdon, Oxon, OX14 4RN

CRC Press is an imprint of Taylor & Francis Group, LLC

ISBN: 978-1-032-97774-4 (hbk)
ISBN: 978-1-032-97919-9 (pbk)
ISBN: 978-1-003-59612-7 (ebk)

DOI: 10.1201/9781003596127

Typeset in Times
by codeMantra

Spherical Cellulose

This book addresses the processing, applications, sustainability aspects, and challenges of spherical cellulose in various size ranges. It also explores themes in cellulose solvents, functional modifications in spherical morphologies, regeneration into spherical structures, and applications with respect to their sizes are discussed, including:

- Current research and advancements to find natural and biocompatible alternatives of synthetic plastics and other carcinogenic materials are summarized as well.
- Covers fundamental and applied knowledge on sustainable and renewable biopolymers and cellulose and their various forms.
- Explains regeneration methods and processes of cellulose to reshape into spherical shapes and varying sizes for specificity in applications.
- Describes the new and ecofriendly approaches to get regenerated cellulose materials and includes pertinent technical feasibilities and economic viabilities.
- Discusses related remediation of the ecosystem via removal of pollutants.

This book is aimed at researchers and graduate students interested in cellulose, nanotechnology, and materials science.

SECTION 3 Cellulose Beads

Editor Biographies

Mohd Shabbir is Assistant Professor in the Department of Applied Sciences, CGC University, Mohali, Punjab, India. Prior to this, he had been working as a postdoctoral fellow in the lab of Professor Xiaogang Luo in the School of Chemical Engineering and Pharmacy, Wuhan Institute of Technology, Wuhan, China. He worked there on regenerated cellulose materials for environmental remediation, sensing, and biomedical applications. He has also worked as Assistant Professor at Sanskriti University, Mathura, NIET, and GLBITM, Greater Noida. He got his doctorate degree in the field of natural dyes application on textiles from Jamia Millia Islamia, New Delhi, India, in 2017. He has published several peer-reviewed scientific research articles of repute and edited six books with international publishers. Biopolymers, cellulose materials, biomaterials for adsorption and catalysis, wastewater treatment, and functional textiles are his current research interests.

Annu serves as Assistant Professor in the School of Mechanical Engineering, Yeungnam University, South Korea. She is an honorary Assistant Professor in the Department of Science and Engineering in Novel Global Community Education Foundation, Australia. Annu secured her doctoral degree in Chemistry from Jamia Millia Islamia (Central University), New Delhi. She has gained further research experience as Research Associate from the Indian Institute of Technology, Delhi. She has authored many scientific research articles in international peer-reviewed journals and book chapters with publishers of international repute. She has edited six books for Elsevier, Wiley, and Taylor & Francis. She is an active member of professional societies like ACS, KSME, APA, and IAAM. Her research focuses on the fabrication and modification of sustainable bionanocomposites, biomaterials, green synthesis of nanoparticles, hybrid nanomaterials, modification in biopolymers and composite nanomaterials for their applications in biomedical, food packaging, nutraceuticals, energy storage devices, biosensors, textiles, and environmental sustainability.

Contributors

Mahapara Abbass
Department of Civil Engineering, Maharishi Markandeshwar Engineering College, Maharishi Markandeshwar (Deemed to be University), Mullana, Ambala, Haryana, India

Oluwatosin David Abodunrin
Euromed University of Fes, UEMF, Fes, Morocco
Department of Biomedical Engineering, College of Engineering, Afe Babalola University, Ado Ekiti, Ekiti State, Nigeria

Shalom Akhai
Department of Mechanical Engineering, Maharishi Markandeshwar Engineering College, Maharishi Markandeshwar (Deemed to be University), Mullana, Ambala, Haryana, India

Ebenezer Oluwatosin Atoyebi
Department of Biomedical Engineering, School of Electrical and Systems Engineering, The Federal University of Technology, Akure, Ondo State, Nigeria
Department of Biomedical Engineering, College of Engineering and Technology, Achievers University, Owo, Ondo State, Nigeria
Department of Biomedical Engineering, Faculty of Technology, University of Ibadan, Ibadan, Oyo State, Nigeria

Rabia Azmat
Department of Zoology, Wildlife and Fisheries, University of Agriculture, Faisalabad, Pakistan

Vishwajeet Bachhar
Center for Environmental Toxin and Emerging-Contaminant Research, Cheng Shiu University, Kaohsiung 833301, Taiwan

Rajesh Bogati
AETs St. John Institute of Pharmacy and Research, Palghar, Maharashtra, India

Jitendra Basrani
Department of Mechanical Engineering, Poornima University, Rajasthan, India

Umesh Bhardwaj
School of Engineering and Technology (SET), CGC University, Mohali, Punjab, India

Naveen Kumar Reddy Bogireddy
Instituto de Ciencias Físicas, National Autonomous University of Mexico (UNAM), Mexico

Temitope T. Dele-afolabi
Institute of Power Engineering (IPE), Universiti Tenaga Nasional (UNITEN), Putrajaya Campus, Jalan IKRAM-UNITEN, Kajang, Selangor, Malaysia

Manisha Duseja
Department of Chemistry, School of Physical Sciences, DIT University, Dehradun, Uttarakhand, India

Kavitha Duraisamy
Department of Food Technology, K S Rangasamy College of Technology, Namakkal, Tamil Nadu, India

Silva Beltrán Eduardo
Instituto de Ciencias Físicas, National Autonomous University of Mexico (UNAM), Mexico

Samson Kolawole Fasogbon
Department of Biomedical Engineering, Faculty of Technology, University of Ibadan, Ibadan, Oyo State, Nigeria

Tahreem Fatima
Department of Zoology, Wildlife and Fisheries, University of Agriculture, Faisalabad, Pakistan

Mir Abdul Gaffar
Department of Wet Process Engineering, Faculty of Textile Chemical Engineering, Bangladesh University of Textiles (BUTEX), Tejgaon, Dhaka, Bangladesh

Naila Ghafoor
Department of Zoology, Wildlife and Fisheries, University of Agriculture, Faisalabad, Pakistan

Aishwarya Gorad
AETs St. John Institute of Pharmacy and Research, Palghar, Maharashtra 401404, India

Heena Gupta
G.L. Bajaj institute of Technology and Management, Greater Noida, India

Rajesh Haldhar
School of Chemical Engineering, Yeungnam University, Gyeongsan, Republic of Korea

Qamar Haneef
Department of Zoology, Wildlife and Fisheries, University of Agriculture, Faisalabad, Pakistan

Athar Adil Hashmi
Department of Chemistry, Jamia Millia Islamia, Okhla, New Delhi, India

Zarqua Hashmi
School of Pharmaceutical Education and Research, Jamia Hamdard, Hamdard Nagar, New Delhi, India

Muhammad Umar Ijaz
Department of Zoology, Wildlife and Fisheries, University of Agriculture, Faisalabad, Pakistan

Nnamdi Chukwunenye Igwe
Euromed University of Fes, UEMF, Fes, Morocco
Department of Mechatronic Engineering, University of Nigeria, Nsukka, Nigeria

Md Rashedul Islam
Department of Yarn Engineering, Faculty of Textile Engineering, Bangladesh University of Textiles (BUTEX), Tejgaon, Dhaka, Bangladesh

Maryam Jameel
Department of Zoology, Wildlife and Fisheries, University of Agriculture, Faisalabad, Pakistan

Vibha Joshi
Department of Chemistry, Graphic Era (Deemed to be University), Dehradun, Uttarakhand 248002, India

Kulbir Kaur
Department of Computer Science and Engineering, Swami Vivekanand Institute of Engineering and Technology, Banur, Punjab, India

Anil Khachi
School of Engineering and Technology (SET), CGC University, Mohali, Punjab, India

Shumaila Kiran
Department of Applied Chemistry, Government College University, Faisalabad Pakistan

Suman Lata
School of Engineering and Technology (SET), CGC University, Mohali, Punjab, India

Muntaha Mahin
Department of Yarn Engineering, Faculty of Textile Engineering, Bangladesh University of Textiles (BUTEX), Tejgaon, Dhaka, Bangladesh

Muhammad Sakander Majid
Department of Zoology, Wildlife and Fisheries, University of Agriculture, Faisalabad, Pakistan

Sachin Kumar Mandotra
Department of Biosciences and Technology (MMEC), Maharishi Markandeshwar (Deemed to be University), Mullana, Ambala, Haryana

Popat Mohite
AETs St. John Institute of Pharmacy and Research, Palghar, Maharashtra, India

Tolulope Motunrayo Omotayo
Department of Biomedical Engineering, Faculty of Technology, University of Ibadan, Ibadan, Oyo State, Nigeria
Department of Biomedical Engineering, College of Engineering and Technology, Achievers University, Owo, Ondo State, Nigeria

Ayodele James Oyejide
Department of Biomedical Engineering, College of Engineering, Afe Babalola University, Ado Ekiti, Ekiti State, Nigeria.
Department of Electrical and Electronics Engineering, Kadir Has University Istanbul, Turkey

Pramod Rakt Patel
Department of Mechanical Engineering, Indian Institute of Technology Jammu, Jammu -181221, India

Riteshkumar Patil
AETs St. John Institute of Pharmacy and Research, Palghar, Maharashtra 401 404, India

Perneta
School of Engineering and Technology (SET), CGC University, Mohali, Punjab, India

Kalai Rajan Poomalaisundar
Department of Food Technology, K S Rangasamy College of Technology, Namakkal, Tamil Nadu, India

Muhammad Zaid Salar
Department of Zoology, Wildlife and Fisheries, University of Agriculture, Faisalabad, Pakistan

Sridevi Sankar
Department of Food Technology, K S Rangasamy College of Technology, Namakkal, Tamil Nadu, India

Quratulann Sattar
Department of Zoology, Wildlife and Fisheries, University of Agriculture, Faisalabad, Pakistan

Rippin Sehgal
Department of Biotechnology, Ambala College of Engineering and Applied Research, Devsthali, Ambala, Haryana, India

Ruby Celsia Arul Selvaraj
Department of Food Technology, K S Rangasamy College of Technology, Namakkal, Tamil Nadu, India

Nithishkumar Sengottaiyan
Department of Food Technology, K S Rangasamy College of Technology, Namakkal, Tamil Nadu, India

Amar Nath Sharma
API Research and Development Unit, Pure and Cure Healthcare Pvt Ltd, Barwala, Haryana, India

Bikram Jit Singh
Department of Biosciences and Technology (MMEC), Maharishi Markandeshwar (Deemed to be University), Mullana, Ambala, Haryana, India

Gulab Singh
Department of Mechanical Engineering, University of Greenwich, London, UK

Manjeet Singh
School of Engineering and Technology (SET), CGC University, Mohali, Punjab, India

Sudarshan Singh
Faculty of Pharmacy, Chiang Mai University, Chiang Mai 50200, Thailand; Office of Research Administration, Chiang Mai University, Chiang Mai 50200, Thailand

Sukhdeep Singh
School of Engineering and Technology (SET), CGC University, Mohali, Punjab, India

Ahamed Ibrahim Syed Maghdoom
Department of Food Technology, K S Rangasamy College of Technology, Namakkal, Tamil Nadu, India

Abu Taha
Department of Chemistry, Jamia Millia Islamia, Okhla, New Delhi, India

Amara Tahir
Department of Zoology, Wildlife and Fisheries, University of Agriculture, Faisalabad, Pakistan

Rajni Verma
Department of Applied Sciences, School of Engineering and Technology (SET), CGC University, Mohali, Punjab, India

Amandeep Singh Wadhwa
Department of Mechanical Engineering, University Institute of Engineering and Technology (UIET), Panjab University, Chandigarh, India

Josue Abraham Lara Zavala
Instituto de Ciencias Físicas, National Autonomous University of Mexico (UNAM), Mexico

Preface

As researchers deeply immersed in the realm of biopolymers and sustainable materials, we have witnessed the remarkable versatility of cellulose across countless applications. Yet, while fibrous, film, and hydrogel forms of cellulose dominate scientific discourse and industrial use, the spherical morphologies ranging from nanoscale particles to micrometer-sized microspheres and millimeter-scale beads remain a specialized yet underexplored frontier. Their unique geometry confers advantages such as elevated surface area, uniform particle size, enhanced porosity, and improved diffusion kinetics, rendering them exceptionally suited for high-performance roles in catalysis, adsorption, drug delivery, environmental remediation, energy storage, and biomedical applications.

The impetus for this book arose from our collective observation that, despite burgeoning research on spherical cellulose regeneration often involving advanced solvents like ionic liquids, NaOH/urea systems, or deep eutectic mixtures, comprehensive syntheses are scarce. Fragmented publications address specific size scales or applications, but no unified resource integrates the foundational science, preparation strategies, size-dependent mechanisms, and cross-disciplinary uses. This gap hinders researchers, engineers, and industry professionals from fully harnessing spherical cellulose's potential, particularly amid growing demands for sustainable, biodegradable alternatives to synthetic materials.

We assembled this volume to provide a definitive, expert-driven compendium. Drawing from leading authorities in biopolymers, materials chemistry, environmental engineering, and biomedical sciences, the 17 chapters systematically cover the field. Through foundational sections, spherical morphologies, environmental sustainability, economic viability, and commercial sources are introduced. Then, through dedicated sections, nano-spherical cellulose (preparation, catalysis, food, and energy applications), cellulose microspheres (wastewater treatment and pharmaceutical delivery), and cellulose beads (separation, drug delivery, and wound healing from waste biomass) are explored.

This work not only consolidates current knowledge but candidly confronts technological challenges: achieving precise size control, developing scalable green regeneration processes, minimizing solvent toxicity and recovery costs, ensuring functionalization efficiency, and navigating regulatory pathways for commercialization. We are profoundly grateful to our contributors for their rigorous scholarship and to the global cellulose community whose innovations laid the groundwork.

We hope this book serves as an invaluable resource for cellulose researchers, materials scientists, environmental specialists, pharmacologists, and students. It aims to inspire advancements in ecofriendly regeneration methods and novel applications, accelerating the transition toward sustainable technologies. We welcome ongoing dialogue as this dynamic field evolves.

1 Spherical Morphologies of Cellulose
Introduction

Vibha Joshi, Vishwajeet Bachhar, Suman Lata, Manisha Duseja, and Rajesh Haldhar

1.1 INTRODUCTION

The most common biopolymer on earth is cellulose, which is the key structural element of the plant cell walls [1, 2]. Cellulose is mostly composed of the linear chains of β-(1→4)-linked with *D*-glucose units, which possesses many remarkable physicochemical properties such as renewability, biodegradability and, mechanical strength [3, 4]. Because of these properties, it is used in a broad range of industries, including the textile industry, paper, packaging, and in making biomedical products. Recent development in the nanotechnology and material sciences have enabled the transformation of cellulose into the different nanoscale structures, out of which the one of the most promising morphologies was the spherical morphology [5, 6]. The spherical cellulose particles are also known as cellulose nanospheres or microspheres that have the various distinct properties such as large surface area, tunable porosity, and shows excellent colloidal stability. These characteristics of the cellulose makes them the ideal for use in various applications such as in drug delivery, catalysis, adsorption, in cosmetics, and is also used as a filler in composite materials [7, 8]. Various spherical cellulose structures that may be synthesized by using a variety of processes such as emulsification, precipitation, spray drying, and self-assembly. The technique and circumstances used have a considerable impact on particle size, surface functioning, and consequent spherical stability. Furthermore, the use of functional groups or composite materials might improve their performance in certain applications [9, 10].

This chapter provides a fundamental review on the spherical morphologies of cellulose, which includes their structural features, synthesis methodologies, and various new emerging applications. It paves the way for a more in-depth examination of advances in this field and implicating their use for current material science and green chemistry.

1.2 EVOLUTION OF CELLULOSE TOWARD THE SPHERICAL MORPHOLOGIES

From the history, cellulose have been used mostly in the forms of fibrous and rod-like formats such as in the form of micro-fibrillated cellulose or cellulose nanocrystals

DOI: 10.1201/9781003596127-1

[11, 12]. However, with the advancement in the nanotechnology, researchers have increasingly created the accurate spherical forms of the cellulose into spherical forms such as nanospheres (10–500 nm), microspheres (micrometer range), and beads (up to millimeter size) [13, 14]. Spherical morphologies are different from the typical fibers because they have an isotropic shape, a high surface-to-volume ratio, and a uniform size. All these important qualities make them more useful for the advanced applications [15, 16]. Two types of preparation strategies have been used.

1.2.1 Top-Down Approaches

The top-down methods use the initial materials such as native cellulose, microcrystalline cellulose, or cotton fibers [17, 18]. These materials are broken down by using acid hydrolysis, enzymatic hydrolysis, or mechanical processing at various conditions such as high acid-to-cellulose ratio, short rod-like pieces lose amorphous portions and self-assemble into single crystalline, spherical cellulose nanospheres (CNS) of ~10–50 nm under certain circumstances [19, 20].

1.2.2 Bottom-Up (Regeneration) Approaches

In the bottom-up method, various techniques, such as the dissolution–regeneration technique, enable the fabrication of micron-scale cellulose beads and porous microspheres. Various parameters such as cooling temperature, stirring rate, solvent, surfactant, and oil-to-water phase ratio may produce cellulose II crystalline forms with diameters ranging from ~5 μm to 1 mm having internal porosity, which can be best suitable for chromatography or adsorbents [21, 22]. These spherical shapes of the cellulose outperform various applications such as medication administration, Pickering emulsions, and pollutant adsorption, where homogeneity, loading capacity, and steady dispersion are crucial. Figure 1.1 represents the schematic representation of the two approaches.

1.3 IMPORTANCE OF SPHERICAL CELLULOSE

Spherical cellulose particles provide a compelling set of physicochemical advantages, which helps in establishing them as a revolutionary material in the sustainable technology [23–25]. Their extremely high surface-to-volume ratio gives them abundant active sites, which are ideal for the adsorption, catalysis, and molecular interactions, which is also seen in regenerated cellulose microspheres, which are designed for enzyme immobilization and chromatographic media, where high surface area and porosity are especially critical. These particles also show the tailorable porosity, with a pore architecture, which ranges from nano to microscale and allows them for accurate encapsulation and controlled release of bioactive chemicals, dyes, or catalysts, which is used in environmental cleaning and biomedical delivery systems. The spherical design also offers better colloidal stability, due to the narrow size distribution and isotropic shapes, which also promotes uniform dispersion in formulations, smooth flow in column operations, and decreased aggregation tendencies, enhancing performance and processing efficiency [26–28].

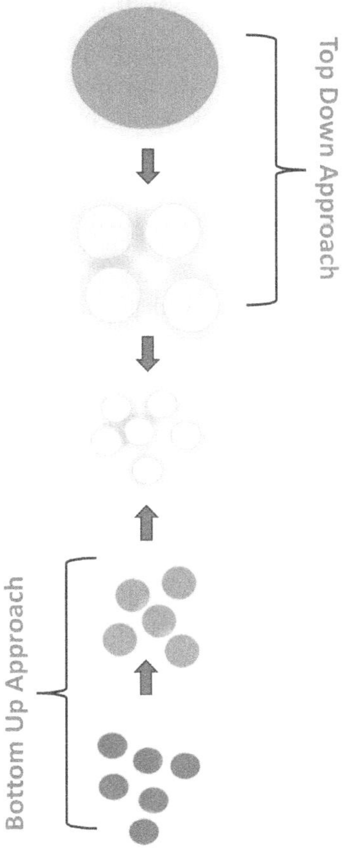

FIGURE 1.1 Diagrammatic representation of the two types of approaches.

Furthermore, cellulose has a chemically flexible surface, which is rich in hydroxyl groups, which allows them for a broad variety of functionalization [29, 30]. Through targeted chemical changes, these spheres may gain binding moieties, stimuli-responsive ligands, catalysts, or antimicrobial coatings, which makes them amenable to targeted adsorption, drug delivery, biosensing, and more. Most importantly, these particles show their non-toxic, biodegradable, and renewable cellulose foundation, which emphasizes their sustainability credentials that are ideal for green chemistry, eco-friendly product design, and circular economy applications in sectors such as medicines, environmental remediation, and energy storage. Overall, the combination of large surface area, controlled porosity, colloidal inertness, and surface tunability elevates spherical cellulose structures to high-performance carriers and functional platforms. They are especially well-suited for applications that need rapid loading and release, selective adsorption, interface stability, and environmental compatibility, which ranges from drug-release systems to adsorbent materials and chromatographic supports [31–33].

1.4 METHODS FOR FABRICATING SPHERICAL CELLULOSE

Several fabrication techniques have been developed to produce spherical cellulose structures, ranging from nanospheres to millimeter-scale beads, which offers a specific control over particle size, porosity, and morphology such as:

1.4.1 Emulsion-Based Techniques

This technique involves dispersion of cellulose solution such as dissolved in TEAOH/urea or ionic liquids into an into an immiscible oil phase, such as cyclohexane, paraffin oil, or sunflower oil, which is used as a surfactants like Span 80 or Span 85 [34, 35]. While stirring, various cellulose-rich droplets forms and solidify into porous, spherical beads in the 10–300 μm range via induced coagulation by acid (e.g., H_2SO_4) or non-solvents like ethanol or butanol. Emulsion-Based Techniques is also used for its simplicity and scalability, which requires less-specialized equipment while offers a good control over the size and distribution. Various parameters like agitation speed, oil-to-water ratio, solvent viscosity, and coagulation conditions allow for the fine-tuning of particle morphology, such as pore size, shell thickness, and internal core shell structure. However, by producing this way, these regenerated cellulose microspheres exhibit high surface area, rapid diffusion pathways, and excellent porosity, which makes them well-suited for various applications, such as drug delivery, adsorption, enzyme immobilization, or chromatographic applications.

1.4.2 Membrane Emulsification with Phase Inversion

It is an advanced and scalable method used where the cellulose solution is pushed through a hydrophobized porous membrane into an oil phase, which forms a droplet, which is then coagulated in ethanol [36, 37]. This also enables exceptionally narrow size distributions and high control over bead diameter (~μm range). Membrane emulsification technique when combined with phase inversion is used for crafting spherical cellulose microspheres with a tight size control (typically 5–20 μm). During this process, a cellulose solution is dissolved in an ionic liquids or organic electrolytes, which is gently forced through a hydrophobic microporous membrane such as porous glass or a polymer with a photolithographic pattern into a continuous oil phase that contains surfactants like Span 80. Under the controlled shear, droplets detach consistently from the pore apertures, which results in extremely monodisperse emulsions. A subsequent phase inversion with an antisolvent such as ethanol results in rapid droplet solidification into clean, spherical cellulose beads. Many researchers use post-treatment cross-linking method for e.g., with glyoxal to increase the strength and to generate sub-10 μm bead sizes with a coefficient of variation under 45%. This low-energy, low-shear technique is ideal for industrial applications and also protects sensitive payloads such as pharmaceuticals or enzymes. By adjusting membrane pore size, flow rates, and shear conditions, it produces uniform, biodegradable cellulose beads for application in food, drug delivery, cosmetics, and environmental cleaning.

1.4.3 Dropping Techniques (Jet Cutting, Syringe, Atomization)

In this technique, syringes, jet cutters, spinning-disk atomizers, or vibrating nozzles are used to extrude cellulose solution as droplets into a coagulation bath, which can be an acid or a non-solvent [38–40]. These droplets solidify into beads, which typically are of sizes from 0.5 to 3 mm. Varying nozzle diameter, flow rate, and bath conditions allows control over bead uniformity and porosity. Various dropping techniques such as jet-cutting, syringe extrusion, and atomization show a robust and scalable way to produce a spherical cellulose bead with well-defined structure and porosity in the range of 0.5–3 mm size. In this method, a cellulose-rich solution is used and prepared in solvents such as NaOH/urea or ionic liquids, which is then extruded through a controlled nozzle system. To generate uniform droplets, simple syringe pumps, jet-cutting devices, spinning-disk atomizers, or vibrating nozzles are used, these droplets fall into a coagulation bath such as in acid solutions like HNO_3 or H_2SO_4, or non-solvents such as ethanol, which induces the rapid phase inversion and solidification into spherical beads. These resulting beads possess a skin core structure where the outer layer initially coagulate fast while the interior layer remains porous, this morphology is tunable via parameters like nozzle diameter, extrusion height, solution viscosity, and coagulation bath concentration and temperature. Various techniques such as jet-cutting or spinning-disk atomization further allows for continuous production. Recent studies have shown that rotating multi-nozzle atomizing systems in ionic liquid solvents can produce micron-scale beads (40–500 μm). These dropping methods have many advantages for large-scale manufacturing as they offer excellent control over bead size distribution and porosity, which are adaptable to include active agents such as enzymes, drugs, cell-laden beads, and require a relatively straightforward setup, which makes them ideal for applications in drug delivery, biocatalysis, chromatography, and adsorbents.

1.4.4 Spray Coagulation and Spray-Drying Technique

Spray coagulation combined with spray drying is a two-step technique with high throughput approach for producing stable, porous cellulose microspheres in the range from ~1 to 3 μm without using any organic solvents or chemical crosslinkers. In this method, a cellulose-aqueous solution, which is produced from NaOH/urea or TEAOH/urea systems and is atomized into a coagulation bath, where the fast-phase inversion produces the gel-like droplets. These droplets are subsequently spray-dried, which rapidly evaporates the solvent and results in intact, hollow, or solid microspheres. This technique also produces consistently sized particles, generally ranges from 1 to 2 μm in diameter [41–43]. While composite variations, such as chitosan-coated cellulose, may reach up to the range of 3 μm. This method is an ecologically safe technique and is suitable with delicate payloads, because it does not use organic solvents or surfactants or need separate crosslinking. The spray-drying step also allows for the co-encapsulation of active substances, which permits medicines or proteins to be loaded during the manufacture, which makes it excellent for drug administration under variable pH circumstances, such as in stomach or intestinal models. Overall, this technique strikes a compromise between scalability, design flexibility, and biocompatibility, making it a viable platform for a wide range of applications, including medicines and catalysis.

1.4.5 Solvent Exchange Technique/Nanoprecipitation (Bottom-Up Regeneration)

This technique is the highly efficient and environmentally friendly method for producing the nano-to-microscale cellulose particles. Firstly, the complete dissolution of cellulose is done in the specialized solvent systems such as in sodium hydroxide/urea, *N*-methylmorpholine-*N*-oxide (NMMO), or in various ionic liquids, which disrupt the extensive hydrogen bonding in native cellulose and allows the polymer chains to disperse at the molecular-scale level [44, 45]. When a non-solvent, such as water or ethanol, is added under controlled conditions, rapid supersaturation occurs, and it results in a thermodynamically induced phase separation. The cellulose chains aggregate and regenerate during this step, which results in the stability of colloidal particles via nucleation and growth mechanisms. After that, the resulting spherical particles usually choose the cellulose II allomorph, which is a thermodynamically more stable form of the regenerated cellulose and shows highly tunable properties such as size, surface area, and porosity, which can be modulated by various variables such as polymer concentration, solvent-to-antisolvent ratio, temperature, and mixing conditions. This bottom-up regeneration approach produces uniform, biocompatible, and biodegradable cellulose spheres without the use of harsh chemicals, crosslinkers, or surfactants. Because of the controllable structure and superior physicochemical properties of these cellulose particles, these particles are widely used in various applications such as in the drug-delivery systems, bioactive agent encapsulation, environmental remediation, tissue engineering scaffolds, and stimuli–responsive materials. Furthermore, the simplicity, scalability, and solvent-recycling potential of this technique makes it very appealing for sustainable material creation in accordance with green chemistry principles [46, 47].

1.4.6 Nanoprecipitation/Emulsion–Diffusion Technique

This technique is a novel bottom-up manufacturing technique, which is widely used in the manufacturing of the homogeneous CNS and nano capsules; these CNS are ideal for high-precision nanotechnology and biomedical applications [48]. The process of making these particles usually starts with the formation of a single or double emulsion system, such as oil-in-water (O/W) or water-in-oil-in-water (W/O/W), in which cellulose or the cellulose derivatives are dissolved in a suitable organic or aqueous solvent phase. After that, the formed emulsion is then subjected to solvent diffusion or evaporation under controlled conditions, which leads to supersaturation and the subsequent nanoprecipitation of the cellulose material. Phase separation occurs at the droplet interface, as the solvent diffuses or evaporates, which allows cellulose chains to self-assemble into well-defined nanostructures. Both nanospheres and nanocapsules are formed with highly controlled morphologies, by just adjusting formulation variables, such as surfactant concentration, solvent type, emulsification energy, and stirring speed. Particle diameter in the range from 100 to 400 nm, shell thickness, surface charge, and encapsulation efficiency can all be precisely tuned. With the help of double emulsion technique, both the hydrophilic and hydrophobic active agents are encapsulated, which make it highly useful for various applications such as in targeted drug delivery, nutraceuticals, and the cosmetics industry. In summary, both techniques nanoprecipitation and emulsion–diffusion provide a strong, clean, and highly controllable pathway for creating the cellulose-based nanostructures with tailored features and putting them at the forefront of the sustainable nanomaterial development for various applications such as biomedicine, diagnostics, controlled-release systems, and functional coatings.

1.4.7 Microfluidic Techniques for Cellulose Sphere Fabrication

Microfluidic droplet-based synthesis technique is a cutting-edge and highly controlled platform for manufacturing cellulose-based micro- and nanospheres with excellent uniformity and accuracy. Unlike the traditional methods and approaches, microfluidic devices are used in microscale fluid channels to precisely control the immiscible fluid streams such as aqueous cellulose solutions and oil phases under the laminar flow conditions [49, 50]. By changing the flow rates, channel design, and fluid viscosity of the material, very monodisperse droplets of different sizes ranging from nanometers to several micrometers may be created. In this method, cellulose or its derivatives are generally dissolved in a continuous or dispersed phase and are formed into droplets utilizing flow-focusing or T-junction geometries on the microfluidic device. Once formed, these droplets solidify through phase inversion by solvent exchange or precipitation or in situ gelation method due to temperature change, pH shift, or ionic crosslinking. The end result is the formation of cellulose micro- or nanospheres with extremely narrow size distributions and customizable internal architectures. One of the most significant advantages of microfluidic techniques is their unparalleled control over droplet formation dynamics, which allows for the precise regulation of particle size, shell thickness (in the case of capsules), porosity, and encapsulation of functional agents. This makes the approach ideal for high-value applications, including medication delivery, diagnostics, microreactors, biosensors, and controlled-release formulations. Although, the capacity to alter and scale in the real time by the microchannel parallelization makes it a feasible option for industrial-scale manufacture of the customized cellulose particles. Although microfluidic systems are more complex and expensive to set up and operate than traditional batch methods, their superior precision, reproducibility, and ability to fabricate structurally sophisticated particles make them a next-generation tool in sustainable materials engineering and biomedical innovation.

1.4.8 Self-Assembly/Top-Down Fragmentation Approaches

This technique combines both controlled degradation and spontaneous rearrangement to produce CNS by providing a clean, template-free route to nanostructure production. In a method of top-down process, the bulk cellulose from the sources such as plant fibers or from microcrystalline cellulose is subjected to precisely controlled hydrolysis either by acidic (typically using sulfuric or hydrochloric acid) or by enzymatic method that results in the fragmentation of cellulose chains into the shorter, rigid rod-like nanocrystals or nanofibrils. Cellulose nanocrystals (CNCs) are rod-shaped cellulose pieces with diameters ranging from 10 to 50 nm. They have high aspect ratios and a reactive surface. At certain conditions, such as controlled drying, sonication, or changes in pH and ionic strength, these nanocrystals spontaneously self-assemble by a substantial hydrogen bonding and van der Waals interactions. This produces compact, spherical nanostructures that do not require any additional templates, surfactants, or organic solvents. The fragmented cellulose unit innate inclination to decrease the surface energy while maximizing their intermolecular contacts drives in the self-assembly process. As a result, highly uniform nanospheres can be formed with the controllable morphology, crystallinity, and surface functionality. This technique is useful for various applications that require green chemistry principles because it eliminates toxic reagents and reduces the environmental impact. Because of their high biocompatibility, renewability, and high mechanical strength, these CNS are ideal for the applications in nanocomposites, drug-delivery systems, emulsion stabilizers, and bioactive coatings. Furthermore, the simplicity, scalability, and environmental friendliness of this technology make it a potential technique for large-scale synthesis of sustainable nanomaterials in medicines, cosmetics, and advanced materials research [51, 52].

1.5 GREEN CHEMISTRY CONSIDERATIONS

Numerous efforts have been undertaken within the frameworks of the sustainable material science and green chemistry principles to develop environmentally friendly solutions for cellulose-based sphere synthesis. These efforts are critical for lowering environmental impact, improving process safety, and helps in promoting widespread industrial adoption. The main considerations are:

1.5.1 Use of Non-Toxic and Renewable Solvents

According to the green chemistry principles, the synthesis of cellulose spheres particles is increasingly based on the non-toxic, renewable solvents such as water, ethanol, deep eutectic solvents (DES), and low-toxicity ionic liquids [53, 54]. These environmentally friendly alternatives eliminate the need for toxic volatile organic compounds, by minimizing environmental and health concerns. Their biodegradability, low volatility, and potential for solvent recovery makes them ideal for long-term and scalable cellulose processing.

1.5.2 Reduction of Energy Consumption

Various processes such as spray-drying, ambient-temperature gelation, and supercritical drying methods are widely used increasingly in the production of cellulose spheres to improve the sustainability of the particles. These techniques are used widely to reduce the thermal and mechanical energy consumption of the product. High-temperature, extended drying, and high-pressure procedures are avoided where necessary. The focus is put on the low-energy mixing, gentle precipitation methods, and process optimization to reduce the total energy footprint while maintaining material quality [55, 56].

1.5.3 Biodegradable and Safe Cross-linking Agents

For the search of the safer and more sustainable methods, the biodegradable and organically produced cross-linking agents are preferred in synthesizing the cellulose sphere. Various natural crosslinkers such as citric acid, genipin, and tannic acid are viable alternatives for the traditional toxicants like glutaraldehyde. These bio-based crosslinkers not only offer the necessary structural stability, but also shows the great biocompatibility that makes them ideal for biomedical, pharmacological, and food applications. These applications support green chemistry that aims at lowering toxicity, increasing biodegradability, and encouraging ecologically responsible material design [57].

1.5.4 Minimization of Chemical Waste and Byproducts

For the synthesis of particles, various techniques such as sustainable and eco-friendly methods are used; these processes are carefully optimized to minimize the volume of the chemical reagents by minimizing waste generation and environmental burden. The main focus is placed on an atom economy, where the reactions are designed to incorporate the maximum proportion of the reactants into the final product. Additionally, the use of renewable, bio-based feedstocks is encouraged to replace petroleum-derived chemicals. These practices not only lower the ecological footprint of cellulose sphere production but also enhance the overall efficiency and sustainability of the process [58, 59].

1.5.5 Process Scalability and Industrial Compatibility

For continuous or semi-continuous manufacturing systems, various techniques are designed to be scalable and adaptable for the production of cellulose spheres. These techniques ensure the consistent product quality and the efficiency of the product at larger volumes. Additionally, these green methods are rigorously evaluated for their economic viability, operational simplicity, and compatibility with the existing industrial infrastructure, facilitating smoother integration into the commercial production of pipelines while maintaining environmental sustainability [60, 61].

1.6 INFLUENCE OF PROCESSING PARAMETERS

The form, size, structural integrity, and functional qualities of the cellulose-based spheres are heavily impacted by processing conditions. Researchers can fine-tune these variables to tailor the final material properties for specific applications ranging from drug delivery to environmental remediation. The key determining elements are:

1.6.1 Solvent Type

The choice of solvent plays a crucial role in determining the cellulose solubility, chain mobility, and regeneration kinetics of the cellulose. Solvents such as ionic liquids, DES, and aqueous NaOH/urea systems can differently affect the formation of cellulose networks, ultimately impacting sphere size, crystallinity, and uniformity.

1.6.2 Cross-Linking Agents

Cross-linking agents play a crucial role in the structural integrity and functional performance of the cellulose spheres. The nature and concentration of the cross-linkers have a direct impact on the mechanical strength, thermal stability, and internal porosity of the finished product. Biodegradable and non-toxic crosslinkers such as citric acid, genipin, or tannic acid are combined to strengthen the cellulose network and also ensure biocompatibility, which makes them ideal for biomedical, pharmaceutical, and food-grade applications. Crosslinker agents must be properly selected and tuned to provide the appropriate balance of the stability and environmental safety [62, 63].

1.6.3 Drying Techniques

Techniques such as spray-drying, freeze-drying, or supercritical drying are used to determine the final textural properties and surface area of the cellulose spheres, which are used for applications in catalysis and controlled release. These methods have a pronounced impact on the surface morphology, internal pore architecture, and in the shrinkage behavior of cellulose spheres [64, 65].

1.6.4 Surfactants and Stabilizers

In the emulsion-based production of cellulose spheres, both surfactants and stabilizers play an important role in providing them with stability and particle homogeneity. Surfactants such as Tween 80, Span 60, and other amphiphilic agents help form droplets, reduce interfacial tension, and help prevent dispersed phases from coalescing. Stabilizers are used to improve the stability of the emulsion by ensuring that the forming spheres are in the same size and shape. The type and amount of these additives are the key factors in determining the particle size distribution, surface smoothness, and overall monodispersity of the particles. Proper selection enables precise control over sphere morphology, which is particularly important in applications that require uniformity, such as drug delivery, cosmetics, and diagnostics [66, 67].

1.7 ADVANTAGES OF CELLULOSE SPHERES

Cellulose-based spherical particles that range in size from micro to nanoscale have emerged as promising material, because of their distinct structural and functional properties. Their production takes use of the inherent benefits of cellulose, a naturally plentiful, biodegradable polymer, while the spherical shape increases surface area, dispersibility, and mechanical stability. These characteristics benefit a broad range of applications in biological, pharmacological, agricultural, cosmetic, and environmental domains. The main benefits of cellulose spheres follow.

1.7.1 High Loading Capacity

For encapsulating and delivering a broad variety of active compounds, the interior porosity and the large surface area of the cellulose spheres make them ideal for use in encapsulating medicines, enzymes, proteins, insecticides, dyes, and heavy metal ions. These contain both hydrophilic and hydrophobic molecules, which allows them multifunctional loading capacity and increasing the effectiveness of medicinal or remediation agents. These characteristics are useful in controlled delivery systems, where great payload efficiency is critical [68].

1.7.2 Controlled and Targeted Release

Cellulose spheres are designed to exhibit the sustained, delayed, or stimuli–responsive release patterns in response to the environmental indicators such as pH, temperature, or enzyme activity. These properties make them excellent for various pharmaceutical medication administration, and also enables longer-lasting therapeutic benefits and lower dose rates. In agriculture, they act as slow-release transporters for fertilizers and insecticides, resulting in more efficient usage and less environmental runoff. The ability to control release kinetics by adjusting cross-linking density, particle size, or surface functionality increases flexibility significantly.

1.7.3 Biocompatibility and Nontoxicity

Because cellulose is naturally sourced and devoid of dangerous chemicals, cellulose spheres have good biocompatibility, low immunogenicity, and nontoxicity, making them suitable for use in biomedical implants, drug carriers, wound healing matrixes, and tissue engineering scaffolds. They degrade into non-toxic byproducts, removing the possibility of bioaccumulation or long-term side effects, which is required for regulatory approval in healthcare and food industries.

1.7.4 Renewable, Biodegradable, and Environmentally Safe

One of the most significant advantages of the cellulose spheres is their conformity with the green chemistry principles and sustainable development goals. Cellulose derived from the plant biomass is renewable, biodegradable, and carbon-neutral, which makes it ideal for reducing the environmental impact of consumer products. Upon disposal, cellulose-based products decompose naturally without leaving toxic residues that help in promoting eco-friendly product life cycles and circular economy models [69].

1.7.5 Structural Tunability and Versatility

Cellulose spheres are produced with the fine control over particle size, surface morphology, porosity, and mechanical strength, and also enables for the customization to suit specific end-use requirements. Functionalization with active groups, additive inclusion, or polymer coating broadens their possibilities across industries such as biosensing, catalysis, cosmetics, and food packaging.

1.8 FUTURE PERSPECTIVES AND APPLICATIONS

The increasing demand for sustainable and functional materials has accelerated research in spherical cellulose. Key application areas include:

1.8.1 Pharmaceuticals

In this sector, the cellulose spheres are used as drug-delivery vehicles, which helps in controlled or in the targeted release properties for various medicinal agents. The biocompatibility and the non-toxic nature make them excellent for application in wound healing, tissue scaffolds, and transdermal administration systems. Future research may focus on stimuli–responsive drug delivery, in which release is triggered by environmental changes like pH, temperature, or enzymes [70].

1.8.2 Environmental Remediation

Cellulose spheres are attractive adsorbents, because of their large surface area and changeable functional groups, which are used for heavy metals, dyes, oils, and other organic contaminants. Future study is projected to provide the functionalized or composite spheres with the improved selectivity and adsorption efficiency, hence this contributes to the clean water technologies and pollution management.

1.8.3 Food Science

Spherical cellulose particles are being investigated in the food sector for their ability to encapsulate flavors, probiotics, essential oils, and preservatives. These spheres protect the sensitive bioactives from the degradation, also increases the shelf life, and also allows for the regulated release during the digestion. This research strives to improve the food safety, functionality, and nutritional delivery.

1.8.4 Biotechnology

Cellulose microspheres and nanospheres are commonly used to immobilize the enzymes, proteins, and microbial cells, which helps in various applications such as biocatalysis, biosensing, and bioprocessing. Their stability, adaptability, and reusability make them the excellent for high-performance biotechnological systems.

1.9 CONCLUSION

This chapter helps in providing an in-depth analysis of the idea, manufacturing methodologies, processing parameters, benefits, and various potential applications of the spherical cellulose morphologies. The change of cellulose from its traditional fibrous structure to well-defined micro- and nanospheres constitutes for a major step forward in the area of sustainable materials and research. Researchers can now precisely customize the size, porosity, and functionality of the cellulose spheres by using a range of synthetic methods, which include the solvent exchange method, nanoprecipitation, microfluidics and self-assembly. The increased interest in cellulose spheres stems from their biodegradability, biocompatibility, renewable origin, and structural plasticity, which makes them appealing for various applications in medicines, food science, biotechnology, and environmental remediation. Their capacity in encapsulating the active agents, distribute compounds in a regulated way, and interact safely with biological systems propels them to the forefront the green and smart material research. In terms of the green chemistry and circular economy principles, spherical cellulose structures stand out as ideal materials for future sustainable technologies. The use of non-toxic solvents, biodegradable crosslinkers, energy-efficient procedures, and waste-reduction methods enhances their environmental appeal. In conclusion, spherical cellulose morphologies are not only a scientific discovery, but also a symbol of a larger transition toward sustainable, renewable, and environmentally friendly material solutions. As technology advances, bio-based spheres will play an increasingly important role in tackling global challenges like as health, the environment, and resource sustainability.

REFERENCES

1. S. Bhaladhare and D. Das, "Cellulose: A fascinating biopolymer for hydrogel synthesis," *J Mater Chem B*, vol. 10, no. 12, pp. 1923–1945, 2022, https://doi.org/10.1039/D1TB02848K.
2. B. Zhang, Y. Gao, L. Zhang, and Y. Zhou, "The plant cell wall: Biosynthesis, construction, and functions," *J Integr Plant Biol*, vol. 63, no. 1, pp. 251–272, Jan. 2021, https://doi.org/10.1111/jipb.13055.
3. B. Joseph, S. V K, C. Sabu, N. Kalarikkal, and S. Thomas, "Cellulose nanocomposites: Fabrication and biomedical applications," *J Bioresour Bioprod*, vol. 5, no. 4, pp. 223–237, 2020, https://doi.org/10.1016/j.jobab.2020.10.001.
4. H. Tu, M. Zhu, B. Duan, and L. Zhang, "Recent progress in high-strength and robust regenerated cellulose materials," *Adv. Mater.*, vol. 33, no. 28, p. 2000682, Jul. 2021, https://doi.org/10.1002/adma.202000682.
5. D. Trache, M. H. Hussin, M. K. M. Haafiz, and V. K. Thakur, "Recent progress in cellulose nanocrystals: Sources and production," *Nanoscale*, vol. 9, no. 5, pp. 1763–1786, 2017, https://doi.org/10.1039/C6NR09494E.
6. B. L. Peng, N. Dhar, H. L. Liu, and K. C. Tam, "Chemistry and applications of nanocrystalline cellulose and its derivatives: A nanotechnology perspective," *Can J Chem Eng*, vol. 89, no. 5, pp. 1191–1206, Oct. 2011, https://doi.org/10.1002/cjce.20554.
7. H. Ullah, H. A. Santos, and T. Khan, "Applications of bacterial cellulose in food, cosmetics and drug delivery," *Cellulose*, vol. 23, no. 4, pp. 2291–2314, 2016, https://doi.org/10.1007/s10570-016-0986-y.
8. Y.-Y. Li, B. Wang, M.-G. Ma, and B. Wang, "Review of recent development on preparation, properties, and applications of cellulose-based functional materials," *Int J Polym Sci*, vol. 2018, no. 1, p. 8973643, Jan. 2018, https://doi.org/10.1155/2018/8973643.
9. S. Soltani, N. Khanian, T. R. Shojaei, N. Asim, Y. Zhao, and T. S. Y. Choong, "Synthesis, characterizations, functionalizations, and biomedical applications of spherical cellulose nanoparticles," in *Handbook of Nanocelluloses: Classification, Properties, Fabrication, and Emerging Applications*, A. Barhoum, Ed., Cham: Springer International Publishing, 2022, pp. 177–200. https://doi.org/10.1007/978-3-030-89621-8_11.

10. A. Stunda-Zujeva, Z. Irbe, and L. Berzina-Cimdina, "Controlling the morphology of ceramic and composite powders obtained via spray drying – A review," *Ceram Int*, vol. 43, no. 15, pp. 11543–11551, 2017, https://doi.org/10.1016/j.ceramint.2017.05.023.
11. M. M. de Souza Lima and R. Borsali, "Rodlike cellulose microcrystals: Structure, properties, and applications," *Macromol Rapid Commun*, vol. 25, no. 7, pp. 771–787, Apr. 2004, https://doi.org/10.1002/marc.200300268.
12. W. Hamad, "On the development and applications of cellulosic nanofibrillar and nanocrystalline materials," *Can J Chem Eng*, vol. 84, no. 5, pp. 513–519, Oct. 2006, https://doi.org/10.1002/cjce.5450840501.
13. J. P. F. Carvalho, A. C. Q. Silva, A. J. D. Silvestre, C. S. R. Freire, and C. Vilela, "Spherical cellulose micro and nanoparticles: A review of recent developments and applications," *Nanomaterials*, vol. 11, no. 10, 2744, Oct. 2021, https://doi.org/10.3390/nano11102744.
14. P. N. Kendre, D. Lokwani, A. Pote, S. Singh, T. J. Jayeoye, and B. G. Prajapati, "Potential technologies to develop cellulose beads and microspheres," in *Regenerated Cellulose and Composites: Morphology-Property Relationship*, M. Shabbir, Ed., Singapore: Springer Nature Singapore, 2023, pp. 145–176. https://doi.org/10.1007/978-981-99-1655-9_6.
15. L. Yang, Z. Zhou, J. Song, and X. Chen, "Anisotropic nanomaterials for shape-dependent physicochemical and biomedical applications," *Chem Soc Rev*, vol. 48, no. 19, pp. 5140–5176, 2019, https://doi.org/10.1039/C9CS00011A.
16. U. Ulusoy, "A review of particle shape effects on material properties for various engineering applications: From macro to nanoscale," Jan. 01, 2023, *MDPI*. https://doi.org/10.3390/min13010091.
17. Q. Qin et al., "'Bottom-up' and 'top-down' strategies toward strong cellulose-based materials," *Chem Soc Rev*, vol. 53, no. 18, pp. 9306–9343, 2024, https://doi.org/10.1039/D4CS00387J.
18. T. C. Mokhena and M. J. John, "Cellulose nanomaterials: New generation materials for solving global issues," *Cellulose*, vol. 27, no. 3, pp. 1149–1194, 2020, https://doi.org/10.1007/s10570-019-02889-w.
19. S. L. Yupanqui-Mendoza and V. Arantes, "Recent advances in cellulose nanospheres: Production, formation mechanisms, properties and applications," *Cellulose*, vol. 31, no. 18, pp. 10609–10649, 2024, https://doi.org/10.1007/s10570-024-06229-5.
20. Y. Habibi, "Key advances in the chemical modification of nanocelluloses," *Chem Soc Rev*, vol. 43, no. 5, pp. 1519–1542, 2014, https://doi.org/10.1039/C3CS60204D.
21. Q. Zhang, W. Zhang, Z. Ding, X. Wang, and J. Huang, "Simple preparation of cellulose microspheres based on a regeneration process," *Int J Biol Macromol*, vol. 311, p. 143352, 2025, https://doi.org/10.1016/j.ijbiomac.2025.143352.
22. Y. Zhang, K. Kobayashi, R. Kusumi, U.-J. Kim, and M. Wada, "Adsorption properties of cellulose beads prepared by emulsion-gelation method and subsequent oxidation," *Colloids Surf A Physicochem Eng Asp*, vol. 682, p. 132859, 2024, https://doi.org/10.1016/j.colsurfa.2023.132859.
23. A. Singh et al., "Unveiling the transformative power of smart cellulosic nanomaterials: Revisiting potential promises to sustainable future," in *Functionalized Cellulose Materials: Sustainable Manufacturing and Applications*, K. K. Kesari, C. Prakash, M. Khalid, and A. Negi, Eds., Cham: Springer Nature Switzerland, 2025, pp. 1–42. https://doi.org/10.1007/978-3-031-76953-5_1.
24. X. Li et al., "An overview of the development status and applications of cellulose-based functional materials," *Cellulose*, vol. 31, no. 1, pp. 61–99, 2024, https://doi.org/10.1007/s10570-023-05616-8.

25. T. Li et al., "Developing fibrillated cellulose as a sustainable technological material," *Nature*, vol. 590, no. 7844, pp. 47–56, 2021, https://doi.org/10.1038/s41586-020-03167-7.
26. E. J. Foster et al., "Current characterization methods for cellulose nanomaterials," *Chem Soc Rev*, vol. 47, no. 8, pp. 2609–2679, 2018, https://doi.org/10.1039/C6CS00895J.
27. J. Y. Zhu et al., "Towards sustainable production and utilization of plant-biomass-based nanomaterials: A review and analysis of recent developments," *Biotechnol Biofuels*, vol. 14, no. 1, p. 114, 2021, https://doi.org/10.1186/s13068-021-01963-5.
28. P. Samyn, A. Barhoum, T. Öhlund, and A. Dufresne, "Review: Nanoparticles and nanostructured materials in papermaking," *J Mater Sci*, vol. 53, no. 1, pp. 146–184, 2018, https://doi.org/10.1007/s10853-017-1525-4.
29. K. Bethke et al., "Functionalized cellulose for water purification, antimicrobial applications, and sensors," *Adv Funct Mater*, vol. 28, no. 23, p. 1800409, Jun. 2018, https://doi.org/10.1002/adfm.201800409.
30. D. Zhao, Y. Zhu, W. Cheng, W. Chen, Y. Wu, and H. Yu, "Cellulose-based flexible functional materials for emerging intelligent electronics," *Adv Mater*, vol. 33, no. 28, p. 2000619, Jul. 2021, https://doi.org/10.1002/adma.202000619.
31. D. Klemm, B. Heublein, H.-P. Fink, and A. Bohn, "Cellulose: Fascinating biopolymer and sustainable raw material," *Angew Chem Int Ed*, vol. 44, no. 22, pp. 3358–3393, May 2005, https://doi.org/10.1002/anie.200460587.
32. B. Thomas et al., "Nanocellulose, a versatile green platform: From biosources to materials and their applications," *Chem Rev*, vol. 118, no. 24, pp. 11575–11625, Dec. 2018, https://doi.org/10.1021/acs.chemrev.7b00627.
33. Q. Yu et al., "Multifunctional composite capsules in drug delivery systems: Bridging pharmaceutical and biomedical applications," *Adv Compos Hybrid Mater*, vol. 8, no. 1, p. 118, 2025, https://doi.org/10.1007/s42114-024-01203-y.
34. S. Taokaew and W. Kriangkrai, "Recent progress in processing cellulose using ionic liquids as solvents," Dec. 01, 2022, *Multidisciplinary Digital Publishing Institute (MDPI)*. https://doi.org/10.3390/polysaccharides3040039.
35. M. Mir and S. Ghasemirad, "Phase inversion emulsification of paraffin oil/polyethylene wax blend in water: A comparison between mixed monomeric and monomeric/gemini surfactant systems," *J Mol Liq*, vol. 359, p. 119315, 2022, https://doi.org/10.1016/j.molliq.2022.119315.
36. J. Lee, K. Y. Cho, E. H. Kim, and H. Lee, "Optimizing cellulose acetate microbead formation through premixed membrane emulsification: Unraveling critical parameters," *Carbohydr Polym Technol Appl*, vol. 8, p. 100545, 2024, https://doi.org/10.1016/j.carpta.2024.100545.
37. C. Charcosset, I. Limayem, and H. Fessi, "The membrane emulsification process—A review," *J Chem Technol Biotechnol*, vol. 79, no. 3, pp. 209–218, Mar. 2004, https://doi.org/10.1002/jctb.969.
38. A. Ali, U. T. Syed, T. S. Bak, and C. A. Quist-Jensen, "Membrane emulsification—A novel solution for treatment and reuse of produced water from oil field," *Membranes (Basel)*, vol. 12, no. 10, Oct. 2022, https://doi.org/10.3390/membranes12100971.
39. H. Bazyar, N. van de Beek, and R. G. H. Lammertink, "Liquid-infused membranes with oil-in-water emulsions," *Langmuir*, vol. 35, no. 29, pp. 9513–9520, Jul. 2019, https://doi.org/10.1021/acs.langmuir.9b01055.
40. A. B. Mapossa, R. K. Tewo, S. S. Ray, W. Mhike, and U. Sundararaj, "A review on thermally induced phase separation technology in the fabrication of microporous polymer membrane devices for sustained-repellent delivery: Crystallization and morphological studies," *J Polym Sci*, vol. 62, no. 23, pp. 5205–5231, Dec. 2024, https://doi.org/10.1002/pol.20240232.

41. S. Wintzheimer et al., "Multifunctional, hybrid materials design via spray-drying: Much more than just drying," *Adv Mater*, vol. 35, no. 47, p. 2306648, Nov. 2023, https://doi.org/10.1002/adma.202306648.
42. Q. Wang et al., "Preparation of cellulose based microspheres by combining spray coagulating with spray drying," *Carbohydr Polym*, vol. 111, pp. 393–399, 2014, https://doi.org/10.1016/j.carbpol.2014.05.002.
43. N. A. Noguez Méndez et al., "Chapter 22- Design and development of pharmaceutical microprocesses in the production of nanomedicine," in *Nanostructures for Oral Medicine*, E. Andronescu and A. M. Grumezescu, Eds., Elsevier, 2017, pp. 669–697. https://doi.org/10.1016/B978-0-323-47720-8.00023-7.
44. R. P. Singh, and K. R. B. Singh, Eds., *Bionanomaterials*. Fundamentals and biomedical applications, IOP Publishing, 2021. https://doi.org/10.1088/978-0-7503-3767-0.
45. H. Zhou et al., "Dissolving-co-catalytic strategy for the preparation of flexible and wet-stable cellulose membrane towards biodegradable packaging," *Int J Biol Macromol*, vol. 275, p. 133454, 2024, https://doi.org/10.1016/j.ijbiomac.2024.133454.
46. D. Khorsandi et al., "Catalytic and biomedical applications of nanocelluloses: A review of recent developments," *Int J Biol Macromol*, vol. 268, p. 131829, 2024, https://doi.org/10.1016/j.ijbiomac.2024.131829.
47. S. Das, B. Ghosh, and K. Sarkar, "Nanocellulose as sustainable biomaterials for drug delivery," *Sens Int*, vol. 3, p. 100135, 2022, https://doi.org/10.1016/j.sintl.2021.100135.
48. C. J. Martínez Rivas et al., "Nanoprecipitation process: From encapsulation to drug delivery," *Int J Pharm*, vol. 532, no. 1, pp. 66–81, 2017, https://doi.org/10.1016/j.ijpharm.2017.08.064.
49. F. Fontana, M. P. A. Ferreira, A. Correia, J. Hirvonen, and H. A. Santos, "Microfluidics as a cutting-edge technique for drug delivery applications," *J Drug Deliv Sci Technol*, vol. 34, pp. 76–87, 2016, https://doi.org/10.1016/j.jddst.2016.01.010.
50. R. Fan et al., "Droplet-based microfluidics for drug delivery applications," *Int J Pharm*, vol. 663, p. 124551, 2024, https://doi.org/10.1016/j.ijpharm.2024.124551.
51. A. Abbasi Moud and A. Abbasi Moud, "Cellulose Nanocrystals (CNC) liquid crystalline state in suspension: An overview," Dec. 01, 2022, *Multidisciplinary Digital Publishing Institute (MDPI)*. https://doi.org/10.3390/applbiosci1030016.
52. S. Yadav, A. K. Sharma, and P. Kumar, "Nanoscale self-assembly for therapeutic delivery," Feb. 25, 2020, *Frontiers Media S.A.* https://doi.org/10.3389/fbioe.2020.00127.
53. E. Carasek, L. Morés, N. da Cunha, G. L. Branco, and A. N. Dias, "Natural deep eutectic solvents and natural solid biosorbents in microextraction techniques of organic compounds based on chromatographic techniques: A critical review," *Adv Sample Prep*, vol. 12, p. 100139, 2024, https://doi.org/10.1016/j.sampre.2024.100139.
54. S. A. Naziri Mehrabani, V. Vatanpour, and I. Koyuncu, "Green solvents in polymeric membrane fabrication: A review," *Sep Purif Technol*, vol. 298, p. 121691, 2022, https://doi.org/10.1016/j.seppur.2022.121691.
55. T. Mai, P.-L. Wang, and M.-G. Ma, "Promising cellulose-based aerogel composites: Preparation methods and advanced applications," *Compos B Eng*, vol. 281, p. 111559, 2024, https://doi.org/10.1016/j.compositesb.2024.111559.
56. S. A. Bahadi, A. H. Hassan, and S. A. Onaizi, "Recent advances in CO_2 adsorption using aerogel materials: A review," *Next Mater*, vol. 7, p. 100388, 2025, https://doi.org/10.1016/j.nxmate.2024.100388.
57. P. Sapuła, K. Bialik-Wąs, and K. Malarz, "Are natural compounds a promising alternative to synthetic cross-linking agents in the preparation of hydrogels?," Jan. 01, 2023, *MDPI*. https://doi.org/10.3390/pharmaceutics15010253.
58. Rijwan, S. Kumar, and S. Kumar, "Green and sustainable approaches in the synthesis of pharmaceutically relevant quinazolinones," *Tetrahedron*, vol. 172, p. 134437, 2025, https://doi.org/10.1016/j.tet.2024.134437.

59. J. C. Slootweg, "Sustainable chemistry: Green, circular, and safe-by-design," *One Earth*, vol. 7, no. 5, pp. 754–758, 2024, https://doi.org/10.1016/j.oneear.2024.04.006.
60. R. Zhang and T. Sun, "Ink-based additive manufacturing for electrochemical applications," *Heliyon*, vol. 10, no. 12, p. e33023, 2024, https://doi.org/10.1016/j.heliyon.2024.e33023.
61. C. Costa and L. Padrela, "Progress on drug nanoparticle manufacturing: Exploring the adaptability of batch bottom-up approaches to continuous manufacturing," *J Drug Deliv Sci Technol*, vol. 111, p. 107120, 2025, https://doi.org/10.1016/j.jddst.2025.107120.
62. S. Hussain and S. S. Maktedar, "Structural, functional and mechanical performance of advanced Graphene-based composite hydrogels," *Results Chem*, vol. 6, p. 101029, 2023, https://doi.org/10.1016/j.rechem.2023.101029.
63. H. Nasution et al., "Hydrogel and effects of crosslinking agent on cellulose-based hydrogels: A review," Sep. 01, 2022, *MDPI*. https://doi.org/10.3390/gels8090568.
64. V. Baudron, P. Gurikov, I. Smirnova, and S. Whitehouse, "Porous starch materials via supercritical-and freeze-drying," *Gels*, vol. 5, no. 1, Mar. 2019, https://doi.org/10.3390/gels5010012.
65. B. Merillas, M. I. Roque, C. M. R. Almeida, M. Á. Rodríguez-Pérez, and L. Durães, "Insights on the influence of the drying method and surface wettability on the final properties of silica aerogels," *Gels*, vol. 11, no. 7, 511, Jul. 2025, https://doi.org/10.3390/gels11070511.
66. E. Tenorio-Garcia, A. Araiza-Calahorra, E. Simone, and A. Sarkar, "Recent advances in design and stability of double emulsions: Trends in Pickering stabilization," *Food Hydrocoll*, vol. 128, p. 107601, 2022, https://doi.org/10.1016/j.foodhyd.2022.107601.
67. R. K. Ameta, K. Soni, and A. Bhattarai, "Recent advances in improving the bioavailability of hydrophobic/lipophilic drugs and their delivery via self-emulsifying formulations," Mar. 01, 2023, *MDPI*. https://doi.org/10.3390/colloids7010016.
68. E. Shamloo et al., "Encapsulation of natural/bio-based materials as nano bio-adsorbents for removal of toxic pollutants from food products: A review," *Carbohydr Polym Technol Appl*, vol. 10, p. 100870, 2025, https://doi.org/10.1016/j.carpta.2025.100870.
69. M. shirzad, A. Salahvarzi, S. Fathi-karkan, A. Rahdar, M. Guettari, and S. Pandey, "Green nanocarriers and biodegradable systems for sustainable drug delivery solutions," *J Drug Deliv Sci Technol*, vol. 111, p. 107208, 2025, https://doi.org/10.1016/j.jddst.2025.107208.
70. J. Gong, L. Hou, Y. C. Ching, K. Y. Ching, N. D. Hai, and C. H. Chuah, "A review of recent advances of cellulose based intelligent-responsive hydrogels as vehicles for controllable drug delivery system," *Int J Biol Macromol*, vol. 264, p. 130525, 2024, https://doi.org/10.1016/j.ijbiomac.2024.130525.

2 Environmental Sustainability through Nanocellulose Applications

Heena Gupta and Perneta

2.1 INTRODUCTION

The rampant environmental concerns have led to modulate many traditional, technological research boundaries. Latest innovations have the potential to redefine the goal of industrial development. Researchers are continuously doing the efforts to go for sustainable development with the use of environment-friendly, green, renewable, non-conventional, and biodegradable raw materials to replace harmful, toxic, and conventional resources [1–4]. Continuous degradation of environment on the global scale has upraised the demand for non-toxic, renewable raw materials. To achieve the goals of sustainable development, researchers are incorporating green or environment-friendly materials in every possible area to minimize environmental damage, conservation of natural resources, and exploration of their nature-friendly recycling properties. Not just the toxic raw materials are the concern of the day, dealing with those during manufacturing, marketing, and disposing are responsible for the contamination of the atmosphere. In a drive toward natural benefits, production using green materials has been taken into account. In recent years, generation of nanomaterials is the example of innovation meeting sustainability. Nanomaterials show a variety of remarkable mechanical, advanced structural, and strengthened features [5,6]. Due to their diversified properties, nanomaterials are being explored in the food sector, wastewater treatment, packaging, coatings, UV protection, polymeric reinforcement, biomedical applications [7], etc. In addition to this, nanomaterials are being used to detect and control greenhouse gas emission and physical, chemical, and biological pollutants in the environment. Among a variety of nanomaterials, nanocellulose is one of the most abundant green materials. Nanocellulose is a prominent representative of a nature-friendly material with remarkable features that make it appropriate for numerous industrial applications [8]. In response to the use of high-value products with low-environmental impact, nanocellulose materials have attracted the attention and research interest of academicians and industrials. These nanomaterials possess high-quality features such as mechanical strength, stiffness, sustainability, biodegradability, and water resistance. Basically, nanocellulose is an

DOI: 10.1201/9781003596127-2

advanced form of the natural plant-based polymer, cellulose. This natural polymer: cellulose is responsible for strengthening the cell wall. For decades we have been exploring the properties of cellulose for the manufacturing of domestic and industrial products of desired features. We can obtain cellulose from a variety of lignocellulosic substances present in plants and non-plant materials. To extract cellulose from any source, one requires proper, precise, and technical methods [9]. Approximately 90% of global pulp requirement is fulfilled by wood pulp obtained from the forests and plantations throughout the world. Along with wood materials, there are many non-wood materials like flax, jute, and kenaf, which consist of a small composition of cellulose that also contributes to 5%–10% of global pulp requirements. The availability of non-wood materials depends on seasonal supply as well. Interestingly, apart from the plant kingdom, cellulose can also be obtained from non-plant materials, such as algae, fungi, bacteria, sea animals. There are so many examples where pure cellulose can be extracted from bacteria [10]. In other words, we can say that with the use of a variety of technological methods, cellulose can be extracted from their numerous sources, which further can be modified into nanocellulose materials. These cellulosic materials derived from natural sources are renewable ones, unlike fossil fuel-based materials. They can be easily broken down by microorganisms, hence lowering the long-term environmental impact related with waste deposition. Manufacturing of these products requires less energy and causes lesser polluting particles in comparison with the manufacturing of synthetic materials.

Nanocellulose plays a significant role in environment remediation in terms of water filtration, removal of dirt particles from contaminated air, and oil spill cleanup [11]. They behave as sustainable agriculture facilitators, reducing the dependence on fossil fuels and the impact on the environment. This chapter gives a brief review about the sources of nanocellulose, their extraction treatment along with processing, limitations, and potential solutions. It also provides numerous applications and utilizations of nanocellulose materials. These modified cellulosic materials have a significant contribution in wastewater treatment and food industry [12]. Though nanocellulose has established utilizations across many sectors but still the limitations exist related to characterization, extraction, modification, manufacturing, cost of production, and regulatory policies. To overcome these shortcomings, one needs advanced, technical knowledge along with assessment tools associated with policy. In contrast, nanocellulose is an environment-friendly product with huge potential for exploration. With additional research and efforts, we can further modify the possibilities with these products and enhance their performance. This chapter focuses on the sustainable manufacturing of nanocellulose, advantages, and limitations of the utilities in every possible sector and their eco-friendly approach. This study is limited to a narrow area—environmental science and engineering, which can guide environmentalists with regard to the manufacturing of new materials to address the latest environmental challenges.

2.2 OVERVIEW OF NANOCELLULOSE

Creating economic and environmental sustainability using nanocellulose is one of the efforts in response to increased public concern and awareness about

TABLE 2.1
Properties or characteristics: nanocellulose

S. No.	Characteristics	Benefits	Ref.
1.	Biodegradability	Extracted from renewable resources; decomposed naturally by microorganisms; safe, non-toxic, and environment-friendly product	[14,15]
2.	Large surface area	Suitable material for adsorption and reaction mechanisms, which increases its reactivity and potential for exploration	[14,15]
3.	Mechanical strength and crystallinity	Significant contributor for the formation of composite materials	[14,15]
4.	Reusability	Potential to reuse many times for adsorption and desorption of particles	[14,15]
5.	Biocompatibility	Ability to involve in tissue culture engineering or biomedical purposes	[14,15]
6.	Tunable surface chemistry	To enhance the efficiency of the material, chemical groups can be altered according to needs such as fire-retardant agents, antimicrobial agents, and conductive polymers	[14,15]
7.	Light weight & high energy density	Reducing the weight of propellant and increasing the range of the projectile fired can be incorporated in military weapons	[14,15]
8.	High viscosity & crosslinking	Surface modifiability for stabilizer, coatings	[14,15]

environmental issues. Moreover, innovation of these materials follows rigorous government policies and regulations. In addition, nanocellulose is a very advanced, efficient, and environment-friendly material in comparison with synthetic plastic products. Ongoing research [13] is likely to address certain limitations associated with nanocellulose products with sustainable credentials and establishing methods that can be further explored for a variety of application possibilities and suitability. From ancient times, forests have been continuous suppliers of plenty of lignocellulosic material for many utilization possibilities. These cellulosic substances are used to manufacture nanocellulose. It has been observed through numerous analyses that due to the large surface-area-to-volume ratio of nanocellulose, these materials possess unique properties (Table 2.1) in comparison with cellulose and its microfibers.

Nanocellulose products are therefore considered as convincing materials for a broad range of applications, including coating, packaging, and transportation in the biomedical industry. Apart from these, nanocellulose can be tuned (shapes, structure, size, dimension, networking of nanofibrils, surface properties) to enhance efficiency of the product to meet the demand of the society (Figure 2.1). Further advancement and innovations in these cellulosic-derived materials open up latest and new applications in various fields such as cardiovascular, visceral surgery, drug delivery systems, medical, and military technology [16,17].

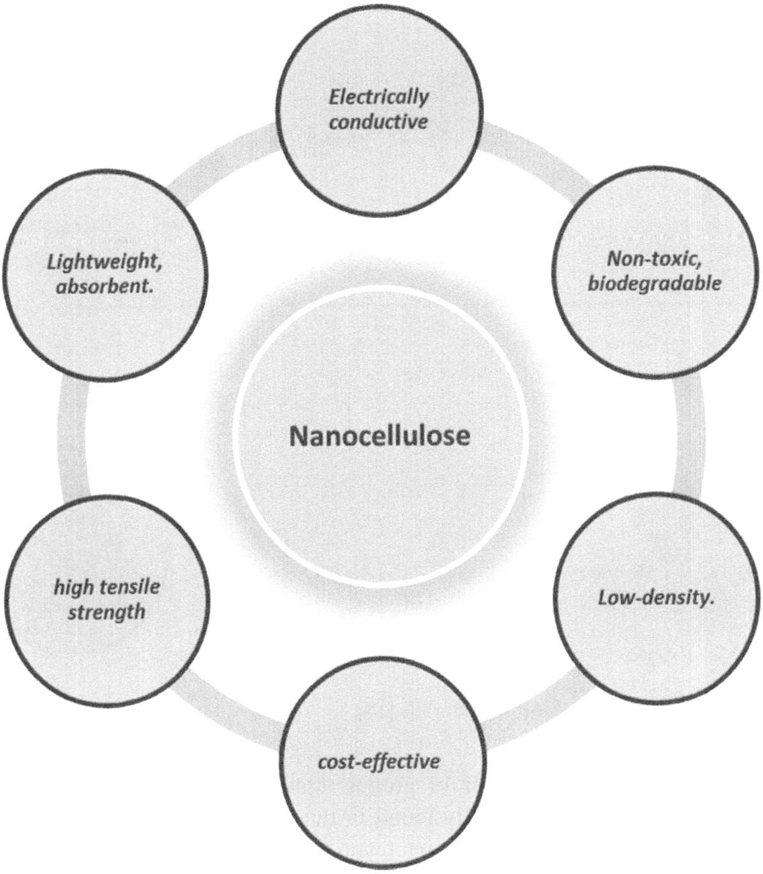

FIGURE 2.1 Properties of nanocellulose.

Nanocellulose (NC) is considered as a promising cellulosic biomaterial having dimensions in nanoscale with high mechanical strength, stiffness, and biodegradability. On the scale of size, crystallinity, tensile strength, NC is classified into three categories (Figure 2.2) such as cellulose nanocrystals (CNCs), bacterial nanocellulose (BNC), and cellulose nanofibers (CNFs). Though these materials have been originated from the same source but exhibit different applications in numerous areas.

Cellulose nanofibrils (CNFs) or ***nanofibrillated cellulose (NFC)*** are nanosized fibrillar structures of a few hundred nanometers or less in size. Typically, it has a width range from 5 to 20 nm. Fibrils above this range comes into the category of ***cellulose microfibrils (CMF).*** Generally, these fibrils are extracted from wood pulp with the help of technical methods such as homogenization, grinding, or microfluidization at a high range of temperature, pressure, and velocity.

Microfibrillated cellulose (MFC) or cellulose microfibril (CMF) is stretchable, entangled, and heterogeneous compared with cellulose nanofibers. It comprises both nano- and micro-size particles. Morphologically, they have different

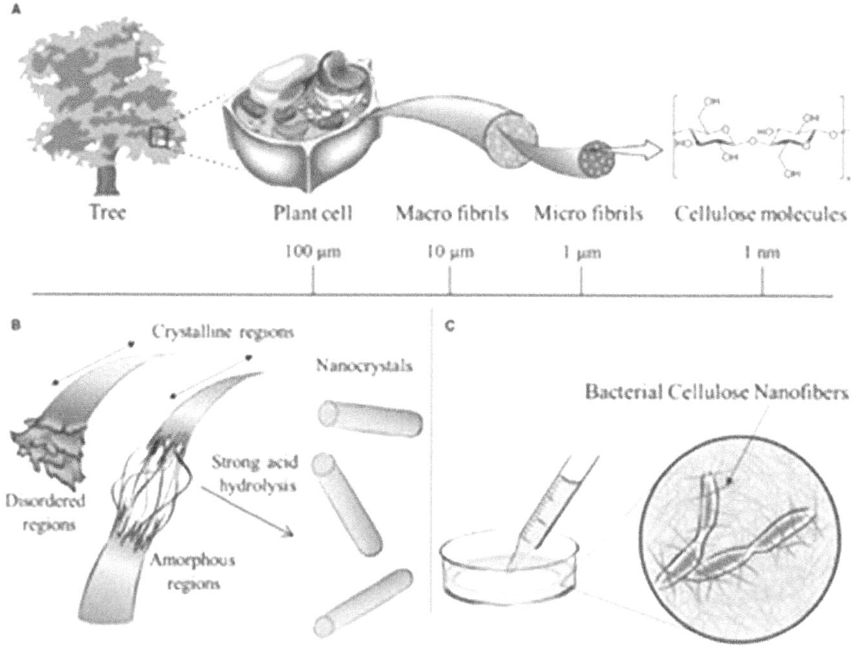

FIGURE 2.2 Types of cellulosic materials [18].

length-to-diameter ratio, number of groups attached to them, surface area, etc. Cellulose microfibrils are naturally found in the innermost surface of the primary cell wall. It possesses actually very interesting phenomena as much as the water is absorbed by the cell, its volume raises and separation of microfibrils starts and formation of new ones begin, which further supports cell strength.

Cellulose nanocrystals (CNCs), or nanocrystalline cellulose (NCC), are sub-microscopic particles with nanosized cellulose fibrils having a typical size of 100–1000 nm. These rod-shaped, crystalline nanoparticles are typically shorter than CNFs. They behave as colloidal stable in water as they are surrounded by negatively charged groups. Nanocrystals (Figure 2.3) have a wide range of applications [19] in several industries with numerous advanced features like high mechanical strength, low cost, electro-magnetic characteristics, and high quality.

Manufacturing of cellulose nanofibers (CNFs) is quite simple and easy in contrast to CNCs as here is no requirement of any chemical splitting procedure, no alteration of structure and cellulose chain length. Indeed, a variety of physical or chemical methods are used for the fabrication of cellulose nanofibers.

Bacterial nanocellulose (BNC) or bacterial cellulose is generally considered as one of the safe products that is being used in industries at a very high rate these days. It is also gaining so much of importance in recent years because of its unique biophysical characteristics such as non-toxic, self-assembling of fibrils, which can be further opted for several biomedical applications [21]. Manufacturing of BNC involves the fermentation of certain bacterial species, including Gram-negative

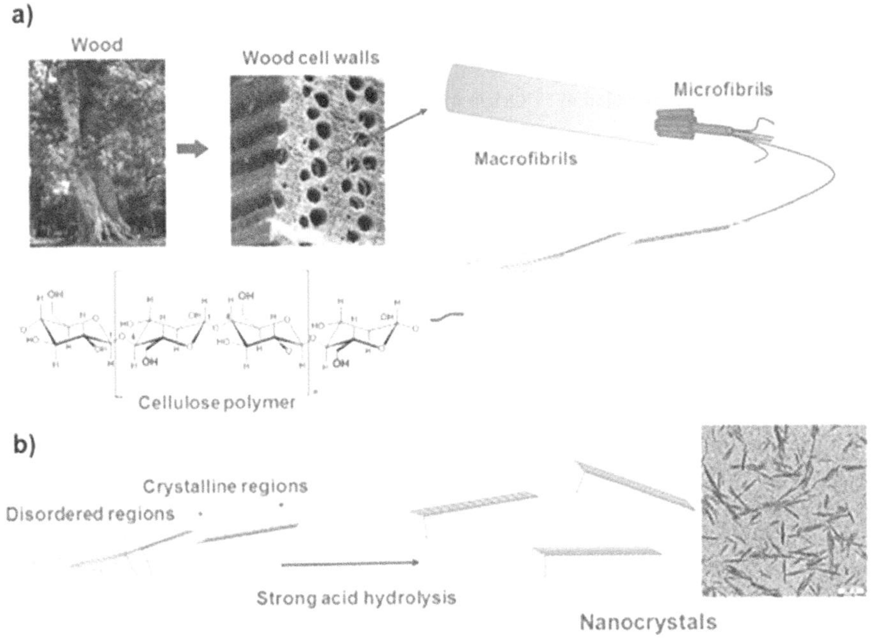

FIGURE 2.3 Manufacturing of nanocrystals [20].

bacteria species such as *Acetobacter* (reclassified as *Komagataeibacter*), *Rhizobium*, *Agrobacterium*, *Pseudomonas*, *Salmonella*, *Alcaligenes*, as well as Gram-positive species such as *Sarcina ventriculi*. Among the number of bacteria, *Acetobacter* is the most significant one. In practice, bacterial nanocellulosic products are generally bigger in size in comparison with their wood or cotton counterparts.

The final size and structure of nanocellulose is ultimately dependent on the raw material and fabrication/isolation methods but generally with a diameter of 10–40 nm when mechanically fibrillated, to 4–10 nm for NCCs, and 2–5 nm for chemical/mechanical extraction. For the manufacturing of polymers, papers, coatings, and membranes, nanocellulose has been gaining preferences these days [22]. Very significantly, bacterial nanocellulose, is being used in implants and tissue engineering. These cellulosic materials can be modified into a number of utilities such as papers, transparent films, aerogels, hydrogels, and even spherical particles. Table 2.2 shows the comparison of a number of products derived from the cellulose, which is being used in a variety of areas, depending on their features and subsequent requirements.

2.3 APPLICATIONS OF NANOCELLULOSE FOR ENVIRONMENTAL SUSTAINABILITY

Recently nanocellulose has expanded its area of utilities because of their significant features like their biocompatibility, renewable nature, mechanical strength, and

TABLE 2.2
Comparison of nanocellulose variant [20]

Variant of nanocellulose	Abbreviation	Origin of raw material	Extraction procedure	Dimensions
Cellulose nanocrystals	CNC	Plant or wood	Acid hydrolysis	Diameter: 20–100 nm Length:1 nm–100 μm
Cellulose nanofiber	CNF	Plant or wood	Mechanical fibrillation	Diameter: 5–50 nm Length: several μm
Bacterial nanocellulose	BNC	Microorganisms	Polymerization and crystallization	Diameter: 2–4 nm Length: 100 nm

cost-effectiveness toward the environmental sustainability. Technological advancement in the field of manufacturing of variants of nanocellulose using aspects and knowledge of physical, chemical, and biological criteria is one of the significant contributions toward dealing with social environmental issues. Nanotechnology is continuously doing efforts to replace toxic, non-renewable, fossil fuel-based materials with nanoproducts. One of the studies [2] says that consistent consumption of nanomaterials can save up to 1.5 billion liters of fuel in one year of the vehicle and lower the carbon emission by more than 5 billion kg per year. Nanocomposites have also made their way to the automobile industry where heavy metal parts are being replaced by highly durable and lightweight raw materials of nanoparticles [23]. Nanotechnology has also been incorporated in the semiconductor manufacturing industry, which exhibits excellent results by preventing greenhouse gases emission and saves billions of dollars. Apart from these, numerous applications [24,25] for nanocellulose are being explored in making hydrophobic fabrication for nanocellulose, manufacturing of antimicrobial products and developing UV barrier, biomedical, electronic, and automobile devices. However, there are still a lot of technical challenges and limitations for the application of nanocellulose to be consumed as a reliable, renewable, and sustainable product for latest and advanced applications. The aim of this study is to explore direct or indirect impact of nanotechnology (nanocellulose) on the environment, specifically to explore pollutants and their remedial methods using nanotechnology.

2.3.1 Air Pollution

Recent advancement in technology coupled with financial growth has led to the direct or indirect impact with significant degradation of the environment. One of the major environmental crises of the current era is air pollution. The inhalation of particulate matter (PM) and biological and chemical pollutants released in the atmosphere is responsible for many respiratory problems in humans and other living organisms [26]. Among the numerous primary and secondary pollutants, PM is one of the most health-hazard causing air pollutants around the world. In other words, they lead to disturb the ecological balance. Nanocellulose has evolved as a novel category of materials having unique and excellent characteristics such as low price, renewability, network-building capability, high aspect ratio, sustainability, low density and weight, biodegradability, high elasticity and mechanical strength, surface modification properties with modulation and stiffness, which can be used for the making air pollution control devices [27]. Owing to their exceptional processability, nanocellulose shows huge potential for advanced applications in the field of aerogels and air filters. Aerogels are solid stuff having interconnected mini pores and a three-dimensional networking framework. Industries have been using silica-based aerogels and chemical air filters from a very long time. Existent or traditional petroleum-based air pollution control equipment are highly stable and non-biodegradable but have sustained in the environment for >1000 years, which is further responsible for environment degradation after disposal as well.

Nanocellulose-based aerogels (NCBA) are a leading category of new generation materials with excellent porosity and absorption potential. Cellulose-derived aerogels with low thermal conductivity, high insulation along with significant flexibility and tensile strength can be opted to replace synthetic polymers and other non-biodegradable polymers for air purification. These aerogels can absorb PM and gas pollutants by making a very fine network. CNF (cellulose nanofibers) is an ideal choice for manufacturing such type of devices as they are good PM absorber. Nanocellulose air filters effectively remove airborne particles (dust, pollen) from airflows by using Sieving (trapping particles between fibers) and inertial impaction (capturing heavier particles) methods [28]. In aerogels, proper absorption and insulation are required, whereas in air filters, filtration of PM is mandatory and nanocellulose satisfies both the requirement. Apart from these, few limitations are associated with the use of **nanocellulosic aerogels** or **air filters**. They can be easily ignited, which can restrict their approach in certain commercial applications. Scientists are continuously doing efforts to further explore the hidden potential or overcome the limitations of these nanofibers.

To monitor air pollution level in the atmosphere, cellulose-based **nanosensors** are being used, which is also a sustainable invention of the decade [29]. They can easily stay in the atmosphere, detect the pollutant level, collect, and store the information to a database based on which remedial treatments can be adopted.

2.3.1.1 Toxic Gases and Heavy Metal Ions

It has been observed by various analyses that exposure to toxic gases, PM, and heavy metal ions leads to health problems, such as respiratory issues, lung cancer, skin cancer, eye infections, and misfunctioning of vital organs. Heavy metal ions cannot be digested by living organisms and decomposed by microorganisms. These ions can be sustained in the nature, air, water, and soil for several years without degradation. In other words, they are non-biodegradable and causes air, water, and soil pollution. Hence, we require some of the equipment that can detect the concentration of the contamination, on the basis of which, control measures can be taken. Industrial emissions along with volcanic eruptions are the main cause of the release of toxic gases and heavy metal ions into the environment. Industrial evolution is the prime most factor for environment degradation. About 2 million tons of waste is collected from the industries every day and approximately 14,000 people have died due to chemical exposure and this number keeps on increasing as the contamination level rises by the minute. The World Health Organization has listed a group of chemicals as toxic for flora and fauna [30]. These chemicals include arsenic, asbestos (silicate minerals), benzene, cadmium, dioxin and dioxin-like substances, inadequate or excess fluoride, lead, mercury, and highly hazardous pesticides. Direct or indirect emission of such chemicals, toxic gases, and heavy metal ions is a threat to human health. Conventional or traditional approach to remove contamination from the environment is very expensive, petroleum-derived and non-biodegradable, which includes chemical oxidation, biodegradation, membrane separation, photocatalysis, liquid extraction, precipitation, electrolytic treatment, electro-coagulation reverse osmosis, and adsorption onto activated carbons. With the use of these methods, only selective removal of chemicals is feasible. Also, incomplete precipitation and the leftover huge amount of sludge are difficult to handle even after disposal. Another restriction with the use of these traditional methods is their high cost too.

In this regard, nanotechnology offers a new path to deal with detection and remediation of these chemical contaminants, toxic gases, and heavy metal ions. The **carbon nanotube (CNT) sensors** can detect the presence of toxic gas molecules in the atmosphere and absorb the same. Several studies claim [31] that in future, these sensors will detect war biochemical gases, air pollutants, and even organic molecules in space. Due to high absorptivity, three-dimensional framework, high precision & sensitivity, selective detection; carbon nanotubes come prior. In other words, we can say nanotubes are high-performer bioadsorbent with strong binding affinity, high specific surface area, and abundance of hydroxyl groups, which are significantly considered in decontamination processes.

Another latest nanotechnological invention is **quantum dots**. This is one of the most accurate, convincing, and advanced nanomaterial-based technology used for the detection of heavy metals. These quantum dots are highly reactive that can detect multiple metals at a time. Zero-dimensional graphene dots are used as heavy metal ion **nanosensors** due to their exceptional features that include brilliant optical properties, high stability, surface area fabrication, and doping. Another example of this category is **carbon quantum dots**, which is extracted from green algae and further used to detect Fe (III) in effluents. The high environmental compatibility of nanobiosensors makes them a perfect choice to be used as attractive fluorescent nanosensors [32].

Detection of heavy metal ions in drinking water can be easily done with the use of cellulose-derived nanocomposites with a core-shell structure having a more advanced surface plasmon resonance bond. They are highly sensitive and can detect the presence of even a very small quantity of heavy metals in drinking water. The presence of any kind of contamination, say heavy metal ions, cadmium, zinc, or lead in drinking water leads to the accumulation of nanocomposites and also changes the wavelength of the adsorption band.

2.3.1.2 Adsorption of Toxic Materials and Gases

Along with the detection or sensing the presence of pollutant/contaminant in the atmosphere, nanocellulose has the potential to remove the same from the environment. With the help of these advance materials, the amount or level of toxicity can be reduced from the surroundings. **Nanocellulose-based carbon nanotubes** (CNTs) are key ingredients for this adsorption due to their high surface area and tunable surface chemistry. Toxic gas can be captured and separated from the environment can be done with nanotechnology. For example, CNTs modified with gold or platinum nanoparticles, graphene, and activated carbon are promising and convincing superior adsorbent to eliminate different kinds of organic and inorganic pollutants from terrestrial and aquatic ecosystems [33,34]. The high adsorption capacity of these materials is specifically because of their porosity and potential to modify surface. Dioxins and their associated compounds (such as polychlorinated dibenzofurans and polychlorinated biphenyls) are extremely toxic to human health and environment. Combustion and oxidation of organic compounds in incinerators are responsible for the production of such carcinogenic materials that affect the entire immune system and fetal growth. Apart from CNTs, activated carbon is also used to remove such dioxin derivatives from the environment in a number of countries like Japan and Europe [35], but the success graph is not very impressive. A number of studies have concluded that the interaction between CNTs and dioxins is much stronger than the interaction between dioxin and activated carbon. These analyses showed the significant removal of such toxic materials via CNTs and nanocomposites.

2.3.1.3 Adsorption of CO_2

In the current decade, the presence of CO and CO_2 in the environment cannot be ignored for a single minute. These affect our immune system and have strong affinity to combine with red blood cells. Binding of CO with RBCs causes the deficiency of O_2 in the body and is responsible for brain hemorrhage. The amount of this carbon-based contamination has been increasing day by day due to excessive industrial phenomena and transportation systems. Along with CO and CO_2, unburnt carbon is also responsible for environmental damage. For years, catalytic converters are being employed in the vehicles to reduce the emissions from internal combustion engines. These converters have the affinity to reduce carbon monoxide, nitrogen oxides, and unburned hydrocarbons in car exhaust. But one of the major or significant limitations associated with the use of these catalytic convertors is the consumption of heavy metals such as palladium, rhodium, and platinum. Efforts are continuously being made to reduce the use of these precious heavy metals in the manufacturing of these convertors with enhanced efficiency. In recent years, many attempts have been made to replace these heavy metals, which are toxic to living beings or to eliminate from the environment by the adsorption. Researchers have developed the nanomaterial that can adsorb particles such as CO_2, CO, and NO_x released during emission from fossil fuel combustion [36,37]. Nanocellulose-based catalytic convertors are not available in the market, yet nanocellulose enhances the efficiency of the other nanomaterials used for adsorption. With a wide range of scientific research, various nanomaterials fused with nanocellulose having strong CO_2 adsorption capabilities have been manufactured apart from activated carbon, zeolite, metal-organic frameworks (MOF), boron nitride etc. The large porosity, surface modification, and suitable thermal and chemical features make them suitable adsorbents of CO_2 under normal conditions. Except nanocellulose, MOFs also show impressive result in CO_2 capturing [38]. It is a promising alternative to other liquid and solid carbon-capture sorbents. It has high nitrogen content with tunable pore size and properties with exceptional adsorption capacities. One of the very famous nanomaterials, ZIF-8 can capture CO_2 either in its molecular cage or in interspatial areas. Nanomaterials such as carbon nanotubes, nanosilica, nanocellulose, MOFs, and porous alumina with unique structural features are considered as perfect adsorbents for such carbon compounds. In recent years, the capturing of CO_2, which is produced by the combustion of fossil fuel draws the attention of environmentalists and researchers. Various technologies for recycling, capturing, adsorption, and desorption have been developed to achieve sustainability. CO_2 emitted from the chemical factories or industries and automobiles can be adsorbed via CNT, nanosilica, and zeolite on a large scale [39].

Nanocellulose also offers promising solutions for the removal of volatile organic compounds from the environment. Various industrial phenomena are responsible for the release of oxides of nitrogen and sulfur, nitric acid, soot formation, polyaromatic compounds, and volatile compounds (VOCs). Environment regulatory policies are clearly focusing on those particles or contaminants that are potentially toxic to living beings. Nanotechnology has been incorporated to purify the air these days [40]. Their catalytic and adsorptive action are effective to eliminate VOCs, sulfur, and nitrogen oxides from the atmosphere under normal circumstances.

2.3.2 Wastewater Treatment

Water is one of the prime requirements of the living beings. It is an inseparable part of every animate life. There will be no survival without water. A number of problems occur if one consumes contaminated water. So, water plays a very important role in everyone's life. Along with availability, purity or quality also affects humankind. Nearly 71% of Earth's surface is occupied by water but to satisfy the needs of humankind, less than 1% is available. This meager quantity or scarcity of fresh water is the reason behind all major water conflicts and crises worldwide. Rapid industrialization and urbanization are responsible for the contamination of surface as well as ground water. Several heavy metal ions and biological contaminants (bacteria, virus, fungi, and protozoa) are being released into the fresh water sources, which degrades the quality of drinking water and makes it unfit for drinking. When the water gets contaminated by organic pollutants, inorganic chemicals, microorganisms, industrial effluents, or any other material, it becomes of no use and is termed as wastewater. This water can't be consumed without any further treatment. These days, wastewater treatment has become a major sustainable initiative. Efforts are continuously being made to conserve the quality of water, protect water supplies, and treat wastewater. This sustainable approach offers selective and efficient ways to treat or purify water, but limitations are also associated with it. Pure and good quality water is important for the survival of all living beings. The current scenario is to approach a sound, cost-effective and environment-friendly wastewater treatment considering all the benefits and limitations. These days, industries are using a few conventional water treatment methods like reverse osmosis, distillation, coagulation–flocculation, bio-sand, and filtration. But these are not effective to remove all heavy metal ions. By selecting the most efficient and cost-effective wastewater purification methods, treated water can be reused.

With the recent advancement in the field of nanotechnology, industries are using nanocellulose for wastewater treatment effectively. This nanotechnology-based adsorption technology is highly efficient and easy to handle. In wastewater treatment, nanocellulose is being consumed for the removal of contaminants or pollutants; microorganisms, heavy metals or particles, inorganic anions, and organic pollutants as they possess prime catalytic and reactive features [41,42]. Nanocellulose also has a large surface area for adsorption or reaction to perform. In addition, it is considered to be highly mobile in solution. Common activities performed via nanocellulose are adsorption for pollutants, nanocatalysis to decompose contaminants, and by including them as nanomembranes for latest separation methods in terms of nanofiltration. Some of the conventional technologies we are using for the primary waste water treatments are adsorption, flocculation, and filtration to date. But we need more specific, more precise low-capital and high-efficiency methods for the purification of wastewater and nanotechnology comes above all.

2.3.2.1 Nanofilters

Nanocellulose is being used as nanofilters in wastewater treatment. Utilization of these nanofilters is a more efficient, capable, and economic way to purify wastewater in comparison with conventional methods at standard temperature and pressure conditions. One of the studies [43] showed that a carbon hollow tube filter made up of radially aligned carbon nanotubes could filter sodium chloride from seawater, reducing sodium from 30.06% to 0.4% and chlorine from 11.44% to 0.17%. The demand for nanofilters has been increasing despite their high cost. The popularity of these nanofilters in the market is due to their unique features such as large surface areas, high water fluxes, tunable pore sizes, antibacterial properties, and the potential to boost water purification methods. In nanofiltration, pathogens like virus, bacteria, pesticides, organic–inorganic pollutants along with calcium and magnesium salts are removed from contaminated water. No further chemicals are required to purify or soften water; hence this method has less impact on the environment. These are effective enough to remove pesticides having inorganic or organic origin along with halogenated solvents. Some of the nanocellulose-based membranes have the potential to alter the nature of highly toxic chlorinated compounds into eco-friendly hydrocarbons and decompose heavy metals such as Ni and Hg, nitrates, and perchlorates. They are also active on radioactive materials of contaminated water.

Nanofiltration is one of the sustainable options for energy saving. A variety of filters are available in the market, depending upon pore size and are called as microfilters, ultrafilters, and nanofilters. In comparison with reverse osmosis, nanofiltration is a low-pressure filtration method that costs less and is very affordable. In addition to this, nanofilters can remove biological contamination from drinking water and agricultural waters.

2.3.2.2 Nanocellulose as Standalone Membrane or Composite Membranes

Water scarcity and pollution intensity worldwide drive the market for the implementation of an effective and environment-friendly approach for water and wastewater treatment and the protection of vital water resources. Nanocellulose-based standalone membranes or composite membranes work very efficiently due to enhanced surface hydrophilicity, selective superior filtration, and superb rejection rates for pollutants along with chemical, mechanical, and thermal stability. These significant features are the necessity of today. These modified nanocellulose-based membranes allow the operation in standard and extreme conditions. The enhanced surface hydrophilicity, selective permeability, higher water flux, high oil rejection percentages (up to 99%), potential to treat dye-contaminated waste water make this fused membrane sustainably suitable and appropriate for oil separation, improve fouling resistance, microfiltration (MF), ultrafiltration (UF), nanofiltration (NF), and reverse osmosis (RO) [44,45]. This advancement is applicable on both oily as well as dry wastewater treatment. Ongoing studies analyze that this is one of the effective solutions for treatment of contaminated water under extreme or harsh conditions. The collective research shows the potential of fused membranes to modify wastewater treatment along with environmental remediation.

2.3.2.3 Carbon Nanotubes

Carbon nanotubes are one of the initial nanoproducts of nanotechnology and came to the light in 1991. These are cylindrical, honeycomb-like structures derived from graphene sheets. Nanotubes are very long, thin, quite stable entities that possess strong and flexible features. Aligned carbon nanotubes can be used for seawater desalination in future. Single-walled carbon nanotubes are being used for bacterial removal whereas multiwalled carbon nanotubes filter exhibited viral removal at low pressure [46]. In addition to this, CNTs can decolorize polluted water. Low energy consumption, tunability, strong antimicrobial activity, antifouling, self-cleaning functions, flexibility, and high tensile strength are those unique features of carbon nanotubes that make them highly efficient in the purification of wastewater. Over conventional methods, nanocellulose-based treatment is more effective and advantageous as it consumes less energy and is also sustainable for the environment. These days, carbon nanotubes have a variety of applications apart from environmental remediation [47]. Carbon nanotubes play an important role in wastewater treatment as well by removing organic matter and tends to clean wastewater more effectively. These are highly stable nanoproducts with high specific surface area. These features make nanotubes a perfect substance for biological treatments and help to speed up the degradation process. Nanotubes are made up of nanofibers that have suitable surface morphology and biocompatibility, which can collect pathogens at the surface and pass clean water. These flexible fibers are responsible for the formation of biofilm on the surface and inside the bed. At the surface, susceptibility to accumulation of microorganisms is very high and quick, but this operation has to be fast to reduce the degree of contamination and nanotubes are the solution.

2.3.2.4 Nano-adsorbents

Excessive industrialization or urbanization has become a huge matter of concern. Variety of pollutants are present in the atmosphere. Different types of pollutants require different treatment. Nanocellulose acts as a sustainable, versatile, and environment-friendly nano-absorbent offering high surface area and pollutant-removal ability. Nano-absorbents are one of the effective solutions for the treatment of organic and inorganic pollutants, which consist of dyes, pesticides, drugs, and organic solvents and metals, fertilizers, acids, alkalis, and salts, respectively. These are effective for environmental remediation with more active sites, tunable functionalities that help in stronger bonding with pollutants, selective adsorption toward heavy metals, etc. Nano-adsorbents are highly appreciated for the adsorption of pollutants from water, air, and soil by leveraging their high reactivity [48]. For the removal of heavy metals from wastewater, nano-absorbents are preferred over conventional methods. These are highly efficient for soil remediation, wastewater treatment, and air pollution control. Nano-absorbents are cost-effective and environment-friendly.

2.3.3 Environmental Catalysts

Overpopulation and excessive industrialization lead to the release of a variety of pollutants into the environment, which disturbs our ecological balance. These contaminants are toxic in nature and harmful for living beings. Hence the immediate degradation of these pollutants is necessary. However, it is very difficult to degrade certain type of pollutants due to their chemical composition, so we need advanced technology for that. Nanocellulose fulfills the criteria and plays a significant role in decomposition. Large surface area, sufficient functional groups, and active sites can easily decompose harmful, toxic pollutants. Nanocellulose acts as a nanocatalyst that can be used for remediation purposes [49]. There are a variety of inorganic, carbon-containing and polymeric nanocatalysts, which are generally being opted for the removal of contaminants from the environment. Adsorption and photocatalytic reduction of heavy metal ions, organophosphorus, chlorinated organic, volatile organic compounds, dyes, and halogenated herbicides are usually preferred over conventional ways of degradation. To air pollution control, wastewater treatment, and related environment remediation, nanocataylsts are being employed [50]. Current status of environment pollution, deteriorated biodiversity, exploitation of natural resources, we need alternative, suitable, advanced technology with low cost and higher efficiency. These days, for wastewater treatment, nanocellulose-based nanocatalysts provide support for metal nanoparticles (NPs), enabling metal-conjugated nanohybrid catalysts due to its versatile surface chemistry and abundant hydroxyl groups. Nanocatalysts are being used for the decomposition of halogenated organic compounds. This type of biodegradation is considered one of the effective methods for the removal of contamination from effluents. Various studies [51] have concluded the reliability of such type of nanomaterials in numerous water conditions. Another sustainable benefit of using nanocatalysts for the degradation of pollutants is their retention. They can be effectively recovered from the effluents after processing by magnetic separation technology. Nanocatalysts are also being employed to control air and water pollution [52]. Hydrocarbon, volatile organic compounds, carbon monoxide, and nitrogen monoxide are present in the environment in abundance due to industrialization and transportation. Nanocatalysts have the potential to decompose such contamination and maintain the environment's sustainability. Traditional catalysts have few limitations associated with them such as high cost and, no standard air-to-fuel ratio. Modern-day nanomaterials act as photocatalysts and antimicrobial catalysts and help to reduce microbial activities. A study says [53] that the fusion of Fe_2O_3 nanoparticles with hydrogen peroxide helps to decompose organic pollutants. In the same manner, a combination of graphene oxide with Fe_2O_3 or with nanotubes shows satisfactory results for adsorption and catalytic purposes in water purification.

Some of the non-inflammatory drugs called diclofenac and ibuprofen can also be treated with nanocatalysts. For photocatalytic degradation, morphologies of catalyst, mass transport processes, visible-light absorption, and surface distribution of active sites are required. In general, it is observed that modern nanocataylsts are being employed for cleaning of the environment or environment remediation.

2.4 CONCLUSIONS

Science and technology play a significant role in environmental remediation via bioremediation, phytoremediation, nanoremediation, and advanced oxidation processes. These methods are being opted by society to eliminate pollutants from soil, air, and water as they offer sustainable, green, and cost-effective alternatives to environmental pollution. Nanotechnology has emerged as one of the advanced tools to resolve a variety of social environmental issues. Air, wastewater, soil, and water nanomaterials can largely detect, degrade, and remove contamination from the environment. Nano-adsorbents, nanofiltration, nanoremediation, nano-photocatalysts, magnetic nanoparticles, and nanosensors are some of the processes used to treat wastewater and pollutant detection. To control environment pollution, nanotechnology is considered as an effective, sustainable, and a greener approach. There are a number of nanomaterials (inorganic, carbonaceous, and polymeric), which are being successfully utilized for environmental remediation. To deal with the contaminants, one needs to be specific. It requires full analysis of the pollutant quality, site of pollutant, composition, and volume for effective remediation.

Considering the efficiencies and potential of nanomaterials, nanocellulose is being opted as an effective means to accomplish the goals of sustainable development. Nanocellulose possesses some unique features such as lightweight, renewable, biocompatible, better mechanical, and thermal properties to perform a significant role toward the production of function-specific nanomaterials. It has been used for the manufacturing of nanocomposites, nano-adsorbents, nanofilters, support for catalytic actions, and nanosensors. Environmental remediation such as wastewater treatment, pollution detection, absorption of heavy metal ions, and toxic gases can be done with nanocellulose-based products as they possess a large surface area, strong mechanical properties, and biodegradability. Apart from environmental protection, nanocellulose is also used for food-packaging due to its non-toxic and barrier properties. Nanocellulose is also explored in the biomedical field but further investigations regarding the toxicity level is yet to be done. Various attempts and research for the manufacturing of advanced nanocellulose-based products are being performed, which is a green bio-based and renewable approach. It can also be concluded that the prospects of nanocellulose as a support to enhance membrane performance in wastewater filtration, environmental remediation, and the development of pollutant sensors/adsorbent and energy devices are also being appreciated. For future development, innovative thoughts/ideas, including the creation of highly efficient, effective, and environment-friendly nanocellulose-based products with renewability for remediation with appropriate cost will be a great challenge. Apart from this, manufacturing of self-cleaning materials is a competitive task. Evaluating the threat and the potential impact of nanomaterials to human health and surrounding is also highly needed. Also, laboratory studies should be implemented in real-case scenario for the remediating number of pollutants such as contaminated water, soil, industrial effluents, sewage, and air.

REFERENCES

1. A. Esmaeili, A. A. Beni, Characterization of PVA/chitosan nano fiber membrane and increasing mechanical properties with cross-linking by heating, *Int. J. Theor. Appl. Mech.* 4 (2019) 26–32.
2. Y. Guo, L. Zhang, F. Zhao, G. Li, G. Zhang, Tribological behaviors of novel epoxy nanocomposites filled with solvent-free ionic SiO_2 nanofluids, *Compos. B Eng.* 215 (2021) 108751.
3. J. Dong, W. Xu, S. Liu, L. Du, Q. Chen, T. Yang, Y. Gong, M. Li, X. Tan, Y. Liu, Recent advances in applications of nonradical oxidation in water treatment: Mechanisms, catalysts and environmental effects, *J. Clean. Prod.* 321 (2021) 128781.
4. M. Rahimi, R. Mehdinavaz, M. Heydarzadeh, A. Hossein, M. Ettelaei, Surface & coatings technology improving biocompatibility and corrosion resistance of anodized AZ31 Mg alloy by electrospun chitosan/mineralized bone allograft (MBA) nanocoatings, *Surf. Coating. Technol.* 405 (2021) 126627.
5. M. J. Moreno-v, F. Rodríguez-f, A. G. L, C. L. Del-toro-s, Sustainable-green synthesis of silver nanoparticles using Safflower (Carthamus tinctorius L.) waste extract and its antibacterial activity, *Heliyon.* 7 (2021) e06923.
6. M. Hosseini, A. H. Abdelrazek, R. Sadri, A. R. Mallah, S. N. Kazi, B. T. Chew, S. Rozali, N. Yusoff, International Journal of Heat and Mass Transfer Numerical study of turbulent heat transfer of nanofluids containing eco- friendly treated carbon nanotubes through a concentric annular heat exchanger, *Int. J. Heat Mass Tran.* 127 (2018) 403–412.
7. M. M. El-Kady, I. Ansari, C. Arora, N. Rai, S. Soni, D. K. Verma, P. Singh, A. El D. Mahmoud, Nanomaterials: A comprehensive review of applications, toxicity, impact, and fate to environment, *J. Mol. Liqs.* 370 (2023) 121046.
8. Z. Gholamvand, D. McAteer, A. Harvey, C. Backes, J. N. Coleman, Electrochemical applications of two-dimensional nanosheets: The effect of nanosheet length and thickness, *Chem. Mater.* 28 (2021) 1–17.
9. P. Phanthong, P. Reubroycharoen, X. Hao, G. Xu, A. Abudula, G. Guan, Nanocellulose: Extraction and application, *Carbon Resour. Convers.* 1 (2018) 32–43.
10. T. Abitbol, A. Rivkin, Y. Cao, Y. Nevo, E. Abraham, T. Ben-Shalom, S. Lapidot, O. Shoseyov, Nanocellulose, a tiny fiber with huge applications, *Curr. Opin. Biotechnol.* 39 (2016) 76–88.
11. R. K. Mishra, A. Sabu, S. K. Tiwar, Materials chemistry and the futurist eco-friendly applications of nanocellulose: Status and prospect, *J. Saudi Chem. Soc.* 22 (2018) 949–978.
12. A. B. Perumal, R. B. Nambiar, J. A. Moses, C. Anandharamakrishnan, Nanocellulose: Recent trends and applications in the food industry, *Food Hydrocoll.* 127 (2022) 107484.
13. K. Dhali, M. Ghasemlou, F. Daver, P. Cass, B. Adhikari, A review of nanocellulose as a new material towards environmental sustainability, *Sci. Tot. Environ.* 775 (2021) 145871.
14. P. Kaur, N. Sharma, M. Munagala, R. Rajkhowa, B. Aallardyce, Y. Shastri, R. Agrawal, Nanocellulose: Resources, physio-chemical properties, current uses and future applications, *Front. Nanotechnol.* 15 (2021) 747329.
15. H. Wei, K. Rodriguez, S. Renneckarbd, P. J. Vikesland, Environmental science and engineering applications of nanocellulose-based nanocomposites, *Environ. Sci.: Nano.* 1 (2014) 302.

16. Md. N. F. Norrrahim, V. F. Knight, N. A. Md. Kasim, F. A. Ujang, N. Janudin, M. A. I. A. Razak, N. A. A. Shah, S. A. Md. Noor, S. H. Jamal, K. K. Ongab, W. Md Z. W. Yunus, Nanocellulose: The next super versatile material for the military, *Mater. Adv.*, 2 (2021) 1485.
17. D. Khorsandi, S. Jenson, A. Zarepour, A. Khosravi, N. Rabiee, S. Iravani, A. Zarrabi, Catalytic and biomedical applications of nanocelluloses: A review of recent developments, *Int. J. Bio. Macromol.* 268 (2024) 131829.
18. D. Trache, A. F. Tarchoun, M. Derradji, T. S. Hamidon, N. Masruchin, N. Brosse, M. H. Hussin, Nanocellulose: From fundamentals to advanced applications, *Front. Chem.* 8 (2020) 1–33.
19. S. K. Misra, K. Pathak, Ind. App. Nanocrystals, Chapter 10- Drug nanocrystals as drug delivery systems, edited by S. Mallakpour, C. M. Hussain, *Micro and Nano Technologies*, Elsevier (2022) 153–178.
20. Y. Zhang, K. Poon, G. Sofia P. Masonsong, Y. Ramaswamy, G. Singh, Sustainable nanomaterials for biomedical applications, *Pharma.* 15 (2023) 922–942.
21. Reshmy R, E. Philip, D. Thomas, A. Madhavan, R. Sindhu, P. Binod, S. Varjani, M. K. Awasthi, A. Pandey, Bacterial nanocellulose: Engineering, production, and applications, *Bioeng.* 12 (2021) 11463–11483.
22. A. Abdel Hakim, R. Mourad, Nanocellulose and its polymer composites: Preparation, characterization, and applications, *Russ. Chem. Rev.* 92 (2023) RCR5076.
23. M. M. Shameem, S. M. Sasikanth, R. Annamalai, R. G. Raman, A brief review on polymer nanocomposites and its applications, *Mater. Today Proc.* 45 (2021) 2536–2539.
24. Md. N. Norizan, S. S. Shazleen, A. H. Alias, F. A. Sabaruddin, M. R. M. Asyraf, E. S. Zainudin, N. Abdullah, Md. S. Samsudin, S. H. Kamarudin, Md. N. F. Norrrahim, Nanocellulose-based nanocomposites for sustainable applications: A review, *Nanomater.* 12 (2022) 3483–3534.
25. Md. N. F. Norrrahim, N. Md. Nurazzi, Md. A Jenol, Md. A. A. Farid, N. Janudin, F. A. Ujang, T. A. T. Yasim-Anuar, S. U. F. S. Najmuddin, R. A. Ilyas, Emerging development of nanocellulose as an antimicrobial material: An overview, *Mater. Adv.* 2 (2021) 3538–3551.
26. M. Saini, M. Joon, S. K. Saini, *Human Health Effects of Particulate Matter*, Airborne Particulate Matter Springer, Singapore (2022) 195–228.
27. S. Sepahvand, H. Kargarzadeh, M. Jonoobi, A. Ashori, S. Ismaeilimoghadam, R. T. Varghese, C. J. Chirayl, B. Azimi, S. Danti, Recent developments in nanocellulose-based aerogels as air filters: A review, *Int. J. Bio. Macromol.* 15 (2023) 125721.
28. S. F. Nitodas, M. Skehan, H. Liu, R. Shah, Current and potential applications of green membranes with nanocellulose, *Membr.* 13 (2023) 694–710.
29. S. Naseem, R. Sajid, M. Nabeel, A. Sadiqa, M. Rizwan, M. R. Zulfiqar, A. Ahmad, D. N. Iqbal, Advancing nanocellulose-based biosensors: Pioneering eco-friendly solutions for biomedical applications and sustainable material replacement, *Int. J. Bio. Macromol.* 309 (2025) 143057.
30. M. K. Nsanya, M. Willkens, R. N. Peck, S. H. Kapiga, E. C. Nyanza, Exposure to toxic chemical elements (Pb, Cd, and Hg) and its association with sustained high blood pressure among secondary school attending adolescents in Northwestern Tanzania, *Environ. Res.* 278 (2025) 121738.
31. S. Saxena, A. K. Srivastava, Chapter 10- Carbon nanotube-based sensors and their application, edited by S. Thomas, Y. Grohens, G. Vignaud, N. Kalarikkal, J. James, *Nano-Optics, Fundamentals, Experimental Methods, and Applications, Micro and Nano Technologies*, Elsevier (2020) 265–291.
32. V. Schroeder, S. Savagatrup, M. He, S. Lin, T. M. Swage, Carbon nanotube chemical sensors, *Chem. Rev.* 119 (2018) 599–663.
33. C. Liné, C. Larue, E. Flahaut, Carbon nanotubes: Impacts and behaviour in the terrestrial ecosystem - A review, *Carbon.* 123 (2017) 767–785.

34. C. Zhang, L. Wu, M.de Perrot, X. Zhao, Carbon nanotubes: A summary of beneficial and dangerous aspects of an increasingly popular group of nanomaterials, *Front Oncol.* 27 (2021) 69381.
35. M. Endo, S. Tsuruoka, G. Ichihara, Carbon Nanotubes in historical and future perspective Summary of an Extended Session at Carbon 2008 in Nagano (JP), *Part Fibre Toxicol.* 10 (2008) 5–21.
36. A. Tripathi, K. Sarkar, S. Patel, U. Mishra, K. P. D. Nigam, D. Panda, K. G. Biswas, Elucidating flow-directed 98% CO_2 absorption using millimeter-sized coiled flow inverters: Nanocellulose-aided sustainable scope, *Chem. Eng. Process.: Process Intensif.* 205 (2024) 110022.
37. F. Baraka, J. Labidi, The emergence of nanocellulose aerogels in CO_2 adsorption, *Sci. Total Environ.* 20 (2024) 169093.
38. S. N. Khan, M. Zhao, F. Abbas, F. M. Enujekwu, Engineering strategies and structural-property relationships of metal-organic frameworks (MOFs) toward enhanced CO_2 adsorption, Engineering strategies and structural-property relationships of metal-organic frameworks (MOFs) toward enhanced CO_2 adsorption, *Chem. Eng. Sci.* 320 (2026) 122626.
39. L. Chavignon, C. Blanc, T. Serodre, M. C. Morris, E. Anglaret, Visible and near infrared absorption and fluorescence studies of the adsorption and coverage of doxorubicin on single-walled carbon nanotubes, *Spectrochim. Acta Part A: Mol. Biomol. Spectrosc.* 347 (2026) 127013.
40. H. Kim, J. Oh, H. Lee, S. Jeong, S. Hwan Ko, Next-generation air filtration nanotechnology for improved indoor air quality, *Chem. Commun.* 61 (2024) 1322–1341.
41. X. Jiao, K. Jia, Y. Yu, D. Liu, J. Zhang, K. Zhang, H. Zheng, X. Sun, Y. Tong, Q. Wei, P. Lv, Nanocellulose-based functional materials towards water treatment, *Carbohydr. Polym.* 350 (2025) 122977.
42. C. Duan, P. Zhang, Min Qi, Recent advances in the extraction of nanocellulose from lignocellulosic waste for wastewater treatment applications, *Int. J. Biol. Macromol.* 304 (2025) 140761.
43. C. R. P. Patel, P. Tripathi, O. N. Srivastava, T. P. Yadav, Filtration of sodium chloride from seawater using carbon hollow tube composed of carbon nanotubes, *Int. J. Smart and Nano Mater.* 5 (2014) 194–206.
44. A. Barhoum, K. Deshmukh, M. L. G. Betancourt, S. Alibakhshi, S. M. Mousavi, A. Meftahi, M. S. K. Sabery, P. Samyn, Nanocelluloses as sustainable membrane materials for separation and filtration technologies: Principles, opportunities, and challenges, *Carbohydr. Polym.* 317 (2023) 121057.
45. P. R. Sharma, S. K. Sharma, B. S. Hsiao, Chapter 5- Nanocellulose in membrane technology for water purification, *Sep. Sci. Technol.* 15 (2022) 69–85.
46. Y. W. Ning, X. Luo, Y. Zhang, P. Tian, Xiao, S. Li, X. Yang, Fan Li, D. Zhang, S. Zhang, Y. Liu, Broad-spectrum nano-bactericide utilizing antimicrobial peptides and bimetallic Cu-Ag nanoparticles anchored onto multiwalled carbon nanotubes for sustained protection against persistent bacterial pathogens in crops, *Int. J. Bio. Macromol.* 265 (2024) 131042.
47. S. S. Ogunsola, M. E. Oladipo, P. O. Oladoye, M. Kadhom, Carbon nanotubes for sustainable environmental remediation: A critical and comprehensive review, *Nano-Struct. Nano-Objects.* 37, 2024, 101099.
48. G. Z. Kyzas, K. A. Matis, Nanoadsorbents for pollutants removal: A review, *J. Mol. Liq.* 203 (2015) 159–168.
49. C. B. Pohrmen, Mrinal, C. Deepika, K. K. Jaiswal, V. Kumar, Amit K. Jaiswal, Green heterogeneous noble nanocatalysts and catalyst-support materials for the production of microalgal biofuels, *Mater. Today Sustain.* 23 (2025) 101226.
50. A. K. Rana, E. Mostafavi, W. F. Alsanie, S. S. Siwal, V. K. Thakur, Cellulose-based materials for air purification: A review, *Ind. Crops and Prod.* 194 (2023) 116331.

51. V. Vaiano, D. Sannino, O. Sacco, Chapter 9- The use of nanocatalysts (and nanoparticles) for water and wastewater treatment by means of advanced oxidation processes, nanotechnology in the beverage industry, edited by A. Amrane, S. Rajendran, T. A. Nguyen, A. A. Assadi, A. M. Sharoba, *Fundamentals and Applications, Micro and Nano Technologies*, Elsevier (2020) 241–264.
52. S. F. Ahmed, M. Mofijur, B. Ahmed, T. Mehnaz, F. Mehejabin, D. Maliat, A. T. Hoang, G. M. Shafiullah, Nanomaterials as a sustainable choice for treating wastewater, *Environ. Res.* 214 (2022) 113807.
53. M. B. de Andrade, A. C. S. Guerra, T. R. T. d. Santos, L. F. Cusioli, R. de Souza Antônio, R. Bergamasco, Simplified synthesis of new GO-α-γ-Fe2O3-Sh adsorbent material composed of graphene oxide decorated with iron oxide nanoparticles applied for removing diuron from aqueous medium, *J. Environ. Chem. Engr.* 8 (2020) 103903.

3 Economic Perspectives and Industrial Regulations about Regenerated Cellulose Materials

Umesh Bhardwaj and Anil Khachi

3.1 INTRODUCTION

The materials industry has seen significant change as a result of the shift to sustainable development. Industries are now looking for renewable and biodegradable materials due to growing concerns about plastic waste, pollution, and excessive use of fossil fuels [1]. Regenerated cellulose is one such choice that has gained a lot of popularity recently [2]. It is regarded as a good, environmentally friendly substitute for many plastics and is made from natural cellulose, the most prevalent organic material found in plants. Using these materials supports international efforts to reduce plastic pollution and lessens the use of non-renewable resources.

Regenerated cellulose is valuable because it is recyclable, renewable, and biodegradable [2]. Unlike traditional plastics, it can break down naturally without producing harmful microplastics. It is also robust, thermally stable, and versatile for a range of uses. Because of these qualities, it can be applied to a variety of fields, including filtration, textiles, packaging, and biomedical applications. New technologies such as closed-loop production and the use of ionic liquids to dissolve cellulose have made its production more efficient and cleaner [3].

Following this introduction, the economic and regulatory concerns surrounding regenerated cellulose are discussed. Cellulose-based materials are becoming a crucial component of creating a circular and ecologically friendly economy as nations and industries strive toward global sustainability goals like the Paris Agreement and the UN Sustainable Development Goals (SDGs) [4].

When scientists discovered how to transform natural cellulose into more workable forms in the late 19th century, regenerated cellulose's history began [5]. The development of rayon, sometimes referred to as "artificial silk," in the 1890s marked a significant turning point because it was significantly less expensive while still having the same appearance and feel as silk [6].

DOI: 10.1201/9781003596127-3

The viscose process emerged as the primary technique for producing regenerated cellulose fibers by the middle of the 20th century [7]. However, carbon disulfide (CS_2), a hazardous chemical that harmed the environment and put workers' health at risk, was used in this process [8]. These issues prompted scientists to look for safer production techniques. As a result, the lyocell method—which dissolved cellulose using *N*-methylmorpholine-*N*-oxide (NMMO)—was developed in the 1980s and 1990s. Because it allowed for cleaner, closed-loop fiber production and decreased hazardous chemicals, this was a significant improvement [9–14].

The use of ionic liquids and other green solvents has advanced further in recent years [15, 16]. These new techniques make it easier to dissolve cellulose, recover solvents more efficiently, and produce less waste. All of these advancements have enhanced regenerated cellulose's industrial scalability and environmental sustainability. The emergence of these materials reflects a global shift away from resource-intensive production and toward circular and eco-friendly manufacturing methods.

3.2 CELLULOSE SOURCES

Cellulose comes from a variety of natural sources, each with its own special qualities and processing needs. Most of the cellulose used today is made from wood pulp, which supplies nearly 90% of the global supply [17]. Hardwoods like birch and eucalyptus and softwoods like pine and spruce differ in their fiber structure, lignin content, and ease of processing. Cotton linters, or the short fibers left over after cotton is ginned, are another important source. Because they contain almost pure cellulose, they are used to make high-end products like specialty papers and regenerated fibers.

More recently, agricultural waste materials can also be used to extract cellulose [18]. Byproducts such as sugarcane bagasse, rice husks, wheat straw, and banana pseudostems provide cheap cellulose and help rural communities while reducing waste [19]. These materials are converted into regenerated cellulose fibers, which reduces forest pressure and turns agricultural waste into something beneficial.

Research is also being done on more recent sources, such as algal and microbial cellulose [20, 21]. *Acetobacter xylinum* and other bacteria produce bacterial cellulose, which is extremely strong, has a fine nanoscale fiber structure, and is highly biocompatible. It is appropriate for biomedical and other cutting-edge applications because of these qualities [22].

3.3 ECONOMIC PERSPECTIVES OF REGENERATED CELLULOSE

The global market for regenerated cellulose fibers has grown significantly over the last ten years due to the need for sustainable and high-performing materials. The market was valued at USD 25 billion in 2024, according to MarketWatch (2024), and is projected to reach USD 40 billion by 2030 [23–25] at an average annual growth rate of about 8%. Because of their robust textile manufacturing infrastructure and easy access to a wealth of raw materials, the Asia-Pacific region—especially China and India—dominates global production. Meanwhile, growing environmental consciousness and the adoption of sustainable fashion practices have made Europe and North America significant consumers.

Tables 3.1–3.3showan overall picture of the global economic status of regenerated cellulose materials. The regenerated cellulose market has been expanding steadily, as Table 3.1 demonstrates. By 2030, it is anticipated to have grown from roughly USD 18 billion in 2022 to USD 35–40 billion. This growth, which has a CAGR of roughly 8%–9%, is indicative of the global shift toward sustainable and renewable materials. The growing market value also demonstrates how regenerated cellulose has spread from its primary application in textiles to its widespread use in filtration, packaging, and medical products.

Table 3.2 highlights that viscose fibers dominate the man-made cellulosic fiber (MMCF) sector, making up nearly 80% of total production. Although modal and lyocell make up smaller shares, their demand is increasing quickly because they offer

TABLE 3.1
Global market size estimates for regenerated cellulose

Year	Estimated Market Size (USD billion)	CAGR/Notes	Source
2022	~18.0 billion	—	[26]
2024	~21.56–21.2 billion	—	[27]
2027	~27.3 billion	~8.7% CAGR (2022–2027)	[28]
2030	~35.44 billion	~8.6% CAGR (2024–2030)	[29]
2032	~42.51 billion	~8.85% CAGR (2024–2032)	[30]
2035	~52.8 billion	~8.65% CAGR (2025–2035)	[31]

TABLE 3.2
Production volumes of Man-Made Cellulosic Fibers—by type

Fiber Type	Approx. Production Volume (Million Tons)	Share of MMCFs	Year	Source
Viscose (Rayon)	~5.8	~80%	2022	[32]
Modal	~0.2	~3% of MMCFs	2022	[33]
Lyocell	~0.3–0.37	~4%–5% of MMCFs	2022/2023	[34]

TABLE 3.3
Regional market share-regenerated cellulose (2023)

Region	Approximate Share of Revenue	Source
Asia-Pacific	~40%	[35]
Europe	~30%	
North America	~15%–20%	

better mechanical properties and have a lower environmental impact. These numbers show that the industry is moving toward cleaner technologies, especially processes like lyocell and ionic-liquid-based production.

Table 3.3 shows how different regions contribute to the global market. The Asia-Pacific region leads with about 40% of total revenue. This is mainly because the region has a strong textile industry, easy access to raw materials, and supportive government policies, especially in China and India. Europe and North America also play important roles, mainly because of strict environmental rules and growing customer interest in sustainable products. Together, these tables show the fast economic growth of the regenerated cellulose sector and its importance in helping the world shift toward a bio-based and circular economy. From an economic point of view, regenerated cellulose fibers allow industries to adopt environmentally friendly materials without losing quality or profit. For example, lyocell fibers hold a premium position in clothing markets because they are soft, breathable, and produced through a cleaner process. Modal fibers, known for strength and comfort, are common in home textiles and activewear. Increasing demand for eco-labeled and sustainable fashion has also boosted market growth.

The future growth of this sector depends on continued technological improvements and compliance with environmental rules. Closed-loop systems that recover solvents and recycle wastewater reduce both costs and environmental harm. Integrating cellulose fiber production with biorefineries—where fibers, biofuels, and biochemicals are produced together—also improves resource efficiency and profitability. Many governments support such efforts through tax benefits, incentives, and research funding that encourage circular economic practices. Companies that focus on transparent supply chains, sustainable forestry, and certifications like OEKO-TEX and FSC gain a strong advantage in the market [36]. These certifications help build trust among consumers and allow companies to charge premium prices, supporting both economic growth and environmental responsibility.

3.4 TECHNOLOGICAL INNOVATIONS IN REGENERATION

Significant technological advancements in the last few decades have changed the process of cellulose regeneration. Newer and cleaner techniques are quickly becoming more significant, even though the conventional viscose process still accounts for the majority of global production. The lyocell process, which employs *N*-methylmorpholine-*N*-oxide (NMMO) in a closed-loop system, is among the most effective inventions [37–39]. Lyocell is one of the most environmentally friendly large-scale fiber production techniques, thanks to this system, which recovers more than 99% of the solvent while drastically reducing emissions and energy consumption.

Beyond lyocell, researchers have developed ionic liquid-based systems, such as 1-ethyl-3-methylimidazolium acetate ([EMIM][OAc]), which can dissolve cellulose directly without chemical modification [40]. These solvents are recyclable, adjustable in their properties, and well suited for sustainable manufacturing. Another promising approach involves NaOH–urea aqueous systems, which provide a low-cost method for dissolving cellulose at sub-zero temperatures [41]. When combined with enzymatic pretreatments, these systems can further improve fiber quality and process efficiency.

Moreover, the introduction of digital and automation technologies is reshaping the cellulose industry [42]. Tools such as process monitoring, real-time data analytics, and predictive control models help improve solvent recovery, optimize spinning conditions, and reduce production waste. The integration of biotechnology, green chemistry, and intelligent process control marks a new era in cellulose regeneration—one that focuses on efficiency, sustainability, and minimal environmental impact.

3.5 INDUSTRIAL REGULATION FOR REGENERATED CELLULOSE

In many nations, safety and environmental laws regulate the manufacture and sale of products made from regenerated cellulose. These rules guarantee that the materials are produced responsibly, safely, and cleanly. The chemicals used in the production of viscose and other cellulose fibers in Europe are strictly regulated by the REACH regulation (Registration, Evaluation, Authorization and Restriction of Chemicals) [43]. By restricting emissions, mandating solvent recovery, and safeguarding employees from dangerous chemicals, the European Chemicals Agency upholds these regulations.

The Environmental Protection Agency (EPA) in the United States carries out similar tasks by enforcing laws such as the Clean Air Act and the Clean Water Act to control pollution and manage chemical waste in an appropriate manner [44]. OSHA also ensures that workers in the cellulose fiber industry are protected from dangerous chemicals such as carbon disulfide and hydrogen sulfide [45]. Together, these steps make production safer for the workers and the environment.

In addition to national laws, international organizations promote sustainable and ethical production. Businesses are encouraged by the International Organization for Standardization (ISO) and the United Nations Environment Program (UNEP) to use resources efficiently, reduce waste, and clearly disclose their environmental impacts [46]. Certification programs such as OEKO-TEX, the EU Ecolabel, the FSC, and the Program for the Endorsement of Forest Certification (PEFC) are in place to ensure that products meet environmentally friendly standards and that raw materials come from forests that are responsibly managed [47].

Policies such as the European Green Deal and the United Nations Sustainable Development Goals also support recycling, waste reduction, and the use of renewable materials. Many countries are now adopting Extended Producer Responsibility (EPR) rules, which require manufacturers to manage their products even after use, including recycling and safe disposal.

When taken as a whole, these policies and programs are changing the regenerated cellulose market. They advocate for transparent supply chains, ethical sourcing, and cleaner technologies. Adhering to these rules enhances a company's standing in the international market, fosters public trust, and protects the environment.

Certifications for sustainability are also crucial for guaranteeing accountability. The PEFC and the FSC attest to the ethical sourcing of wood pulp. Environmental and social standards are met during manufacturing, thanks to textile-focused certifications like the EU Ecolabel, the Global Organic Textile Standard, and the OEKO-TEX Standard 100 [48]. Verification of sustainability claims is aided by international trade and labeling standards such as ISO 14024 and ISO 26000 [49]. EPR programs, which hold manufacturers accountable for the full life cycle of their products, including post-consumer waste, are being expanded in many nations. By rewarding businesses that use cleaner technologies, these changing regulations not only safeguard the environment but also promote innovation.

3.6 SUSTAINABILITY, CIRCULAR ECONOMY, AND INDUSTRIAL APPLICATIONS

The sustainable production of regenerated cellulose materials now revolves around the idea of the circular economy. Waste reduction, resource efficiency, and product reuse are prioritized in a circular system. This entails creating procedures for cellulose production that minimize water use, reuse waste pulp, and achieve solvent recovery rates close to 100%. Regenerated cellulose fibers have substantially lower carbon footprints than polyester and nylon, according to Life Cycle Assessment studies. For example, Lyocell produces no harmful byproducts and uses less energy and water. Additionally, improved strength-to-weight ratios provided by cellulose-based composites and nanocellulose materials are creating new opportunities in the biomedical, automotive, and packaging industries. From advanced biocomposites used in aerospace and medicine to nonwoven textiles, hygiene products, and industrial filters, industrial applications are still growing. Their adaptability and compatibility with contemporary sustainability frameworks provide an economic advantage. Cellulose fibers are becoming essential components of many businesses' green product portfolios as they implement bio-based branding strategies.

3.7 CHALLENGES AND FUTURE PROSPECTS

The broad use of regenerated cellulose materials is hampered by a number of issues, despite impressive technological advancements. The significant financial outlay needed for green manufacturing infrastructure is one of the main obstacles [50]. Setting up solvent recovery and closed-loop systems requires large upfront expenditures. Additionally, prices are still being driven down by competition from inexpensive synthetic fibers like polyester. The management of the raw material supply chain presents another difficulty. Strict forestry certifications must be followed in order to source wood pulp sustainably, which could reduce supply and raise prices. Collecting, shipping, and preprocessing cellulose made from agricultural waste presents logistical challenges. Nonetheless, there is still a bright future ahead. The market for sustainable fiber is anticipated to be dominated by regenerated cellulose materials due to rising consumer awareness, government incentives, and the global shift to carbon neutrality [51]. New developments include the investigation of bioengineered microbial cellulose for high-performance applications, the integration of nanocellulose and lignin composites, and the 3D printing of cellulose-based materials. Overcoming technological and economic obstacles will require international cooperation between academia, business, and policymakers. Adoption and market growth can be further accelerated by lifecycle-based policy designs, sustainability reporting, and open innovation platforms.

3.8 CONCLUSION

Materials made from regenerated cellulose represent a harmonious fusion of technological innovation, environmental responsibility, and economic feasibility. Their increasing importance in contemporary industries is a reflection of the global movement away from conventional, linear production systems and toward more circular and sustainable models. Regenerated cellulose fibers provide a high-performance, biodegradable, and renewable substitute for traditional petrochemical-based materials as environmental regulations tighten and sustainability becomes a key component of corporate strategies.

Economically speaking, the market for regenerated cellulose is still growing due to continued research, advancements in technology, encouraging government regulations, and growing consumer demand for environmentally friendly goods. The industry's momentum is also anticipated to be strengthened by regulatory initiatives to promote green manufacturing and harmonize international standards.

Regenerated cellulose materials are an example of how industries can move toward sustainability without compromising performance or profitability by fusing scientific innovation, economic opportunity, and environmental policy. They represent a significant advancement in technology as well as a step in the direction of a more accountable and resource-efficient global economy.

REFERENCES

1. Operamolla, A., 2019. Recent advances on renewable and biodegradable cellulose nanopaper substrates for transparent light-harvesting devices: Interaction with humid environment. *International Journal of Photoenergy*, 2019(1), p.3057929.
2. Tu, H., Zhu, M., Duan, B. and Zhang, L., 2021. Recent progress in high-strength and robust regenerated cellulose materials. *Advanced Materials*, 33(28), p.2000682.
3. Wang, H., Gurau, G. and Rogers, R.D., 2012. Ionic liquid processing of cellulose. *Chemical Society Reviews*, 41(4), pp.1519–1537.
4. Verkerk, P.J., Hassegawa, M., Van Brusselen, J., Cramm, M., Chen, X., Imparato Maximo, Y. and Koç, M., 2021. *Forest products in the global bioeconomy: Enabling substitution by wood-based products and contributing to the Sustainable Development Goals*. Food & Agriculture Org.

5. Woodings, C., 2001. A brief history of regenerated cellulosic fibers. In *Regenerated cellulose fibers* (pp.1–21). Boca Raton, FL, Boston, MA, New York, Washington, DC: CRC Press.
6. Luft, M.G., 1925. Rayon-man-made silk. *Journal of Chemical Education*, 2(10), p.864.
7. Wilkes, A.G., 2001. The viscose process. In *Regenerated cellulose fibres* (pp.37–61). Boca Raton, FL, Boston, MA, New York, Washington, DC: CRC Press.
8. Sieja, K., von Mach-Szczypiński, J. and von Mach-Szczypiński, J., 2017. Health effect of chronic exposure to carbon disulfide (CS_2) on women employed in viscose industry. *Medycyna Pracy. Workers' Health and Safety*, 69(3), pp.329–335.
9. Perepelkin, K.E.E., 2007. Lyocell fibres based on direct dissolution of cellulose in N-methylmorpholine N-oxide: Development and prospects. *Fibre Chemistry*, 39(2), pp.163–172.
10. Rosenau, T., Potthast, A., Adorjan, I., Hofinger, A., Sixta, H., Firgo, H. and Kosma, P., 2002. Cellulose solutions in N-methylmorpholine-N-oxide (NMMO)–degradation processes and stabilizers. *Cellulose*, 9(3), pp.283–291.
11. Sawiak, M., Souto, B.A., Lawson, L., Lo, J. and Dolez, P.I., 2025. Recovery of N-Methylmorpholine N-Oxide (NMMO) in lyocell fibre manufacturing process. *Fibers*, 13(1), p.3.
12. Rosenau, T. and French, A.D., 2021. N-Methylmorpholine-N-oxide (NMMO): Hazards in practice and pitfalls in theory. *Cellulose*, 28(10), pp.5985–5990.
13. Hytönen, E., Sorsamäki, L., Kolehmainen, E., Sturm, M. and von Weymarn, N., 2023. Lyocell fibre production using NMMO–A simulation based techno-economic analysis. *BioResources*, 18(3), pp.6384–6411.
14. Rosenau, T., Potthast, A., Milacher, W., Adorjan, I., Hofinger, A. and Kosma, P., 2005. Discoloration of cellulose solutions in N-methylmorpholine-N-oxide (Lyocell). Part 2: Isolation and identification of chromophores. *Cellulose*, 12(2), pp.197–208.
15. Zhao, H., Xia, S. and Ma, P., 2005. Use of ionic liquids as 'green' solvents for extractions. *Journal of Chemical Technology & Biotechnology: International Research in Process, Environmental & Clean Technology*, 80(10), pp.1089–1096.
16. Earle, M.J. and Seddon, K.R., 2000. Ionic liquids. Green solvents for the future. *Pure and Applied Chemistry*, 72(7), pp.1391–1398.
17. Marinho, E., 2025. Cellulose: A comprehensive review of its properties and applications. *Sustainable Chemistry for the Environment*, 11, p.100283.
18. Bian, H., Gao, Y., Luo, J., Jiao, L., Wu, W., Fang, G. and Dai, H., 2019. Lignocellulosic nanofibrils produced using wheat straw and their pulping solid residue: From agricultural waste to cellulose nanomaterials. *Waste Management*, 91, pp.1–8.
19. Oyedeji, S., Patel, N., Krishnamurthy, R. and Fatoba, P. O., 2024. Agricultural wastes to value-added products: Economic and environmental perspectives for waste conversion. In *Biowaste to value-added products: Economics and technologies* (pp. 215248). Cham: Springer.
20. Gunnison, D. and Alexander, M., 1975. Basis for the susceptibility of several algae to microbial decomposition. *Canadian Journal of Microbiology*, 21(5), pp.619–628.
21. Gunnison, D. and Alexander, M., 1975. Basis for the resistance of several algae to microbial decomposition. *Applied Microbiology*, 29(6), pp.729–738.
22. Nicu, R., Ciolacu, F. and Ciolacu, D.E., 2021. Advanced functional materials based on nanocellulose for pharmaceutical/medical applications. *Pharmaceutics*, 13(8), p.1125.
23. Stenton, M., Houghton, J.A., Kapsali, V. and Blackburn, R.S., 2021. The potential for regenerated protein fibres within a circular economy: Lessons from the past can inform sustainable innovation in the textiles industry. *Sustainability*, 13(4), p.2328.
24. Zainul Armir, N.A., Zulkifli, A., Gunaseelan, S., Palanivelu, S.D., Salleh, K.M., Che Othman, M.H. and Zakaria, S., 2021. Regenerated cellulose products for agricultural and their potential: A review. *Polymers*, 13(20), p.3586.

25. Woodings, C. ed., 2001. *Regenerated cellulose fibres*. Boca Raton, FL, Boston, MA, New York, Washington, DC: Elsevier.
26. Fortune Business Insights, Regenerated Cellulose Market Size, Share, Growth & Trends (2024–2032), https://www.fortunebusinessinsights.com/regenerated-cellulose-market-102020, Retrieved on November 11, 2025.
27. Mordor Intelligence, Regenerated Cellulose Market Report – Growth, Trends, and Forecasts (2024–2029), https://www.mordorintelligence.com/industry-reports/regenerated-cellulose-market, Retrieved on November 11, 2025.
28. Market Research Future (MRFR), Regenerated Cellulose Market Research Report – Global Forecast till 2030, https://www.marketresearchfuture.com/reports/regenerated-cellulose-market-3098, Retrieved on November 11, 2025.
29. Precedence Research, Regenerated Cellulose Market – Global Industry Analysis (2024–2033), Retrieved on November 11, 2025.
30. MarketWatch, Regenerated Cellulose Market Overview, Demand, and Forecast Report (2024–2030), https://www.marketwatch.com/press-release/regenerated-cellulose-market-forecast-2024-2030, Retrieved on November 11, 2025.
31. Research and Markets, Regenerated Cellulose Market Size, Share & Forecast Report (2024–2031), https://www.researchandmarkets.com/reports/5712357/regenerated-cellulose-market, Retrieved on November 11, 2025.
32. Textile Exchange, Materials Market Report 2023- Virgin Man-Made Cellulosic Fibers (MMCFs) Production Facts and Figures, November 11, 2025.
33. Textile Exchange, Materials Market Report 2024 – Global MMCF Production and Share by Type, November 11, 2025.
34. Fortune Business Insights, Man-Made Cellulose Fiber Market Size, Share & Forecast 2024–2032, https://www.fortunebusinessinsights.com/man-made-cellulose-fiber-market-102759, November 11, 2025.
35. Verified Market Reports, Regenerated Cellulose Fiber Market Size, Expansion, SWOT & Forecast 2033 – Regional Contribution in 2023, Retrieved on November 11, 2025.
36. Walter, C. and Pethile, D., 2025. Regulatory standards and certifications for sustainable textiles. In *Advances in textile materials and processing techniques for sustainability* (pp. 413–451). Singapore: Springer Nature Singapore.
37. Sawiak, M., Souto, B.A., Lawson, L., Lo, J. and Dolez, P.I., 2025. Recovery of N-Methylmorpholine N-Oxide (NMMO) in lyocell fibre manufacturing process. *Fibers*, 13(1), p.3.
38. Plakantonaki, S., Tournis, I., Zacharopoulos, N., Kouvelos, E., Sapalidis, A.A. and Athanasekou, C., 2025. Introducing reverse osmosis and membrane distillation in the recovery process of N-Methylmorpholine N-Oxide in cellulose fiber production. *Recycling*, 10(5), p.194.
39. Reipsar, A., 2020. Modelling of lyocell fibre production and technical analysis of solvent recovery concepts.
40. Tan, X., Li, X., Chen, L. and Xie, F., 2016. Solubility of starch and microcrystalline cellulose in 1-ethyl-3-methylimidazolium acetate ionic liquid and solution rheological properties. *Physical Chemistry Chemical Physics*, 18(39), pp.27584–27593.
41. Korhonen, O., 2020. Short-fiber reinforced all-cellulose composites using aqueous NaOH solution.
42. Särkkä, T., Gutiérrez-Poch, M. and Kuhlberg, M., 2018. Technological transformation in the global pulp and paper industry: Introduction. In *Technological transformation in the global pulp and paper industry 1800–2018: Comparative perspectives* (pp. 1–10). Cham: Springer International Publishing.
43. Bergkamp, L. ed., 2013. *The European Union REACH regulation for chemicals: Law and practice*. Oxford: OUP Oxford.

44. Act, C.A., Act, R. and Act, R., 2006. Environmental protection agency (EPA). Report on Carcinogens, 168.
45. Sapuan, S.M., Ilyas, R.A. and Asyraf, M.R.M., 2021. Occupational safety and health administration in composite industry. In *Safety and health in composite industry* (pp. 229–252). Singapore: Springer Singapore.
46. Clark, G., 2007. Evolution of the global sustainable consumption and production policy and the United Nations Environment Programme's (UNEP) supporting activities. *Journal of Cleaner Production*, 15(6), pp.492–498.
47. Almeida, L., 2014. Ecolabels and organic certification for textile products. In *Roadmap to sustainable textiles and clothing: Regulatory aspects and sustainability standards of textiles and the clothing supply chain* (pp. 175–196). Singapore: Springer Singapore.
48. Lemes, P.G., Zanuncio, J.C., Jacovine, L.A., Wilcken, C.F. and Lawson, S.A., 2021. Forest stewardship council and responsible wood certification in the integrated pest management in Australian forest plantations. *Forest Policy and Economics*, 131, p.102541.
49. Minkov, N., Lehmann, A., Winter, L. and Finkbeiner, M., 2020. Characterization of environmental labels beyond the criteria of ISO 14020 series. *The International Journal of Life Cycle Assessment*, 25(5), pp.840–855.
50. Koh, J.M., 2018. *Green infrastructure financing*. Cham: Springer International Publishing.
51. Chi, T., Frattali, A., Liu, H. and Chen, Y., 2023. Regenerated Cellulose Fibers (RCFs) for future apparel sustainability: Insights from the US consumers. *Sustainability*, 15(6), p.5404.

4 Commercial Sources of Cellulose

Technological Challenges and Economic Viabilities in the Global Perspective

Sukhdeep Singh, Jitendra Basrani, Gulab Singh, Pramod Rakt Patel, and Manjeet Singh

4.1 INTRODUCTION

Cellulose, the world's most abundant organic polymer, forms the structural basis of plant cell walls. This linear chain of glucose units is organized into a hierarchical structure, giving it significant strength and versatile chemistry. Long used in paper and textiles, cellulose is now a key feedstock for the bioeconomy. Its renewability, biodegradability, and ease of modification make it a prime candidate to replace fossil-fuel-based materials, from plastics to composites, aligning it with the goals of sustainable innovation (Habibi et al., 2010; Moon et al., 2011).

The commercial importance of cellulose is underscored by the size and trajectory of its markets. Demand drivers include packaging (especially the shift toward biodegradable and recyclable packaging), textile fibers, specialty cellulose derivatives for pharmaceuticals and personal care, and rapidly growing segments such as micro- and nanocellulose for advanced materials. Market analyses suggest substantial growth in both bulk cellulose and higher value derivatives, reflecting the dual nature of cellulose markets: large volumes for traditional applications and rapidly increasing value for specialty products (Data Bridge Market Research, 2023).

The industrial production of cellulose, the foundational polymer for the pulp, paper, and burgeoning bioeconomy, relies primarily on three distinct feedstock families. Each category presents a unique profile of advantages, limitations, and logistical considerations, shaping its role within the global supply chain. The most established and dominant source is **wood-based feedstocks**, which form the backbone of the traditional pulp and paper industry. This category is divided predominantly between softwood and hardwood species, each contributing specific fibrous qualities to the final product. The wood-based model benefits from over a century of development, resulting in sophisticated forestry management practices, integrated milling operations, and efficient global logistics that enable significant economies of scale.

DOI: 10.1201/9781003596127-4

However, this mature system faces increasing pressures. Wood-based supply chains are sensitive to land-use conflicts, biodiversity conservation concerns, and the substantial environmental footprint of conventional chemical pulping processes, notably the kraft and sulfite methods, which are energy intensive and generate significant emissions (Klemm et al., 2005).

As an alternative, **agricultural residues** offer a compelling pathway to valorize waste streams. Crop byproducts such as straw, sugarcane bagasse, corn stover, and rice husks represent a vast, often underutilized, reservoir of lignocellulosic material. Their use as cellulose feedstock provides a dual benefit: it reduces the environmental hazards associated with open-field burning, a common disposal method that causes severe air pollution, and it creates an additional revenue stream for the agricultural sector. Many of these residues also possess a high cellulose content, making them theoretically excellent raw materials. Nevertheless, their utilization is hampered by significant logistical hurdles. These materials are inherently geographically dispersed, seasonally available, and often have variable composition and high levels of contamination (such as soil and ash), which complicates collection, storage, and preprocessing economics (Kumar et al., 2020).

The third major family encompasses non-wood plants and specialty fibers. This diverse group includes cotton linters, hemp, flax, and fast-growing grasses like bamboo. These sources are often characterized by their high cellulose purity and, in some cases, superior fiber length and strength. Certain non-wood plants, particularly hemp and bamboo, exhibit favorable agronomic profiles, such as rapid-growth cycles and high yield per hectare. Bamboo, in particular, is receiving considerable attention, especially in Asia, as a rapid-growth feedstock for pulp and specialty cellulose. Its fast renewability and promising market growth are positioning it as a significant fiber source for the future (Yashi Paper, 2023a, 2023b; Verified Market Reports, 2023). While often requiring specialized processing infrastructure, these specialty fibers cater to high-value markets and offer a complementary stream to traditional wood and agricultural residue sources.

Each feedstock class comes with different implications for processing, product spectrum, and regional competitiveness. For instance, wood continues to dominate in regions with extensive forest resources and industrial pulp capacity, while bagasse and other residues are attractive in major agricultural economies (e.g., sugarcane producing regions) where co-location with mills lowers transport costs (Zhao, X., Zhang, L., & Liu, D. 2019).

The core technical challenge across feedstocks is cost-effective and sustainable fractionation: separating cellulose from hemicellulose, lignin, and extractives with sufficient yield, purity, and minimal environmental burden.

a. **Conventional pulping**. Kraft and sulfite pulping are mature, high-yield processes but involve substantial chemical and energy inputs and produce effluents that require careful management. Although these processes remain cost-effective at scale, their environmental footprint and suitability for specialty, high-purity cellulose are limited.
b. **Emerging pretreatments and green solvents**. Organosolv, ionic liquids, and deep eutectic solvents (DES) represent promising routes to gentler

fractionation and easier lignin recovery. These approaches can produce higher-purity cellulose streams and enable valorization of co-products, but they face practical hurdles: solvent cost, solvent recovery and recycling, corrosion and material compatibility, and scale-up challenges that impact overall techno-economic viability. Recent perspectives emphasize solvent recovery and lignin chemistry as the pivotal engineering and economic constraints for ionic-liquid-based biorefineries (Melikoglu, M., 2025).

c. **Enzymatic and biological approaches**. Enzymatic hydrolysis and tailored cellulase systems offer lower-temperature processing and higher selectivity, which is attractive for producing specialty cellulose and fermentable sugars in integrated biorefineries. Advances in cellulase engineering and production technologies aim to reduce enzyme costs and increase activity, but enzyme economics and process integration remain critical barriers for many commercial pathways (Ranjan, R et al. 2023).

d. **Nanocellulose production**. Microfibrillated and nanocrystalline cellulose open new high-value markets (e.g., composites, barrier coatings, rheology modifiers). However, producing nanocellulose at competitive cost and consistent quality requires energy-efficient fibrillation, controlled chemical treatments, and standardized characterization methods—all active areas of industrial R&D and pilot demonstrations. Market reports forecast rapid growth for nanocellulose, reflecting both industrial interest and the promise of diversified end uses (Global Microp and Nanocellulose Market Report 2024–2035).

e. **Economic and regional considerations**. Economic viability hinges on factors such as raw material proximity to processing capacity, modality of transport, scale of operations, and policy support (subsidies, R&D incentives, and waste-valorization schemes). Developed economies typically invest in advanced processing and higher-value cellulose derivatives, facilitated by existing capital and R&D ecosystems (Lynd et al., 2017; Kumar et al., 2020; Popp et al., 2021). Conversely, emerging economies often supply substantial raw cellulose (e.g., agricultural residues, non-wood fibers) but must balance economic benefits against land-use impacts and investment needs for clean processing technologies (Carus et al., 2015). For example, sugarcane-rich regions have a natural synergy for bagasse utilization in pulping and energy, while bamboo utilization is particularly relevant in certain Asian contexts where bamboo pulp production has scaled significantly (Yashi Paper, 2023a, 2023b). Price volatility, global trade flows, and competition from cheaper synthetic polymers add additional uncertainty (Rana et al., 2021). Achieving competitiveness for cellulose-based alternatives therefore requires either cost parity at scale (through process intensification and feedstock logistics) or premium pricing enabled by superior performance, biodegradability, or regulatory drivers that favor bio-based materials (IEA Bioenergy, 2022).

f. **Sustainability, circularity, and policy levers**. Sustainability considerations permeate all feedstock choices. Avoiding deforestation, maintaining soil health (when harvesting agricultural residues), and minimizing water

> and chemical footprints are essential (IEA Bioenergy, 2023). Circular approaches—recycling paper and textile cellulose streams, integrating biorefineries that valorize lignin and hemicellulose, and designing products for end-of-life recovery—can dramatically improve life-cycle outcomes and material economics (Clark et al., 2016). Policy interventions (carbon pricing, waste management regulations, public procurement policies favoring bio-based materials, and R&D funding) will strongly influence the pace and direction of cellulose commercialization (European Commission, 2020; IEA Bioenergy, 2022).

This chapter examines the principal commercial sources of cellulose—wood, agricultural residues, and non-wood plants—and evaluates the technological and economic constraints that influence their adoption at scale. The aim is to synthesize technical considerations (extraction/pulping, purification, and value-added conversion), supply chain and economic drivers (raw material availability, processing and logistics costs, competition from synthetic substitutes), and sustainability dimensions (land use, carbon footprint, and circularity). By framing the discussion from a global perspective, the chapter highlights regional strengths and trade-offs, and suggests pathways where innovations—technical and policy—can improve viability and sustainability.

4.2 COMMERCIAL SOURCES OF CELLULOSE

Cellulose feedstocks for commercial use fall into three practical categories: wood-based sources (industrial forestry and pulps), agricultural residues (byproducts such as bagasse and straw), and non-wood plant sources and specialty fibers (cotton linters, hemp, bamboo, etc.). Each class differs in availability, fiber chemistry, processing needs, environmental footprint, and regional competitiveness. Global cellulose production remains dominated by wood-based sources, with agricultural residues and marine biomass contributing smaller but steadily increasing shares, as illustrated in Figure 4.1. The relative contribution and cellulose content of these sources is highlighted in Table 4.1, in which wood-based, agricultural, and non-wood cellulose are compared. The following sections analyze these classes in detail and present regional case studies and quantitative context.

4.2.1 Wood-Based Cellulose

Wood remains the dominant source of cellulose globally, contributing nearly 180–200 million tons annually, primarily from softwoods and hardwoods. Softwoods, rich in long fibers, are preferred in packaging and construction-grade pulp, whereas hardwoods with shorter fibers are dominant in printing and specialty paper.

4.2.1.1 Softwood vs. Hardwood Pulp: Fiber Characteristics and Implications

Wood pulp remains the backbone of the global cellulose supply chain. Softwoods (e.g., pine, spruce) produce long, strong fibers that confer superior tensile strength and are preferred for printing and high-strength paper grades. Hardwoods (e.g.,

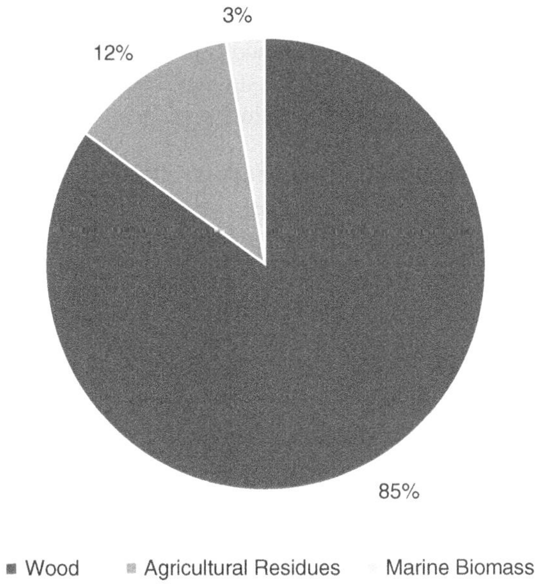

FIGURE 4.1 Global cellulose production from major sources.

TABLE 4.1
Major commercial sources of cellulose and their estimated global production volumes (compiled from FAO 2022a, 2022b; Klemm et al. 2018)

Source Type	Examples	Global Production (Million Tons/ Year)	Cellulose Content (%)	Key Applications
Wood-based	Softwood, hardwood	180–200	40–50	Pulp & paper, textiles, packaging
Agricultural residues	Wheat straw, corn stover, bagasse	100–120	30–40	Biofuels, pulp, composites
Non-wood plants	Cotton, hemp, bamboo, kenaf	25–30	35–95 (cotton highest)	Specialty papers, textiles, bioplastics

eucalyptus, birch) give shorter, finer fibers that suit smoother papers and many specialty applications. The chemical composition (lignin/hemicellulose ratios) and fiber geometry strongly influence pulping efficiency, bleaching demands, and end-product properties (Klemm et al., 2005; Sixta, 2006). These intrinsic differences also affect yield: hardwoods typically give higher pulp yield on a dry-mass basis, while softwoods often yield papers with better wet and dry strength; decision/industrial implication: mill choice of feedstock depends on the target product mix: newsprint and packaging favor long-fiber blends for strength, while specialty and fine papers may prioritize hardwoods or purified cellulose streams.

4.2.1.2 Major Producers and Global Supply Chains

Global wood pulp production is concentrated in regions with extensive commercial forestry and integrated pulp & paper industries. Recent statistical reporting indicates a multi-million ton annual industry: wood pulp production rose substantially in the early 2020s, with combined wood and recycled pulp supply supporting a large global paper and board sector. Europe, North America, China, Brazil, and Canada are among the largest producers, benefiting from vertical integration (forest → mill → transport) and established logistics that enable economies of scale (UNECE, 2022; CEPI, 2023b).

4.2.1.3 Economic and Environmental Trade-Offs

Wood-based pulping (kraft, sulfite) is mature and cost-effective at scale, but it carries environmental burdens: energy and chemical consumption, odor and effluent management, and land-use impacts when forestry is not sustainably managed. Certification schemes (FSC, PEFC) and mill-level investments in recovery boilers and closed-loop chemical recovery mitigate impacts, but the initial capital and operating complexity remain barriers in regions lacking infrastructure and regulatory frameworks (FAO, 2021).

4.2.2 Agricultural Residues

Agricultural residues are attractive because they are often abundant, low-cost, and represent valorization of a previously under-utilized waste stream. However, residue use raises unique logistical and processing challenges. Agricultural byproducts such as wheat straw, corn stover, and bagasse provide a renewable and low-cost cellulose stream, especially in emerging economies. However, the contribution varies significantly across regions. For instance, Brazil's sugarcane bagasse dominates Latin America, while India utilizes rice and wheat straw extensively. A regional comparison of forestry and agricultural cellulose production is summarized in Table 4.2.

4.2.2.1 Key Residue Types and Properties

Common residues used for cellulose extraction include sugarcane bagasse, straws (rice, wheat), corn stover, and husks (e.g., rice husk). These materials typically contain significant cellulose fractions but also higher proportions of hemicellulose and silica (especially rice husk), which affect pulping chemistry and equipment wear. Residue composition is more variable than wood, and contaminants (soil, stones) can increase preprocessing needs (Goodman, 2020).

4.2.2.2 Logistical and Seasonal Constraints

Residues are highly seasonal and geographically dispersed. Collection, densification (baling), transport distance, and storage (moisture control, biodegradation)

TABLE 4.2
Regional comparison of cellulose production from forestry and agriculture sectors (compiled from FAO 2022a, 2022b)

Region	Primary Source	Estimated Share of Global Output (%)	Notable Producers
North America	Softwood & hardwood	25–30	USA, Canada
Europe	Hardwood & agricultural	20–25	Finland, Sweden, Germany
Asia-Pacific	Bamboo, bagasse, cotton	35–40	China, India, Indonesia
Latin America	Sugarcane bagasse	10–15	Brazil, Argentina
Africa	Agricultural residues	<5	South Africa, Kenya, Egypt

are critical determinants of economic viability. The capital intensity of collection networks and the low bulk density of many residues mean that profitability often depends on co-location with processing facilities (e.g., sugar mills, biorefineries) (Kumar et al., 2020).

4.2.2.3 Processing Approaches and Technical Issues

Pulping of residues often uses adapted chemical pulping or soda-anthraquinone processes and pretreatments (alkaline extraction, steam explosion, acid hydrolysis) to remove hemicellulose and facilitate delignification. Bagasse is relatively easier to pulp (lower lignin than some woods), making it a frequent candidate for integrated sugar mill-paper mill operations. However, residues can require more bleaching and generate different by-product streams that need management (Sixta, 2006; Goodman, 2020).

4.2.2.4 Environmental and Social Considerations

Using residues reduces open-field burning and associated air pollution (a major benefit in many developing countries). However, removing residues from fields can reduce soil organic matter and long-term fertility unless sustainable harvesting fractions and soil management practices are applied (FAO, 2021; Popp et al., 2021).

4.2.3 Non-Wood Plant Sources and Specialty Fibers

This category includes cotton linters, hemp, bamboo, and other fast-growing grasses or specialty plants. These sources are often higher in cellulose content and can satisfy niche or high-purity demands. Cotton linters, with 90%–95% cellulose content, are among the purest sources but are cost-intensive. Yield per hectare for selected crops is detailed in Table 4.3, highlighting bamboo and kenaf as promising high-yield options.

4.2.3.1 Cotton Linters: Purity, Strategic Importance, and Supply Risk

Cotton linters (short fibers left on seed cotton after ginning) are a high-purity cellulose source historically used for high-grade papers, cellulose derivatives (e.g.,

TABLE 4.3
Cellulose yield per hectare for selected crops (data compiled from Austin et al., 2024; Lam et al., 2003; Vivas, 2024)

Crop/Source	Average Yield (tons/ha)	Cellulose Content (%)	Effective Cellulose Yield (tons/ha)
Cotton linters	2–2.5	90–95	1.8–2.3
Sugarcane bagasse	70–100 (wet biomass)	35–40	25–30
Wheat straw	3–5	35–40	1.0–2.0
Bamboo	20–30	40–50	8–12

nitrocellulose, cellulose ethers), and specialty applications. Global supply is geographically concentrated where cotton ginning is large (notably China and India). Recent geopolitical and defense-industry concerns have highlighted supply vulnerabilities for critical applications (e.g., nitrocellulose for propellants), prompting policy attention in some regions to diversify sources or expand domestic capacity (Financial Times, 2024).

4.2.3.2 Bamboo: Rapid Renewability and Regional Scaling

Bamboo is a fast-growing grass with high cellulose yield per hectare and has been developed at scale in China as a pulp source. China dominates bamboo pulp production, with millions of tons of bamboo pulp produced annually and an established industrial base for bamboo-based paper and specialty cellulose products (Yashi Paper, 2023a, 2023b). Bamboo's rapid-growth cycle and high biomass yield make it attractive for regions with suitable agro-climatic conditions, though pulping must address silica content and specific chemical treatments (FAO, 1998).

4.2.3.3 Hemp: High Fiber Quality and Emerging Markets

Industrial hemp yields strong bast fibers rich in cellulose and has attracted interest for textiles, composites, and specialty papers. The global hemp fiber market is growing rapidly (multiple market analyses project strong CAGR), driven by demand for sustainable textiles and building materials (Spherical Insights, 2024). Nonetheless, regulatory variability (especially in Europe and certain countries) creates uncertainty for rapid scaling; policy shifts can meaningfully alter production economics and supply (Grand View Research, 2025). When analyzed in terms of land-use efficiency, comparative yields of cellulose per hectare demonstrate that certain crops significantly outperform traditional sources, thereby offering more sustainable alternatives (Figure 4.2). Table 4.4 shows the comparison of the key performance indicators of major cellulose feedstocks, emphasizing production efficiency, processing requirements, and cost competitiveness to guide industry-scale adoption decisions.

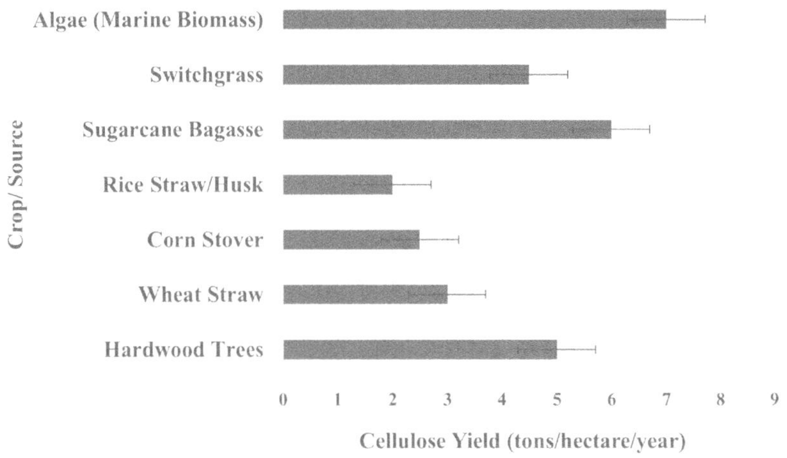

FIGURE 4.2 Comparative yield of cellulose per hectare (Biomass Connect, 2023).

TABLE 4.4
Comparative data table—feedstock metrics (summary)

Feedstock	Typical Cellulose Content (dry wt, %)*	Key Producing Regions	Main Advantages	Primary Challenges
Softwood wood pulp	40%–45% cellulose (whole wood)	Canada, Scandinavia, Russia, USA	Long fibers → strong paper; mature supply chains	High capital mills; forestry sustainability concerns (UNECE)
Hardwood wood pulp	40%–50% cellulose	China, Brazil, EU, USA	Higher yield; fine fiber for speciality papers	Bleaching chemicals; variable lignin types (UNECE)
Sugarcane bagasse	40%–50% (cellulose + hemicellulose)	Brazil, India, Thailand	Waste valorization; co-location with sugar mills	Seasonal supply; collection logistics; preprocessing (PMC)
Straw/corn stover	30%–45% (varies)	Asia (rice/wheat), USA (corn)	Large volumes; reduces open burning	Low density, high logistics cost; silica (rice) (PMC)

(Continued)

TABLE 4.4 (*Continued*)
Comparative data table—feedstock metrics (summary)

Feedstock	Typical Cellulose Content (dry wt, %)*	Key Producing Regions	Main Advantages	Primary Challenges
Cotton linters	90%+ cellulose (very high purity)	China, India, USA	High purity → specialty cellulose & derivatives	Supply concentrated; strategic dependence (Financial Times, 2024)
Bamboo	40%–60% (species dependent)	China, SE Asia	Rapid growth; high biomass yield	Pulping adaptation; silica in some species (Yashi Paper, 2023a, 2023b)
Hemp fiber	50%–70% cellulose (bast fraction)	Europe, China, N. America	Strong fibers; low input crop	Regulatory uncertainty; processing infrastructure (Spherical Insights, 2024)

*Typical ranges are approximate and feedstock- and variety-dependent; values are intended for comparative context rather than exact specification.

4.3 TECHNOLOGICAL CHALLENGES IN CELLULOSE PRODUCTION

Despite its abundance and renewability, the large-scale utilization of cellulose is hindered by a range of technological barriers. These include energy- and chemical-intensive extraction methods, limited scalability of green processing technologies, and challenges in ensuring consistency and quality across diverse raw material sources. As industries increasingly seek sustainable alternatives to petroleum-derived polymers, addressing these hurdles is critical for achieving both economic competitiveness and environmental responsibility (Figure 4.3).

4.3.1 Extraction and Purification

The initial phase of isolating cellulose from lignocellulosic biomass is a critical determinant of the final material's quality, yield, and environmental footprint, encompassing a range of techniques from established industrial methods to novel emerging alternatives. This process, which must effectively separate the robust crystalline cellulose from the surrounding matrix of lignin and hemicellulose, is evolving to meet higher demands for sustainability, purity, and efficiency across various industries.

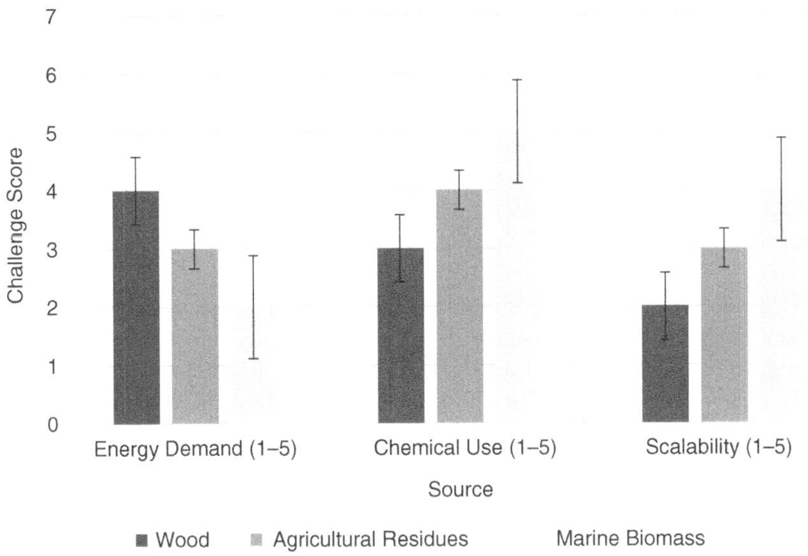

FIGURE 4.3 Technological challenges in cellulose extraction and processing.

4.3.1.1 Conventional Pulping Methods

The kraft process remains the dominant industrial method, accounting for nearly 80% of global pulp production (FAO, 2022a, 2022b). It involves cooking wood chips in sodium hydroxide and sodium sulfide solutions, producing strong fibers but generating large volumes of black liquor waste. Although chemical recovery systems partially offset costs, the process is highly energy intensive. The sulfite process, once widely used for dissolving pulp (viscose rayon, cellulose acetate), has declined due to chemical recovery limitations and environmental issues related to sulfur emissions. Mechanical pulping offers higher yields (up to 95%) but results in lower-quality fibers with high lignin content, unsuitable for long-lasting products.

4.3.1.2 Emerging Alternatives

Emerging alternatives to these conventional methods are being developed to address their environmental and efficiency shortcomings, including organosolv pulping, which utilizes organic solvents like ethanol or acetic acid under pressure to effectively fractionate lignocellulosic biomass into relatively pure streams of cellulose, lignin, and hemicellulose; while this method offers the significant advantages of easier, higher-quality lignin recovery for valorization and reduced water pollution compared with kraft systems, the high costs associated with the distillation and recovery of organic solvents to make the process economically viable remain a significant bottleneck for widespread industrial adoption. Another promising alternative is the use of ionic liquid (IL) systems, which are salts in a liquid state that can dissolve cellulose directly, enabling homogeneous chemical processing and modification; however, the industrial scalability of this method is currently hindered by the high cost of the ionic liquids themselves, considerable challenges in achieving near-total solvent recyclability, and lingering toxicity concerns regarding the environmental impact of certain ILs. Furthermore, physical pre-treatment methods like steam explosion—which uses high-pressure steam to explosively decompress and break apart biomass structure—and ammonia fiber expansion (AFEX)—which uses liquid ammonia to swell and disrupt crystalline cellulose—are primarily used to enhance the accessibility of cellulose for subsequent enzymatic hydrolysis in bioethanol plants, but their application remains limited and not yet widespread within traditional pulp industries focused on fiber production rather than sugar conversion.

a. Organosolv pulping: Uses organic solvents (ethanol, acetic acid) to fractionate lignocellulosic biomass. While offering easier lignin recovery and reduced pollution, costs of solvent recovery remain a bottleneck.
b. IL systems: ILs can dissolve cellulose directly, enabling homogeneous processing. However, high cost, recyclability issues, and toxicity concerns hinder industrial adoption.
c. Steam explosion and AFEX: Pre-treatment methods for agricultural residues that enhance cellulose accessibility for enzymatic hydrolysis. Scalable in bioethanol plants but not yet widespread in pulp industries.

4.3.2 Processing Innovations

Advances in enzymatic hydrolysis, ionic liquids, and green solvents are transforming cellulose extraction into a more sustainable process. However, significant differences exist in crystallinity, morphology, and end-use suitability of cellulose types. A detailed comparison is shown in Table 4.5.

4.3.2.1 Enzymatic Hydrolysis

Cellulase enzymes can selectively break down lignocellulosic biomass into fermentable sugars, which are further utilized for nanocellulose or bioethanol. Although enzymatic processes are environmentally benign, high enzyme costs and long

TABLE 4.5
Comparison of properties of different forms of cellulose

Type of Cellulose	Morphology/ Structure	Crystallinity (%)	Typical Applications
Microcrystalline (MCC)	Granular, micro-scale	50–60	Pharmaceuticals, food additives, cosmetics
Nanocrystalline (NCC)	Rod-like, nano-scale	70–90	Nanocomposites, packaging films, coatings
Bacterial cellulose (BC)	3D nanofiber network	90	Medical implants, wound dressings, membranes

Note. Data on morphology, crystallinity, and applications compiled from Spence et al. (2011) and Azizi Samir et al. (2005).

reaction times limit economic viability. Advances in immobilized enzymes and genetically engineered microbial strains offer potential breakthroughs (Singh, R., et al. 2022).

4.3.2.2 Green Solvents

The pursuit of green solvents for cellulose processing has led to significant advancements, with DES standing out as particularly promising due to their affordability, biocompatibility, and tunability, as they can be formulated from natural compounds like choline chloride and urea to efficiently break the extensive hydrogen-bonding network of cellulose through the action of hydrogen bond donors and acceptors. Similarly, aqueous NaOH/urea systems operate as a green solvent by rapidly cooling the cellulose solution to create a transient, frozen complex that prevents polymer reaggregation, allowing for dissolution at low temperatures; however, this process demands strict control over concentration, temperature, and regeneration conditions to be effective and avoid premature gelation. Transitioning to nanocellulose production, these bio-based nanomaterials (CNC, CNF) offer an exceptional strength-to-weight ratio, superior barrier properties, and inherent biodegradability, making them ideal for advanced applications in packaging, electronics, and biomedical devices (Klemm, D., et al. 2018). However, their commercialization is hampered by significant production challenges: CNC production traditionally relies on harsh sulfuric acid hydrolysis, which introduces issues of reactor corrosion, potential cellulose degradation, and the generation of large volumes of neutralization effluent that poses a serious environmental disposal problem. Conversely, CNF production primarily through mechanical methods like high-pressure homogenization or microfluidization is notoriously energy intensive, consuming up to 30,000 kWh/ton due to the immense forces required to delaminate cellulose fibrils, presenting a severe economic and sustainability barrier. Furthermore, a critical lack of standardization in key material properties—such as particle size distribution, degree of crystallinity, and surface chemistry-across different production batches and manufacturers creates inconsistency, complicates downstream processing, and hinders the reliable integration of nanocellulose into large-scale industrial applications, thus slowing its market adoption despite its tremendous potential.

4.3.3 Quality and Standardization

The inherent variability in raw materials presents a fundamental challenge for the cellulose industry, as the chemical and physical properties of cellulose—including its purity, crystallinity, and degree of polymerization—vary significantly depending on the plant species, its specific growth conditions, and the subsequent processing method employed; for instance, cellulose derived from rice straw is characterized by a high ash and silica content that complicates processing, while cotton linters are prized for their ultra-pure, long-fibered cellulose, a variability that critically complicates the blending of feedstocks and poses substantial quality assurance hurdles in highly regulated industries like pharmaceuticals, where excipient purity is tightly controlled and any inconsistency can jeopardize product safety and efficacy (Future Journal of Pharmaceutical Sciences, 2025).

This natural inconsistency is exacerbated by significant standardization issues, where while established benchmarks like the ISO 5264 standard for beating and refining tests and various TAPPI standards provide reliable frameworks for assessing traditional paper-grade pulp, they lack comprehensive and universally accepted specifications for novel and advanced materials such as nanocellulose and cellulose-based biocomposites, leading to problematic inconsistencies in the measurement and reporting of key parameters like crystallinity index, degree of polymerization, and nanofibril dimensions across different laboratories, which in turn creates uncertainty, hinders reliable product certification, and ultimately slows broader market acceptance. These challenges are viewed through distinct regional perspectives: in Europe, there is a strong and coordinated focus on developing rigorous EN standards for bio-based products, an effort supported and funded by large EU frameworks like Horizon 2020 to ensure quality and foster trust; in contrast, China's rapid expansion of its viscose and dissolving pulp industry has sometimes been accompanied by variable and inconsistent product quality, drawing criticism from international buyers who require uniform material performance; meanwhile, in the United States, the Food and Drug Administration (FDA) mandates strict compliance and extensive documentation for cellulose used in food, pharmaceuticals, and dietary supplements, creating formidable market barriers for imported non-wood cellulose sources from regions that lack similarly validated and recognized production and testing standards.

4.4 ECONOMIC VIABILITY AND MARKET DYNAMICS

The commercial potential of cellulose-based products is not determined solely by technological feasibility, but also by their economic competitiveness and adaptability within global markets. Factors such as raw material availability, supply chain logistics, production costs, and market competition from synthetic alternatives strongly influence the industrial utilization of cellulose. Moreover, global disparities in infrastructure, investment, and regulatory frameworks create distinct opportunities and challenges across regions. This section explores the cost drivers, global market trends, and investment-policy frameworks that shape the commercial viability of cellulose.

4.4.1 Cost Drivers

a. **Raw Material Availability**
The cost of cellulose production is heavily influenced by local feedstock availability. Regions with abundant forest resources, such as Scandinavia and North America, benefit from lower feedstock costs due to established supply chains (FAO, 2022a, 2022b; UNECE, 2023). By contrast, countries reliant on imports, such as China, incur higher costs due to transport and tariffs (Pöyry, 2021). From an economic perspective, a comparison of production costs and infrastructural requirements highlights clear differences in the viability of each cellulose source (Figure 4.4).
b. **Logistics and Infrastructure**
Collection, storage, and transportation of agricultural residues add significant costs, especially in developing economies. For example, in India, the

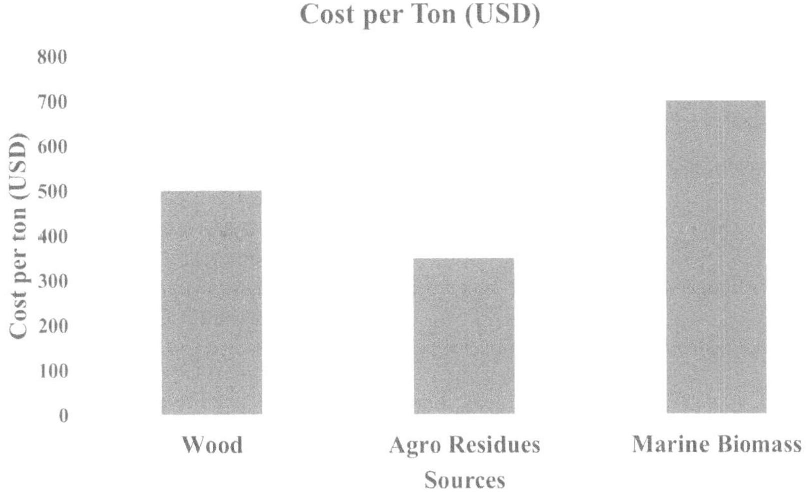

FIGURE 4.4 Economic viability comparison of cellulose sources.

cost of collecting rice straw can exceed the value of the cellulose extracted, discouraging private investment (ICAR, 2022). Efficient biomass aggregation models (as in Brazil's sugarcane sector) help mitigate these challenges.

c. **Processing Costs**
Traditional pulping requires large capital-intensive plants, with per-ton costs dominated by chemicals and energy inputs. Emerging green technologies, though environmentally preferable, often increase per-ton costs due to high enzyme or solvent expenses (Klemm et al., 2018; Zhang et al., 2021).

d. **Competition from Alternatives**
Synthetic polymers (e.g., polyethylene, polypropylene) and recycled paper compete directly with cellulose products. Plastics are often cheaper, while recycled paper reduces demand for virgin pulp in packaging and printing sectors (MarketsandMarkets, 2023; Ellen MacArthur Foundation, 2022).

The major cost drivers and their impact on cellulose economics are summarized in Table 4.6, highlighting raw material dependency, processing costs, and policy influences as critical determinants of overall viability.

4.4.2 Global Market Analysis

The global cellulose market is geographically diverse. **North America and Europe** dominate high-value cellulose derivatives such as microcrystalline cellulose (MCC) and nanocellulose, benefiting from strong R&D ecosystems and advanced processing technologies. Conversely, **Asia-Pacific**—led by China, India, and Indonesia—accounts for the largest share of raw cellulose production, driven by vast agricultural and forestry outputs. Latin America, particularly Brazil, plays a crucial role in agricultural cellulose through its massive sugarcane bagasse production (Klein and Luna, 2022). Meanwhile, Africa, despite its vast agricultural base, remains a **minor contributor** due to limited infrastructure, technology adoption, and investment.

TABLE 4.6
Global cost drivers in cellulose production and utilization

Cost Driver	Impact on Production Costs	Examples/Notes
Raw material availability	High—affects supply chain	Seasonal crop residues, forestry yield
Processing costs	High—chemicals, energy intensive	Kraft pulping, enzymatic hydrolysis
Logistics & transport	Medium—bulk material handling	Bagasse transport in Brazil, straw in India
Competition with synthetics	High—petroleum-based polymers	Plastics, rayon, synthetic fibers
Policy & subsidies	Variable—reduces effective cost	EU green deal incentives, US bioeconomy acts

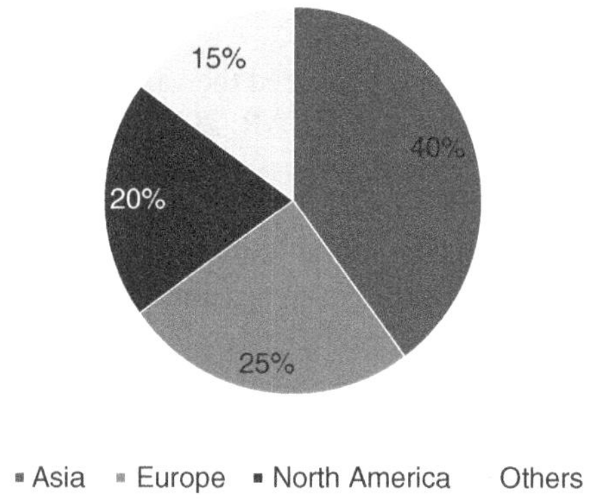

FIGURE 4.5 Regional distribution of cellulose utilization.

Furthermore, regional utilization patterns underscore that Asia and Europe remain the largest markets for cellulose applications, while North America shows steady but moderate demand (Figure 4.5).

4.4.2.1 Regional Distribution

The global cellulose industry exhibits distinct regional strengths and challenges. North America, historically strong in high-quality paper and specialty derivatives, is pivoting toward bioplastics and pharmaceutical applications as demand for traditional paper wanes. Europe is leveraging policy initiatives like the EU Green Deal to lead in sustainable markets, investing heavily in nanocellulose, bio-packaging, and circular economy models. The Asia-Pacific region, dominated by China, is the largest consumer, importing massive pulp volumes for textiles and packaging, while India explores cost-sensitive, agro-residue-based production. Meanwhile, parts of Latin America, particularly Brazil and Chile, compete as low-cost producers, thanks to fast-growing eucalyptus plantations, though this raises ongoing sustainability concerns about monoculture farming.

4.4.2.2 Demand Sectors

a. **Packaging**: Fastest-growing segment, driven by bans on single-use plastics and consumer demand for biodegradable alternatives.
b. **Textiles (viscose, rayon)**: Major growth in Asia, particularly China and India.
c. **Pharmaceuticals and Food**: MCC (microcrystalline cellulose) used in tablets, capsules, and food thickeners represents a high-margin market, dominated by North America and Europe.
d. **Advanced Applications**: Nanocellulose for electronics, biomedical devices, and 3D printing remains a niche but high-potential market.

Looking ahead, projections suggest robust global market growth for cellulose-based industries, with demand expected to rise sharply through 2040 (Figure 4.6). Table 4.7 summarizes regional production capacities, utilization patterns, and sectoral applications, providing insights into the current scale and direction of cellulose commercialization worldwide.

4.4.3 Investment and Policy Frameworks

Government incentives, subsidies, and R&D investments are vital in determining the economic attractiveness of cellulose-based industries. For instance, the European Union's Green Deal promotes bio-based materials, while the U.S. Bioeconomy Act provides R&D funding for biomass conversion. Similarly, India's National Bio-Energy Mission supports the valorization of agricultural residues.

Such frameworks not only reduce production costs but also encourage private investment, particularly in biorefineries that integrate cellulose with biofuels and other co-products.

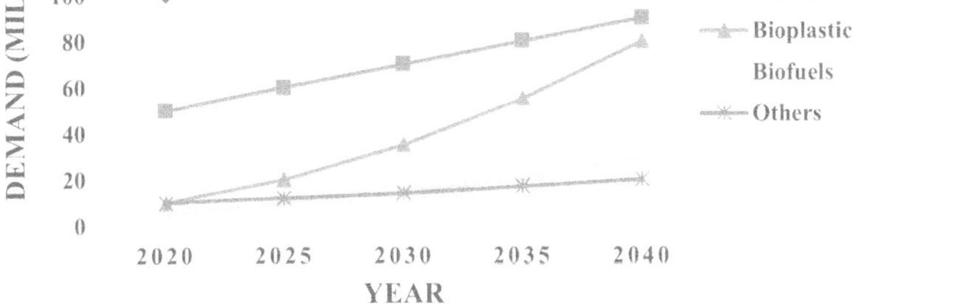

FIGURE 4.6 Global market trends and projected growth (ResearchAndMarkets.com, 2023).

TABLE 4.7
Global cellulose market overview (ResearchAndMarkets.com, 2023)

Region	Production (Million Tons)	Consumption (Million Tons)	Key Demand Sectors	Growth Drivers
North America	55	45	Packaging, pharma, MCC	Innovation in bio-based products
Europe	35	40	Nanocellulose, packaging, pharma	EU Green Deal, circular economy
Latin America	45	20	Tissue, fine paper	Low-cost eucalyptus pulp exports
Asia-Pacific	65	80	Textiles, packaging, viscose	Rising population, industrialization
Others (Africa, Middle East)	8	10	Packaging, imports	Growing consumer base, limited local production

4.4.3.1 Subsidies and Incentives

a. Governments worldwide are increasingly recognizing cellulose as a strategic resource in the bioeconomy transition.
b. EU: Horizon 2020 and Green Deal programs fund cellulose R&D, including pilot-scale nanocellulose plants.
c. India: Provides subsidies for biomass aggregation and mandates 10% co-firing of agro-residues in thermal power plants, indirectly promoting cellulose valorization.
d. China: Offers tax incentives for viscose producers but faces criticism for weak environmental regulations.

4.4.3.2 R&D Investments

a. Global corporations (e.g., Stora Enso, UPM, Suzano) are investing in biorefineries to diversify beyond pulp and paper.
b. Venture capital interest in nanocellulose startups has grown, particularly in North America, for applications in sustainable packaging and biomedical materials.

4.4.3.3 Regulatory Impacts

a. Bans on single-use plastics (e.g., EU 2021 directive, India 2022 ban) are accelerating cellulose packaging adoption.
b. FDA/EMA regulations for pharmaceutical cellulose derivatives create high entry barriers but ensure stable demand.

4.5 SUSTAINABILITY AND CIRCULAR ECONOMY APPROACHES

The sustainable utilization of cellulose depends not only on its renewable origin but also on how effectively production systems address ecological impacts and resource efficiency. Traditional pulp and paper industries have long been associated with deforestation, high water consumption, and chemical effluents. However, the transition toward a circular bioeconomy is opening new pathways for cellulose utilization, emphasizing waste valorization, recycling, and biorefinery integration. This section explores the environmental trade-offs associated with cellulose production and outlines circular economy strategies that can enhance long-term viability. A comprehensive summary of these sustainability strategies, including their potential benefits and implementation challenges, is presented in Table 4.8.

4.5.1 Environmental Trade-Offs

The environmental trade-offs of cellulose production are multifaceted, beginning with the critical choice of sourcing, where wood-based cellulose from large-scale forest harvesting has historically contributed to significant deforestation and biodiversity loss, particularly in vulnerable tropical regions like Indonesia and the Amazon, though certified forestry schemes such as the FSC and the PEFC aim to mitigate these impacts by ensuring sustainable harvesting practices, while the alternative of using agricultural residues like wheat straw or bagasse reduces direct deforestation risks and prevents the harmful practice of open-field burning, yet diverting these residues away from their traditional uses for soil enrichment and livestock feed may

TABLE 4.8
Sustainability Approaches in Cellulose Production

Approach	Description	Benefits	Challenges
Agro-residue utilization	Using crop residues (e.g., wheat straw, corn stover, bagasse) as cellulose feedstock	Reduces agricultural waste, low-cost resource	Variability in composition, seasonal availability
Closed-loop processing	Recycling solvents and chemicals in pulping and extraction	Lower environmental impact, reduced chemical demand	High upfront investment, need for process control
Biotechnological methods	Enzymatic hydrolysis or microbial fermentation for cellulose isolation	Energy-efficient, minimal toxic byproducts	Slower processing times, scalability issues
Marine biomass cultivation	Using seaweed and algae as renewable cellulose sources	Fast growth, no arable land use, high-yield potential	Processing complexity, regional adaptation
Circular economy integration	Valorization of cellulose byproducts into biofuels, bioplastics, and chemicals	Enhances resource efficiency, multiple revenue streams	Requires advanced infrastructure and investment

create new ecological trade-offs by impacting soil health and agricultural sustainability (Worku et al. 2023).

Furthermore, the carbon footprint and energy use vary considerably by source and processing method, as conventional kraft pulping emits significant amounts of CO_2 but achieves partial mitigation through the combustion of black liquor for energy, agro-residue pulping is generally less energy intensive but requires additional preprocessing steps for silica removal, and advanced nanocellulose production is notoriously energy-heavy unless more efficient hybrid enzyme-mechanical methods are employed by Zhao et al. (2019).

The issue of water pollution and effluents is also paramount, with conventional pulping consuming vast quantities of water (30–70 m^3 per ton of pulp) and generating effluents rich in pollutants like lignin, chlorinated organics, and high COD/BOD loads, a problem that has been reduced in Europe and North America by transitioning to elemental chlorine-free and totally chlorine-free bleaching technologies, though the adoption of these cleaner technologies remains limited in major producing regions across Asia and Africa.

A case study in sustainable practices is exemplified by Scandinavian countries like Finland and Sweden, which have set industry benchmarks through substantial investments in closed-loop water systems, energy-efficient mill technologies, and rigorously certified forestry management, resulting in production facilities that exhibit some of the lowest carbon footprints globally while also pioneering integrated biorefinery concepts where waste streams like lignin and hemicellulose are valorized into marketable fuels and chemicals, as documented in studies by Robertson et al. (2018) and RSC Sustainability (2024).

4.5.2 Closed-Loop Systems

The development of closed-loop systems in the cellulose industry is a critical component of the circular economy, exemplified by the widespread recycling of cellulose waste, particularly in paper recycling, which stands as one of the most established examples with global recycling rates exceeding 60% (CEPI, 2023a), though a key limitation exists as recovered fibers are widely used for packaging, yet their quality degrades after five to seven cycles due to shortening and hornification, prompting innovations in fiber reinforcement with nanocellulose to extend recycling potential, a practice with regional disparities where Europe achieves some of the highest rates exceeding 70% supported by stringent EU directives on waste management, while in countries like India and China, recycling rates are growing but remain fundamentally limited by inadequate waste segregation infrastructure (Haile et al. 2021).

Beyond mechanical recycling, a more holistic approach is found in biorefinery integration, where cellulose production is increasingly linked to models that fractionate lignocellulosic biomass into multiple high-value streams, directing cellulose toward traditional products like paper and textiles as well as advanced materials like nanocellulose, while simultaneously valorizing lignin for adhesives, carbon fibers, and bio-based chemicals, and hemicellulose for bioethanol and food additives, an integrated approach that drastically reduces waste, maximizes economic returns from the same biomass input, and fully aligns with circular economy principles.

This concept is further expanded through industrial symbiosis, creating interconnected systems as seen in Brazil where sugarcane bagasse biorefineries integrate power generation, ethanol production, and pulp manufacturing to create a nearly zero-waste system, and in Japan, where companies like Spiber and Asahi Kasei are pioneering advanced systems for cellulose regeneration from textile waste, producing new fibers while significantly reducing the landfill burden. A compelling case study of this closed-loop innovation is found in Canada, where pilot-scale projects are successfully converting paper mill sludge—a waste product with high disposal costs—into valuable nanocellulose for use in coatings and composites, demonstrating a practical model where waste streams are redefined as feedstocks, thereby simultaneously enhancing economic viability and sustainability by addressing waste disposal problems and creating new revenue streams from upcycled materials (Ozola 2019).

4.6 CASE STUDIES AND FUTURE PROSPECTS

Cellulose has evolved far beyond its conventional role in the paper and textile industries, moving toward high-value applications across biomedical, energy, and environmental sectors. For example, nanocellulose-based wound dressings and scaffolds are already demonstrating promise in biomedicine, while cellulose nanofibers are being investigated for next-generation supercapacitors and bio-based batteries. Similarly, in the packaging sector, cellulose films are increasingly being developed to replace plastics, with future prospects involving active packaging that integrates sensors to monitor food freshness. In environmental domains, cellulose membranes are emerging as cost-effective tools for water filtration and oil spill remediation, with ongoing research on CO_2 capture further expanding their role in sustainability. These emerging and

TABLE 4.9
Innovative Applications and Future Prospects of Cellulose

Application Area	Current Uses	Future Prospects
Biomedical	Wound dressings, scaffolds, drug delivery	Smart biomaterials with AI-enabled diagnostics, tissue engineering
Packaging	Biodegradable films, food packaging	Nanocellulose-based active packaging with sensors for freshness detection
Energy storage	Cellulose nanofibers in supercapacitors	High-capacity bio-based batteries and hybrid energy devices
Electronics	Flexible substrates, conductive composites	Transparent cellulose nanofiber displays, wearable electronics
Construction	Reinforcement in composites and insulation	Carbon-negative cellulose composites for sustainable building
Environmental	Water filtration membranes, oil spill cleanup	Nanocellulose for CO_2 capture, large-scale environmental remediation

future opportunities are systematically mapped in Table 4.9, which highlights current applications alongside anticipated technological advancements.

The commercial development of cellulose has been shaped by a diverse range of industrial practices, regional strategies, and technological innovations. Examining global case studies offers valuable insights into both the successes and the persistent challenges of cellulose utilization, while also highlighting directions for future development.

One of the most widely cited success stories in the pulp and paper sector is Metsä Fiber's Bioproduct Mill in Äänekoski, Finland. Recognized as the world's largest bioproduct mill, it produces approximately 1.3 million tons of pulp annually. Beyond pulp production, the facility embodies the principles of the circular economy by operating as an energy self-sufficient system, generating bioenergy from black liquor and other side streams. Moreover, it integrates downstream valorization by producing biogas, tall oil, and emerging nanocellulose products. This example demonstrates the feasibility of coupling large-scale cellulose production with environmental sustainability and economic diversification **(Metsä Group, 2022).**

In contrast, Brazil's Suzano Pulp and Paper offers a case study rooted in agricultural and silvicultural efficiency. Brazil has emerged as a global leader in short-fiber pulp production, owing to its eucalyptus plantations that mature in as little as seven years—considerably faster than boreal forests. Through investments in genetic improvements and precision agriculture, Suzano has achieved a sustainable forestry model that aligns with both economic growth and environmental stewardship (Suzano, 2023). This case illustrates how natural endowments, when paired with scientific innovation, can shape global competitiveness.

Asia, particularly Japan, provides an example of high-value innovation. Companies such as Nippon Paper and Oji Holdings have been pioneers in scaling nanocellulose for advanced applications in automotive composites and food packaging. Supported by government subsidies and the establishment of the Cellulose Nanofiber Business Promotion Consortium, Japan has transformed laboratory-scale research into industrial initiatives. However, challenges remain in reducing production costs and achieving international market penetration, underscoring the gap between innovation and commercialization (Spence et al., 2011).

In India, efforts to valorize agricultural residues into pulp highlight a different trajectory. Small-scale pulp mills in states such as Punjab and Uttar Pradesh utilize wheat and rice straw, thereby providing an alternative to residue burning—a practice that contributes significantly to regional air pollution. Supported by government-backed initiatives such as the National Bio-Energy Mission, these mills not only mitigate environmental hazards but also provide economic opportunities for rural communities. Nevertheless, technical challenges, such as the high silica content of straw, continue to limit scalability and product quality.

The regional distribution of cellulose innovation reveals distinct trajectories. Europe has placed sustainability at the core of its cellulose strategy, prioritizing eco-labels, life-cycle assessments, and low-carbon production under the Horizon Europe framework. This has positioned Europe as a leader in the circular bioeconomy. North America, in contrast, has consolidated its pulp and paper industry, focusing on high-value derivatives such as cellulose acetate and microcrystalline cellulose for pharmaceutical applications. Asia-Pacific, led by China, is experiencing rapid growth in viscose and regenerated cellulose fibers, though sustainability concerns persist due to coal-based energy reliance. Meanwhile, Africa is emerging as a frontier for bamboo-based cellulose, with Ethiopia and Kenya developing capacity through United Nations Industrial Development Organization projects.

Looking ahead, cellulose holds significant promise in advanced applications. In the biomedical field, nanocellulose scaffolds are being explored for tissue engineering and wound dressings. In electronics, transparent cellulose-based substrates are being investigated for flexible displays and conductive films. In sustainable packaging, cellulose barrier coatings offer alternatives to petroleum-based plastics, particularly in food and beverage sectors. Furthermore, cellulose derivatives are showing potential in energy devices, including supercapacitors and solid-state electrolytes.

The integration of digital technologies will further shape the cellulose industry. Artificial intelligence offers opportunities to optimize pulping and enzymatic processes by reducing energy consumption and chemical inputs. Similarly, blockchain technologies are being explored to enhance transparency and sustainability within cellulose supply chains, ensuring certified sourcing and eco-label compliance. From an economic perspective, the global demand for cellulose is projected to rise from approximately 160 million tons in 2023 to nearly 220 million tons by 2035 (FAO, 2024). This growth will be fueled by policies promoting single-use plastic bans and increased consumer demand for sustainable products. However, the scalability of nanocellulose production and the development of cost-competitive closed-loop recycling systems will remain critical for long-term viability.

Across these global case studies, a common theme emerges: the integration of forestry, agriculture, and industrial innovation is essential for building sustainable cellulose value chains. Regional disparities in technological adoption remain significant, but collaborative frameworks are beginning to bridge these gaps. The implementation of circular economy principles—particularly waste valorization and recycling—offers a viable pathway for cellulose to maintain its competitive edge in a global bio-based economy.

In conclusion, cellulose stands at the nexus of tradition and innovation. Its future success will depend on the ability of industries, policymakers, and researchers to work collectively, balancing economic competitiveness with ecological responsibility. The next phase of cellulose utilization is likely to be defined not merely by production volumes, but by the ability to create high-value, sustainable, and technologically advanced applications that align with global transitions toward a bioeconomy.

4.7 CONCLUSION

Cellulose, the most abundant renewable biopolymer on Earth, remains at the forefront of efforts to build a sustainable bioeconomy, and its versatility—spanning traditional paper and textiles to emerging applications in nanomaterials, biomedical devices, and energy systems—demonstrates both its historical significance and its future potential, yet the pathway to maximizing cellulose's value is far from straightforward. From the analysis presented in this chapter, several key conclusions can be drawn, the first of which is that the commercial sources of cellulose—wood, agricultural residues, and non-wood plants—each present distinct advantages and limitations, and while wood-based cellulose dominates global supply, agricultural residues and specialty fibers such as cotton, hemp, and bamboo offer sustainability benefits, especially when integrated into circular economy frameworks. A second conclusion is that the technological challenges associated with cellulose production remain a central barrier, where conventional pulping technologies are energy- and chemical-intensive, while advanced approaches such as enzymatic treatments, ionic liquids, and nanocellulose production require further cost reductions and process optimization to achieve industrial-scale adoption, meaning overcoming these barriers will demand innovations in process engineering, material science, and digital tools. Third, it is clear that economic viability and market dynamics shape cellulose's future trajectory as much as technology does, as raw material costs, logistics, and global trade patterns dictate competitiveness, while synthetic alternatives continue to provide strong competition, but nevertheless, government incentives, R&D subsidies, and the rising demand for eco-friendly materials are creating opportunities for cellulose to consolidate its role in global markets. Fourth, sustainability considerations are inseparable from the discussion of cellulose, since deforestation, carbon emissions, and land-use trade-offs pose serious challenges, but the adoption of closed-loop systems, recycling initiatives, and biorefinery integration can transform cellulose production into a model for sustainable industry, and these circular economy approaches not only mitigate ecological impacts but also unlock new streams of economic value through waste valorization. Finally, global case studies illustrate that success in cellulose commercialization is context-dependent, where Finland's bioproduct mills exemplify integration of scale and sustainability, Brazil demonstrates agro-silvicultural efficiency, Japan highlights high-value innovation in nanocellulose, and India showcases the potential of agricultural residue utilization, and together, these examples emphasize that diverse pathways can contribute to a shared goal of sustainable cellulose utilization.

To fully realize the potential of cellulose within a sustainable bioeconomy, a multifaceted and concerted set of actions is required. First and foremost, it is essential to strengthen interdisciplinary collaboration by fostering closer cooperation between material scientists, chemical engineers, economists, and policymakers to holistically address the multifaceted challenges of cellulose production and commercialization. Concurrently, there must be a significant and targeted investment in green technologies, focusing on scaling up enzymatic processing, green solvents, and nanocellulose production through dedicated funding for R&D, pilot plants, and industrial demonstration projects. Governments have a critical role to play by promoting robust policy frameworks for sustainability, using subsidies, tax incentives, and certification systems to actively reward eco-friendly practices while discouraging unsustainable deforestation and energy-intensive processes. Furthermore, enhancing global supply chain transparency through the adoption of blockchain and AI-enabled tools is necessary to ensure traceability, certify sustainable sourcing, and optimize logistics. A fundamental shift is needed to prioritize circular economy models, where industries focus on waste valorization, the recycling of cellulose-based products, and integration into biorefineries to ensure long-term economic and environmental viability. It is also crucial to encourage regional specialization, enabling countries to leverage their natural advantages—such as bamboo in Africa, agro-residues in India, and eucalyptus in Brazil—to develop cellulose industries that are uniquely tailored to their local contexts. Finally, to sustain this progress, we must foster education and awareness by ensuring that academic institutions and industry training programs are equipped to prepare a workforce capable of addressing cellulose's technical and economic challenges through innovation-driven solutions.

Cellulose represents more than a raw material, it is a cornerstone for building the bio-based future. By combining scientific innovation, economic pragmatism, and sustainability principles, cellulose can transition from being primarily a structural biopolymer to a strategic resource for 21st-century industries. The challenges are formidable, but with coordinated global action, cellulose has the potential to redefine materials science, reshape industrial ecosystems, and contribute significantly to the transition toward a circular and sustainable global economy.

REFERENCES

Austin, D., A. Bourguignon, and J. Laird. 2024. Kenaf: Opportunities for an Ancient Fiber Crop. *Agronomy* 14 (7): 1542. https://doi.org/10.3390/agronomy14071542.

Azizi Samir, M. A. S., F. Alloin, and A. Dufresne. 2005. Review of Recent Research into Cellulosic Whiskers, Their Properties and Their Application in Nanocomposite Field. *Biomacromolecules* 6 (2): 612–626. https://doi.org/10.1021/bm0493685.

Biomass Connect. 2023. *Hemp as a Biomass Crop*. Last modified October 18, 2023. Accessed October 27, 2025. https://www.biomassconnect.org/technical-articles/hemp-as-a-biomass-crop/.

Carus, M., A. Raschka, and S. Piotrowski. 2015. *The Climate Impact of Bio-based Materials*. nova-Institute. https://bio-based.eu/climate.

Confederation of European Paper Industries (CEPI). (2023a). *European Paper Recycling Council (EPRC) Monitoring Report 2022*. Brussels, Belgium. https://www.cepi.org/wp-content/uploads/2023/09/EPRC-Monitoring-Report-2022_Final.pdf

CEPI (Confederation of European Paper Industries). 2023b. *Key Statistics 2022- European Paper Industry*. https://www.cepi.org/statistics.

Clark, J. H., R. Luque, and A. S. Matharu. 2016. Green Chemistry, Biofuels, and Biorefineries. *Annual Review of Chemical and Biomolecular Engineering* 3: 183–207. https://doi.org/10.1146/annurev-chembioeng-062011-081014

Data Bridge Market Research. 2023. *Global Cellulose Market - Industry Trends and Forecast to 2029*. Accessed October 27, 2025. https://www.databridgemarketresearch.com/reports/global-cellulose-market.

Ellen MacArthur Foundation. 2022. *Circular Economy in the Paper and Packaging Sector*.

European Commission. 2020. *A New Circular Economy Action Plan for a Cleaner and More Competitive Europe*. https://eur-lex.europa.eu/legal-content/EN/TXT/?uri=CELEX%3A52020DC0098.

FAO (Food and Agriculture Organization of the United Nations). 1998. *Resources of Non-Wood Fibre in China*. https://www.fao.org/4/w7734e/w7734e04.htm.

FAO (Food and Agriculture Organization of the United Nations). 2021. *Sustainable Wood and Non-Wood Biomass Utilization*. https://www.fao.org.

FAO (Food and Agriculture Organization of the United Nations). 2022a. *FAO Yearbook of Forest Products 2022*. Rome: FAO. https://www.fao.org/forestry/statistics.

FAO (Food and Agriculture Organization of the United Nations). 2022b. *Pulp and Paper Capacities: Survey 2021–2026*.

Financial Times. 2024. *European Defence Groups Warn over Reliance on Chinese Cotton Used in Gunpowder*. February 8, 2024. https://www.ft.com/content/23807ef8-fc6b-41c9-ae7b-9c9ad3a27e82.

Garg, T., S. Arora, and R. Pahwa. 2025. Cellulose and Its Derivatives: Structure, Modification, and Application in Controlled Drug Delivery. *Future Journal of Pharmaceutical Sciences* 11 (1): 76.

Goodman, B. A. (2020). Utilization of Waste Straw and Husks from Rice Production: A Review. *Journal of Bioresources and Bioproducts,* 5(3), 143–162. https://doi.org/10.1016/j.jobab.2020.07.001

Grand View Research. 2025. *Hemp Fiber Market Size, Share & Trends Analysis Report*. Accessed October 27, 2025. https://www.grandviewresearch.com/industry-analysis/hemp-fiber-market-report.

Habibi, Y., L. A. Lucia, and O. J. Rojas. 2010. Cellulose Nanocrystals: Chemistry, Self-Assembly, and Applications. *Chemical Reviews* 110 (6): 3479–3500. https://doi.org/10.1021/cr900339w.

Haile, A., G. G. Gelebo, T. Tesfaye, W. Mengie, M. A. Mebrate, A. Abuhay, and D. Y. Limeneh. 2021. Pulp and Paper Mill Wastes: Utilizations and Prospects for High Value-Added Biomaterials. *Bioresources and Bioprocessing* 8: 35. https://doi.org/10.1186/s40643-021-00385-3.

ICAR (Indian Council of Agricultural Research). 2022. *Utilization of Crop Residues in India: Challenges and Strategies*.

IEA Bioenergy. 2022. *Annual Report 2021*. Last modified April 2022. https://www.ieabioenergy.com/wp-content/uploads/2022/04/IEA-Bioenergy-Annual-Report-2021.pdf.

IEA Bioenergy. 2023. *Environmental Sustainability - Bioenergy Review 2023*. Accessed October 27, 2025. https://www.ieabioenergyreview.org/environmental-sustainability/.

Klein, H. S., and Luna, F. V. 2022. The Development of a Modern Cellulose Industry in South America. *Latin American Research Review* 57(4): 753–774. doi:10.1017/lar.2022.65

Klemm, D., B. Heublein, H. P. Fink, and A. Bohn. 2005. Cellulose: Fascinating Biopolymer and Sustainable Raw Material. *Angewandte Chemie International Edition* 44 (22): 3358–3393. https://doi.org/10.1002/anie.200460587.

Klemm, D., et al. 2018. "Nanocellulose as a Natural Source for Groundbreaking Applications in Materials Science: Today's State and Future Trends." *Materials Today* 21 (7): 720–748.

Kumar, M., A. Parmar, R. Chandra, and P. Chowdhary. 2020. Agricultural Residues as Biomass Feedstocks for Bioethanol Production: Challenges and Opportunities. *Renewable and Sustainable Energy Reviews* 120: 109648. https://doi.org/10.1016/j.rser.2019.109648.

Kumar, V., et al. 2020. "Energy-Efficient Hybrid Approach Combining Enzymatic Pretreatment and Mechanical Fibrillation for Nanocellulose Production: The Finnish Experience." *Industrial Crops and Products* 148: 112274.

Lam, T. B. T., K. Iiyama, and B. A. Stone. 2003. Structural Characteristics of Cell Walls of Kenaf (Hibiscus cannabinus L.). *Journal of Wood Science* 49 (4): 255–261. https://doi.org/10.1007/s10086-002-0469-7.

Lynd, L. R., P. J. Weimer, W. H. van Zyl, and I. S. Pretorius. 2017. "Microbial Cellulose Utilization: Fundamentals and Biotechnology." *Microbiology and Molecular Biology Reviews* 66 (3): 506–577. https://doi.org/10.1128/MMBR.66.3.506-577.2002.

MarketsandMarkets. 2023. *Cellulose Market by Application, Type, and Region - Global Forecast to 2028*.

Melikoglu, M. 2025. Organosolv Lignin: A Comprehensive Review of Pretreatment Advancements and Biorefinery Integration. *Next Materials* 9: 101136.

Metsä Group. 2022. *The Äänekoski Bioproduct Mill*. Accessed October 27, 2025. https://www.metsagroup.com/metsafibre/about-metsafibre/pulp-production/aanekoski-bioproduct-mill/.

Moon, R. J., A. Martini, J. Nairn, J. Simonsen, and J. Youngblood. 2011. Cellulose Nanomaterials Review: Structure, Properties and Nanocomposites. *Chemical Society Reviews* 40 (7): 3941–3994. https://doi.org/10.1039/c0cs00108b.

Ozola, Z. U. 2019. Paper Waste Recycling: Circular Economy Aspects. *Environmental and Climate Technologies* 23: 261–268. https://doi.org/10.2478/rtuect-2019-0094.

Popp, J., S. Kovács, J. Oláh, and P. Balogh. 2021. The Role of Biomass Residues in a Sustainable Bioeconomy: Circular Economy Perspectives. *Sustainability* 13 (14): 7834. https://doi.org/10.3390/su13147834.

Pöyry. 2021. *Global Pulp Market Intelligence Report*. AFRY Management Consulting.

Rana, D., R. Fangueiro, and D. Oliveira. 2021. Natural Fibers for Textile Reinforcement: Economics, Performance and Sustainability. *Textile Research Journal* 91 (11): 1272–1286.

Ranjan, R., R. Rai, S. B. Bhatt, and P. Dhar. 2023. Technological Road Map of Cellulase: A Comprehensive Outlook to Structural, Computational, and Industrial Applications. *Biochemical Engineering Journal* 198: 109020.

ResearchAndMarkets.com. 2023. *Global Cellulose Derivatives Industry Research Report 2023–2030: Increasing Use in Various End-Use Verticals and Growing Demand for Biodegradable and Eco-Friendly Products*. Last modified August 3, 2023. Accessed October 27, 2025. https://www.businesswire.com/news/home/20230803140427/en/.

Robertson, A., et al. 2018. Cellulosic Biofuel Contributions to a Sustainable Energy Future: Choices and Outcomes. *Science* 359 (6378): eaal2324. https://doi.org/10.1126/science.aal2324.

RSC Sustainability. 2024. *Unraveling the Climate Neutrality of Wood Derivatives and Biopolymers*.

Singh, R., et al. 2022. Advances in Cellulase Technology for Efficient Hydrolysis of Lignocellulosic Biomass: A Review. *Bioresource Technology* 344: 126254.

Sixta, H., ed. 2006. *Handbook of Pulp*. Weinheim: Wiley-VCH Verlag GmbH.

Spence, K. L., R. A. Venditti, O. J. Rojas, Y. Habibi, and J. J. Pawlak. 2011. A Comparative Study of Energy Consumption and Physical Properties of Microfibrillated Cellulose Produced by Different Processing Methods. *Cellulose* 18 (4): 1097–1111. https://doi.org/10.1007/s10570-011-9533-z.

Spherical Insights. 2024. *Global Hemp Fiber Market Size to Exceed USD 41.36 Billion by 2033*. Last modified April 2024. Accessed October 27, 2025. https://www.globenewswire.com/news-release/2024/04/22/2867140/0/en/Global-Hemp-Fiber-Market-Size-To-Exceed-USD-41-36-Billion-By-2033-CAGR-Of-22-09.html.

Suzano. 2023. *Annual and Sustainability Report 2022*. Suzano Pulp and Paper. https://www.suzano.com.br/news/suzano-reports-record-operating-cash-generation-of-r22-6-billion-in-2022.

UNECE (United Nations Economic Commission for Europe). 2022. *Forest Products Annual Market Review 2021–2022*. https://unece.org.

UNECE (United Nations Economic Commission for Europe). 2023. *Forest Sector Outlook Study*.

Verified Market Reports. 2023. *Global Specialty Cellulose Market Report - Trends, Forecast and Competitive Analysis*. Accessed October 27, 2025. https://www.verifiedmarketreports.com.

Vivas, R., and C. Hernández. 2024. Economic Analysis of Bamboo Plantations and Bio-Based Applications. *Biofuels, Bioproducts and Biorefining* 18 (2): 2634. https://doi.org/10.1002/bbb.2634.

Worku, L. A., A. Bachheti, R. K. Bachheti, C. E. Rodrigues Reis, and A. K. Chandel. 2023. Agricultural Residues as Raw Materials for Pulp and Paper Production: Overview and Applications on Membrane Fabrication. *Membranes* 13 (2): 228. https://doi.org/10.3390/membranes13020228.

Yashi Paper. 2023a. *Bamboo Pulp Market and Sustainability Outlook.* Accessed October 27, 2025. https://www.yashipaper.com.

Yashi Paper. 2023b. *China's Bamboo Pulp Papermaking Industry Is Moving Towards Modernization and Scale.* Accessed October 27, 2025. https://www.yashipaper.com/news/chinas-bamboo-pulp-papermaking-industry-is-moving-towards-modernization-and-scale/.

Zhang, L., et al. 2021. Recent Advances in Green Solvents for Biomass Processing. *Green Chemistry* 23: 1342–1359.

Zhao, Qingjian, Sheng Ding, Zuomin Wen, and Anne Toppinen. 2019. Energy Flows and Carbon Footprint in the Forestry-Pulp and Paper Industry. *Forests* 10 (9): 725. https://doi.org/10.3390/f10090725.

Section 1

Nano-Spherical Cellulose

5 Preparation and Characterization of Nanospherical Cellulose

Ebenezer Oluwatosin Atoyebi, Oluwatosin David Abodunrin, Samson Kolawole Fasogbon, Ayodele James Oyejide, and Temitope T. Dele-Afolabi

5.1 INTRODUCTION

5.1.1 Background on Cellulose Nanomaterial

Cellulose is Earth's most abundant organic polymer, made up of repeating β-(1→4) linked D-glucose units (Figure 5.1), with a density of 1.59 g cm^{-3} (Sun, 2005; Ghasemi et al., 2017; Ge et al., 2022). It forms the structural backbone of plant cell walls and is also found in bacteria, algae, and fungi. The extensive intra- and intermolecular hydrogen bonding in cellulose gives rise to its impressive tensile strength, biodegradability, and renewability (Cheng et al., 2023; Sharma et al., 2019). Historically, cellulose has been central to industries such as paper, textiles, pharmaceuticals, and packaging.

In recent years, nanotechnology has enabled the creation of nanocellulose, defined as cellulose materials with at least one dimension below 100 nm (Tofanica et al., 2025). Three main types of nanocellulose exist. First, cellulose nanocrystals (CNCs): rigid rod-like particles, typically 3–20 nm in width and 100–500 nm (up to 2000 nm) in length (Channab et al., 2024; Rana et al., 2021). They possess high crystallinity (50%–90%), axial stiffness (~100–150 GPa), and low thermal expansion (Cheng et al., 2023). Second, cellulose nanofibrils (CNFs) are flexible, web-like fibers with

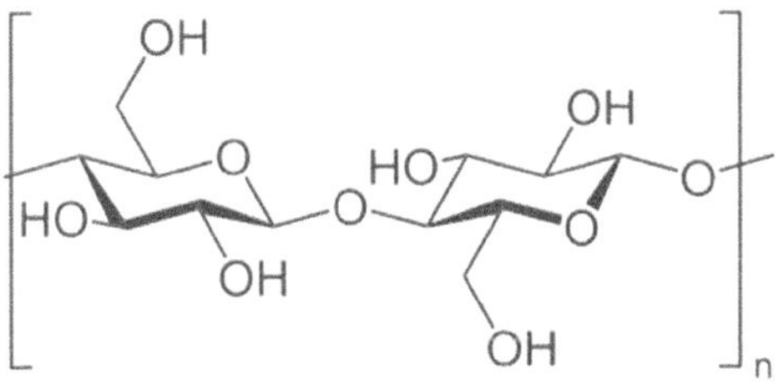

FIGURE 5.1 Structure of cellulose.

DOI: 10.1201/9781003596127-6

diameters of 4–100 nm and lengths up to several microns (Tofanica et al., 2025). CNFs exhibit high aspect ratios, shear-thinning rheology, and form strong fibrous networks. Lastly, bacterial nanocellulose (BNC) is synthesized by microorganisms into highly crystalline, three-dimensional fiber mats with Young's moduli reaching 15 GPa (Cheng et al., 2023; Randhawa et al., 2022).

Nanocellulose combines the biodegradability, biocompatibility, and versatility of bulk cellulose with nanoscale properties such as high surface area, exceptional mechanical strength, and the ability to form films, gels, and aerogels (Cheng et al., 2023). Also, production techniques vary. CNCs are produced via acid hydrolysis of cellulose to alter the amorphous regions (Tofanica et al., 2025; Kumar & Kumar, 2022). They are derived through mechanical fibrillation, often preceded by enzymatic or TEMPO-mediated oxidation (Kumar & Kumar, 2022), while BNC is generated through microbial cultivation (Kumar & Kumar, 2022). Each structural form and production method influences the physicochemical properties of nanocellulose.

In addition, nanocellulose finds applications in composites, barrier films and packaging, biomedical hydrogels and scaffolds, drug delivery carriers, electronics, sensors, and water purification (Dele-Afolabi et al., 2023a, 2023b; Rahman et al., 2021; Parvej, 2016). Its renewable origin and tunable characteristics promote high-performance, eco-friendly alternatives to traditional materials (Niknafs et al., 2024; Casalini et al., 2023; Jo et al., 2022; Meftahi et al., 2022; Carvalho et al., 2021). The mentioned areas of applications are discussed in details in Section 5.4. The overview of this chapter is presented in Figure 5.2.

5.1.2 Significance of Nanospherical Cellulose (NSC)

NSC represents a unique subclass within the broader family of nanocellulose, distinguished by its spherical morphology and nanoscale dimensions. Produced through techniques such as acid hydrolysis, emulsification, or self-assembly, NSC particles

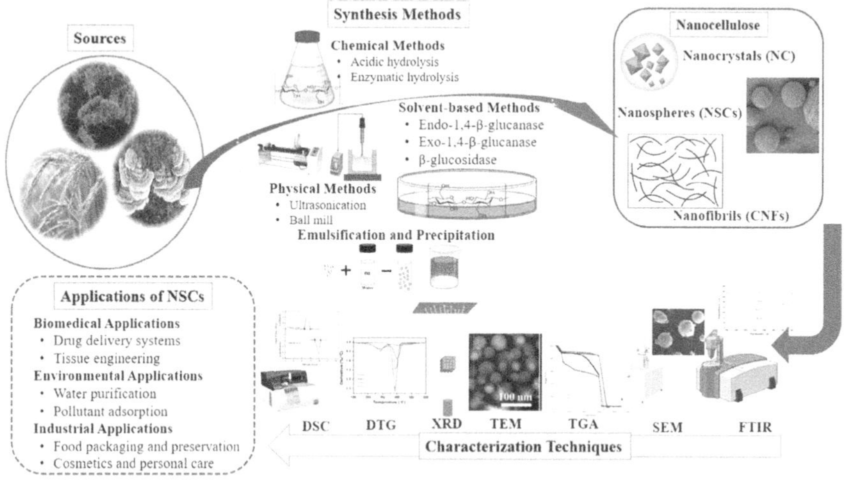

FIGURE 5.2 Graphical abstract.

typically range from 20 to 500 nm and present a uniform, isotropic structure (Phuong et al., 2022; Carvalho et al., 2021). This spherical geometry confers distinct physicochemical properties that set NSC apart from its fibrous or rod-like counterparts, making it particularly valuable for biomedical, environmental, and industrial applications (Khine & Stenzel, 2020).

A defining advantage of NSC is its high specific surface area. The spherical structure maximizes surface-to-volume ratio, thereby providing more interaction sites per unit mass compared with elongated nanocellulose forms (Herreros-Lopez, 2016; You et al., 2025). This enhanced surface area directly contributes to improved adsorption, surface modification, and drug delivery performance, relevant in pharmaceutical applications, where a larger surface area supports higher drug loading capacity and facilitates controlled release profiles (El Allaoui et al., 2023; Janmohammadi et al., 2023; Parvej, 2016).

Uniformity and colloidal stability further distinguish NSC. Unlike nanofibrils, which can entangle and unpredictably alter flow properties, spherical nanoparticles disperse evenly in liquid media (Gobi et al., 2024). This attribute enhances formulation stability, making NSC a superior choice for applications requiring consistent rheology, such as cosmetics, pharmaceuticals, and food products (Wang et al., 2025; Meftahi et al., 2022; Tang et al., 2021). Moreover, their resistance to agglomeration improves shelf life and industrial processability (Phuong et al., 2022).

In biomedical engineering, the spherical form of NSC enhances performance in targeted drug delivery, tissue engineering scaffolds design, and injectable systems. Its consistent size supports smoother circulation in biological fluids, improves cellular uptake, and reduces mechanical irritation in tissues, which are its advantages over fibrous nanocellulose forms (Alebachew et al., 2025; El Allaoui et al., 2023; Mujtaba et al., 2023). Combined with biocompatibility and biodegradability, NSC is an attractive material for applications such as cancer therapy, genetic delivery, and tissue regeneration (Niknafs et al., 2024). In addition, the abundant surface hydroxyl groups on NSC enable versatile chemical modifications, such as esterification, etherification, oxidation, or graft polymerization, tailoring hydrophilicity, charge, and reactivity for specific needs in composites, films, hydrogels, and coatings (Mujtaba et al., 2023; Dong et al., 2015).

From a sustainability perspective, NSC aligns with global efforts for renewable, biodegradable materials. It is typically derived from biomass sources like wood pulp, cotton, or agricultural residues such as palm residues, offering a low-carbon alternative to synthetic nanomaterials (You et al., 2025; Abitbol et al., 2023; Mehanny et al., 2021). The biodegradability of NSC minimizes microplastic pollution, and recent advancements in green synthesis and scalable processes are lowering production costs (Kendre et al., 2023). These combined features position NSC as a promising material for high-performance applications. Elaborate discussions on the significance of NSC is presented in Section 5.4.

5.2 SYNTHESIS METHODS FOR NSC

There are different methods by which NSCs can be synthesized, broadly categorized as both chemical and physical methods. Typically, these methods are combined based on the feedstock, the intended applications and the desired quality of the NSC being synthesized. This section discusses both the methods.

5.2.1 Chemical Methods

5.2.1.1 Acid Hydrolysis

Acid hydrolysis is the most commonly used approach for the synthesis of NSC (Vasconcelos et al., 2017; Dufresne, 2012). This method involves the cleavage of glycosidic bonds by hydrogen ions (H^+) that penetrates the molecules of the amorphous cellulose, which then results in the release of individual crystallites. While the source of the cellulose can determine the quality of the yield, by tuning the conditions of hydrolysis reaction, such as the reaction time, temperature, type of acid, and its concentration, NSC endowed with improved mechanical and physical properties can be obtained (Vasconcelos et al., 2017). For instance, a prolonged period of hydrolysis and elevated temperature have been associated with decreased crystallinity of NSC, in addition to significant modifications of their physical and morphological properties (Vasconcelos et al., 2017; Dufresne, 2012). It is known that a decreasing crystallinity of biomaterials impacts crucial properties, including enhancing water uptake and porosity, improvement of biodegradation, cell attachment, and proliferation, which are vital to successful tissue engineering applications, especially soft tissues (Atoyebi et al., 2023; Rahman et al., 2021).

In the acid hydrolysis of cellulose, strong inorganic acids such as sulfuric acid (H_2SO_4) and hydrochloric acid (HCl) are widely employed, owing to their high proton-donating capacity, efficient disruption of amorphous cellulose regions, and ability to selectively cleave glycosidic bonds (Pawcenis et al., 2022; Tian et al., 2022). Depending on the desired characteristics, high acidic concentration employed during hydrolysis can yield NSC with enhanced amorphousness or crystallinity (Pawcenis et al., 2022; Wang et al., 2007). When HCl is employed, nanospherical crystals with relatively low surface charge are typically produced, with limited stability in aqueous systems (Pawcenis et al., 2022; Wang et al. (2007). Due to the weak electrostatic repulsion between the particles, such nanosphere cellulose easily aggregates in suspension. On the other hand, H_2SO_4 attaches sulfate ester groups with a negative charge ($-OSO_3^-$) on the surface of the nanosphere cellulose (Vasconcelos et al., 2017; Wang et al., 2007). These groups enhance colloidal stability by increasing surface charge density, thus preventing particles agglomeration. Nevertheless, the presence of sulfate groups have been implicated for their ability to reduce the thermal stability of the nanosphere cellulose (Pawcenis et al., 2022; Araki et al., 1998). This predisposes them to degradation at relatively lower temperatures (Martínez-Sanz et al., 2011), which may be relevant to applications such as tissue engineering in which biodegradation of scaffolds over a period of implantation is vital to the regeneration of tissues (Atoyebi et al., 2023). Since hydrolysis with either HCl or H_2SO_4 impact nanosphere cellulose differently, studies have suggested a combined hydrolysis by both acids to harness their advantages, including thermal stability (Corrêa et al., 2010; Vasconcelos et al., 2017; Wang et al., 2007).

5.2.1.2 Enzymatic Hydrolysis

Unlike the acidic hydrolysis in which strong acids are employed, enzymatic hydrolysis is an eco-friendly and highly selective method of hydrolyzing cellulose into nanoparticles (Stankova et al., 2025; Meyabadi & Dadashian, 2012). This approach involves the application of cellulolytic enzymes (cellulases and hemicellulases) to cleave β-1,4-glycosidic linkages present in cellulose under mild conditions (Tsai et al., 2014; Cai et al., 2023; Zhang et al., 2021). Typically, this involves three main types of enzymes secreted by the fungus *Trichoderma reesei*, which are endoglucanase, exoglucanase, and β-glucosidase (Beldman et al., 1985; Rani et al., 2014; Stankova et al., 2025). Endo-1,4-β-d-glucanase (EG, EC 3.2.1.4) randomly cleave internal β-1,4 bonds within the amorphous regions of cellulose chains (Levy et al., 2002; Stankova et al., 2025). Also, exo-1,4-β-d-glucanase or cellobiohydrolase (CBH, EC 3.2.1.91) acts on the chain ends to release cellobiose units as products. CBH1 (or Cel7A) and CBH2 (or Cel6A) are two examples of this exo-acting enzyme. Particularly, CBH1 cleaves the reducing ends, in contrast to CBH2, which cleaves the non-reducing ends of the cellulose polymers. The cellobiose by-products are then hydrolyzed by the β-glucosidases into glucose monomers (Xia et al., 2019; Mishra et al., 2019). These enzymes work synergistically to progressively reduce the degree of polymerization of cellulose without damaging its crystalline structure.

5.2.2 Solvent-Based (SB) Methods

SB methods involve the dissolution of cellulose in solvents, followed by regeneration or precipitation to yield NSC (Liebert, 2010; Nawaz et al., 2019). These methods are highly valuable for processing NSC by enhancing control of its particle morphology and surface chemistry, which is essential to their applications in different fields (Hauru et al., 2012). Unlike other methods, such as acid or enzymatic hydrolysis in which cellulose chains are depolymerized, SB techniques work by dissolving the polymer matrix, followed by precipitation under precise conditions to achieve uniform spherical particles (Heinze & Koschella, 2005; Medronho & Lindman, 2014a; Nawaz et al., 2019).

To dissolve cellulose by SB methods, a wide variety of solvents have been explored by researchers, including *N,N*-dimethylacetamide/lithium chloride (DMAc/LiCl), *N*-methylmorpholine *N*-oxide (NMMO), ionic liquids (ILs) such as 1-ethyl-3-methylimidazolium acetate ([EMIM][OAc]), and deep eutectic solvents (Potthast et al., 2002; Liebert, 2010; Hauru et al., 2012; Ghasemi et al., 2017). Typically, these solvents disrupt the hydrogen bonds in cellulose, which enables it to dissolve without derivatization, and the cellulose can then be precipitated in an anti-solvent medium such as acetone, water, or ethanol (Medronho & Lindman, 2014a; Nawaz et al., 2019; Ge et al., 2022). Further processing during regeneration under suitable agitation, temperature, and concentration conditions yields spherical nanoparticles. The spherical morphology of cellulose particles produced by SB methods can be modulated by adjusting the type of solvent, polymer concentration, and precipitation kinetics ((Heinze & Koschella, 2005; Nosar et al., 2016). In addition, this can also be achieved by introducing surfactants or emulsifying agents (Goddard, 1994; Hoffmann et al., 1996). Furthermore, its surface reactivity for specific applications such as drug loading, enzyme immobilization, or biodegradable film reinforcement can be improved via post-precipitation modifications (Hoffmann et al., 1996; Zhu et al., 2001; D'Addio & Prud'homme, 2011; Novotna et al., 2013; Silva Filho et al., 2013; Saeed et al., 2022).

However, SB processes are not without limitations. Many cellulose solvents are toxic, expensive, or difficult to recycle, which makes the method less sustainable on a large scale (Nawaz et al., 2019). Additionally, high viscosity of cellulose solutions and challenges with complete solvent removal may affect processing efficiency and product purity (Medronho & Lindman, 2014a; Medronho & Lindman, 2014b). Nevertheless, SB methods are highly relevant in research for their ability to yield precise NSC with tuned properties and high potential for drug delivery systems, hydrogel matrixes, and controlled release tissue engineering scaffolds in biomedical applications (Guo et al., 2012; Nosar et al., 2016; Parvej, 2016; Nawaz et al., 2019; Queiroz et al., 2021; Chen et al., 2022).

5.2.3 Emulsion and Precipitation Techniques

Emulsion and precipitation techniques are bottom-up methods used for producing NSC particles (Chin et al., 2018; Carvalho et al., 2021). Typically, these techniques involve the dispersion of a cellulosic substrate, such as carboxymethyl cellulose (CMC), CA, EC, and regenerated cellulose, into another immiscible phase, thereby forming an emulsion (Yan et al., 2015; Joshy et al., 2017; Abbaspoor et al., 2019; Cordt et al., 2020). This is then followed by the precipitation of cellulose into spherical nanoparticles (Carvalho et al., 2021). The process is widely adopted for biomedical and pharmaceutical applications due to its compatibility with bioactive compounds and ease of surface functionalization (Carvalho et al., 2021).

There are a variety of emulsion and precipitation techniques that can be categorized as emulsion solvent evaporation (ESE), anti-solvent precipitation (ASP), and inverse emulsion and cross-linking based on the type of phase interaction and stabilization mechanism involved. In ESE, cellulosic substrates are first dissolved in a suitable organic solvent. This solution is then emulsified into an aqueous phase containing surfactants like polyvinyl alcohol (PVA) to create oil-in-water emulsions (Sousa et al., 2019; Li et al., 2019; Nguyen et al., 2024). Through solvent evaporation or diffusion, the cellulose precipitates within the droplets, thereby forming solid nanospheres (Jia et al., 2016; Yan et al., 2017; Li et al., 2019), which are often smooth, uniform, and capable of encapsulating drugs or bioactives for controlled release (Li et al., 2019; O'Donnell et al., 2020; Wang et al., 2020; Zhu et al., 2022; Samanta et al., 2022).

Unlike in the ESE where direct emulsion with solvent diffusion applies, the ASP is based on liquid–liquid phase separation and does not require the application of surfactants (Anandharamakrishnan et al., 2014; Zhao et al., 2025a). In this method, cellulose is dissolved in a good solvent, such as DMAc, or ILs, and is then added dropwise into a non-solvent or anti-solvent phase, commonly water or ethanol (Harmeet et al., 2016; Carvalho et al., 2021; Zhao et al., 2025a). This initiates the precipitation of cellulose into spherical nanoparticles (Carvalho et al., 2021; Zhao et al., 2025a). ASP is particularly attractive due to its simplicity and ability to produce nanospheres without requiring toxic cross-linking agents or surfactants. The ASP relies on the rapid supersaturation and nucleation of cellulose as it diffuses into the anti-solvent, with particle morphology being strongly affected by concentration gradients, diffusion rates, and mixing conditions (Dengning et al., 2014; Zhang et al., 2022; Zhao et al., 2025a).

In some cases, especially for hydrophilic cellulose derivatives, water-in-oil inverse emulsions are used (Wu et al., 2021; Than-ardna et al., 2023). Typically, the aqueous cellulose phase is dispersed in an organic continuous phase, and chemical cross-linking agents such as glutaraldehyde and epichlorohydrin are employed to stabilize the spherical droplets (Namazi et al., 2012; Subedi et al., 2016; Huang et al., 2017; Wang et al., 2024). This results in the formation of cross-linked cellulose nanospheres, which are suitable for encapsulating both hydrophilic and hydrophobic drugs (Lu & Hsieh, 2010). However, care must be taken to remove residual cross-linking agents to ensure biocompatibility.

5.2.4 Physical Methods

Unlike chemical methods in which reagents that may interfere with structure of NSCs are employed, physical methods rely on mechanical force, heat, and physical interactions to manipulate cellulose particle morphology into spherical forms. Generally, by-products of physical methods are environmentally friendly, which makes them relevant in biological applications. *Some* of the physical techniques are discussed in this section.

5.2.4.1 Ball Milling

Ball milling is a mechanical, top-down technique employed in cellulose preparation for reducing its particle size, thereby increasing its surface area for various applications. Typically, this technique involves the use of grinding media (usually ceramic or steel balls) in a rotating chamber to apply impact and shear force to break down larger cellulose feedstock into smaller particles (Chhabra & Basavaraj, 2019). For biological applications, ceramic grinding media and the rotating chamber could be preferred over steel to limit the risk of contamination as a result of friction.

The ability of the ball mill technique to break down cellulose to particle sizes up to nanoscale as well as decrystallizing them are advantages that make it relevant to processing NSC for biological applications (Zhao et al., 2006; Avolio et al., 2012; Ling et al., 2019). In applications such as tissue engineering and drug delivery systems, decreasing crystallization of cellulose is associated with improved processability, increased surface area, higher drug loading capacity, and enhanced water absorption that impacts its characteristics such as biodegradability and bioavailability. Ball mill serves as an effective pre-treatment to enhance the efficiency of subsequent processes such as acid hydrolysis and ASP, which ultimately shape the particles into spherical forms (Yabushita et al., 2014).

5.2.4.2 Ultrasonication

In ultrasonication, high frequency sound waves (about 20 kHz) are employed to break down cellulose fibers into nanoparticles, typically between 100 and 300 nm diameter (Meyabadi et al., 2014). During this process, acoustic cavitation generated by using an ultrasound sonicator induces shear forces and microjetting, which breaks down the cellulose. As shown in Figure 5.3, the process involves an ultrasound sonicator that generates the ultrasonic frequency delivered to the sample in suspension via a probe connected to it. Parameters such as amplitude, duration, and temperature

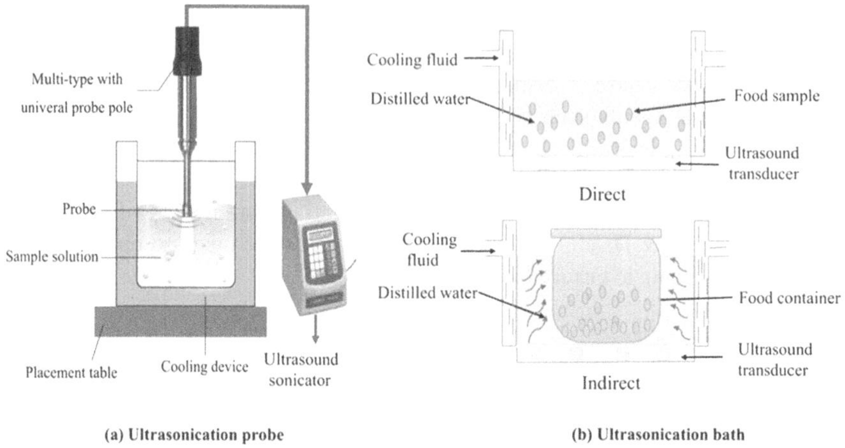

FIGURE 5.3 Ultrasonication (Kumar et al., 2023).

significantly influence the resulting particle sizes and diameters (Ponce-Reyes et al., 2014). Moreover, ultrasonication can be combined with pre-heated and/or partially hydrolyzed cellulose that helps to further refine the cellulose nanospheres (Keerati-U-Rai et al., 2009; Hemmati et al., 2019).

5.3 CHARACTERIZATION TECHNIQUES FOR NSC

Due to its unique properties, including large surface area, biodegradability, adjustable surface chemistry, and recognizable spherical morphology, a thorough characterization is necessary in order to fully comprehend and utilize the potential of NSC (Dele-Afolabi et al., 2023a). The physical, chemical, morphological, and thermal characteristics of NSC are assessed using a variety of methods. These include thermal analysis techniques like differential scanning calorimetry (DSC) and thermogravimetric analysis (TGA) for evaluating thermal stability; Fourier-transform infrared spectroscopy (FT-IR) and X-ray diffraction (XRD) for structural and chemical composition; and microscopy methods like transmission electron microscopy (TEM) and scanning electron microscopy (SEM) for morphological analysis. When combined, these methods offer vital information about the structure–property relationships of NSC, enabling its efficient application in a variety of fields. The various techniques are discussed in this section.

5.3.1 Morphological Characterization

The cellulose nanosphere (CN) from rice straw (RS), corncob (CC), pineapple leaf (PL), and peel (PP) is shown in Figure 5.4 along with its appearance and microstructural morphology (Romruen et al., 2022). All of the samples had a white powder look; cellulose is represented by this color. The FESEM results showed that the extracted CN had a uniformly distributed spherical shape. RS-CN, CC-CN, PL-CN, PP-CN, and COM-CN have diameter ranges of 28–82, 2–44, 8–42, 11–36, and 28–127 nm,

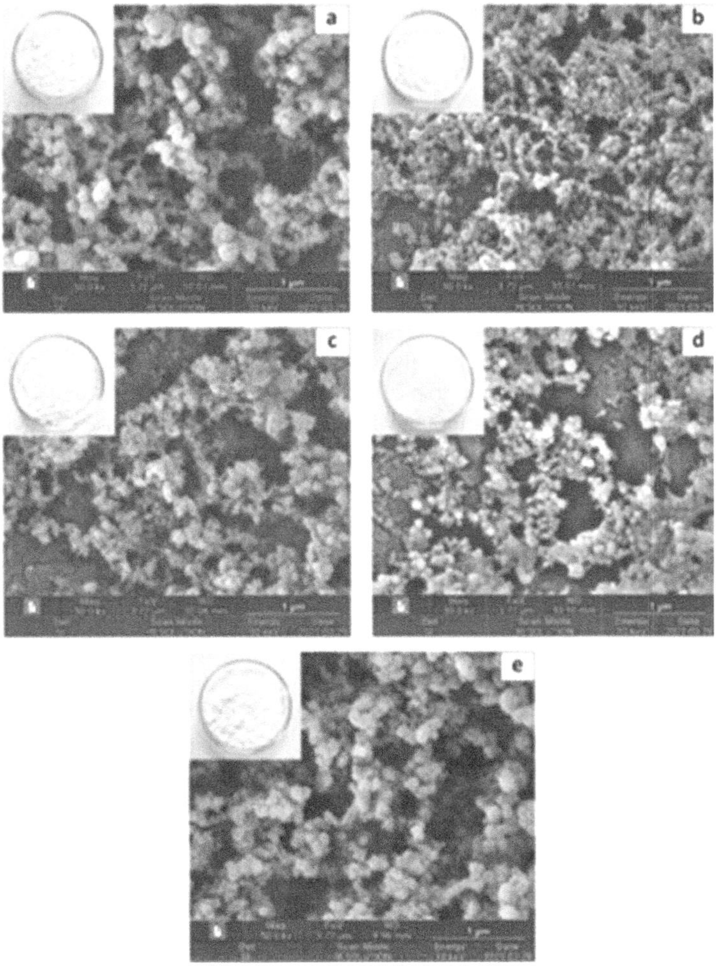

FIGURE 5.4 SEM images of cellulose nanospheres separated from various sources compared with commercial cellulose: (a) rice straw, (b) corncob, (c) pineapple leaf, (d) pineapple peel, and (e) commercial cellulose (magnification 50 k×) (Romruen et al., 2022).

respectively. Ultrasonication facilitates the hydrolysis process by increasing cellulose hydroscopicity, facilitating acid molecular penetration, improving reactant-to-reactant contact, and accelerating the rate of reaction. The shape of CN demonstrates some agglomeration caused by the drying process, indicating the presence of a strong cohesive force inside the nanomaterials. The NSC has larger aspect ratios and a larger specific surface area than nanocellulose of the same size. Therefore, NSC has a higher potential than rod-shaped nanocellulose to improve the crystallinity and heat stability of a polymer matrix at the same nanocellulose concentration.

Another study described the enzymatic hydrolysis of discarded cotton fibers and sonication treatment to produce cellulose nanoparticles with a spherical shape (Meyabadi et al., 2014). A drop of highly diluted suspension of nanoparticles was put on a glass slide and examined after freeze-drying in order to produce a clearer image. The detected nanoparticles were spherical in shape and had a diameter of less than 100 nm. As discussed in Section 5.2.1.1, the most popular technique for creating cellulose nanoparticles is acid hydrolysis; however, mineral acids are not environmentally friendly and result in by-products. Additionally, mineral acids are somewhat linked to challenging maintenance issues as well as handling, shipping, and process control issues. Consequently, as compared with acid hydrolysis, the enzymatic hydrolysis approach employed in this study is preferred.

SEM was used to analyze the morphology of the PVA/nanocellulose films made from cotton linen and linter (Ibrahim et al., 2010). According to the findings, a porous structure was seen in the film that was created when the percentage of linen nanocellulose was increased to 60%. This suggested that the tensile strength was decreased when the percentage of linen nanoparticles was increased to 40% and 60%. In the meantime, an increase in the amount of cotton linter nanocellulose caused holes to grow in the PVA, which explains why the tensile strength decreased. The spherical shape of the generated cellulose nanoparticles was confirmed by the TEM data. Although the cellulose linen-derived particles had varied shapes, they were mostly spherical in shape. A practical and easy method of creating cellulose nanoparticles with diameters ranging from 570 to 60 nm was made possible by the application of a facile multi-stage process (Zhang et al., 2007). High-resolution SEM pictures of the nanocrystalline cellulose showed that the 190 nm cellulose particles were spherical in shape, with an average diameter of 85 nm, after being acidic or sonicated for three hours. According to supporting information from TEM pictures of aggregated cellulose nanoparticles, the particles were mostly spherical in shape and had average aspect ratios between 0.91 and 1.10.

All of the CNC particles that Wang et al. (2007) produced were spherical in nanodimension, and there was no apparent change in the CNC sample particle size throughout treatments. The sample was designated CNC-a when the inner suspension pH was approximately 5.0. After more than a month of constantly washing CNC-a with distilled water until the suspension was almost neutral, CNC-b was created. After stirring the CNC-a suspension with a 1.0% (w/w) NaOH solution to lower the pH to about 9.0, it was left to stand for 48 hours at room temperature. After that, the suspension was neutralized with distilled water to get the CNC-c sample. A specified volume of CNC-a suspension was added, stirring, to a 20-fold 1.0% NaOH solution to create the sample CNC-d. The mixture was then left to rest for 48 hours at room temperature before being washed with distilled water to achieve neutrality.

TEM images showed some degree of CNC particle aggregation, most likely as a result of water evaporation. The average diameter of the spherical CNC, as determined by TEM images, was roughly 60 nm, significantly lower than the microcrystalline cellulose size of 20 mm (Figure 5.5). The formation of the spherical CNC particles depended on the combined acid composition and ratio. However, the spherical CNC was formed in large part by the ultrasonic treatment. High-resolution SEM (HRSEM) was used elsewhere to examine the morphology of microcrystalline cellulose (MCC) and dodecane-encapsulated spherical cellulose containers (SCC-D) (Tzhayik et al., 2014). According to HRSEM analysis, MCC (Avicel) has an irregular

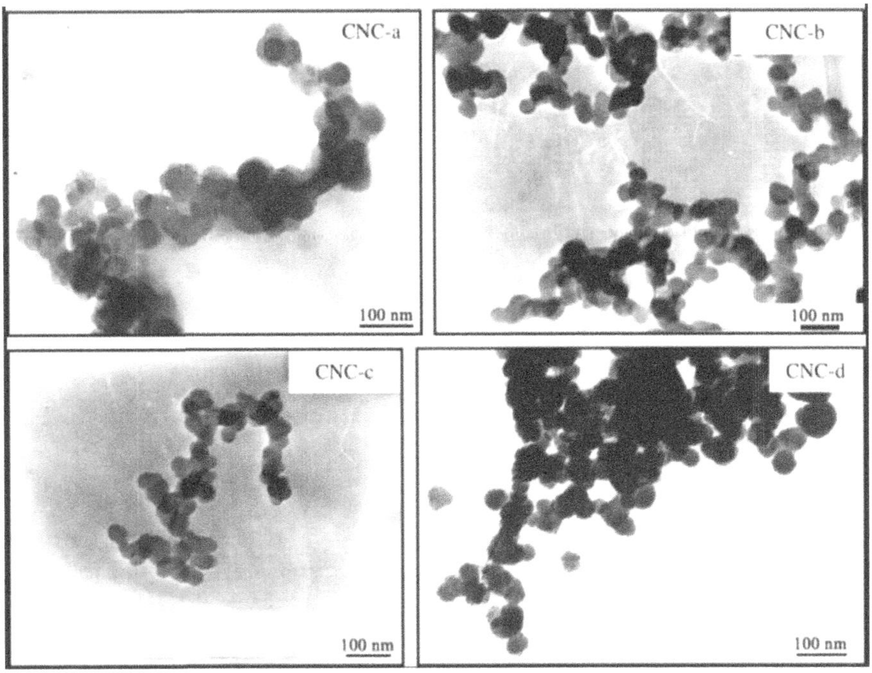

FIGURE 5.5 TEM images of the spherical cellulose nanocrystals (CNC) samples with different treatments (Wang et al., 2007).

shape and a particle size of 50 μm. On the other hand, the SCC-Ds had an average diameter of 52 ± 3 nm and were shaped as smooth surface spherical particles. In the meantime, the NSC showed signs of aggregating, and after a few days, they produced cellulose microspheres that ranged in diameter from 1 to 10 μm.

In a different study by Lu and Hsieh (2010), the cellulose microfibrils are clearly visible in the SEM pictures of cellulose powder prior to acid hydrolysis (Figure 5.6a and b). These original cellulose microfibrils swelled and separated into considerably smaller crystalline cellulose products in three different forms: rods (Figure 5.6c and d), spheres (Figure 5.6e and f), and porous networks (Figure 5.6g and h) after being hydrolyzed by acid at 45°C. Spherical nanocrystals were the most prevalent form, whereas porous networks were the least common. The standard techniques of centrifugation and filtration were unable to separate these three distinct types of nanocrystals. The TEM picture of the rods of CNC revealed lengths of 200–400 nm and diameters of less than 10 nm (Figure 5.6d). Spherical CNC were the predominant by-products of acid hydrolysis and freeze-drying. These cellulose spheres (Figure 5.6e) may have originated from the process of turning cotton fibers into filter paper or from self-assembled short cellulose rods through interfacial hydrogen bonding. SEM (Figure 5.6e) revealed that the spherical CNC were loosely packed, and TEM (Figure 5.6f) revealed that the isolated spherical CNC had sizes ranging from 10 to 100 nm.

The TEM result showed that the NSC produced by acid hydrolysis, followed by lipase-catalyzed esterification with 3-mercaptopropionic acid, ranged in size from

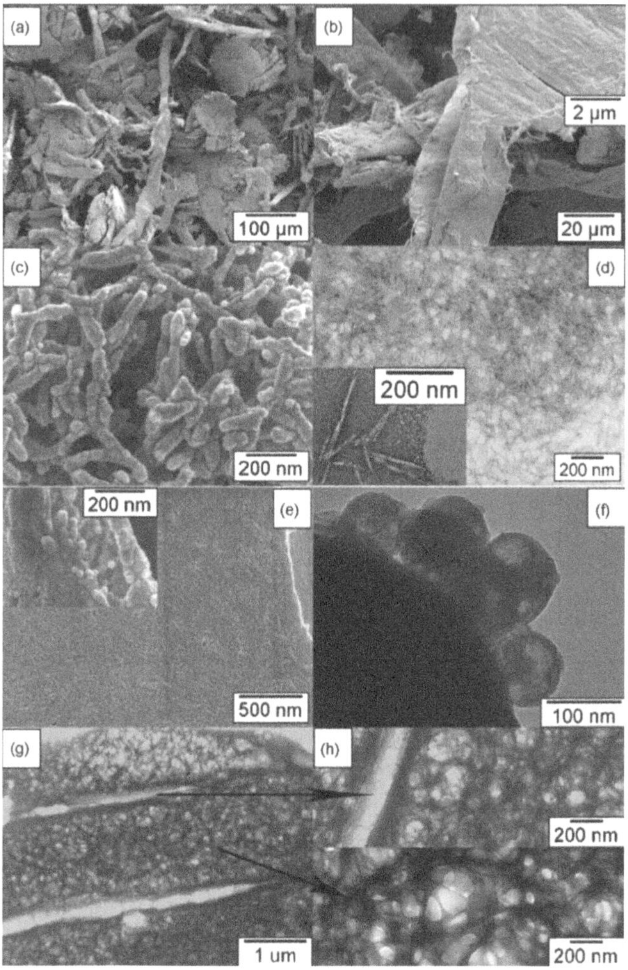

FIGURE 5.6 SEM images of cellulose (a and b) powder and SEM (left, except g with TEM) and TEM (right) images of nanocrystals in the forms of (c and d) rods, (e and f) spheres, (g and h) porous network (Lu & Hsieh, 2010).

45 to 75 nm (Ram & Chauhan, 2018). According to the SEM analysis of the spherical nano-sized cellulose sample made by enzymatically breaking down cellulose with a single cellulase, the result was spherical particles about 30 nm in size. These were mixed with irregular rod-like particles that were hundreds of nanometers long; these could be the intermediate state of the final spherical particles product (Chen et al., 2018). Even though the enzyme concentration of 200 u/ml was quite high and the reaction conditions had already been tuned throughout the sample preparation phase, there were still a lot of uneven strip particles.

After being extracted from jujube seeds using a sulfuric acid hydrolysis process and alkaline and bleaching treatments, the CNCs exhibited a diameter of 27–34 nm (Maheri et al., 2022). The presence of sulfur impurities in CNCs was believed to be caused by the production of negatively charged sulfate groups on the CNC surface following the completion of the acid hydrolysis, based on the EDX spectrum. The TEM analysis revealed that the CNC particles had an evenly distributed, nanoscale, spherical shape that shrank to the point of evaporation. Their enormous specific surface area, van der Waals forces, and tiny diameters were the causes of this (Lu et al., 2013). The resulting spherical morphology appeared to be mostly caused by pre-mechanical and mechanical techniques, as well as the outcomes of dialysis and ultrasonic treatments during acid hydrolysis. With a mean size of 27.37 nm, the resulting spherical nanocrystals were the same as the findings of FESEM analysis. According to the information obtained, it was possible to verify the effectiveness of the conditions used during the acid hydrolysis treatment of jujube seed fibers as well as the presence of distinct nanocrystals in the aqueous suspension.

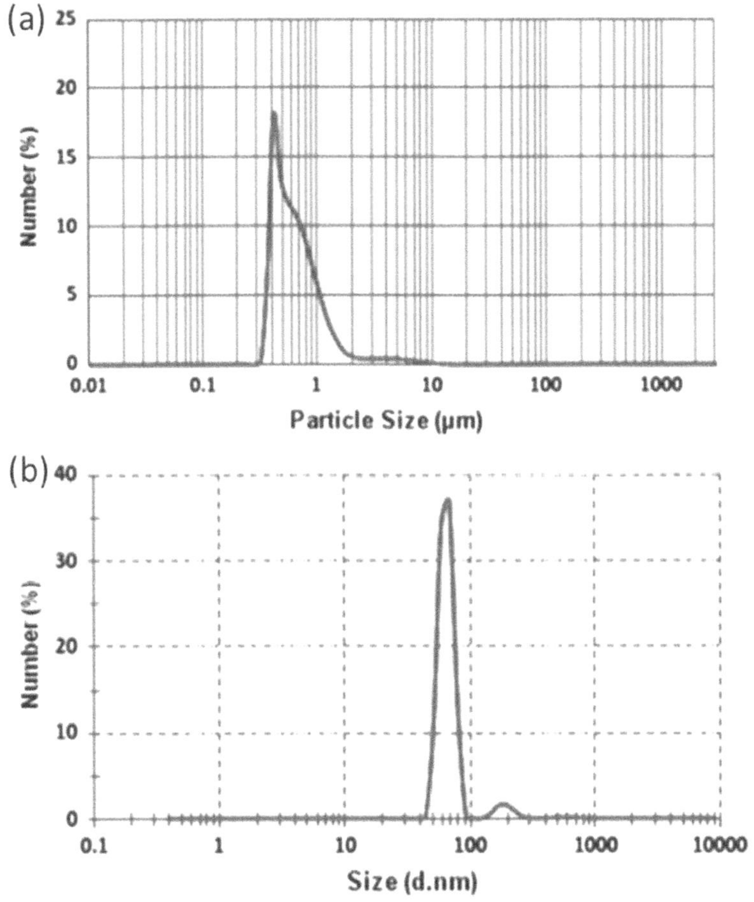

FIGURE 5.7 Particle size distributions of cellulose from (a) hydrolyzed cotton and (b) cotton nanoparticles (Meyabadi et al., 2014).

5.3.2 Particle Size and Distribution Analysis

Figure 5.7 displays the findings of light scattering investigations on the distribution and particle size of the cellulose from hydrolyzed cotton and the corresponding nanoparticles (Meyabadi et al., 2014). The figures demonstrated a noticeable shift in the particle size distribution toward lower values for the nanoparticles compared with hydrolyzed cotton. A particle size analysis of the hydrolyzed cotton suspension revealed a bimodal distribution with an average of 0.417 and 3.802 μm for each peak. The average particle size, d (0.5) = 0.526 μm hydrolyzed cotton suspension, showed that half of the particles were smaller than that size. Based on the particle size distribution of the nanoparticle suspension, the final mixture had two different particle sizes, roughly 70 nm and 200 nm. However, the majority of the particles were between 40 and 90 nm, and there were very few particles between 100 and 300 nm. A steady suspension resulted from the nanoparticles' lower size when compared with hydrolyzed cotton.

Depending on the initial cellulose type, the result of the acid hydrolyzed cellulose suspension had two distinct particle sizes, according to particle size analysis (Ibrahim et al., 2010). The average sizes of the nanocellulose derived from linen and cotton linter were 10.9 nm and 5.9 nm, respectively. According to further research on image morphometry, the particle area of the nanocellulose derived from cotton linter was 16.35 μm^2 at a distance of 15.76 μm, whereas the area of the nanocellulose derived from linen was 29.41 μm^2 at a distance of 19.76 μm.

A Malvern 3000 Zetasizer was used to analyze the particle size of the hydrolyzed cellulose suspension, and the results showed that the final combination had two distinct particle sizes with averages of roughly 500 nm and 70–200 nm (Zhang et al., 2007). Subsequent investigation revealed that in order to improve product yield, particle size, and dispersion, the original cellulose fibers could be mechanically refined to pass through a 20-mesh Wiley mill before pre-treatment. Nevertheless, this approach did not completely eradicate the cellulose nanoparticles' bimodal distribution. The bimodal particle size distribution appeared during ultrasonication, most likely due to the uneven dispersion of the suspension during the treatment, according to a detailed examination of the hydrolysis process. This problem was resolved by mechanical stirring during ultrasonic treatment. A reliable, high-yielding procedure required first mechanically refining the cellulose fibers, then hydrolyzing them with ultrasonic acid while swirling mechanically. After being sonicated for approximately ten minutes, the obtained microspheres returned to their initial nanometer size as measured by DLS (average diameter of 47 ± 3 nm) and the Nanosight apparatus (20–100 nm particle distribution with an average size of 41 ± 18 nm) (Tzhayik et al., 2014). There was good agreement between the DLS results and the size determined by the HRSEM measurements. The cellulose microspheres were determined to be aggregates of cellulose nanospheres. On the other hand, whether kept in a sealed bottle at ambient temperature or in the refrigerator (4°C), SCCs remained stable for over six months. A globular web of interwoven fibers with an approximate width of 40 nm and an unknown length was visible when the cellulose containers were lyophilized.

CNCs produced at various centrifugal speeds had widely differing particle sizes (Ren et al., 2022). Alkali dispersion at pH 9.0 was used to characterize the sample particle sizes after CNCs were separated at different centrifugal speeds. The samples had a lot of rod-like CNCs when the rotational speed was between 500 and 2500 rpm, and the sample size decreased as the rotational speed increased. CNCs resembling rods and shorter than 1.0 μm in length were developed at 1000 rpm. When the rotating speed reached 3000 rpm, pure spherical CNCs were generated, with the particle size mainly distributed between 24 and 76 nm. The CNCs produced at 4000 rpm were slightly larger than those obtained at 3000 rpm, measuring 70 nm, as a result of the re-aggregation of the CNCs caused by the elevated centrifugation speed. When big particles were eliminated by alkali dispersion, 1000 rpm was obviously the ideal separation condition for rod-like CNCs, while 3000 rpm produced pure spherical CNCs with lower sizes.

As a result of having an average zeta potential of −41.6 mV, more intense negative charges were seen throughout the case of CNCs derived from jujube seeds (Maheri et al., 2022). Since sulfate groups are typically carried to the surface during the hydrolysis process by H_2SO_4, which results in the induction of greater negative charges against HCl and other mineral acids, this outcome was rather anticipated (Abu-Danso et al., 2017). Furthermore, because the obtained value was greater than 15 mV, which is thought to be the minimal stability for identifying the start of compaction, it was concluded that the CNC suspension had good dispersity and sufficient stability.

5.3.3 Chemical and Structural Analysis

5.3.3.1 FT-IR Spectroscopy

The physicochemical properties of lingo-cellulosic materials can be investigated using the straightforward method of FT-IR spectroscopy. This process produced the CNCs depicted in Figure 5.8 and demonstrated the hydrolysis of untreated chemical compounds, both alkaline and acidic (Maheri et al., 2022). Furthermore, the hydrophilic behavior of the samples was evident from the band width in the 3500–3300 cm^{-1} region, which was connected to the OH groups present in their fundamental composition. The apparent adsorption peak between 3400 and 3000 cm^{-1} was caused by the elongation and bending of OH cellulose groups, which mirror the reflection of hydrophilic molecules. These peaks (3430 cm^{-1}) gradually decreased throughout the spectra when the fibers were chemically treated. This was due to the removal of lignin and hemicellulose during the treatment. The intensity of the peak at 894 cm^{-1}, which demonstrated the purity of the CNCs, was created by C-H vibrations. Additionally, the wavenumber of 1426 cm^{-1} indicates the flexural vibration of cellulose, whereas the band at 3430 cm^{-1} is linked to the stretching of the samples' pre-existing C–H and O–H.

The peak in the treatment spectrum at 1742 cm^{-1} is associated with the uronic acetylene ester groups of hemicellulose or the ester bond of the carboxylic group of ferulic and *p*-coumaric acid in lignin and/or hemicellulose (Satyamurthy & Vigneshwaran, 2013). Because hemicellulose and lignin were significantly removed, the peak that was visible in the acid hydrolysis spectrum decreased and subsequently disappeared from the CNCs spectra as a result of acid extraction. The C–H attraction was responsible for the peak at 2899–2926 cm^{-1}. Although some of them were

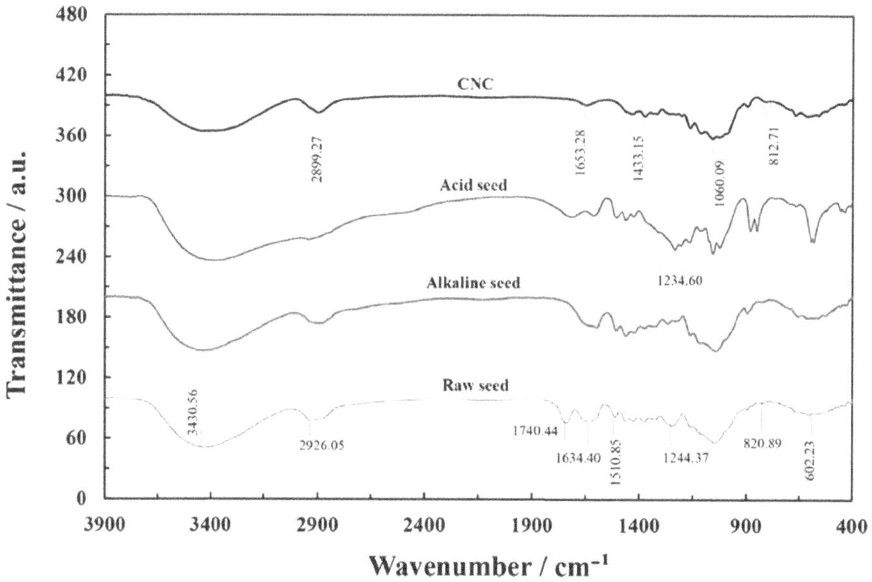

FIGURE 5.8 Fourier-transform infrared spectra of hydrolyzed seed, alkaline seed, untreated jujube seed, and CNCs (Maheri et al., 2022).

omitted, the peaks seen in the samples at 1634–1653, 1510, and 1595 cm^{-1} were also suggestive of the aromatic ring in lignin and adsorbed water. The spectra at 1250 cm^{-1} are considered an external feature of CO since the aryl group in lignin was removed from the CNCs. Across the spectrum of treated samples, the strips were more intense than cellulose because the crystalline content of the strips rose when lignin was eliminated. In actuality, the treated samples had stronger peaks that were ascribed to the C–H cellulose group than the crude ones in the 1400–1300 cm^{-1} region. The observed signal at 1234 cm^{-1} demonstrated that the acid hydrolysis of the hydrolyzed sample resulted in the OH replacing the sulfur groups (Bano & Negi, 2017; El Achaby et al., 2018; Xu et al., 2018).

FT-IR was used to confirm the reaction between the PVA and nanocelluloses (Ibrahim et al., 2010). A band at 3429 cm^{-1} was ascribed to the PVA hydroxyl group O–H stretching vibration. All of the hydroxyl groups in PVA and nanocelluloses are engaged in the reaction, as evidenced by the O–H stretching for the films generated, which are represented by a broad band at around 3290 cm^{-1}. For the generated PVA/nanocelluloses, a distinct band at about 1250 cm^{-1} indicated the presence of an acetyl C=O group on the PVA backbone. Around 2931 cm^{-1}, the peak caused by the aliphatic C–H stretching vibrations emerged. The carbonyl (C=O) stretching vibration of the a-keto carbonyl present in the nanocelluloses may be the cause of the absorption peak seen at about 1722 cm^{-1}, thereby validating the reaction between the nanocelluloses and the PVA. The PVA spectrum included the CH stretch band at about 1423 cm^{-1}. Additionally, the bands at about 1429 cm^{-1} were linked to CH in the nanocellulose CH group plane deformation. In contrast, the C–O stretching vibrations of the aliphatic primary and secondary alcohols in nanocellulose were involved in the bands at about 1033 cm^{-1}.

The FT-IR spectra of produced cellulose nanospheres (SCC) and MCC are shown in comparison in Figure 5.9 (Tzhayik et al., 2014). In general, the spectra do not differ significantly. At 3333 and 3341 cm^{-1}, the cellulose alcohol groups that were H-bonded were exhibited distinctive stretching vibrations in both MCC and SCC. This implies that hydroxyl groups and H bonds were present prior to and following the cross-linking of cellulose nanospheres, and that the formation of cellulose spheres did not eliminate or weaken these hydrogen bonds. It appears that the sonication promoted the breaking and the establishment of new H bonds during the spheres' development. The peak at 2902 cm^{-1} in the MCC curve indicates the stretching vibration of the C–H bonds of glucose units, whereas the bending vibration of these bonds absorbs at 1428 cm^{-1}. The distinctive peaks of the MCC at 1638 cm^{-1} and the SCCs at 1640 cm^{-1} are probably associated with the bending mode of water molecules that are intimately bonded. MCCs and SCCs have similar spectral properties in the 800–1200 cm^{-1} region of the MCC fingerprint. The existence of all these distinct peaks attests to the fact that the shell of the container is made entirely of cellulose and that the sonochemical creation of SCCs did not substantially alter the fundamental structural units of cellulose.

The crystalline I_β characteristic peaks, OH stretching at 3270 cm^{-1} and OH out-of-plane bending at 710 cm^{-1}, were found in the Fourier-Transform Analogous spectra of untreated cotton, hydrolyzed cotton, and the corresponding nanoparticles (Meyabadi et al., 2014). It was clear that the enzymatic hydrolysis and ultrasound treatment retained the cotton cellulose I_β allomorph. Overall, following the hydrolysis and ultrasonication, every band held its place. Spherical CNCs were constantly cleaned as washing periods increased, and variations in the FT-IR spectrum

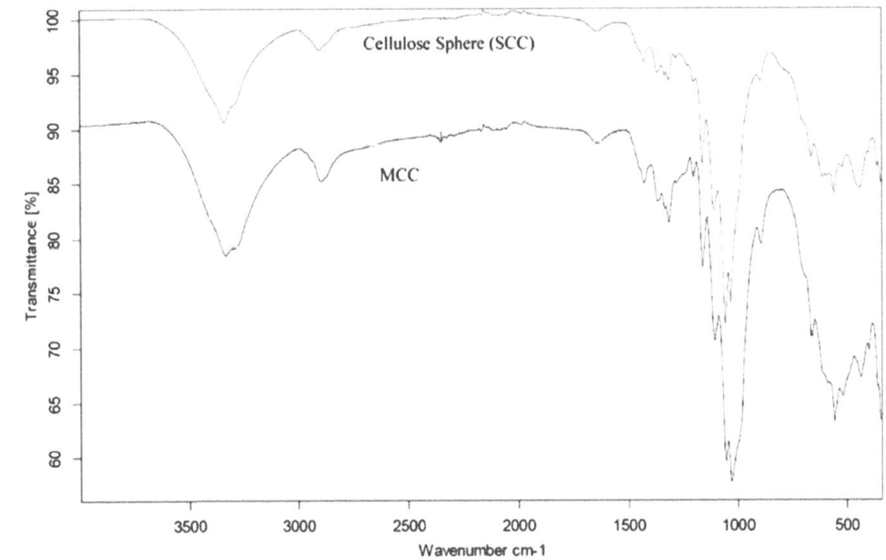

FIGURE 5.9 Fourier-transform infrared spectra of microcrystalline cellulose and cellulose nanosheres (Tzhayik et al., 2014).

absorption peak intensity were observed (Ren et al., 2022). In the FT-IR spectra, the OH peak was located at 3345 and 1638 cm^{-1}. Remarkably, more OH was present at the 3345 and 1638 cm^{-1} peaks. The tensile vibration peaks of CH and CO cyclics were situated near the peak at 1031 cm^{-1}. The peak at 665 cm^{-1} was linked to the C–OH out-of-plane bending peak. The two primary FT-IR discrepancies observed by Lu and Hsieh (2010) in their study are the sulfate peak of the CNC from the esterification reaction at 1205 cm^{-1} and the stronger peak of the adsorbed water in the cellulose granules at 1641 cm^{-1}. The existence of sulfate groups was further verified by the EDS, which showed that the CNCs had a weak sulfur peak that made up only 0.85 wt% of the total, whereas the carbon peak accounted for 64.38 wt%. According to this atomic ratio, approximately one out of every 12 anhydroglucopyranose units became esters through a reaction with sulfuric acid. This supports the idea that the resulting sulfate groups were mostly found on the crystal surfaces and significantly enhanced charge repulsion-based nanocrystal separation.

By identifying the functional groups present, FT-IR analysis was used to examine the composition of the isolated CN (Romruen et al., 2022). The similarity of the diffraction spectra of all the samples showed that the chemical functionality of the separated CN was unaffected by the shape of the source material. The removal of the lignin residue was shown by the lack of the peak at 1541 cm^{-1} (aromatic CC = benzene ring stretching). This occurred as a result of the cellulose-lignin linkages being seriously disrupted by the acid hydrolysis process. The fact that the spectra of pure cellulose and NSC are similar is one important conclusion that can be made from these investigations (Ram & Chauhan, 2018). Therefore, there was no indication of cellulose sulfonation. This demonstrates that when SNC was formed by acid hydrolysis, the structural composition remained unchanged. The FT-IR spectrum of SNC-3-MPA (3-mercaptopropionic acid) showed three prominent absorption bands: one at 3336 cm^{-1} (–O–H stretching), one at 2901 cm^{-1} (C–H symmetrical stretching), two bands at 2472 and 2359 cm^{-1} that are attributed to the thiol group (S–H) absorption, and a band at 1720 cm^{-1} that was attributed to the ester C=O group that emerged after the esterification reaction.

5.3.3.2 XRD

Through chemical treatment, samples and CNCs were separated from jujube seeds, and XRD analyses were used to determine their size and crystallization index (CI). Observations showed that cellulose is divided into two regions: crystalline and amorphous. Hemicellulose and lignin make up the amorphous portion (Alemdar & Sain, 2008). Figure 5.10 displays the XRD patterns of the jujube seed CNCs that were produced, hydrolyzed, alkalized, and untreated (Maheri et al., 2022). The crystalline manner of cellulose, which included a sharp diffraction peak at $2\theta = 22.6°$ (002), indicated the high perfection of crystal lattice in accordance with XRD patterns, whereas the appearance of two weak peaks in $2\theta = 15.4°$ (101) and 34.5° (004) referred to the typical pattern of cellulose I (Nagalakshmaiah et al., 2016). It was found that this curve had a peak at roughly 22.6°, which is suggestive of a typical cellulose network topology and also represents amorphous phase diffusion capabilities. The systematic elimination of cellulosic polysaccharides and the cessation of the amorphous zone were the causes of the greater crystallization level of CNCs (Wang et al., 2019).

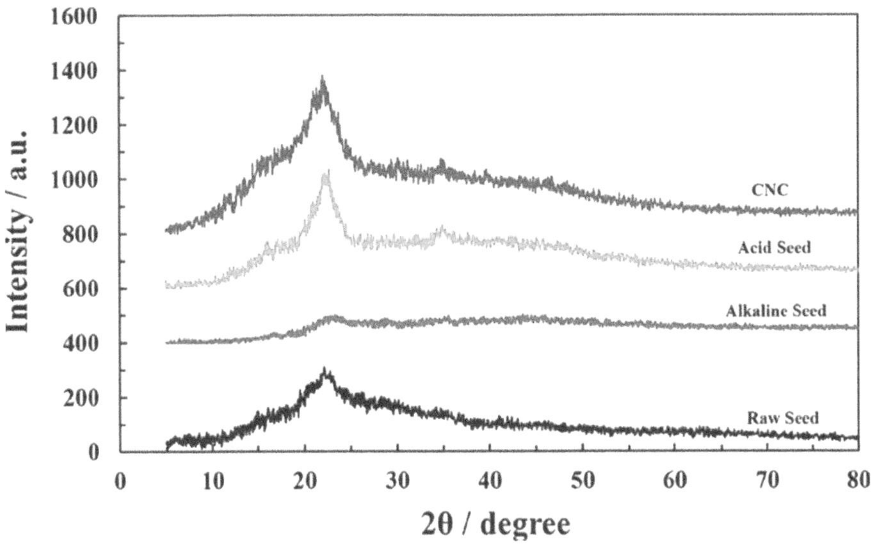

FIGURE 5.10 XRD patterns of hydrolyzed seed, alkaline seed, untreated jujube seed, and CNCs that were synthesized (Maheri et al., 2022).

However, the grade of crystallinity is determined by the type of chemical process and conditions that are applied to the material. Numerous scholars have looked into the phenomenon of a particular increase in crystallization after executing an acid treatment (Marett et al., 2017). Due to the resulting rise in CI, it was suggested that lignin, hemicellulose, and pollutants be totally eliminated from untreated jujube seeds. Additionally, it can be said that the hydrolysis and other treatments did not change the crystal structure of the cellulose in any way. The crystal size, which appears to be relative, was determined to be 26.59 nm CNC using the Scherer equation (Gu et al., 2017).

Using XRD diffraction analysis, cellulose I was detected from the freeze-dried 0.237 mm beginning cellulose, fibers that had been pre-treated with NaOH–DMSO, and cellulose nanoparticles with a mean diameter of 80 nm. Cellulose II was identified for the pre-treated cellulose and nanocellulose, which had an average diameter of 80 nm (Zhang et al., 2007). Following the alkaline pre-treatment and the initial acid hydrolysis, cellulose II was produced. This micrometer-scale cellulose exhibited amorphous areas that are selectively hydrolyzed. Subsequent acidic sonication converted the micrometer-scale cellulose into nanocrystalline cellulose.

It was evident from their X-ray curves that commercial MCC and CNC samples had comparable XRD patterns (Wang et al., 2007). This observation demonstrated that cellulose I of MCC had retained its crystalline structure after hydrolysis with a mixed acid (Majdanac et al., 1991). The degree of crystallinity further supported this view. CNC samples have a somewhat lower crystallinity than MCC samples. One explanation for the high surface amorphous ratio in CNC was that they had a high specific surface area. Consequently, the various thermal degradation behaviors in CNC were caused by differences in the outermost structure of nanocrystalline particles, notably the presence of sulfate groups. The diffraction profiles of the resulting CN and MCC are comparable, although during the cellulose container formation process, there were some variations in peak intensity (Figure 5.11a) (Tzhayik et al., 2014). The MCC peaks have stronger XRD peak intensities than the SCCs. The crystallinity of the SCCs was not exactly the same as that of the original MCC.

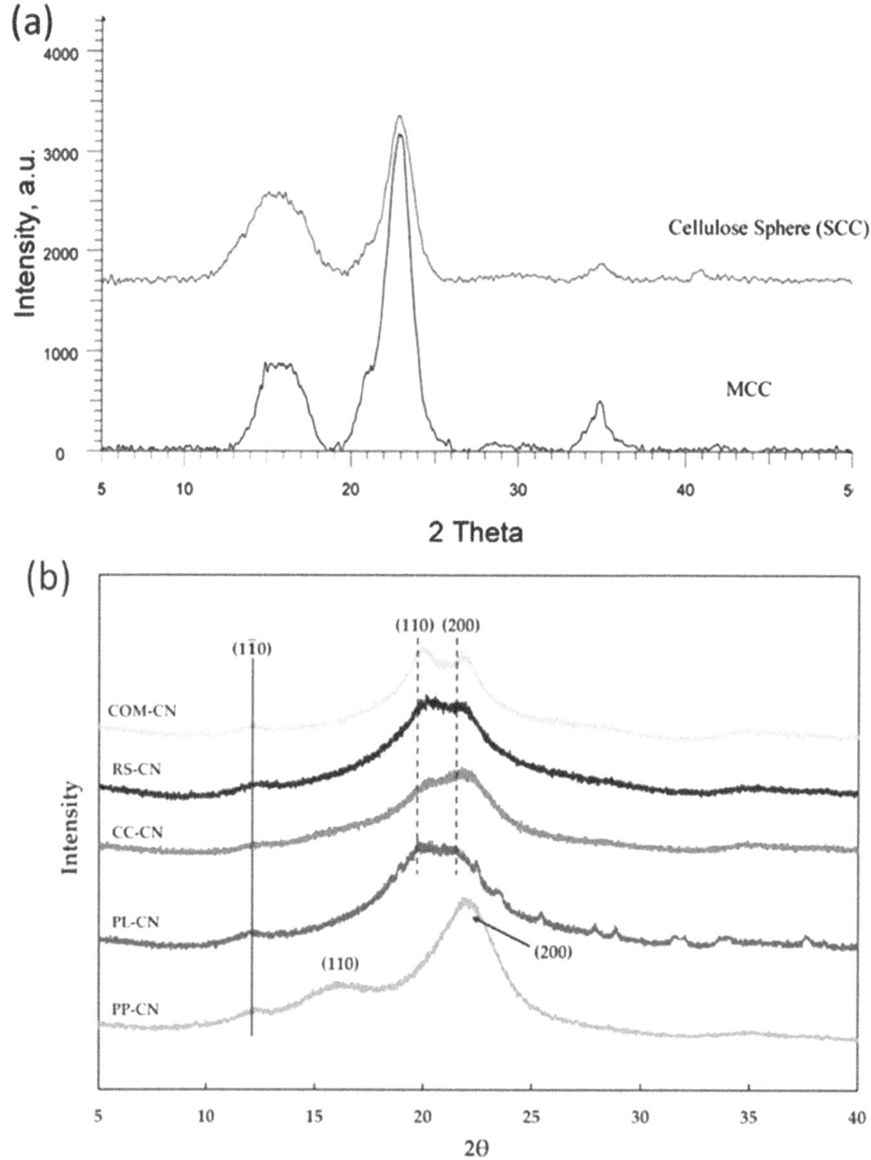

FIGURE 5.11 XRD patterns of (a) microcrystalline cellulose and cellulose nanospheres (Tzhayik et al., 2014), and (b) cellulose nanospheres isolated from different agricultural by-products compared with commercial cellulose (Romruen et al., 2022).

Compared with MCC, SCCs are more amorphous, and the ultrasonic effect is crucial in lessening the sample crystalline nature. However, as Zhang and colleagues concluded (Zhang et al., 2013), these XRD profile alterations showed that the ultrasonic primarily affected the particle size of the cellulose chains and had no discernible effect on the supramolecular ordering of MCC.

The XRD spectra of CN, which was separated from a number of agricultural by-products, such as RS, CC, PL, and PP, are shown in Figure 5.11b (Romruen et al., 2022). RS-CN, CC-CN, PL-CN, and COM-CN have measured diffraction peaks at 12.1°, 20.0°, and 21.8°, respectively. The peak that is displayed corresponds to the crystallographic planes of $(1\bar{1}0)$, (110), and (200) and indicates the cellulose-II structure (French, 2014). Thus, these results verified that the acid hydrolysis process produced the cellulose-II crystal structure. On the other hand, PP-CN displays the peaks at 2θ at approximately 12.1°, 16.0°, and 22.0° accordingly.

While the peak at 12.1° suggests the presence of a trace amount of cellulose II, the peaks at 16.0° and 22.0° corresponded to the distinctive diffraction maximum of cellulose I. The XRD-assessed CI showed how the mechanical and physical properties of the materials changed. The CI values for RC-CN, CC-CN, PL-CN, PP-CN, and COM-CN were 58.43, 73.58, 64.75, 43.98, and 63.23%, respectively. The increased conversion of cellulose I to cellulose II in the CI indicates the preferred breakdown of cellulose amorphous regions during sulfuric acid hydrolysis. However, the loss of the crystallization zone as a result of the hydrogen bonds between the cellulose macromolecules breaking could be the cause of the CI decline.

In the diffraction patterns of treated materials (hydrolyzed cotton and the nanoparticles), a little increase in the peak intensity of the (002) plane ($2\theta = 22.5°$) indicated more perfection of the crystal lattice in the (002) plane as compared with untreated cotton (Meyabadi et al., 2014). Additionally, for treated samples, the peak of the (101) plane ($2\theta = 14.7°$) became somewhat more intense. Following the hydrolysis and ultrasonication procedures, there were no discernible changes in the CI values of the cellulose. Naturally, there was only a slight variation in the CI of the samples. Additionally, the cotton that was not treated and the cotton that was treated with enzymes had crystallinity indices of 80.46% and 82.80%, respectively. Due to the quick removal of amorphous cellulose close to the surface caused by the difference in reactivity between crystalline and amorphous cellulose, there was a minor rise in CI. Additionally, because the amount of surface amorphous cellulose was so small in comparison with bulk amorphous cellulose, its loss did not significantly alter the overall crystallinity. Notably, the technique used in this investigation preserved the crystalline structure of cellulose I in spherical nanoparticles. However, the alternative techniques caused the crystalline structure of the nanoparticles to transform into cellulose II (Li et al., 2001; Zhang et al., 2007).

Three cellulose I characteristic peaks were seen in the XRD of cellulose powder and nanocrystals at $2\theta = 14.7°$, 16.4°, and 22.6°. Higher perfection of the crystal lattice in the (2 0 0) plane compared with original cellulose was indicated by the sharpening of the 22.6° peak of the (2 0 0) plane of CNC (Lu & Hsieh, 2010). In addition to becoming stronger, the peak for the $(1\bar{1}0)$ plane ($2\theta = 14.7°$) separated from the 110 ($2\theta = 16.4°$) peaks for CNC. After the acid hydrolysis and freeze-drying operation, the CNC became more crystalline than the original cellulose while maintaining the cellulose Iβ crystalline structure, according to the XRD data.

5.3.4 Thermal Stability Analysis

The thermal properties of extracted CN from several agricultural by-products (RS, corncob [CC], PL, and peel [PP]) were examined using a TGA-DSC combination testing system (Romruen et al., 2022). For its potential use as a reinforcing material in bionanocomposites, the assessment of thermal stability is essential. The inherent characteristics of the samples and the molecular interactions among the different macromolecules determine the polymer thermal stability. Figure 5.12 displays the TGA thermogram, DTG, and DSC curve results, while Table 5.1 provides a summary of the analysis' findings. The isolated CN from RS, CC, PL, PP, and COM had degradation temperatures (T_{max}) of 314°C, 318°C, 323°C, 327°C, and 313°C, respectively. The slight variation suggests that the CNSs produced have comparable compositions and levels of thermal stability. Depending on the type of raw material, the ultimate residues after thermal degradation remained at 600°C. The largest final

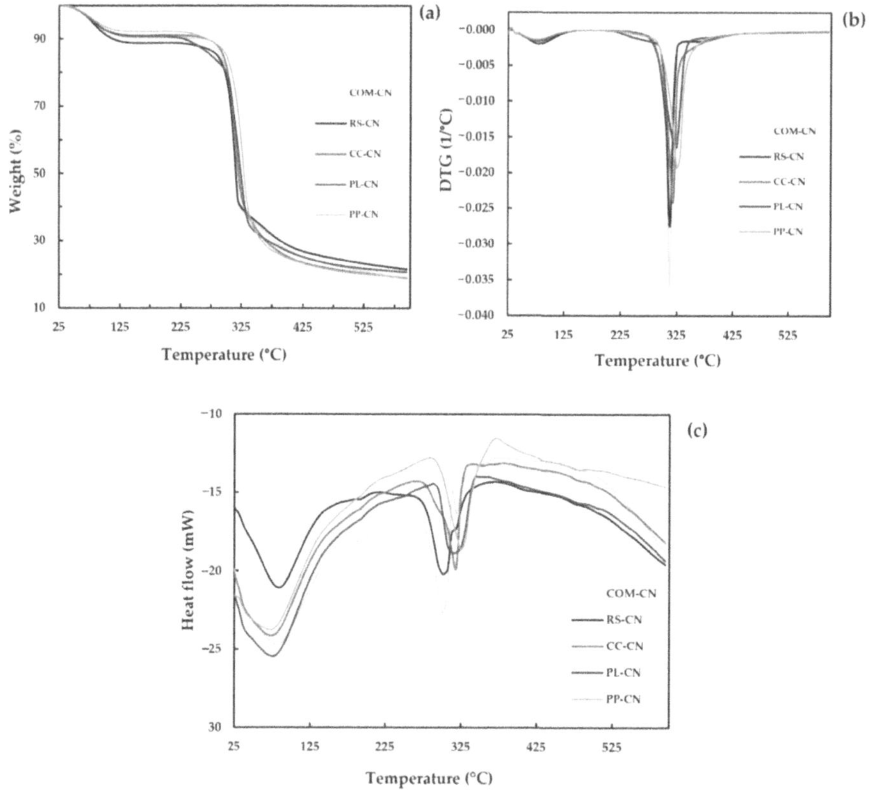

FIGURE 5.12 Cellulose nanospheres extracted from various sources are compared with commercial cellulose using TGA (a), DTG (b), and DSC (c) curves of the following are cellulose nanospheres: RS-CN (rice straw), CC-CN (corncob), PL-CN (pine leaves), PP-CN (pine peels), and COM-CN (cellulose nanosphere hydrolyzed from commercial cellulose) (Romruen et al., 2022).

TABLE 5.1
Comparing the thermal characteristics of cellulose nanospheres from various sources to commercial cellulose (Romruen et al., 2022)

Source of Cellulose Nanosphere	TGA Analysis			DSC Analysis		
	T_{onset} (°C)	T_{max} (°C)	Residue at 600°C (%)	T_{onset} (°C)	T_{max} (°C)	ΔH (J/g)
Rice straw	304.22	314.67	27.54	275.94	302.00	177.04
Corncob	301.60	318.00	25.38	305.13	317.33	133.55
Pineapple leaf	295.62	323.33	25.03	295.42	318.00	120.44
Pineapple peel	302.25	327.67	23.34	296.10	327.33	191.44
Commercial cellulose	305.49	313.67	25.29	276.24	302.33	176.14

residue (27.54%) is displayed by RS-CN, which is followed by CC-CN (25.38%), COM-CN (25.29%), PL-CN (25.03%), and PP-CN (23.34%), in that order. The relatively small weight losses demonstrate that the robust cellulose structure could withstand high temperatures.

The differences in final residues between the CNs may be influenced by differences in crystallinity and chemical structure. The sulfate ester groups functioned as a flame retardant in the acid hydrolysis process to create CN, which led to a high final residue of CN. The low final residue of the PP-CN may vary depending on the type of cellulose crystal structure. With the exception of PP-CN, which appeared in cellulose type I, extracted CN was found to be cellulose type-II based on the results of XRD examination. The crystal structure of cellulose II exhibited remarkable heat stability. DSC curves were used to assess the evolving thermomolecular behaviors of isolated CNs, as seen in Figure 5.12c. The findings show that the degradation trends of each sample were comparable. An endothermic peak was initially seen between 30°C and 130°C, indicating that the absorbed heat energy was used for the evaporation of water. At 200°C, a little second endothermic peak appeared, indicating the start of cellulose breakdown. Around 300°C, an endothermic peak was observed, which is indicative of the decarboxylation and depolymerization of cellulose. The DSC analysis was in agreement with the TGA investigation, as evidenced by the striking similarity between the obtained DSC values and the TGA data.

In other cases, the typical degradation temperatures of untreated cotton, hydrolyzed cotton, and nanoparticles were not significantly altered by the hydrolysis or sonication of cotton fibers (Meyabadi et al., 2014). The shape-related deterioration trends seem to be consistent across all samples. Concurrent degradation processes such as depolymerization, dehydration, and breakdown of the glycosyl rings, followed by the production of a charred residue, produced the dominating peak between 300°C and

350°C. The oxidation and breakdown of the charred residue into low molecular weight gaseous compounds was also the cause of the peak above 350°C. Due to the decreased crystallinity of the cotton nanoparticles in comparison with the other samples, the decomposition process began at a lower temperature of roughly 250°C. Because of the increased surface area exposure to heat, the high surface area of nanoparticles significantly reduced their thermal stability. Moreover, the breakdown at lower temperatures would demonstrate a quicker heat transfer in the cellulose nanoparticles. Additionally, those with higher crystallinity had more burned residue. From 11.09% for the untreated cotton to 25.21% and 7.34% for the hydrolyzed cotton and nanoparticles, respectively, the char yield rose. These nanoparticles exhibited higher thermal stability than spherical nanoparticles produced by acid hydrolysis. Figure 5.13 displays the TG and DTG curves for various CNC. In the low temperature (<120°C) range, all CNC samples experienced a slight weight loss, which was caused by the evaporation of absorbed water (Wang et al., 2007). The TG and DTG graphs show that in the high temperature range (>150°C), the degradation behaviors of CNC samples differed from one another. The DTG curves showed two different pyrolysis processes, and the degradation of CNC samples with distinguishable acid sulfate groups (CNC-a and CNC-b) occurred over a wider temperature range. Between 150°C and 290°C, one process took place, and between 320°C and 500°C, another. The thermal breakdown of CNC significantly changed to a higher temperature and took place within a limited temperature range when the acid sulfate groups were neutralized by NaOH solution. Even though sample CNC-c still displayed two pyrolysis processes, the second process predominated because the first process, which occurred between 220°C and 260°C, was so tiny. Only one pyrolysis process was visible in the degradation of CNC-d following treatment with the higher content NaOH solution (1.0%).

Notably, samples containing acid sulfate groups had significantly more char residue at 500°C than samples neutralized with a 1.0% NaOH solution. Char residue generation was aided by sulfuric acid, a well-known dehydration catalyst (Kim et al., 2001; Scheirs et al., 2001). The acid-catalyzed dehydration was hindered when the hydrogen ion (H^+) was swapped out for the alkaline ion (Na^+). Furthermore, the yield of char residue was reduced because alkaline ions with a highly polar field could serve as a catalyst for the full hemolytic breakdown of glycosidic units (Antal et al., 1988; Piskorz et al., 1989). Additionally, it was noted that CNC-a and CNC-b decomposed at lower temperatures and exhibited comparable DSC curves (Figure 5.13c).

The endothermic peak, which peaked at about 185°C, matched the first fitting peak of primary pyrolysis, which is the dehydration reaction mediated by acid sulfate groups (Kim et al., 2001). When acid sulfate groups were neutralized by NaOH solution, the CNC disintegration in DSC curves changed dramatically to the higher temperature, suggesting a significant increase in their thermal stability. An endothermic peak above 300°C was visible in the CNC-d DSC curve, indicating the pyrolysis process. Sample CNC-c showed two distinct endothermic peaks between 230°C and 240°C. This finding verified that the minor breakdown did, in fact, take place prior to the main pyrolysis process.

In other cases, the thermal behaviors of the CNC differed dramatically from those of the original cellulose granules (Lu & Hsieh, 2010). The original cellulose

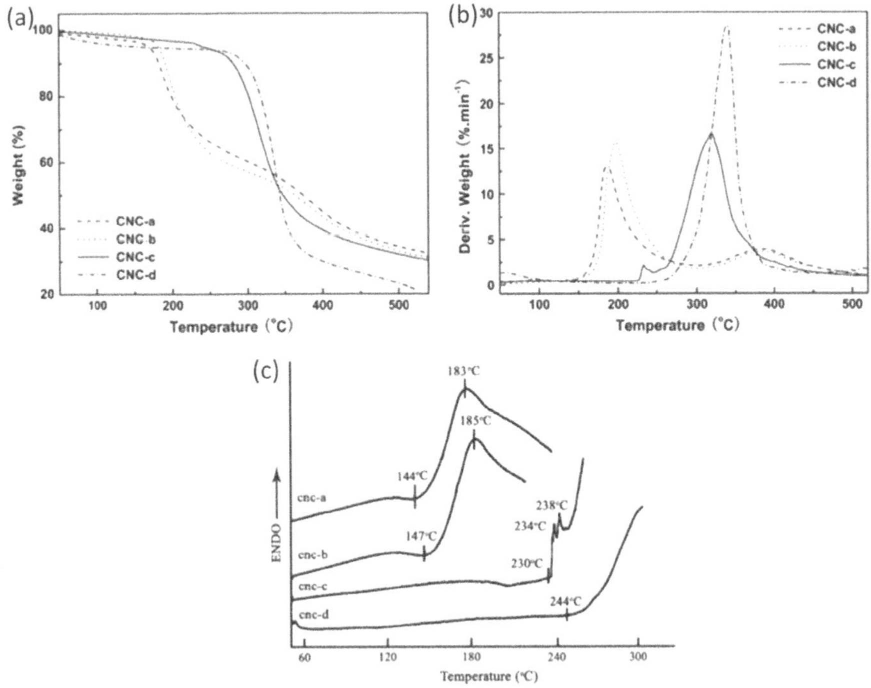

FIGURE 5.13 (a) TG, (b) DTG, and (c) DSC curves of spherical cellulose nanocrystal samples (Wang et al., 2007).

decomposed as it usually does, with an onset temperature somewhat above 300°C and a significant mass loss that left only 2.87% ash at 600°C. However, in both DSC and TGA, the cellulose nanocrystals displayed slower thermal transitions that began at a lower temperature of about 150°C. The CNCs lost over 40% of their mass between 150°C and 300°C, and then another 30% between 300°C and 600°C. At 600°C, they had a much greater residue of around 30%. Different decomposition–gasification processes may be responsible for these significant thermal behavioral changes between the original cellulose powders and the CNCs. At approximately 180°C, cellulose is known to break down into levoglucosan (1,6-anhydro-β-d-glucopyranose), which effectively gasifies at 300°C. A different breakdown mechanism, possibly direct solid-to-gas phase transitions aided by surface sulfate groups, is suggested by the appearance of small endotherms at 160°C and 210°C and the gradual mass loss of the CNCs in the 150°C–300°C frequency range. Thermal analysis (TGA) was used to evaluate the effect of functionalization on the thermal degradation behavior of cellulose, spherical nanocellulose (SNC), and spherical nanocellulose-3-mercaptopropionic acid (SNC-3 MPA) (Ram & Chauhan, 2018). The initial decomposition temperature (IDT) of pure cellulose is roughly 300°C, per published studies. The same was hydrolyzed with sulfuric acid and then reduced to 150°C. The benefit of combined hydrolysis (HCl and H_2SO_4) in this work was that the IDT of the adsorbent is 220°C, while that of the SNC was significantly higher at 260°C. In contrast to both SNC and cellulose, the latter had a similar end breakdown temperature and a far higher amount of residue left over.

The extracted CNCs and untreated jujube seed were analyzed for the factor of thermal stability under various conditions using the TGA method (Maheri et al., 2022). Several studies have demonstrated that using a temperature range of 200°C–450°C can significantly break down cellulosic materials (Alemdar & Sain, 2008). At lower temperatures (between 25°C and 150°C), slight weight loss is typically an indication of moisture evaporation and removal. Because cellulose, lignin, and hemicellulose have various chemical structures, ligno-cellulosic compounds showed breakdown at varying temperatures. It was found that the hemicellulose degradation process began at 250°C and progressed to 315°C. Additionally, the untreated jujube seed showed a peak in decomposition and attained the highest mass loss rate at 294°C. In the instance of hemicellulose, it was assumed that the reduced thermal stability was due to the presence of acetyl groups.

It was clearly expected that hemicellulose degradation would be identified before lignin and cellulose decomposition. Cellulose began to break down around 290°C, which is indicative of point T_m, and continued to do so until it reached 470°C. Additionally, at 468°C and 404°C, respectively, the untreated sample and the CNCs that were produced saw the highest rate of weight loss. In comparison with the untreated sample, the CNCs sample may have had a lower maximum degradation temperature due to the size of the prepared CNCs, which may have led to the induced drop in thermal stability of CNCs following chemical and mechanical processes (Fisher et al., 2002). This may result from the chemical process removal of most of the lignin and hemicellulose, as well as from increased crystallinity of the cellulose and increased surface area exposure to heat (Wang et al., 2007). MCC had a maximum pyrolysis temperature (T_{max}) of 338.9°C, while spherical CNCs had a T_{max} of 309.4°C, which was slightly lower (Ren et al., 2022). The 211°C –378°C range was where the majority of the mass loss of spherical CNCs occurred. Spherical CNCs had a lower initial weight loss temperature than MCC due to their smaller particle size, decreased crystallinity, and larger specific surface area, which speeds up heat transmission. CNCs produced by enzymatic hydrolysis have good thermal stability and an extensive range of potential uses in the solar energy industry because they have higher starting pyrolysis temperatures (T_0) and maximum pyrolysis temperatures (T_{max}) than those made by acid hydrolysis.

5.4 APPLICATIONS OF NSC

5.4.1 Biomedical Applications

5.4.1.1 Drug Delivery

NSC has gained considerable attention as a sustainable and effective biopolymer for drug delivery (Shuya, 2023; Lukova et al., 2023). With its diameter typically ranging from 20 nm to several hundred nanometers in diameter, its spherical shape, extensive surface area, and rich hydroxyl chemistry support versatile drug loading and surface functionalization (Carvalho et al., 2021; Aljabali et al., 2022; Dehkhoda et al., 2022).

Its biocompatible and biodegradable nature also makes it well-suited as a therapeutic carrier (Khine & Stenzel, 2020).

A major asset of NSC-based systems is their ability to regulate drug loading and release, often achieved through both physical encapsulation and chemical grafting of drug molecules onto the cellulose surface (Carvalho et al., 2021). For instance, Ning et al. (2022) created spherical cellulose–titania hybrids loaded with triclosan, demonstrating sustained release over several hours following the Higuchi diffusion model, indicative of a controlled, diffusion driven release mechanism. Their formulation retained its antimicrobial effect against *E. coli* and *S. aureus* after release. Similarly, Işıtan et al. (2024) produced ~300 nm NSC from recycled cotton that achieved approximately 99% doxorubicin entrapment efficiency and sustained release over 34 days, exhibiting effective cytotoxicity against A549 lung cancer cells.

In addition, stimulus–responsive NSC carriers further exhibit intelligent behavior. For instance, earlier studies employing magnetically active CMC cores achieved dexamethasone release under specific pH conditions, highlighting potential for on-demand drug delivery (Cai et al., 2017). More so, recently, curcumin-loaded NSC nanoparticles showed accelerated release at acidic pH (~5.5) and demonstrated therapeutic effects in breast cancer cell cultures (Aboushoushah et al., 2024). Also, NSC platforms accommodate a wide range of payloads, such as bacterial cellulose-based capsules that have been used to facilitate colon-specific curcumin delivery (Aguiar et al., 2025; Shi et al., 2024) and NSC hydrogels adapted as protective vectors for siRNA and DNA, achieving gene delivery with minimal toxicity (Valataba et al., 2024). NSC-based systems also enhance topical, transdermal, and oral drug delivery. For example, triclosan–TiO_2 patches deliver topical antibacterial activity, insulin-laden cellulose hydrogels enable oral administration, and injectable NSC hydrogels offer localized, controlled drug release (Thala et al., 2025; Agafonov et al., 2018).

5.4.1.2 Tissue Engineering and Regenerative Medicine

Due to its high porosity, tuneable stiffness, and naturally abundant hydroxyl groups, NSC has emerged as a promising scaffold material that closely mimics the extracellular matrix (ECM) (Janmohammadi et al., 2023). As reviewed by Mirtaghavi et al. (2020), three-dimensional NSC hydrogels and aerogels derived from plant and bacterial sources provide hydrated networks that support cell adhesion, proliferation, and differentiation, making them ideal for wound dressings, artificial skin, and implantable scaffolds.

In bone tissue engineering, NSC reinforces composite scaffolds by enhancing mechanical strength and promoting osteogenesis. Niknafs et al. (2024) incorporated bioactive glass nanoparticles into bacterial cellulose–silk fibroin scaffolds, yielding enhanced mechanical stiffness and upregulated osteogenic markers such as Runx2 and alkaline phosphatase in human adipose-derived stem cells. Keshvardoostchokami et al. (2020) also demonstrated that nanobioglass–silk fibroin–CMC scaffolds supported hMSC adhesion and osteogenic differentiation, showing increased ALP activity, biomineralization, and Runx2 expression. In soft tissue and cartilage repair as well, hydrogels made from CMC or hydroxypropyl methylcellulose (HPMC) have been successfully injected to support chondrocyte growth and ECM deposition under physiological conditions (Prasad et al., 2023). In addition, NSC-based bio-inks have been used in 3D bioprinting of auricular and meniscal tissues, demonstrating cell compatibility and ECM formation (Lin et al., 2022).

Furthermore, for neural tissue engineering, Bordoni et al. (2020) found that 3D CNF or bacterial nanocellulose scaffolds enhanced adhesion, differentiation, and network formation of PC12 and neuroblastoma cells. Additionally, wood-derived NFC hydrogels with shear-thinning properties have successfully supported 3D cultures of stem and hepatic cells, offering ECM-like environments suitable for drug testing and regenerative applications (Merivaara et al., 2022). Wound healing also benefits from NSC-based materials. Porous nanocellulose hydrogels cross-linked with gelatin or chitosan create moist environments that support angiogenesis and fibroblast activity, accelerating skin regeneration (Wei et al., 2021; Dele-Afolabi et al., 2023a). Functionalization with chitosan or silver nanoparticles also enhances antimicrobial efficacy, making these hydrogels especially useful in burn treatment (Wei et al., 2021).

5.4.2 Environmental Applications

5.4.2.1 Water Purification

In areas of sustainable and effective adsorbent for environmental remediation, NSC has gained relevance, thanks to its exceptional surface properties, including a high specific surface area, abundant hydroxyl functionality, and tunable porosity (Dele-Afolabi et al., 2023a). These properties make NSC ideal for both adsorption processes and filtration membranes (Gobi et al., 2024).

Derived from renewable biomass, NSC offers a green alternative to synthetic polymers used in water treatment (Carvalho et al., 2021), with heavy metal adsorption ranking among its most impactful applications. Modified NSC materials, through methods such as carboxymethylation or TEMPO oxidation, are capable of binding toxic metal ions, including lead, cadmium, chromium, and copper. Recent reviews report removal efficiencies reaching nearly 99% within minutes (Lemessa et al., 2024). These functionalized adsorbents not only display high affinity but also commendable reusability, positioning NSC as a competitive biomaterial in heavy metal remediation (Lemessa et al., 2024). Membrane fabrication with NSC further expands its usage. By casting spherical cellulose particles into membrane forms (using vacuum filtration, electrospinning, or freeze-drying), researchers are developing multifunctional filters spanning microfiltration to reverse osmosis. Such membranes benefit from adjustable pore size and surface chemistry, enabling simultaneous removal of bacteria, viruses, organic dyes, and metal ions (Castro & Abejón, 2024). For example, integrating NSC with cellulose acetate or blending with other biopolymers produces dense yet permeable films suited for clean water use (Saberi Riseh et al., 2023).

In addition, antibacterial and antifouling capabilities are realized by embedding silver or titanium dioxide nanoparticles into NSC membranes. A study on CNC/Ag nanocomposite membranes reported 99.4% *E. coli* reduction and a high flux recovery ratio (92.6%) against humic acid fouling (Xu et al., 2020). These composite films resist biofilm formation, making them suitable for medical-grade water purification and drinking water standards (Lemessa et al., 2024). Organic contaminant adsorption is another strong suit (Lu & Yuan 2018). Hydrophobic modifications to NSC enhance its capacity to sequester dyes, oil droplets, pesticides, and pharmaceuticals from wastewater, where organic pollutants are difficult to eliminate through traditional methods (Rana et al., 2021).

5.4.2.2 Pollutant Adsorption

A key advantage of NSC-based materials lies in their biodegradability and eco-friendliness. Unlike many synthetic polymers, NSC does not create secondary pollution and can be regenerated through mild treatments such as pH shifts or salt washes (Hassanisaadi et al., 2023; Shahzad et al., 2023). Moreover, surface chemistry modifications like succinic anhydride treatment enhance metal-binding efficiencies by enabling chelation or ion-exchange mechanisms (Sejie et al., 2023). The nanoscale spherical morphology of NSC offers a high surface area-to-volume ratio, adjustable surface chemistry, and abundant hydroxyl functionality, enabling efficient adsorption of a broad spectrum of water pollutants, ranging from heavy metals to organic compounds (Gobi et al., 2024).

One of the most impactful applications of NSC involves the removal of heavy metals from contaminated water bodies (Dele-Afolabi et al., 2023b; Lu & Yuan 2018). Through surface modifications such as carboxylation, sulfonation, or amination, NSC develops strong binding sites for toxic ions, including lead, cadmium, chromium, arsenic, and mercury (Marimuthu et al., 2023; Jamshaid et al., 2017). Functionalized NSC adsorbents can achieve greater than 90% removal efficiency within minutes, with performance strongly influenced by pH, contact duration, and pollutant concentration (Gobi et al., 2024; Lemessa et al., 2024). Beyond inorganic toxins, NSC excels in adsorbing organic pollutants. Electrostatic attraction, π–π interactions, and hydrophobic forces enable effective capture of dyes, pharmaceuticals, and pesticides (Salahshoori et al., 2024). For example, carboxylated cellulose nanocrystals capturing methylene blue and rhodamine B, while MOF-enhanced composites removed antibiotics like tetracycline and ciprofloxacin with high efficiency (Salahshoori et al., 2024).

Moreso, advanced NSC-based adsorbents often combine with biochar or metal-organic frameworks (MOFs) to enhance pollutant uptake. For instance, Zr-MOF/bacterial cellulose aerogels achieved tetracycline capacities around 318 mg/g, while Cu-based MOF composites in a biopolymer matrix reached over 180 mg/g in tetracycline adsorption (Abdelhamid & Mathew, 2022). These studies also demonstrated that adsorption follows Langmuir isotherm and pseudo-second-order kinetics, suggesting monolayer chemisorption mechanisms with high selectivity and recyclability.

5.4.3 Industrial Application

5.4.3.1 Food Packaging and Preservation

NSC nanoparticles significantly enhance tensile strength and stiffness while reducing oxygen and water vapor transmission by reinforcing biodegradable polymers such as PLA. This is a key factor in extending shelf life. Karkhanis et al. (2021) demonstrated that PLA films containing just 1 wt% CNCs achieved a 40% longer shelf life for moisture-sensitive crackers, largely due to their improved moisture barrier properties (Wongthanaroj et al., 2022). Complementary work has shown that nanocellulose coatings on paperboard markedly increase resistance to grease, oil, and oxygen permeation (Lahti, 2019).

In addition to passive protection, NSC enables active and intelligent packaging. Incorporating bioactive components—such as essential oils, tannins, or antimicrobial silver/TiO_2 nanoparticles—into NSC films has proven effective in delaying spoilage. Casalini et al. (2023) reported nanocellulose films with thyme and oregano essential oils preserved raspberries' firmness and reduced weight loss over 12 days at 1°C. Pickering emulsion-based films with lemon or cinnamon essential oils also show strong antimicrobial and antioxidant performance (Cai et al., 2025). Also, meantime, hydrogels and pH-sensitive indicator films offer real-time freshness signals (Das et al., 2023).

Furthermore, NSC supports edible coating applications for fresh produce. A composite of chitosan and nanocellulose infused with essential oils extended strawberry shelf life, reducing weight loss and decay by over 70% over seven days at 20°C (Dong et al., 2015). Recent LCNF–chitosan–raspberry extract coatings similarly enhanced strawberry quality, mitigating fungal damage and size loss during storage (Rabasco-Vílchez et al., 2024). Production methods, including extrusion, solution casting, spray drying, and roll-to-roll coating, enable scalable manufacturing of NSC films and coatings, while advances in multilayer coatings, such as nanocellulose combined with hydrophobic biowax or shellac, create films with superior moisture and gas barrier properties suitable for industrial adoption (Jo et al., 2022).

5.4.3.2 Cosmetics and Personal Care

As a natural stabilizer, thickener, and emulsifier, NSC enhances texture and product performance. Its high surface area and shear-thinning properties enable smooth, consistent creams, gels, and hair-care products (Wang et al., 2025; Meftahi et al., 2022). For instance, CNCs combined with PLA create foundation liquids that are easy to apply and gentle to skin, reducing the need for harsh cleansers (Tang et al., 2021).

In makeup, NSC provides smooth, dry finishes and helps absorb excess oil, ensuring longevity. Porous spherical cellulose microparticles like CELLUFLOW™ reduce dye migration in lipsticks, enhancing wear and minimizing skin staining (You et al., 2025). Beyond texture and longevity, NSC acts as an active carrier for functional ingredients. Bacterial nanocellulose sheet masks can deliver vitamins, hyaluronic acid, or green tea extracts slowly into skin—boosting hydration, elasticity, and repair (You et al., 2025; Almeida et al., 2021). NSC hydrogels and composite films also allow controlled release of actives, improving bioavailability and sustained effectiveness (Salimi et al., 2024).

Biodegradable microbeads represent another notable application in this domain. Cellulosic microspheres, such as those developed at the University of Bath, offer a sustainable alternative to plastic, breaking down into harmless sugars in water and confronting ocean pollution without compromising exfoliant performance (Kendre et al., 2023). Similarly, CELLUFLOW™ delivers biodegradable slip and absorbency benefits comparable to plastic exfoliants (You et al., 2025). Importantly, safety and sustainability are strong selling points for NSC. Nanocellulose is non-toxic, non-inflammatory, and biocompatible upon topical application and aligns with consumer demand for green beauty solutions (Zagórska-Dziok & Sobczak, 2020). As an eco-friendly ingredient derived from plant or bacterial biomass, NSC supports the industry shift toward environmentally conscious formulations.

5.4.4 Emerging and Potential Applications

NSC transcends its established applications, entering a realm of transformative innovation across materials science, biomedicine, and sustainable technologies. Its unique combination of mechanical robustness, tunable surface chemistry, and environmental compatibility positions it as a key material at the frontier of scientific advancement (Carvalho et al., 2021).

One exciting development is in bioelectronics and conductive materials. For example, sulfated CNC incorporated as templates for green synthesis of PEDOT produce mixed ionic–electronic conductors with notable conductivity (~5 s cm^{-1}) suitable for organic electrochemical transistors (Matura et al., 2024). Similarly, that PANI polymerized on nanocellulose to create flexible, biodegradable composites with conductivities up to 90 mS cm^{-1} has been demonstrated (Zhao et al., 2025b; Chen et al., 2020). Such composites combine the best of both mechanical stability of cellulose and the electrical functionality of the polymers, making them ideal for sensors, supercapacitors, and wearable electronics.

In gas capture and separation, NSC aerogels functionalized with MOFs or amino groups show selective CO_2 adsorption from gas streams (Adegoke et al., 2022). These lightweight aerogels possess large surface areas and adjustable chemistries, presenting low-cost, eco-friendly options for carbon capture technologies. The role of NSCs in developing antimicrobial surfaces and airborne pathogen control is especially timely. A sprayable NSC-based coating reportedly achieved rapid inactivation of SARS-CoV-2 (Stanislas et al., 2022; Otto & De Villiers, 2020), reducing droplet transfer on high-touch surfaces, signifying potential use in clinics, public transport, and healthcare products.

The optical and photonic domain has seen the emergence of structurally colored NSC microparticles. By self-assembling CNCs into cholesteric patterns, researchers produced vibrant pigments that reflect red, green, and blue light without chemical dyes. These "photonic pigments" offer sustainable options for colorants, anti-counterfeit inks, and smart coatings responsive to their environment (Dong et al., 2023). In neural repair, the development of 3D-printed piezoelectric cellulose composites has been ground breaking, combining cellulose scaffolds with KNN ceramics using directional freeze casting (Liu et al., 2025; Pîrvu et al., 2024). In spinal cord injury models, these scaffolds guided neural stem-cell growth and provided electromechanical stimulation to promote regeneration (Luo et al., 2019). Such biodegradable scaffolds restore structural and electrical pathways in neural tissue while eliminating the need for surgical removal. Other future directions include cellulose-stabilized Pickering emulsions, which may serve as fat substitutes or drug carriers in food, cosmetics, and hygiene sectors (Wang et al., 2025; Guo & Jiang, 2025).

5.5 CONCLUSION

A thorough review of the main preparation and characterization methods used in the investigation of NSC, a unique class of nanocelluloses with spherical form and exceptional physicochemical properties, is presented in this chapter. Understanding the structure–property relationships of NSC is essential for optimizing its performance in cutting-edge applications like drug delivery systems, nanocomposites, tissue engineering scaffolds, and sustainable packaging, and the chapter highlights the significance of a comprehensive preparation and characterization in this regard. The chapter is organized according to the structure and main types of preparation and characterization techniques. The physical, chemical, and enzymatic methods used for preparing NSC are discussed. In addition, the functions of high-resolution imaging methods like TEM and SEM in assessing particle size, distribution, and surface morphology are examined. The chapter also examined NSC particle size and size distribution, which are important factors affecting its stability, surface area, physical behavior, and functional performance in a range of applications. Methods like XRD are described in depth for evaluating the polymorphic forms of cellulose and the crystallinity index. The chapter also discusses Raman spectroscopy and FT-IR spectroscopy, which are used to confirm chemical changes and identify distinctive functional groups. The importance of TGA and DSC in assessing thermal stability, transition temperatures, and degradation—critical for processing and end-use applications—is discussed in this chapter.

REFERENCES

Abbaspoor, S., Ashrafi, A., Abolfarsi, R. (2019). Development of self-healing coatings based on ethyl cellulose micro/nano-capsules. *Surface Engineering*, 35, 273–280.

Abdelhamid, H. N., and Mathew, A. P. (2022). Cellulose–metal organic frameworks (CelloMOFs) hybrid materials and their multifaceted applications: A review. *Coordination Chemistry Reviews*, 451, 214263.

Abitbol, T., Kubat, M., Brännvall, E., Kotov, N., Johnson, C. M., Nizamov, R., Nyberg, M., Miettunen, K., Nordgren, N., Stevanic, J. S., and Guerreiro, M. P. (2023). Isolation of mixed compositions of cellulose nanocrystals, microcrystalline cellulose, and lignin nanoparticles from wood pulps. *ACS Omega,* 8(24): 21474–21484.

Aboushoushah, S. F., Abaza, S. F., Elbialy, N. S., and Mohamed, N. (2024). Synergistic effect of "methyl cellulose-dextran" on oral curcumin delivery via casein nanomicelle: Fabrication, characterization, and cancer therapeutic efficacy assessment. *Cellulose*, 31(16), 9731–9755.

Abu-Danso, E., Srivastava, V., Sillanpää, M., and Bhatnagar, A. (2017). Pretreatment assisted synthesis and characterization of cellulose nanocrystals and cellulose nanofibers from absorbent cotton. *International Journal of Biological Macromolecules*, 102, 248–257.

Adegoke, K. A., Oyedotun, K. O., Ighalo, J. O., Amaku, J. F., Olisah, C., Adeola, A. O., ... and Conradie, J. (2022). Cellulose derivatives and cellulose-metal-organic frameworks for CO_2 adsorption and separation. *Journal of CO_2 Utilization*, 64, 102163.

Agafonov, A. V., Evdokimova, O. L., Svensson, F. G., Håkansson, S., Seisenbaeva, G. A., and Kessler, V. G. (2018). Hybrid drug delivery patches based on spherical cellulose nanocrystals and colloid titania—synthesis and antibacterial properties. *Nanomaterials* (2079–4991), 8(4), 228.

Aguiar, A. C., Bianchi, J. R., Lopes, J. H., Ferreira, F. V., and Lona, L. M. F. (2025). Nanocellulose-based capsules with pH responsiveness for colon-targeted curcumin delivery. *ACS Applied Nano Materials*, 8(4), 2033–2045.

Alebachew, A. W., Dong, Y., Abdalkarim, S. Y. H., Wu, C., and Yu, H. Y. (2025). Recent progress of multifunctional nanocellulose-based pharmaceutical materials: A review. *International Journal of Biological Macromolecules*, 306, Part 2, 141427.

Alemdar, A., and Sain, M. (2008). Isolation and characterization of nanofibers from agricultural residues–Wheat straw and soy hulls. *Bioresource Technology*, 99(6), 1664–1671.

Aljabali, A. A. A., Obeid, M. A., Rezigue, M. M., Alqudah, A., Charbe, N. B., Chellappan, D. K., Mishra, V., Pardhi, D. M., Dureja, H., Gupta, G., Prasher, P., Dua, K., Barhoum, A., and Tambuwala, M. M. (2022). Nanocelluloses as a novel vehicle for controlled drug delivery. In: Barhoum, A. (ed) *Handbook of Nanocelluloses*, 507–542. Cham: Springer.

Almeida, T., Silvestre, A. J., Vilela, C., and Freire, C. S. (2021). Bacterial nanocellulose toward green cosmetics: Recent progresses and challenges. *International Journal of Molecular Sciences*, 22(6), 2836.

Anandharamakrishnan, C. (2014). Liquid-based nanoencapsulation techniques. In: Anandharamakrishnan, C. (ed) *Techniques for Nanoencapsulation of Food Ingredients. SpringerBriefs in Food, Health, and Nutrition*, 29–41. New York: Springer.

Antal, G. V. J. M., Sezkely, T., Till, F., and Jakab, E. (1988). Simultaneous thermogravi metric-mass spectrometric studies of the thermal decomposition of biopolymers. *Energy Fuels*, 2, 267–272.

Araki, J., Wada, M., Kuga, S., and Okano, T. (1998). Flow properties of microcrystalline cellulose suspension prepared by acid treatment of native cellulose. *Colloids and Surfaces A: Physicochemical and Engineering Aspects*, 142(1), 75e82.

Atoyebi, E. O., Oyejide, A. J., Dele-Afolabi, T. T., Azmah Hanim, M. A., and Ojo-Kupoluyi, O. J. (2023). Scaffold modeling advancement in biomaterials application. In: Hashmi, S. (ed) *Comprehensive Materials Processing (Second Edition)*, 56–71. Amsterdam: Elsevier.

Avolio, R., Bonadies, I., Capitani, D., Errico, M., Gentile, G., and Avella, M. (2012). A multitechnique approach to assess the effect of ball milling on cellulose. *Carbohydrate Polymers*, 87(1), 265–273.

Bano, S., and Negi, Y. S. (2017). Studies on cellulose nanocrystals isolated from groundnut shells. *Carbohydrate Polymers*, *157*, 1041–1049.

Beldman, G., Searle-Van Leeuwen, M. F., Rombouts, F. M., and Voragen, F. G. J. (1985). The cellulase of Trichoderma viride. *European Journal of Biochemistry*, *146*(2), 301–308.

Bordoni, M., Karabulut, E., Kuzmenko, V., Fantini, V., Pansarasa, O., Cereda, C., and Gatenholm, P. (2020). 3D printed conductive nanocellulose scaffolds for the differentiation of human neuroblastoma cells. *Cells*, 9(3), 682.

Cai, C., Zhang, C., Li, N., Liu, H., Xie, J., Lou, H., Pan, X., Zhu, J. Y., and Wang, F. (2023). Changing the role of lignin in enzymatic hydrolysis for a sustainable and efficient sugar platform. *Renewable and Sustainable Energy Reviews*, 183, 113446.

Cai, F., Duan, Z., Yu, D., Song, Z., and Lu, P. (2025). Antimicrobial packaging activity enhancement by lemon essential oil pickering emulsion stabilized with nanocellulose microgel particles. *Food Packaging and Shelf Life*, 47, 101439.

Cai, Y., Yuan, F., Wang, X., Sun, Z., Chen, Y., Liu, Z., … and Wang, S. (2017). Synthesis of core–shell structured Fe_3O_4 carboxymethyl cellulose magnetic composite for highly efficient removal of Eu (III). *Cellulose*, 24, 175–190.

Carvalho, J. P. F., Silva, A. C. Q., Silvestre, A. J. D., Freire, C. S. R., and Vilela, C. (2021). Spherical cellulose micro and nanoparticles: A review of recent developments and applications. *Nanomaterials*, 11(10), 2744.

Casalini, S., Baschetti, M. G., Cappelletti, M., Guerreiro, A. C., Gago, C. M., Nici, S. and Antunes, M. D. (2023) Antimicrobial activity of different nanocellulose films embedded with thyme, cinnamon, and oregano essential oils for active packaging application on raspberries. *Frontiers in Sustainable Food Systems* 7, 1190979.

Castro, K., and Abejón, R. (2024). Removal of heavy metals from wastewaters and other aqueous streams by pressure-driven membrane technologies: An outlook on reverse osmosis, nanofiltration, ultrafiltration and microfiltration potential from a bibliometric analysis. *Membranes*, 14(8), 180.

Channab, B. E., El Idrissi, A., Essamlali, Y., and Zahouily, M. (2024). Nanocellulose: Structure, modification, biodegradation and applications in agriculture as slow/controlled release fertilizer, superabsorbent, and crop protection: A review. *Journal of Environmental Management*, 352, 119928.

Chen, C., Xi, Y., and Weng, Y. (2022). Recent advances in cellulose-based hydrogels for tissue engineering applications. *Polymers*, 14(16), 3335.

Chen, S., Yang, K., Leng, X., Chen, M., Novoselov, K. S., and Andreeva, D. V. (2020). Perspectives in the design and application of composites based on graphene derivatives and bio-based polymers. *Polymer International*, 69(12), 1173–1186.

Chen, X. Q., Deng, X. Y., Shen, W. H., and Jia, M. Y. (2018). Preparation and characterization of the spherical nanosized cellulose by the enzymatic hydrolysis of pulp fibers. *Carbohydrate Polymers*, 181, 879–884.

Cheng, W., Zhu, Y., Jiang, G., Cao, K., Zeng, S., Chen, W., … and Yu, H. (2023). Sustainable cellulose and its derivatives for promising biomedical applications. *Progress in Materials Science*, 138, 101152.

Chhabra, R., and Basavaraj, M. G. (2019). Particle size reduction and enlargement. In: Chhabra, R., and Basavaraj, M. G. (eds) *Coulson and Richardson's Chemical Engineering (Sixth Edition)*, 2a, 205–280. Oxford: Butterworth-Heinemann.

Chin, S. F., Jimmy, F. B., and Pang, S. C. (2018). Size controlled fabrication of cellulose nanoparticles for drug delivery applications. *Journal of Drug Delivery Science and Technology*, 43, 262–266.

Cordt, C., Meckel, T., Geissler, A., and Biesalski, M. (2020). Entrapment of hydrophobic biocides into cellulose acetate nanoparticles by nanoprecipitation. *Nanomaterials*, 10, 2447.

Corrêa, A. C., Teixeira, E. M., Pessan, L. A., and Mattoso, L. H. C. (2010). Cellulosenanofibers from curauá fibers. *Cellulose*, 17, 1183–1192.

D'Addio, S. M., and Prud'homme, R. K. (2011). Controlling drug nanoparticle formation by rapid precipitation. *Advanced Drug Delivery Reviews*, 63(6), 417–426.

Das, P. P., Kalyani, P., Kumar, R., and Khandelwal, M. (2023). Cellulose-based natural nanofibers for fresh produce packaging: Current status, sustainability and future outlook. *Sustainable Food Technology*, 1(4), 528–544.

Dehkhoda, S., Bagheri, M., Heydari, M., Rabieh, S. (2022). Extraction of carboxylated nanocellulose from oat husk: Characterization, surface modification and in vitro evaluation of indomethacin drug release. *International Journal of Biological Macromolecules*, 212(2022), 165–171.

Dele-Afolabi, T. T., Azmah Hanim, M. A., Ojo-Kupoluyi, O. J., and Atoyebi, E. O. (2023a). Chapter 25- Biobased materials for wastewater treatment. In: Ahmed, S., and Annu (eds) *Advanced Applications of Biobased Materials*, pp. 593–624. Cambridge, MA: Elsevier.

Dele-Afolabi, T. T., Azmah Hanim, M. A., Ojo-Kupoluyi, O. J., and Atoyebi, E. O. (2023b). Chitosan nanocomposites as wound healing materials: Advances in processing techniques and mechanical properties. *Pertanika Journal of Science & Technology*, 31(1), 543–575. https://doi.org/10.47836/pjst.31.1.32.

Dengning, X., Yong, G., and Fude, C. (2014). Application of precipitation methods for the production of water-insoluble drug nanocrystals: Production techniques and stability of nanocrystals. *Current Pharmaceutical Design*, 20(3), 408–435(28).

Dong, F., Li, S. J., Liu, Z. M., Zhu, K. X., Wang, X. L., and Jin, C. D. (2015). Improvement of quality and shelf life of strawberry with nanocellulose/chitosan composite coatings. *Bangladesh Journal of Botany*, 44, 709–717.

Dong, Y., Zhao, H., Jin, H., Liu, S., Cheng, Q., Wang, L., and Li, Y. (2023). Eco-friendly cellulosic photonic pigments with tunable structural color for anti-counterfeiting and decorative applications. *Polymer*, 282, 126165.

Dufresne, A. (2012). *Nanocellulose: From Nature to High Performance Tailored Materials*. Berlin/Boston: Walter de Gruyter GmbH & Co KG.

El Achaby, M., El Miri, N., Hannache, H., Gmouh, S., and Aboulkas, A. (2018). Production of cellulose nanocrystals from vine shoots and their use for the development of nanocomposite materials. *International Journal of Biological Macromolecules*, 117, 592–600.

El Allaoui, B., Benzeid, H., Zari, N., el kacem Qaiss, A., and Bouhfid, R. (2023). Functional cellulose-based beads for drug delivery: Preparation, functionalization, and applications. *Journal of Drug Delivery Science and Technology*, 88, 104899.

Fisher, T., Hajaligol, M., Waymack, B., and Kellogg, D. (2002). Pyrolysis behavior and kinetics of biomass derived materials. *Journal of Analytical and Applied Pyrolysis*, 62(2), 331–349.

French, A. D. (2014). Idealized powder diffraction patterns for cellulose polymorphs. *Cellulose*, 21(2), 885–896.

Ge, W., Shuai, J., Wang, Y., Zhou, Y., and Wang, X. (2022). Progress on chemical modification of cellulose in "green" solvents. *Polymer Chemistry*, 13(3), 359–372.

Ghasemi, M., Tsianou, M., and Alexandridis, P. (2017). Assessment of solvents for cellulose dissolution. *Bioresource Technology*, 228, 330–338.

Gobi, M., Kumar, A., Singh, J., Singh, S., and Ramamurthy, P. C. (2024). Nanocellulose-based adsorption for the removal of heavy metal from wastewater—A review. *Water Conservation Science and Engineering*, 9(1), 24.

Goddard, E. D. (1994), Polymer/surfactant interaction—Its relevance to detergent systems. *Journal of American Oil Chemists' Society*, 71, 1–16.

Gu, J., Hu, C., Zhong, R., Tu, D., Yun, H., Zhang, W., and Leu, S. Y. (2017). Isolation of cellulose nanocrystals from medium density fiberboards. *Carbohydrate Polymers*, 167, 70–78.

Guo, K., and Jiang, J. (2025). Boronate-polyol networks reinforce cellulose-stabilized pickering emulsions. *Carbohydrate Polymers*, 366, 123905.

Guo, Y., Wang, X., Shu, X., Shen, Z., and Sun, R. C. (2012). Self-assembly and paclitaxel loading capacity of cellulose-graft-poly (lactide) nanomicelles. *Journal of Agricultural and Food Chemistry*, 60(15), 3900–3908.

Harmeet, K., Virender, K., Krishan, K., Sandeep, R., Parveen, K., Jasbir, S. (2016). Polymer particulates in drug delivery. *Current Pharmaceutical Design*, 22(19), 2761–2787(27).

Hassanisaadi, M., Riseh, R. S., Rabiei, A., Varma, R. S., and Kennedy, J. F. (2023). Nano/micro-cellulose-based materials as remarkable sorbents for the remediation of agricultural resources from chemical pollutants. *International Journal of Biological Macromolecules*, 246, 125763.

Hauru, L. K. J., Hummel, M., King, A. W. T., Kilpeläinen, I., and Sixta, H. (2012). Role of solvent parameters in the regeneration of cellulose from ionic liquid solutions. *Biomacromolecules* 13(9), 2896–2905.

Heinze, T., and Koschella, A. (2005). Solvents applied in the field of cellulose chemistry - A mini review. *Polímeros: Ciência e Tecnologia*, 15(2), 84–90.

Hemmati, F., Jafari, S. M., and Taheri, R. A. (2019). Optimization of homogenization-sonication technique for the production of cellulose nanocrystals from cotton linter. *International Journal of Biological Macromolecules*, 137, 374–381.

Herreros-Lopez, A. N. A. (2016). *Synthesis and Characterization of Novel Nanomaterials Based on Nanocrystalline Cellulose*. https://hdl.handle.net/11368/2907985.

Hoffmann, H., Kästner, U., Dönges, R., and Ehrler, R. (1996). Gels from modified hydroxyethyl cellulose and ionic surfactants. *Polymer Gels and Networks*, 4(5–6), 509–526.

Huang, S., Wang, Z., Shen, J., Wu, R., Zhao, H., Wang, Y., Wang, Y., and Xia, Y. (2017). Surface functionalization of cellulose nanocrystals with polymeric ionic liquids during phase transfer, *Carbohydrate Polymers*, 157(2017), 1426–1433. https://doi.org/10.1016/j.carbpol.2016.11.024.

Ibrahim, M. M., El-Zawawy, W. K., and Nassar, M. A. (2010). Synthesis and characterization of polyvinyl alcohol/nanospherical cellulose particle films. *Carbohydrate Polymers*, 79(3), 694–699.

Işıtan, A., Pasquardini, L., Bersani, M., Gök, C., Fioravanti, S., Lunelli, L., ... and Koluman, A. (2024). Sustainable production of microcrystalline and nanocrystalline cellulose from textile waste using HCl and NaOH/Urea treatment. *Polymers*, 17(1), 48.

Jamshaid, A., Hamid, A., Muhammad, N., Naseer, A., Ghauri, M., Iqbal, J., ... and Shah, N. S. (2017). Cellulose-based materials for the removal of heavy metals from wastewater–an overview. *ChemBioEng Reviews*, 4(4), 240–256.

Janmohammadi, M., Nazemi, Z., Salehi, A. O. M., Seyfoori, A., John, J. V., Nourbakhsh, M. S., and Akbari, M. (2023). Cellulose-based composite scaffolds for bone tissue engineering and localized drug delivery. *Bioactive Materials*, 20, 137–163.

Jia, Y., Zhai, X., Fu, W., Liu, Y., Li, F., Zhong, C., 2016. Surfactant-free emulsions stabilized by tempo-oxidized bacterial cellulose. *Carbohydrate Polymers*, 151, 907–915.

Jo, H. M., Kim, D. H., Lee, S. H., and Lee, J. Y. (2022). Multi-layer barrier coating technology using nano-fibrillated cellulose and a hydrophobic coating agent. *BioResources*, 17(4), 6222.

Joshy, K. S., Snigdha, S., George, A., Kalarikkal, N., Pothen, L. A., and Thomas, S. (2017). Core–shell nanoparticles of carboxy methyl cellulose and compritol-PEG for antiretroviral drug delivery. *Cellulose*, 24, 4759–4771.

Karkhanis, S. S., Stark, N. M., Sabo, R. C., and Matuana, L. M. (2021). Potential of extrusion-blown poly(lactic acid)/cellulose nanocrystals nanocomposite films for improving the shelf-life of a dry food product. *Food Packaging and Shelf Life*, 29, 100689.

Keerati-U-Rai, M., and Corredig, M. (2009). Effect of dynamic high pressure homogenization on the aggregation state of soy protein. *Journal of Agricultural and Food Chemistry*, 57(9), 3556–3562.

Kendre, P. N., Lokwani, D., Pote, A., Singh, S., Jayeoye, T. J., and Prajapati, B. G. (2023). Potential technologies to develop cellulose beads and microspheres. In Shabbir, M. (ed) *Regenerated Cellulose and Composites: Morphology-Property Relationship*, 145–176. Singapore: Springer Nature Singapore.

Keshvardoostchokami, M., Majidi, S. S., Huo, P., Ramachandran, R., Chen, M., and Liu, B. (2020). Electrospun nanofibers of natural and synthetic polymers as artificial extracellular matrix for tissue engineering. *Nanomaterials*, 11(1), 21.

Khine, Y. Y., and Stenzel, M. H. (2020). Surface modified cellulose nanomaterials: A source of non-spherical nanoparticles for drug delivery. *Materials Horizons*, 7(7), 1727–1758.

Kim, D. Y., Nishiyama, Y., Wada, M., and Kuga, S. (2001). High-yield carbonization of cellulose by sulfuric acid impregnation. *Cellulose*, *8*, 29–33.

Kumar, G., Upadhyay, S., Yadav, D. K., Malakar, S., Dhurve, P., and Suri, S. (2023). Application of ultrasound technology for extraction of color pigments from plant sources and their potential bio-functional properties: A review. *Journal of Food Process Engineering*, 46(6), e14238.

Kumar, R., and Kumar, G. (2022). Nanocellulose: Fascinating and sustainable nanomaterial for papermaking. In *Nanotechnology in Paper and Wood Engineering* (pp. 389–407). Cambridge: Elsevier.

Lahti, J. (2019). Nanocellulose and polylactic acid based multilayer coatings for barrier applications. In *Biennial TAPPI European PLACE Conference, Porto (Portugal)* (pp. 446–455). TAPPI Press.

Lemessa, G., Chebude, Y., and Alemayehu, E. (2024). Synthesis of cellulose-based nano composites as sustainable adsorbents for adsorption removal of heavy metals from wastewater. A review. *African Journal of Water Conservation and Sustainability*, 10(1), 1–12.

Levy, I., Shani, Z., and Shoseyov, O. (2002). Modification of polysaccharides and plant cell wall by endo-1,4-β-glucanase and cellulose-binding domains. *Biomolecular Engineering*, 19(1), 17–30.

Li, X. F., Ding, E. Y., and Li, G. K. (2001). A method of preparing spherical nano-crystal cellulose with mixed crystalline forms of cellulose I and II*. *Chinese Journal of Polymer Science (World Scientific Publishing Company)*, 19(3), 291–296.

Li, Y., Zhao, X., Liu, Y., Yang, J., Zhang, Q., Wang, L., Wu, W., Yang, Q., Liu, B. (2019). Melatonin loaded with bacterial cellulose nanofiber by Pickering-emulsion solvent evaporation for enhanced dissolution and bioavailability. *International Journal of Pharmaceutics*, 559(2019), 393–401. https://doi.org/10.1016/j.ijpharm.2019.01.059.

Liebert, T. (2010). Cellulose solvents – remarkable history, bright future. Cellulose solvents: For analysis, shaping and chemical modification. Chapter 1, *ACS Symposium Series*, 1033, 3–54.

Lin, K. H., Enomae, T., and Chang, F. C. (2019). Cellulose nanocrystal isolation from hardwood pulp using various hydrolysis conditions. *Molecules (Basel, Switzerland)*, 24(20): 3724.

Lin, L., Jiang, S., Yang, J., Qiu, J., Jiao, X., Yue, X., … and Zhang, L. (2022). Application of 3D-bioprinted nanocellulose and cellulose derivative-based bio-inks in bone and cartilage tissue engineering. *International Journal of Bioprinting*, 9(1), 637.

Ling, Z., Wang, T., Makarem, M., Santiago Cintrón, M., Cheng, H. N., Kang, X., … and French, A. D. (2019). Effects of ball milling on the structure of cotton cellulose. *Cellulose*, 26, 305–328.

Liu, L., Huang, P., Xu, S., Chen, X., Fu, K., Li, J., … and Wu, S. (2025). Self-polarized cellulose nanofiber-reinforced PVDF-based piezoelectric composites via direct-ink-writing 3D printing for pressure sensing and energy harvesting. *Sensors and Actuators A: Physical*, 381, 116084.

Lu, H., Gui, Y., Zheng, L., and Liu, X. (2013). Morphological, crystalline, thermal and physicochemical properties of cellulose nanocrystals obtained from sweet potato residue. *Food Research International*, 50(1), 121–128.

Lu, P., and Hsieh, Y. L. (2010). Preparation and properties of cellulose nanocrystals: Rods, spheres, and network. *Carbohydrate Polymers*, 82(2), 329–336.

Lu, Y., and Yuan, W. (2018). Superhydrophobic three-dimensional porous ethyl cellulose absorbent with micro/nano-scale hierarchical structures for highly efficient removal of oily contaminants from water. *Carbohydrate Polymers*, 191(2018), 86–94. https://doi.org/10.1016/j.carbpol.2018.03.018.

Lukova, P., Katsarov, P., and Pilicheva, B. (2023). Application of starch, cellulose, and their derivatives in the development of microparticle drug-delivery systems. *Polymers*, 15(17), 3615.

Luo, H., Cha, R., Li, J., Hao, W., Zhang, Y., and Zhou, F. (2019). Advances in tissue engineering of nanocellulose-based scaffolds: A review. *Carbohydrate Polymers*, 224, 115144.

Maheri, H., Hashemzadeh, F., Shakibapour, N., Kamelniya, E., Malaekeh-Nikouei, B., Mokaberi, P., and Chamani, J. (2022). Glucokinase activity enhancement by cellulose nanocrystals isolated from jujube seed: A novel perspective for type II diabetes mellitus treatment (In vitro). *Journal of Molecular Structure*, 1269, 133803.

Majdanac, L. D., Poleti, D., and Teodorovic, M. J. (1991). Determination of the crystallinity of cellulose samples by x-ray diffraction. *Acta Polymerica*, 42(8), 351–357.

Marett, J., Aning, A., and Foster, E. J. (2017). The isolation of cellulose nanocrystals from pistachio shells via acid hydrolysis. *Industrial Crops and Products*, 109, 869–874.

Marimuthu, T., Chee, C. Y., and Sulaiman, N. M. N. (2023). A review on the use of cellulose nanomaterials for wastewater remediation of heavy metal ions. *International Journal of Environmental Science and Technology*, 20(3), 3421–3436.

Martínez-Sanz, M., Lopez-Rubio, A., and Lagaron, J. M. (2011). Optimization of thenanofabrication by acid hydrolysis of bacterial cellulose nanowhiskers. *Carbohydrate Polymers*, 85, 228–236.

Matura, K., D'Orsi, R., Spagnuolo, L., Mayr, F., Cobet, M., Putz, C., … and Tekoglu, S. (2024). Nanocrystalline cellulose-based mixed ionic–electronic conductor for bioelectronics. *Journal of Materials Chemistry C*, 12(41), 16701–16713.

Medronho, B., and Lindman, B. (2014). Brief overview on cellulose dissolution/regeneration interactions and mechanisms. *Advances in Colloid and Interface Science*, 222, 502–508.

Medronho, B., and Lindman, B. (2014). Competing forces during cellulose dissolution: From solvents to mechanisms. *Current Opinion in Colloid & Interface Science*, 19(1), 32–40.

Meftahi, A., Samyn, P., Geravand, S. A., Khajavi, R., Alibkhshi, S., Bechelany, M., and Barhoum, A. (2022). Nanocelluloses as skin biocompatible materials for skincare, cosmetics, and healthcare: Formulations, regulations, and emerging applications. *Carbohydrate Polymers*, 278, 118956.

Mehanny, S., Abu-El Magd, E. E., Ibrahim, M., Farag, M., Gil-San-Millan, R., Navarro, J., El Habbak, A., and El-Kashif, E. (2021). Extraction and characterization of nanocellulose from three types of palm residues. *Journal of Materials Research and Technology*, 10, 526–537.

Merivaara, A., Koivunotko, E., Manninen, K., Kaseva, T., Monola, J., Salli, E., … and Yliperttula, M. (2022). Stiffness-controlled hydrogels for 3D cell culture models. *Polymers*, 14(24), 5530.

Meyabadi, T. F., and Dadashian, F. (2012). Optimization of enzymatic hydrolysis of waste cotton fibers for nanoparticles production using response surface methodology. *Fibers and Polymers,* 13, 313–321.

Meyabadi, T. F., Dadashian, F., Sadeghi, G. M. M., and Asl, H. E. Z. (2014). Spherical cellulose nanoparticles preparation from waste cotton using a green method. *Powder Technology*, 261, 232–240.

Mirtaghavi, A., Luo, J., and Muthuraj, R. (2020). Recent advances in porous 3D cellulose aerogels for tissue engineering applications: A review. *Journal of Composites Science*, 4(4), 152.

Mishra, S., Goyal, D., Kumar, A., and Dantu, P. K. (2019). Biotechnological Applications of β-Glucosidases in Biomass Degradation. In: Yadav, A., Singh, S., Mishra, S., Gupta, A. (eds) *Recent Advancement in White Biotechnology through Fungi. Fungal Biology*, 257–281. Cham: Springer.

Mujtaba, M., Negi, A., King, A. W., Zare, M., and Kuncova-Kallio, J. (2023). Surface modifications of nanocellulose for drug delivery applications; a critical review. *Current Opinion in Biomedical Engineering*, 28, 100475.

Nagalakshmaiah, M., Mortha, G., and Dufresne, A. (2016). Structural investigation of cellulose nanocrystals extracted from chili leftover and their reinforcement in cariflex-IR rubber latex. *Carbohydrate Polymers*, 136, 945–954.

Namazi, H., Fathi, F., and Heydari, A. (2012). Nanoparticles based on modified polysaccharides. In: Hashim, A. A. (ed) *The Delivery of Nanoparticles*, 149–184. London: IntechOpen.

Nawaz, H., Zhang, J., Tian, W., Wu, J., Zhang, J. (2019). Chemical modification of cellulose in solvents for functional materials. In: Han, B., Wu, T. (eds) *Green Chemistry and Chemical Engineering. Encyclopedia of Sustainability Science and Technology Series*, 1–34. New York: Springer.

Nguyen, V. T., Pham, N. A. T., Duong, T. H. T., Nguyen, T. V., Pham, L. H., Ly, P. H., Nguyen, D. M., Thuy, N. T. N., Hoang, D. (2024). High-performance hydrophobic aerogel based on nanocellulose, graphene oxide, polyvinyl alcohol, and hexadecyltrimethoxysilane: Structure, properties, and applicability. *Journal of Environmental Chemical Engineering*, 12(4), 113215.

Niknafs, B., Meskaraf-Asadabadi, M., Hamdi, K., and Ghanbari, E. (2024). Incorporating bioactive glass nanoparticles in silk fibroin/bacterial nanocellulose composite scaffolds improves their biological and osteogenic properties for bone tissue engineering applications. *International Journal of Biological Macromolecules*, 266, 131167.

Ning, L., Jia, Y., Zhao, X., Tang, R., Wang, F., and You, C. (2022). Nanocellulose-based drug carriers: Functional design, controllable synthesis, and therapeutic applications. *International Journal of Biological Macromolecules*, 222, 1500–1510.

Nosar, M. N., Salehi, M., Ghorbani, S., Beiranvand, S. P., Goodarzi, A., and Azami, M. (2016). Characterization of wet-electrospun cellulose acetate based 3-dimensional scaffolds for skin tissue engineering applications: Influence of cellulose acetate concentration. *Cellulose*, 23, 3239–3248.

Novotna, K., Havelka, P., Sopuch, T., Kolarova, K., Vosmanska, V., Lisa, V., Svorcik, V., and Bacakova, L. (2013). Cellulose-based materials as scaffolds for tissue engineering. *Cellulose*, 20, 2263–2278.

O'Donnell, K. L., Oporto-Velásquez, G. S., and Comolli, N. (2020). Evaluation of acetaminophen release from biodegradable poly (vinyl alcohol) (pva) and nanocellulose films using a multiphase release mechanism. *Nanomaterials*, 10(2), 301. https://doi.org/10.3390/nano10020301.

Otto, D. P., and De Villiers, M. M. (2020). Layer-by-layer nanocoating of antiviral polysaccharides on surfaces to prevent coronavirus infections. *Molecules*, 25(15), 3415.

Parvej, K. H. (2016). *Interaction of ciprofloxacin with micro-and modified nano-crystalline cellulose.* http://lib.buet.ac.bd:8080/xmlui/bitstream/handle/123456789/4554/Full%20Thesis.pdf?sequence=1&isAllowed=y.

Pawcenis, D., Leśniak, M., Szumera, M., Sitarz, M., and Profic-Paczkowska, J. (2022). Effect of hydrolysis time, pH and surfactant type on stability of hydrochloric acid hydrolyzed nanocellulose. *International Journal of Biological Macromolecules*, 222(Part B), 1996–2005.

Phuong, H. T., Thoa, N. K., Tuyet, P. T. A., Van, Q. N., and Hai, Y. D. (2022). Cellulose nanomaterials as a future, sustainable and renewable material. *Crystals*, 12(1), 106.

Pîrvu, C. I., Sover, A., and Abrudeanu, M. (2024). Participation of polymer materials in the structure of piezoelectric composites. *Polymers*, 16(24), 3603.

Piskorz, J., Radlein, D. S. A., Scott, D. S., and Czernik, S. (1989). Pretreatment of wood and cellulose for production of sugars by fast pyrolysis. *Journal of Analytical and Applied Pyrolysis*, 16(2), 127–142.

Ponce-Reyes, C. E., Chanona-Perez, J. J., Garibay-Febles, V., Palacios-Gonzalez, E., Karamath, J., Terres-Rojas, E., and Calderon-Dominguez, G. (2014). Preparation of cellulose nanoparticles from agave waste and its morphological and structural characterization. *Revista Mexicana de Ingeniería Química* 13(3), 897–906.

Potthast, A., Rosenau, T., Buchner, R., Röder, T., Ebner, G., Bruglachner, H., Sixta, H., and Kosma, P. (2002). The cellulose solvent system N,N-dimethylacetamide/lithium chloride revisited: The effect of water on physicochemical properties and chemical stability. *Cellulose*, 9, 41–53.

Prasad, A. S., Wilson, J., and Thomas, L. V. (2023). Designer injectable matrices of photocrosslinkable carboxymethyl cellulose methacrylate based hydrogels as cell carriers for gel type autologous chondrocyte implantation (GACI). *International Journal of Biological Macromolecules*, 224, 465–482.

Queiroz, B. G., Ciol, H., Inada, N. M., and Frollini, E. (2021). Hydrogel from all in all lignocellulosic sisal fibers macromolecular components. *International Journal of Biological Macromolecules*, 181, 978–989.

Rabasco-Vílchez, L., Jiménez-Jiménez, F., Possas, A., Morcillo-Martín, R., and Pérez-Rodríguez, F. (2024). Evaluation of the effect of nanocellulose edible coating on strawberries inoculated with Aspergillus flavus through image analysis. *Lwt*, 191, 115697.

Rahman, M. M., Howlader, A., Rahman, E., Roy, H., Hosen, M. M., Rahman, M., Sarkar, S. D., Roy, C. K., Bin Imran, A., and Firoz, S. H. (2021). Development of functionalized nanocrystalline cellulose-based polyelectrolytes with high water uptake. *Polymer Journal*, 53(2021), 913–921.

Ram, B., and Chauhan, G. S. (2018). New spherical nanocellulose and thiol-based adsorbent for rapid and selective removal of mercuric ions. *Chemical Engineering Journal*, 331, 587–596.

Rana, A. K., Mishra, Y. K., Gupta, V. K., and Thakur, V. K. (2021). Sustainable materials in the removal of pesticides from contaminated water: Perspective on macro to nanoscale cellulose. *Science of the Total Environment*, 797, 149129.

Randhawa, A., Dutta, S. D., Ganguly, K., Patil, T. V., Patel, D. K., and Lim, K. T. (2022). A review of properties of nanocellulose, its synthesis, and potential in biomedical applications. *Applied Sciences*, 12(14), 7090.

Rani, V., Mohanram, S., Tiwari, R., Nain, L., and Arora, A. (2014). Beta-glucosidase: Key enzyme in determining efficiency of cellulase and biomass hydrolysis. *Journal of Bioprocessing & Biotechnology*, 5(1), 197.

Ren, R. W., Chen, X. Q., and Shen, W. H. (2022). Preparation and separation of pure spherical cellulose nanocrystals from microcrystalline cellulose by complex enzymatic hydrolysis. *International Journal of Biological Macromolecules*, 202, 1–10.

Romruen, O., Kaewprachu, P., Karbowiak, T., and Rawdkuen, S. (2022). Isolation and characterization cellulose nanosphere from different agricultural by-products. *Polymers*, 14(13), 2534.

Saberi Riseh, R., Gholizadeh Vazvani, M., Hassanisaadi, M., and Skorik, Y. A. (2023). Micro-/nano-carboxymethyl cellulose as a promising biopolymer with prospects in the agriculture sector: A review. *Polymers*, 15(2), 440.

Saeed, U., Abdullah, T., and Al-Turaif, H. (2022). Surface morphology and biochemical characteristics of electrospun cellulose nanofibril reinforced PLA/PBS hollow scaffold for tissue engineering. *Fibers and Polymers,* 23, 2539–2548.

Salahshoori, I., Wang, Q., Nobre, M. A., Mohammadi, A. H., Dawi, E. A., and Khonakdar, H. A. (2024). Molecular simulation-based insights into dye pollutant adsorption: A perspective review. *Advances in Colloid and Interface Science*, 333, 103281.

Salimi, M., El Idrissi, A., Channab, B. E., Essamlali, Y., Firouzabadi, A. G., Beygi, M., … and Motamedi, E. (2024). Cellulose-based controlled release fertilizers for sustainable agriculture: Recent trends and future perspectives. *Cellulose*, 31, 10679–10726.

Samanta, A. P., Ali, M. S., Orasugh, J. T., Ghosh, S. K., Chattopadhyay, D. (2022). Crosslinked nanocollagen-cellulose nanofibrils reinforced electrospun polyvinyl alcohol/methylcellulose/polyethylene glycol bionanocomposites: Study of material properties and sustained release of ketorolac tromethamine, *Carbohydrate Polymer Technologies and Applications*, 3(2022), 100195.

Satyamurthy, P., and Vigneshwaran, N. (2013). A novel process for synthesis of spherical nanocellulose by controlled hydrolysis of microcrystalline cellulose using anaerobic microbial consortium. *Enzyme and Microbial Technology*, 52(1), 20–25.

Scheirs, J., Camino, G., and Tumiatti, W. (2001). Overview of water evolution during the thermal degradation of cellulose. *European Polymer Journal*, 37(5), 933–942.

Sejie, F. P., Matshwele, J. T., Nareetsile, F. M., and Obuseng, V. C. (2023). Nanocellulose surface modification reactions and their influence on the adsorption of heavy metal ions from water-a review. *Chemical Review and Letters*, 6(3), 245–255.

Shahzad, A., Ullah, M. W., Ali, J., Aziz, K., Javed, M. A., Shi, Z., … and Yang, G. (2023). The versatility of nanocellulose, modification strategies, and its current progress in wastewater treatment and environmental remediation. *Science of the Total Environment*, 858, 159937.

Sharma, A., Thakur, M., Bhattacharya, M., Mandal, T., and Goswami, S. (2019). Commercial application of cellulose nano-composites–A review. *Biotechnology Reports*, 21, e00316.

Shi, Y., Zhang, Y., Zhu, L., Miao, Y., Zhu, Y., and Yue, B. (2024). Tailored drug delivery platforms: Stimulus-responsive core–shell structured nanocarriers. *Advanced Healthcare Materials*, 13(1), 2301726.

Shuya, L. (2023). Advances in drug delivery applications of modified bacterial cellulose-based materials. *Frontiers in Bioengineering and Biotechnology*, 11(2023), 1252706.

Silva Filho, E. C., Santos Júnior, L. S., Silva, M. M. F., Fonseca, M. G., Santana, S. A. A., and Airoldi, C. (2013). Surface cellulose modification with 2-aminomethylpyridine for copper, cobalt, nickel and zinc removal from aqueous solution. *Materials Research*, 16, 79–84.

Sousa, S., Costa, A., Silva, A., and Simões, R. (2019). Poly(lactic acid)/Cellulose films produced from composite spheres prepared by emulsion-solvent evaporation method. *Polymers*, 11(1), 66.

Stanislas, T. T., Bilba, K., de Oliveira Santos, R. P., Onésippe-Potiron, C., Savastano Junior, H., and Arsène, M. A. (2022). Nanocellulose-based membrane as a potential material for high performance biodegradable aerosol respirators for SARS-CoV-2 prevention: A review. *Cellulose*, 29(15), 8001–8024.

Stankova, L., Kutova, A., Doubkova, M., Kvitek, O., Sedlar, A., Svorcik, V., and Bacakova, L. (2025). Cellulases and their promising use in tissue engineering and cell sheet technology. *Materials & Design*, 254, 114111.

Subedi, G., Shrestha, A. K., and Shakya, S. (2016). *Study of effect of different factors in formulation of micro and nanospheres with solvent evaporation technique. PHARMSCI-3–182.*

Sun, C. (2005). True density of microcrystalline cellulose. *Journal of Pharmaceutical Sciences*, 94(10), 2132–2134.

Tang, J., He, H., Wan, R., Yang, Q., Luo, H., Li, L., and Xiong, L. (2021). Cellulose nanocrystals for skin barrier protection by preparing a versatile foundation liquid. *ACS Omega*, 6(4), 2906–2915.

Thala, K., Machina, A., Parveen, S., and Ramana, L. N. (2025). Reusable and non-invasive TiO2-based photodynamic transdermal patch (RPT) for treating MDR-negative bacteria strain and promote wound healing through a synergistic approach of ROS-induced RNS. *Colloids and Surfaces B: Biointerfaces*, 245, 114301.

Than-ardna, B., Weder, C., and Manuspiya, H. (2023). Superhydrophilic bacterial cellulose membranes efficiently separate oil-in-water emulsions. *Journal of Materials Science*, 58, 5086–5103.

Tian, W., Gao, X., Zhang, J., Yu, J., and Zhang, J. (2022). Cellulose nanosphere: Preparation and applications of the novel nanocellulose. *Carbohydrate Polymers*, 277(2022), 118863.

Tofanica, B. M., Mikhailidi, A., Fortună, M. E., Rotaru, R., Ungureanu, O. C., and Ungureanu, E. (2025). Cellulose nanomaterials: Characterization methods, isolation techniques, and strategies. *Crystals*, 15(4), 352.

Tsai, C. T., and Meyer, A. S. (2014). Enzymatic cellulose hydrolysis: Enzyme reusability and visualization of β glucosidase immobilized in calcium alginate. *Molecules (Basel, Switzerland)*, 19(12), 19390–19406.

Tzhayik, O., Pulidindi, I. N., and Gedanken, A. (2014). Forming nanospherical cellulose containers. *Industrial & Engineering Chemistry Research*, *53*(36), 13871–13880.

Valatabar, N., Oroojalian, F., Kazemzadeh, M., Mokhtarzadeh, A. A., Safaralizadeh, R., and Sahebkar, A. (2024). Recent advances in gene delivery nanoplatforms based on spherical nucleic acids. *Journal of Nanobiotechnology*, 22(1), 386.

Vasconcelos, N. F., Feitosa, J. P. A., Portela da Gama, F. M., Morais, J. P. S., Andrade, F. K., Filho, M. S., and Rosa, M. F. (2017). Bacterial cellulose nanocrystals produced under different hydrolysis conditions: Properties and morphological features. *Carbohydrate Polymers,* 155(2017), 425–431.

Wang, L., Wang, C., Wang, L., Zhang, Q., Wang, Y., and Xia, X. (2020). Emulsion electrospun polylactic acid/apocynum venetum nanocellulose nanofiber membranes with controlled sea buckthorn extract release as a drug delivery system. *Textile Research Journal*, 91 (9–10), 1046–1055.

Wang, N., Ding, E., and Cheng, R. (2007). Thermal degradation behaviors of spherical cellulose nanocrystals with sulfate groups. *Polymer*, 48(12), 3486–3493.

Wang, Q., Chen, J., Zhao, L., Wang, S., Li, P., Du, K. (2024). A sphere-on-sphere structure: Cu nanoparticles grown on cellulose microspheres for the removal of environmental contaminants. *Industrial Crops and Products*, 213(2022), 118401.

Wang, X., Zhong, C., Zhong, Y., Fan, Z., Liu, Z., Xu, P., … and Liu, J. (2025). Impressive merits of nanocellulose driving sustainable beauty. *Carbohydrate Polymers*, 353, 123270.

Wang, Z., Yao, Z., Zhou, J., He, M., Jiang, Q., Li, S., … and Luo, S. (2019). Isolation and characterization of cellulose nanocrystals from pueraria root residue. *International Journal of Biological Macromolecules*, 129, 1081–1089.

Wei, Z., Wu, C., Li, R., Yu, D., and Ding, Q. (2021). Nanocellulose based hydrogel or aerogel scaffolds for tissue engineering. *Cellulose*, 28(12), 7497–7520.

Wongthanaroj, D., Jessmore, L. A., Lin, Y., Bergholz, T. M., Stark, N. M., Sabo, R. C., and Matuana, L. M. (2022). Sustainable and eco-friendly poly (Lactic acid)/cellulose nanocrystal nanocomposite films for the preservation of oxygen-sensitive food. *Applied Food Research*, 2(2), 100222.

Wu, Z., Sun, H., Xu, Z., Chi, H., Li, X., Wang, S., Zhang, T., and Zhao, Y. (2021). Underwater mechanically tough, elastic, superhydrophilic cellulose nanofiber-based aerogels for water-in-oil emulsion separation and solar steam generation. *ACS Applied Nano Materials*, 4(9), 8979–8989.

Xia, J., Yu, Y., Chen, H., Zhou, J., Tan, Z., He, S., … and Li, X. (2019). Improved lignocellulose degradation efficiency by fusion of β-glucosidase, exoglucanase, and carbohydrate-binding module from Caldicellulosiruptor saccharolyticus. *BioResources*, 14(3), 6767–6780.

Xu, C., Chen, W., Gao, H., Xie, X., and Chen, Y. (2020). Cellulose nanocrystal/silver (CNC/Ag) thin-film nanocomposite nanofiltration membranes with multifunctional properties. *Environmental Science: Nano*, 7(3), 803–816.

Xu, K., Liu, C., Kang, K., Zheng, Z., Wang, S., Tang, Z., and Yang, W. (2018). Isolation of nanocrystalline cellulose from rice straw and preparation of its biocomposites with chitosan: Physicochemical characterization and evaluation of interfacial compatibility. *Composites Science and Technology*, 154, 8–17.

Yabushita, M., Kobayashi, H., Hara, K., and Fukuoka, A. (2014). Quantitative evaluation of ball-milling effects on the hydrolysis of cellulose catalyzed by activated carbon. *Catalysis Science & Technology*, 4(8), 2312–2317.

Yan, C. F., Yu, H. Y., Yao, J. M. (2015). One-step extraction and functionalization of cellulose nanospheres from lyocell fibers with cellulose II crystal structure. *Cellulose*, 22, 3773–3788.

Yan, H., Chen, X., Song, H., Li, J., Feng, Y., Shi, Z., Wang, X., Lin, Q., (2017). Synthesis of bacterial cellulose and bacterial cellulose nanocrystals for their applications in the stabilization of olive oil pickering emulsion. *Food Hydrocolloids*, 72, 127–135.

You, X., Zhou, Y., Sun, R., Pei, X., Zhou, H., Deng, D., … and Tan, Y. (2025). Biodegradable microbeads for personal care products and cosmetics. *Green Chemistry*, 27, 2823–2845.

Zagórska-Dziok, M., and Sobczak, M. (2020). Hydrogel-based active substance release systems for cosmetology and dermatology application: A review. *Pharmaceutics*, 12(5), 396.

Zhang, J., Elder, T. J., Pu, Y., and Ragauskas, A. J. (2007). Facile synthesis of spherical cellulose nanoparticles. *Carbohydrate Polymers*, 69(3), 607–611.

Zhang, J., Tian, Z., Ji, X., and Zhang, F. (2022). Fabrication mechanisms of lignin nanoparticles and their ultraviolet protection ability in PVA composite film. *Polymers*, 14(19), 4196.

Zhang, Q., Benoit, M., Vigier, K. D. O., Barrault, J., Jégou, G., Philippe, M., and Jérôme, F. (2013). Pretreatment of microcrystalline cellulose by ultrasounds: Effect of particle size in the heterogeneously-catalyzed hydrolysis of cellulose to glucose. *Green Chemistry*, 15(4), 963–969.

Zhang, Q., Lu, Z., Su, C., Feng, Z., Wang, H., Yu, J., and Su, W. (2021). High yielding, one-step mechano-enzymatic hydrolysis of cellulose to cellulose nanocrystals without bulk solvent. *Bioresource Technology*, 331, 125015.

Zhao, H., Kwak, J. H., Wang, Y., Franz, J. A., White, J. M., and Holladay, J. E. (2006). Effects of crystallinity on dilute acid hydrolysis of cellulose by cellulose ball-milling study. *Energy & Fuels*, 20(2), 807–811.

Zhao, W., Zhou, Y., Zhang, G., Li, Y., Liao, X., Lai, G., Jiang, Y., Jia, S., Su, Z., Qi, J., and Zhang, S. (2025a). Efficient and size-controllable method and mechanism for preparing cellulose nanospheres. Biomacromolecules 2025 26 (4), 2665–2674.

Zhao, X., Zhang, Y., Lv, Z., Ding, Z., Wang, Y., Yang, H., ... and Chen, H. (2025b). Nanocellulose and its composites toward flexible supercapacitor. Advanced Materials Technologies, 10(10), 2401709.

Zhu, L., Feng, L., Luo, H., Dong, R., Wang, M., Yao, G., and Chen, J. (2022). Characterization of polyvinyl alcohol-nanocellulose composite film and its release effect on tetracycline hydrochloride. *Industrial Crops and Products*, 188(Part B), 115723.

Zhu, L., Vijay, K., and Gilbert, S. B. (2001). Examination of oxidized cellulose as a macromolecular prodrug carrier: Preparation and characterization of an oxidized cellulose-phenylpropanolamine conjugate. *International Journal of Pharmaceutics*, 223(1–2), 35–47.

6 Catalytic Nano-Spherical Cellulose Materials

(a) Organic Synthesis Technologies and (b) Catalytic Degradation Technologies

Vishwajeet Bachhar, Vibha Joshi, Manisha Duseja, and Rajesh Haldhar

6.1 INTRODUCTION

Nano-spherical cellulose materials have appeared as groundbreaking tools in catalysis because of diverse physical and chemical characteristics. For instance, these nanomaterials are derived from renewable resources that not only show larger surface area and versatility but also support and develop green and sustainable technologies [1,2]. The spherical morphology of the cellulose nanomaterials at the nanoscale significantly increases the reactivity, stability, and compatibility with many catalytic reactions.

The latest development in the field of nanotechnology and green chemistry have eased the development of nano-spherical cellulose as catalytic material. Applications that require efficient chemical transformations in mild environments show promise for these structures [3,4]. Nano-spherical cellulose is used in catalysis in two major areas such as catalytic degradation and organic synthesis reaction technologies. Nano-spherical cellulose is a strong platform for immobilizing various catalytic agents, like metal nanoparticles, organocatalysts, and enzyme mimetics, in organic synthesis [5,6]. In addition to being reusable, these hybrid catalysts show enhanced activity and selectivity in various organic reactions like coupling reactions, oxidation, and reduction.

It is feasible to alter the surface of nano-spherical cellulose chemically to add certain functional groups that can aid in the attachment of the active catalytic species as shown in Figure 6.1. The synthesis of the variety of composite catalysts has been made feasible by approaches like surface grafting, in-situ reduction, and covalent bonding [7]. These catalysts work very efficiently in reactions that are important for the manufacture of pharmaceuticals, renewable fuels, and fine chemicals.

DOI: 10.1201/9781003596127-7

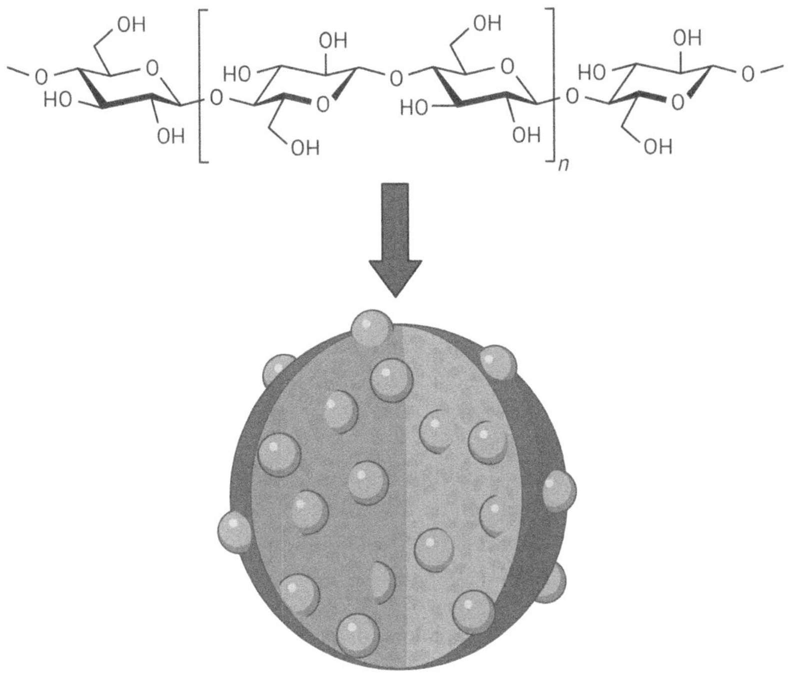

FIGURE 6.1 Conversion of cellulose into cellulose nanospheres via various methods.

Furthermore, they provide benefits such as affordability, reusability, and environmental safety while often working on a par with conventional catalysts. They are thus being investigated more, and they are potential substitutes for traditional metal oxide and polymer-based catalytic systems [8,9].

The function of nano-spherical cellulose materials in environmental catalysis, particularly for catalytic degradation technologies, is equally important. Synthesis of the efficient and sustainable remediation approach is essential for solving the worldwide problem of environmental pollution, specifically the toxic contamination of water supplies with dyes, pesticides, medicines, and industrial discharge. Materials based on nano-spherical cellulose is promising in this area by acting as redox active or photocatalysts [10–12]. These cellulose structures can promote the degradation of organic pollutants under light irradiation when doped with photocatalytic agents like iron oxide (Fe_2O_3), zinc oxide (ZnO), or titanium dioxide (TiO_2). This helps create safer and cleaner habitats [13–15].

The recent advances in research and technical advancement in the field of catalytic nano-spherical cellulose materials are thoroughly reviewed in this chapter. In both organic synthesis and environmental remediation, it methodically discusses the synthesis plans, functionalization methods, and performance indicators of these materials. The mechanistic understanding of their catalytic activities, the difficulties encountered in their actual use, and the future possibilities for their optimization and scale-up are given particular attention.

6.2 STRUCTURAL AND CHEMICAL PROPERTIES OF NANO-SPHERICAL CELLULOSE

6.2.1 Morphology and Surface Characteristics

These materials are characterized by their highly well-defined and uniform spherical morphology, which usually ranges from 20 to 200 nm [16,17]. Various kinds of fabrication methods like acid hydrolysis, emulsion templating, or self-assembly methods, allowing precise control over particle size and shape for achieving nanoscale structures [15].

Moreover, surface-to-volume ratio is enhanced by the spherical geometry of these nanocellulose particles, and it is important for the catalytic and adsorption-based studies and applications. Larger surface area allows for a greater number of active sites to be exposed, which facilitates increase in the interaction with target molecules, reactants, or contaminants.

Moreover, the nano-spherical cellulose surface can be easily modified because of many hydroxyl groups present on the backbone of the cellulose. These functional groups enable chemical grafting, surface functionalization, or nanoparticle immobilization, which enhance the catalytic efficiency, selectivity, and reactivity of the material.

Furthermore, these show an excellent dispersibility in aqueous and organic solvents, preventing aggregation and maintains their active surface area. The uniformity in the morphological structure also ensures constant behavior across applications, like heterogeneous catalysis, biosensing, drug delivery, and environmental remediation.

Conclusively, the structural morphology and properties of the surface of nano-spherical cellulose make it an ideal contender for various kinds of catalytic and functional material applications, which owe to its structural uniformity, modifiability, and high-performance surface properties.

6.2.2 Surface Chemistry

The hydroxyl (–OH) functional groups that are abundant on the surface of nano-spherical cellulose (NSC) provide it considerable chemical reactivity and adaptability for surface modification. An extensive array of hydroxyl functional groups in cellulose nanospheres enable a wide variety of chemical derivatization techniques. Methods like carboxylation, sulfonation, phosphorylation, amination, and esterification enhance the surface properties of nanocellulose for certain catalytic applications [18,19].

These chemical modifications are necessary to enhance the anchoring and dispersion of catalytic species, such as metal nanoparticles, metal oxides, enzymes, or organocatalysts. For example, carboxylated or aminated surfaces provide sites for covalent bonding or electrostatic interactions with metal ions or enzyme cofactors to boost catalytic efficiency.

These alterations change the material's hydrophilicity or hydrophobicity, making it suitable for reactions in both organic and aqueous phases. Figure 6.2 shows various surface modification of cellulose nanospheres.

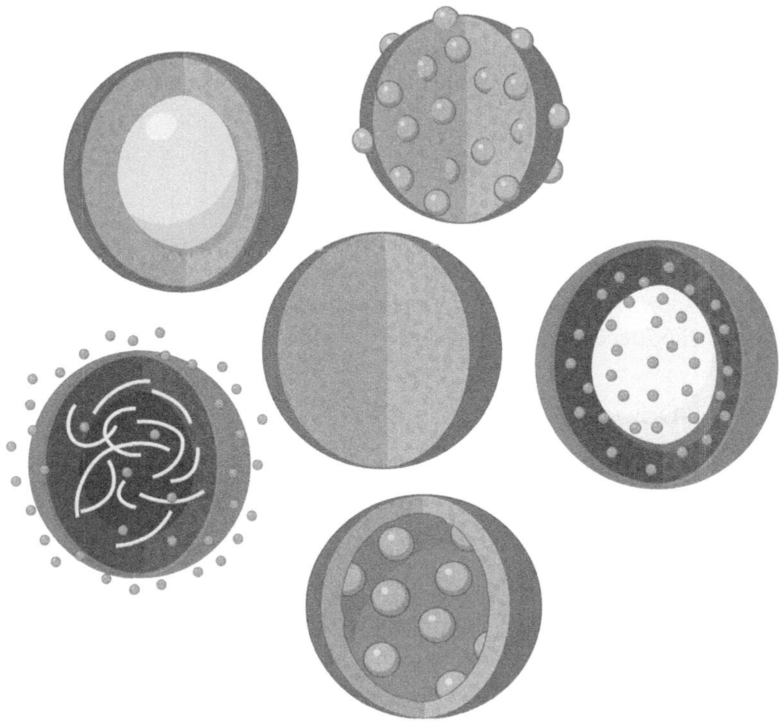

FIGURE 6.2 Various types of surface modification of cellulose nanospheres.

6.2.3 Thermal and Chemical Stability

One of the major benefits of NSC in catalysis is its improved thermal and chemical stability and this property significantly expands its applicability in harsh processing environments. The nano-spherical form of cellulose shows enhanced thermal stability due to properties like decreased crystallinity, uniformity of the particle size, and better heat dissipation across its large surface area, even when native cellulose tends to break down at temperatures over 250°C [2].

Its heat resistance is further increased by additional crosslinking or inorganic functionalization, such as doping with metal oxides (such as TiO_2, ZnO, or FeO_4). At high temperatures, these hybrid materials not only maintain the cellulose's structural integrity but also provide synergistic catalytic activity. In high-temperature catalytic processes, such as gas-phase reactions, polymerization, and oxidative degradation, the thermal stability is very advantageous.

In terms of chemical resistance, NSC is relatively inert in a wide pH range. Its robust performance in acidic and alkaline media allows it to serve as a catalyst support or reactive intermediate in diverse chemical transformations, such as ester hydrolysis, transesterification, or redox reactions. Chemical modification and composite formation further enhance this stability, making the material compatible with corrosive environments, polar solvents, and oxidizing agents.

Overall, the characteristic thermal and chemical flexibility of NSC makes it a perfect platform for heterogeneous catalysis, especially in industrial settings where durability and resistance to harsh conditions are critical for continuous operation and material longevity.

6.3 SYNTHESIS OF NSC

NSC materials have received a lot of interest for their customizable surface area, biocompatibility, and for their catalytic potential. Their synthesis is basically classified into top-down, bottom-up, and green synthesis techniques with each providing distinct benefits, depending on application and environmental factors [20,21].

6.3.1 Top-Down Approaches

Top-down methods involve the breaking down of bulk cellulose into nanoscale spherical forms by using physical or chemical treatments.

- **Acid Hydrolysis**: This is one of the widely used processes in which various powerful mineral acids such as sulfuric or hydrochloric acid are used to selectively hydrolyze the amorphous areas of the cellulose by yielding the crystalline nanospheres. For minimizing excessive deterioration, the acid content and the reaction time must be carefully controlled throughout the process.
- **Ultrasonication**: For the ultrasonication method, high-energy ultrasonic waves are used to induce cavitation in the cellulose dispersions by breaking down the cellulose into nano-sized spherical particles. This method is often used in combination with chemical treatments to enhance uniformity.
- **Mechanical Milling**: In this technique, cellulose fibers are used to high-energy ball-milling or jet-milling technique to reduce them into the nanoscale spheres. It is a solvent-free, scalable, and energy-intensive method that is suitable for large-scale production.

6.3.2 Bottom-Up Approaches

Bottom-up approaches create NSC structures from molecular or nanoscale materials through controlled assembly.

- **Emulsion Templating**: In emulsion templating, cellulose solutions are disseminated into emulsions such as oil-in-water or water-in-oil, by which then they solidify or coagulate and result in the creation of homogeneous nanospheres. Surfactants and stabilizers are critical for regulating particle size and shape.
- **Sol–Gel Synthesis**: This method includes the converting of cellulose sol into a gel phase by generally using the crosslinking or gelation chemicals, and then drying to produce nanospheres. It enables the introduction of new functional groups or active agents.

- **Microfluidic Synthesis**: Advanced microfluidic devices can accurately regulate droplet production and particle size, which results in a highly controllable and repeatable approach for producing monodisperse cellulose nanospheres.

6.3.3 Green Synthesis Routes

Green synthesis emphasizes eco-friendly routes by avoiding harsh chemicals and helps in employing renewable resources or mild conditions.

- **Use of Plant Extracts**: Phytochemicals present in plant extracts act as reducing and the stabilizing agents and help in facilitating the transformation of cellulose into nanospheres under benign conditions.
- **Enzymatic Hydrolysis**: Cellulolytic enzymes are preferentially split into cellulose chains to form nanostructured materials. This approach is incredibly specialized, gentle, and environmentally friendly.
- **Microwave-Assisted Synthesis**: Microwave irradiation enhances reaction kinetics and uniform heating by accelerating the conversion of cellulose into nanospheres while reducing energy consumption and processing time [22,23].

6.4 FUNCTIONALIZATION TECHNIQUES

For NSC, post-synthesis functionalization is critical for adjusting its catalytic activity, stability, and compatibility in a variety of applications, such as organic transformations and environmental remediation.

6.4.1 Surface Grafting

With the help of surface grafting, one can add additional functional groups or polymeric layers to cellulose nanospheres to increase their reactivity and dispersibility.

- **Silanization**: Various silane-coupling agents such as APTES and GPTMS are used for grafting organosilanes into the hydroxyl-rich cellulose surfaces that provide the anchor points for further chemical modifications or nanoparticle binding.
- **Polymer Brushes**: "Grafting from" or "grafting to" techniques are used to enable the attachment of polymer chains onto NSC by improving solubility, mechanical stability, and catalytic compatibility of the cellulose [24,25].

6.4.2 Metal Nanoparticle Immobilization

For imparting the catalytic activity for the redox and organic syntheses reactions, metal nanoparticles are embedded into cellulose nanospheres.

- **In-situ Reduction Using Green Agents**: Various metal salts such as $AgNO_3$ and $AuCl_4$ are reduced directly on the cellulose surfaces by using eco-friendly reducing agents such as ascorbic acid or plant extracts. This technique ensures uniform distribution and also minimizes environmental hazards.

- **Electroless Plating**: This technique is a chemical deposition method where the metal ions are reduced and then they are plated onto the surface of the cellulose without any external power source. This technique also yields a robust metallic layer, which is also suitable for catalytic applications.

6.4.3 Enzyme Immobilization

Enzyme immobilization of NSC serves as a versatile platform, which also helps in enhancing their stability, reusability, and catalytic efficiency.

- **Physical Adsorption**: In this technique enzymes adhere to the cellulose surface through non-covalent interactions such as by van der Waals or electrostatic force. This technique is simple and, when it is reversible, provides a widely used high enzyme activity.
- **Covalent Bonding**: In cellulose nanospheres, various functional groups such as aldehyde, carboxyl forms a stable covalent bond with the enzyme residue that enhances durability and also helps in minimizing the leaching during the reactions.
- **Entrapment**: Various enzymes are encapsulated within the polymeric or gel-like matrix on the cellulose surface that provides a protective microenvironment and also sustained activity over time [26–28].

6.5 CATALYTIC APPLICATIONS IN ORGANIC SYNTHESIS

The science of organic synthesis has been advanced with the usage of NSC in catalytic systems. They provide structures for metal nanoparticles and enzymes due to the large surface area, versatility of surface chemistry, and biocompatibility. These hybrid systems support a variety of organic transformations under moderate, eco-friendly conditions, following the concepts of green chemistry [29,30].

6.5.1 Oxidation Reactions

Various cellulose-supported gold (Au) and platinum (Pt) nanoparticles have shown excellent effectiveness in the oxidation of alcohols to aldehydes or ketones. The high dispersion of metal nanoparticles on the surface of the cellulose confirms maximum catalytic exposure and reduces the need for extra loading, which helps in sustainable and cheaper processes [31,32].

Moreover, these catalysts are used for green oxidation of aldehydes, by using oxidants like H_2O_2 under harsh conditions. These systems minimize the environmental impact by side-stepping harmful reagents and offer high selectivity and conversion rates and make them ideal for various chemical synthesis.

6.5.2 Reduction Reactions

An important process in the synthesis of colors, pharmaceuticals, and agrochemicals is the reduction of nitroarenes to their equivalent anilines. Cellulose–silver composite nanomaterials have shown excellent catalytic activity. Electron transport and reaction kinetics are increased by the stability and nanoscale dispersion of Ag particles on cellulose [33].

Moreover, alkenes and carbonyl compounds are efficiently hydrogenated via catalysts based on cellulose. These reduction methods offer a high-activity, recyclable, and eco-friendly substitute to conventional metal oxide supports when applied to bio-based materials using supported metal catalysts.

6.5.3 C–C Bond Formation

Cross-coupling reactions like Suzuki-Miyaura coupling have been advanced significantly from the use of cellulose nanospheres, which is loaded with palladium. These composite catalysts aid in effective C–C formation of bond with outstanding yields and reduced leaching, which highlights their strength and potential for reuse [34].

Moreover, for Suzuki coupling, cellulose-supported composite catalysts are used in the Heck and Sonogashira reactions. Structural and chemical characteristics of cellulose nanosphere help in the stabilization of metal catalysts, which aids in efficient coupling under harsh reaction conditions, and conclusively contributes to the sustainability of synthetic techniques.

6.5.4 Esterification and Transesterification

It is found that cellulose nanospheres play a significant role in esterification and transesterification reactions, specifically in areas like biodiesel production. Surface functionalization and thermal stability permit for the immobilization of many acidic or basic catalytic species, which is required for these transformations [35].

One of the most important uses is the immobilization of lipase enzymes onto cellulose nanospheres. These bio-catalytic systems increase enzyme stability and reusability, which facilitates the effective transesterification of triglycerides to methyl esters. Using such green catalytic systems not only enhances reaction efficiency but also supports the development of sustainable biofuel technologies.

6.6 PHOTOCATALYTIC DEGRADATION OF DYES

The photocatalytic synthesis of synthetic dyes y utilizing NSC, which is supported by a catalyst, is a green and effective method for environmental cleanup. This method uses semiconductor materials such as TiO_2 or ZnO when adsorbed on the cellulose nanospheres to improve light absorption, charge separation, and catalytic activity.

- **TiO_2- or ZnO-Doped Cellulose Nanospheres**: When TiO_2 or ZnO is doped onto the surface of cellulose nanospheres, it significantly improves their photocatalytic efficiency due to the synergistic effect of the semiconductor's photocatalytic properties and high dispersion/stabilization, which is offered by the cellulose matrix. Cellulose nanospheres here act as a structural support by facilitating the uniform distribution of nanoparticles and also preventing the aggregation. Moreover, the biopolymeric surface can be chemically modified to improve the interaction with dye molecules, thus enhancing degradation kinetics.

- **Visible Light and Solar Irradiation Activity**: The enhanced and the modified TiO_2 or ZnO nanocatalysts on cellulose nanospheres may be developed to extend their various activities into the visible light spectrum by the metal and non-metal doping or the surface sensitization. This enables the photocatalytic process to operate well in natural sunlight by making it both economically and the ecologically feasible [36]. Dye molecules are adsorbed onto the catalyst surface and also undergo oxidative breakdown into non-toxic byproducts, such as CO_2 and H_2O, via reactive oxygen species (ROS) formed during light irradiation.

This sustainable method reveals the ability of NSC materials to enable improved photocatalytic systems, particularly for wastewater treatment, environmental detoxification, and dye effluent remediation, thereby contributing to circular and clean technology development.

6.7 DEGRADATION OF PESTICIDES AND PHARMACEUTICALS

Various persistent organic contaminants, such as pesticides and the pharmaceutical residues, contaminate water bodies by endangering both the environment and human health. NSC materials that are functionalized with the catalytically active components have showed the great potential for aiding the breakdown by improved oxidation processes.

- **Fenton-Like Reactions Catalyzed by Fe-Based Cellulose Hybrids**: Various cellulose nanospheres containing iron-based species such as Fe_3O_4, Fe_2O_3, or Fe (II)/Fe (III) complexes function as the heterogeneous catalyst for various reactions such as Fenton and Fenton-like reactions. These reactions create hydroxyl radicals (•OH) in the presence of hydrogen peroxide. These radicals are very reactive, capable of converting complex pesticide and converting medicine compounds into harmless end products. The cellulose matrix provides a biocompatible and dispersible platform by increasing catalytic efficiency, stability, and reusability of the hybrid material.
- **Reactive Oxygen Species Generation**: In the ROS generation, Fe-based cellulose hybrids are used catalytically to activate the H_2O_2 or persulfates by generating the ROS such as hydroxyl radicals, sulfate radicals ($SO_4^{-\bullet}$), and superoxide ($O_2^{\bullet-}$) radicals [37]. These ROS can cause the non-selective oxidation of the pharmaceutical contaminants by promoting mineralization. Because of catalyst systems, these moderate reaction conditions and customizable redox characteristics make it ideal for practical applications in environmental remediation.

6.8 ADSORPTIVE–CATALYTIC HYBRID SYSTEMS

Adsorptive–catalytic hybrid systems are used to improve the overall efficiency of contaminant removal, which is based on the NSC that has been developed by combining the pollutant capture with in-situ catalytic degradation.

- **Dual Function as Adsorbent and Catalyst**: NSC materials have significant adsorptive interactions with a variety of organic contaminants because of their porous structure and the adjustable surface chemistry and nano-sphericity. When these are equipped with catalytic components such as metal oxides, enzymes, or doped semiconductors, these systems perform two functions: first they concentrate pollutants by adsorption and then they catalytically break them down [38].
- **Synergistic Removal and Degradation of Pollutants**: Both adsorption and catalysis work together to improve the removal efficiency and the degradation rate. Here the adsorbed pollutants are retained near the catalytic sites, which improves the reaction kinetics and also reduces the secondary contamination. This method is very useful for treating low-concentration contaminants in the complicated aqueous matrixes, such as in pharmaceutical wastewater and the pesticide runoff.

6.9 MECHANISTIC INSIGHTS

For optimizing the performance and extending the applicability in organic synthesis and environmental remediation, it is necessary to understand the catalytic mechanisms for NSC systems.

6.9.1 Electron Transfer and Activation Mechanisms

Catalytic potential of cellulose-based nanomaterials often centers on electron donor–acceptor interactions. With the enhancement of the charge transfer between the catalyst and the reactant, these interactions promote the commencement of reactions and start the activation of substrates. Radical-based processes, as they produce the ROS that is responsible for the breakdown of organic molecules, predominate in many oxidation reactions. When metals or metal oxides are mixed with nanocellulose, these routes are enhanced by stabilizing transient radical species.

6.9.2 Synergistic Effects

Metal–support interactions have a major role in improving the catalytic performance. The cellulose matrix offers a stabilizing and dispersing platform for metal nanoparticles, which prevents the aggregation and facilitates the electron transfer. The cooperative catalysis using co-catalysts, like enzymatic or metallic, can also facilitate synergistic effects that boost selectivity and turnover rates. Dual or cascade reactions can occur in a single catalytic setup due to these complex systems.

6.10 PERFORMANCE EVALUATION

Quantifying catalytic performance is an important parameter for benchmarking the effectiveness and practicality of cellulose-based catalysts in various chemical reactions.

6.10.1 Catalytic Activity

Catalytic activity is majorly examined via reaction kinetics, as it provides details about the reaction rate under specific conditions. Turnover frequency and turnover number are the two major parameters, which shows the catalyst's efficacy and durability. Increased TOF and TON values is the indication of the catalyst's potential to facilitate multiple reaction cycles with minor degradation or deactivation [39].

6.10.2 Selectivity and Yield

Both selectivity and product yield are influenced by various parameters like surface modifications on the catalyst. Functionalization of cellulose nanospheres can increase active site exposure and direct product pathways. Furthermore, the choice of the solvent and pH of the reaction medium suggestively affect the interaction between reactants and catalysts, which alters with the selectivity and conclusively enhances the overall yields.

6.11 CHALLENGES AND LIMITATIONS

Although NSC materials show increasing promise in catalytic degradation technologies, several obstacles are still needed to be overcome before they can be widely used and adopted in the real world.

6.11.1 Scalability

Synthesis of NSC is mainly done by using sophisticated or green approaches and often faces scalability issues. But nowadays, most current methods are optimized for laboratory-scale production and are not easily adaptable to industrial settings [40]. Various factors such as long processing times, solvent use, and stringent reaction conditions can limit the throughput. While green synthesis methods such as by using plant extracts or enzymatic reactions are more sustainable, they are also cost-effective and also process reproducibility that remain key hurdles for scale-up.

6.11.2 Stability and Leaching

Stability of the catalyst and metal leaching are major concerns in repeated or in long-term applications. Metal nanoparticles or active sites are immobilized on cellulose nanospheres that may leach into the environment during catalytic cycles by reducing the performance and raising toxicity concerns. Moreover, various exposure to the harsh reaction conditions such as pH, temperature, and oxidative species can deteriorate the structural integrity of the hybrid material that leads to the loss of catalytic efficiency over time.

6.11.3 Recyclability and Regeneration

Many of the cellulose-based catalysts are reusable as well as their recyclability is also limited by the deactivation mechanisms, such as fouling, structural collapse, or aggregation of the active sites. So, effective washing and regeneration protocols are majorly required to restore catalytic performance, but repeated treatment can also compromise the functional integrity of the cellulose matrix, while designing more robust materials with self-regeneration capability is an active area of research.

6.12 FUTURE PERSPECTIVES

For NSC catalysts, future research is very necessary, which should focus on the innovation, integration, and industrial applications, with an emphasis on the sustainability and the multifunctionality of the material.

6.12.1 Smart and Multifunctional Catalysts

The smart and multifunctional catalysts are the stimuli–responsive systems that may activate or deactivate the response to the environmental cues such as light, pH, or temperature that are being developed as a result of the advancements in material design. These "smart catalysts" might provide exact control over parameters of the reaction. Additionally, combining these two functions, such as sensing and catalysis at the same time, will create a new possibility for the targeted degradation and for the real-time pollution monitoring [41].

6.12.2 Hybrid Nanocomposites

The efficiency of the NSC catalysts in the hybrid nanocomposite materials may be greatly improved by combining them with the sophisticated nanomaterials such as graphene, metal-organic frameworks, and carbon dots. These combinations may also increase the charge separation, electron mobility, and also the surface reactivity that leads to higher photocatalytic and Fenton-like efficiencies. Such hybrids also allows for the modification of material characteristics to suit particular environmental or biological applications.

6.12.3 Industrial Applications

To attain practical significance, research must go from material synthesis to reactor design and then to process integration. The addition of NSC catalysts to flow-through catalytic reactors provides a scalable solution for wastewater treatment, pharmaceutical effluent remediation, and also for pesticide degradation. These inventions can promote circular economy frameworks in which trash is not only eliminated but also converted into useful byproducts.

6.13 CONCLUSION

NSC materials are one of the most exceptional family of green catalytic materials with a wide range of use in organic synthesis and environmental remediation applications. These are used in extremely effective, selective, and reusable catalytic systems due to their chemical characteristics and physical adaptability. These materials have the potential and have great importance in the development of sustainable technologies, an eco-friendlier chemical industry, and international initiatives to develop a circular bioeconomy as research advances.

REFERENCES

1. Phuong, H.T., Thoa, N.K., Tuyet, P.T.A., Van, Q.N., and Hai, Y.D. (2022) Cellulose nanomaterials as a future, sustainable and renewable material. *Crystals (Basel)*, **12** (1), 106.
2. Khorsandi, D., Jenson, S., Zarepour, A., Khosravi, A., Rabiee, N., Iravani, S., and Zarrabi, A. (2024) Catalytic and biomedical applications of nanocelluloses: A review of recent developments. *Int J Biol Macromol*, 268, 131829.
3. Thomas, B., Raj, M.C., B, A.K., H, R.M., Joy, J., Moores, A., Drisko, G.L., and Sanchez, C. (2018) Nanocellulose, a versatile green platform: From biosources to materials and their applications. *Chem Rev*, 118 (24), 11575–11625.
4. de Oliveira, M.L., Rodrigues, L.M., Silva da Veiga, M.A.M., and Souza, L.R.R. (2024) Advancing nanomaterial synthesis: Harnessing green chemistry for sustainable innovation. *Green Analytical Chemistry*, 11, 100148.
5. Bilal, M., Khaliq, N., Ashraf, M., Hussain, N., Baqar, Z., Zdarta, J., Jesionowski, T., and Iqbal, H.M.N. (2023) Enzyme mimic nanomaterials as nanozymes with catalytic attributes. *Colloids Surf B Biointerfaces*, 221, 112950.
6. Geleto, S.A., Ariti, A.M., Gutema, B.T., Abda, E.M., Abiye, A.A., Abay, S.M., Mekonnen, M.L., and Workie, Y.A. (2023) Nanocellulose/Fe3O4/Ag nanozyme with robust peroxidase activity for enhanced antibacterial and wound healing applications. *ACS Omega*, 8 (51), 48764–48774.
7. Mujtaba, M., Negi, A., King, A.W.T., Zare, M., and Kuncova-Kallio, J. (2023) Surface modifications of nanocellulose for drug delivery applications: A critical review. *Curr Opin Biomed Eng*, 28, 100475.
8. Hassaan, M.A., El-Nemr, M.A., Elkatory, M.R., Ragab, S., Niculescu, V.-C., and El Nemr, A. (2023) Principles of photocatalysts and their different applications: A review. *Top Curr Chem*, 381 (6), 31.
9. Melián-Cabrera, I. (2021) Catalytic materials: Concepts to understand the pathway to implementation. *Ind Eng Chem Res*, 60 (51), 18545–18559.
10. Ningthoujam, R., Singh, Y.D., Babu, P.J., Tirkey, A., Pradhan, S., and Sarma, M. (2022) Nanocatalyst in remediating environmental pollutants. *Chemical Physics Impact*, 4, 100064.
11. González-González, R.B., Parra-Saldívar, R., Alsanie, W.F., and Iqbal, H.M.N. (2022) Nanohybrid catalysts with porous structures for environmental remediation through photocatalytic degradation of emerging pollutants. *Environ Res*, 214, 113955.
12. Rana, R., Ahmad, W., Mishra, A., Kumar, S., and Ahmad, A. (2025) Nanocellulose/nanocellulose-based membranes in wastewater treatment: A sustainable path forward for environmental protection. *Food Chem X*, 29, 102654.
13. Chen, D., Cheng, Y., Zhou, N., Chen, P., Wang, Y., Li, K., Huo, S., Cheng, P., Peng, P., Zhang, R., Wang, L., Liu, H., Liu, Y., and Ruan, R. (2020) Photocatalytic degradation of organic pollutants using TiO2-based photocatalysts: A review. *J Clean Prod*, 268, 121725.
14. Batchamen Mougnol, J.B., Waanders, F., Ntwampe, S.K.O., Fosso-Kankeu, E., and Al Alili, A.R. (2022) Synthesis of eco-friendly ZnO-based heterophotocatalysts with enhanced properties under visible light in the degradation of organic pollutants. *Environmental Systems Research*, 11 (1), 25.
15. V, D., B, V., and Sebastian, A. (2023) Photocatalytic degradation of model pollutants using ZnO/Ag3PO4 heterostructure thin films under visible and solar irradiation. *Opt Mater (Amst)*, **138**, 113646.

16. Joshi, V., Bachhar, V., Negi, A., Mishra, S.S., and Duseja, M. (2025) Green synthesis of silver nanoparticles using Pachygone laurifolia extract and pharmacological evaluation for antioxidant, antidiabetic, and antibacterial activity. *J Mol Struct*, **1342**, 142730.
17. Joshi, V., Bachhar, V., Bhatia, A., Rom, T., Duseja, M., and Shukla, R.K. (2025) Green synthesis of multifunctional AgFe nanoparticles using Piper chaba extract: Evaluation of antioxidant, antidiabetic, and antibacterial activities. *Colloids Surf A Physicochem Eng Asp*, **725**, 137619.
18. Peng, S., Luo, Q., Zhou, G., and Xu, X. (2021) Recent advances on cellulose nanocrystals and their derivatives. *Polymers (Basel)*, **13** (19), 3247.
19. Tavakolian, M., Jafari, S.M., and van de Ven, T.G.M. (2020) A review on surface-functionalized cellulosic nanostructures as biocompatible antibacterial materials. *Nanomicro Lett*, **12** (1), 73.
20. Thakur, V., Guleria, A., Kumar, S., Sharma, S., and Singh, K. (2021) Recent advances in nanocellulose processing, functionalization and applications: A review. *Mater Adv*, 2 (6), 1872–1895.
21. Singh, S., Bhardwaj, S., Tiwari, P., Dev, K., Ghosh, K., and Maji, P.K. (2024) Recent advances in cellulose nanocrystals-based sensors: A review. *Mater Adv*, 5 (7), 2622–2654.
22. Magalhães, S., Fernandes, C., Pedrosa, J.F.S., Alves, L., Medronho, B., Ferreira, P.J.T., and Rasteiro, M. da G. (2023) Eco-friendly methods for extraction and modification of cellulose: An overview. *Polymers (Basel)*, **15** (14), 3138.
23. Villagrán, Z., Anaya-Esparza, L.M., Velázquez-Carriles, C.A., Silva-Jara, J.M., Ruvalcaba-Gómez, J.M., Aurora-Vigo, E.F., Rodríguez-Lafitte, E., Rodríguez-Barajas, N., Balderas-León, I., and Martínez-Esquivias, F. (2024) Plant-based extracts as Reducing, Capping, and Stabilizing Agents for the Green Synthesis of Inorganic Nanoparticles. *Resources*, **13** (6), 70.
24. Frka-Petesic, B., Parton, T.G., Honorato-Rios, C., Narkevicius, A., Ballu, K., Shen, Q., Lu, Z., Ogawa, Y., Haataja, J.S., Droguet, B.E., Parker, R.M., and Vignolini, S. (2023) Structural Color from Cellulose Nanocrystals or Chitin Nanocrystals: Self-Assembly, Optics, and Applications. *Chem Rev*, 123 (23), 12595–12756.
25. Kassie, B.B., Getahun, M.J., Azanaw, A., Ferede, B.T., and Tassew, D.F. (2024) Surface modification of cellulose nanocrystals for biomedical and personal hygiene applications. *Int J Biol Macromol*, 282, 136949.
26. Califano, V., and Costantini, A. (2020) Immobilization of cellulolytic enzymes in mesostructured silica materials. *Catalysts*, **10** (6), 706.
27. Maghraby, Y.R., El-Shabasy, R.M., Ibrahim, A.H., and Azzazy, H.M.E.-S. (2023) Enzyme immobilization technologies and industrial applications. *ACS Omega*, 8 (6), 5184–5196.
28. Prabhakar, T., Giaretta, J., Zulli, R., Rath, R.J., Farajikhah, S., Talebian, S., and Dehghani, F. (2025) Covalent immobilization: A review from an enzyme perspective. *Chemical Engineering Journal*, 503, 158054.
29. Wieszczycka, K., Staszak, K., Woźniak-Budych, M.J., Litowczenko, J., Maciejewska, B.M., and Jurga, S. (2021) Surface functionalization – The way for advanced applications of smart materials. *Coord Chem Rev*, 436, 213846.
30. Malik, S., Singh, J., Goyat, R., Saharan, Y., Chaudhry, V., Umar, A., Ibrahim, A.A., Akbar, S., Ameen, S., and Baskoutas, S. (2023) Nanomaterials-based biosensor and their applications: A review. *Heliyon*, 9 (9), e19929.
31. Dikshit, P.K., Kumar, J., Das, A.K., Sadhu, S., Sharma, S., Singh, S., Gupta, P.K., and Kim, B.S. (2021) Green synthesis of metallic nanoparticles: Applications and limitations. *Catalysts*, **11** (8), 902.
32. Hutchings, G.J., and Edwards, J.K. (2012) Chapter 6- Application of gold nanoparticles in catalysis, In: Roy L. Johnston; J.P. Wilcoxon (eds) *Frontiers of Nanoscience*, vol. 3, Elsevier, pp. 249–293.

33. Zhang, S., Xu, X., Ye, Z., Liu, Y., Wang, Q., Chen, Q., Jiang, Y., Qi, J., Tian, D., Xu, J., Wu, F., Feng, X., and Lu, Y. (2023) A large-nanosphere/small-nanosphere (cellulose/silver) antibacterial composite with prominent catalytic properties for the degradation of p-nitrophenol. *Appl Surf Sci*, 608, 155192.
34. Riva, L., Nicastro, G., Liu, M., Battocchio, C., Punta, C., and Sacchetti, A. (2022) Pd-loaded cellulose nanoSponge as a heterogeneous catalyst for Suzuki–Miyaura coupling reactions. *Gels*, **8** (12), 789.
35. Atofarati, E.O., and Enweremadu, C.C. (2026) A review of sustainable biofuel catalysts and additive technologies using nanoparticles. *Fuel*, 403, 136100.
36. El Mchaouri, M., Mallah, S., Abouhajjoub, D., Boumya, W., Elmoubarki, R., Essadki, A., Barka, N., and Elhalil, A. (2025) Engineering TiO2 photocatalysts for enhanced visible-light activity in wastewater treatment applications. *Tetrahedron Green Chem*, 6, 100084.
37. Van, N.T.T., Phan, A.N., Cuong, V.C., Van, N.T.T., Thanh, H.G.-T., Khai, N.Q., Tri, N., Nguyen, T.-T., Bui, X.-T., and Huynh, K.P.H. (2023) Enhanced heterogeneous Fenton degradation of p-nitrophenol by Fe3O4 nanoparticles decorated cellulose aerogel from banana stem. *Environ Technol Innov*, **30**, 103041.
38. Bader, Q.A., Al-Sharify, Z.T., Dhabab, J.M., Zaidan, H.K., Rheima, A.M., Athair, D.M., Joseph, T.M., and kianfar, E. (2024) Cellulose nanomaterials in oil and gas industry and bio-manufacture: Current situation and future outlook. *Case Studies in Chemical and Environmental Engineering*, 10, 100993.
39. Zhou, Q., Liao, L., Zhou, H., Li, D., Tang, D., and Yu, F. (2022) Innovative strategies in design of transition metal-based catalysts for large-current-density alkaline water/seawater electrolysis. *Materials Today Physics*, 26, 100727.
40. Pradhan, D., Jaiswal, A.K., and Jaiswal, S. (2022) Emerging technologies for the production of nanocellulose from lignocellulosic biomass. *Carbohydr Polym*, 285, 119258.
41. Bachhar, V., Joshi, V., Bhatia, A., Rom, T., Duseja, M., and Shukla, R.K. (2025) Green synthesis of AgFe bimetallic nanoparticles from calyptocarpus vialis plant extract for enhanced catalytic reduction of 4-NP, antioxidant and antibacterial activities. *J Environ Chem Eng*, 13 (3), 116829.

7 Functional Roles of Nano-Spherical Cellulose in Food Systems

Muhammad Umar Ijaz, Amara Tahir, Quratulann Sattar, Muhammad Zaid Salar, Naila Ghafoor, and Shumaila Kiran

7.1 INTRODUCTION

Cellulose is one of the most common natural polymers found on earth and is mainly extracted from plant cell walls. Cellulose was originally isolated and identified by French chemist Anselme Payen in 1838, who named it cellulose, distinguishing it as a separate component of plant tissues (Payen, 1838). Due to biocompatibility and environmental sustainability, cellulose derivatives and cellulose have been receiving widespread attention and are found to have far-reaching implications. Planned measures have to be adopted to avoid wastage and deterioration in perishable foods. Here, an appropriate choice of packaging material can play a crucial role in avoiding the spoilage of foods (Wsoo et al., 2020). Due to its low price, renewability and good physicochemical and mechanical properties, cellulose has gained popularity in the world market, especially in the food and textile industry (Hua et al., 2014).

A huge quantity of waste from farming, such as fruits, vegetables, cotton straw, and forest residues, needs to undergo recycling. (Nascimento et al., 2018). Because of their distinctive properties, biodegradability, and superior physical properties, as well as their superior mechanical properties, cellulose, nanocellulose (NC) and their derivatives attract attention for use in food packaging. Due to the recent advancements in nanotechnology, nano materials, i.e., metal oxides, and nanocrystals, have been of considerable interest for use across all sectors, including food packaging, pharmaceuticals, and cosmeceuticals (Oun and Rhim, 2016).

In this context, cellulose-based nanomaterials have garnered much attention as biocompatible, renewable, and eco-friendly substitutes (Abdul Khalil et al., 2016). Cellulose is the utmost renewable and most abundant raw polymer on Earth, the annual production of this biopolymer is estimated to be between 10^{10} and 10^{11} tonne (Habibi et al., 2010)). Cellulose nanofibers (CNF) are derived from numerous materials like wood, ramie hemp, flax, cotton, sisal, jute (Liu et al., 2021), wood remains from forests (Moriana et al., 2016), farm waste (Ingle et al., 2014), waste from fruits (Hiasa et al., 2014; Ferrer et al., 2017), and bacterial cellulose (BC) (Reiniati et al., 2017). Nanocellulose has numerous appealing properties, including biodegradability,

DOI: 10.1201/9781003596127-8

sustainability, superior mechanical properties, high aspect ratio (ca. 70), high surface area (ca. 150 m^2/g), light density, less environmental impact, less cost, and tunable surface properties (Pickering et al., 2016). These properties make them more useful as nanofillers in food-packaging film for enhancing mechanical, and moisture, a properties (Bettaieb et al., 2015).

7.2 SYNTHESIS AND CHARACTERIZATION OF NANO-SPHERICAL CELLULOSE

7.2.1 Methods Used for Nanostructured Cellulose (NSC) Preparation

Nanoparticles (NPs) are synthesized via physical (Abou El-Nour et al., 2010), chemical (González et al., 2019), and biological (Rafique et al., 2017) processes. The green synthesis strategy has been popularized due to its low toxicity and reduced use of chemicals. The distinctiveness of nanoparticles may be studied by microscope and spectroscopic methods to assess particle size, morphology, and surface plasmon resonance (Chandrasekharan et al., 2022). AgNPs have attracted much attention for their surface plasmon resonance, biocompatibility, electrical conductivity, antioxidant properties, and chemical stability. These features allow them to be used in treating viruses, bacteria (Salar et al., 2015), fungi (Velmurugan et al., 2015), and cancer (Ghramh et al., 2020). NPs have found widespread applications in anti-corrosive coatings, sensors, delivery agents for drugs, solar cells, water purification, agriculture, consumer goods, and catalysis (Khan et al., 2014). Some immobilization methods are pad-dry-cure, in-situ synthesis, printing, layer-by-layer deposition, solution immersion, and sonochemical techniques (Fernandes et al., 2020). AgNPs have been successfully applied to a wide range of textile materials, such as cotton, PC blend, wool, silk, polyester, regenerated cellulose, and polyamide. AgNPs have been reported to permeate biological membranes, penetrate into cells, and accumulate in organs like brain, heart, liver, spleen, lungs, and kidneys, which can impact physiological processes (Syafiuddin, 2019).

7.2.2 Isolation Methods of Nanocellulose

Nanocellulose is produced from lignocellulosic materials by various methods including acid hydrolysis (Liu et al., 2008), mechanical, chemical/TEMPO-mediated oxidation, ammonium persulfate (Oun and Rhim, 2016), and cellulose regeneration (Wang et al., 2015). The characteristics of nanocellulose are different, depending on the cellulose raw material and the process of separation. Nanocellulose can be derived directly from purified cellulose sources like cotton fiber and bacterial cellulose (BC), or from a lignocellulosic biomass, like grasses, with or without the removal of lignin (Liu et al., 2016).

7.2.2.1 Mechanical Approaches

Mechanical methods are used to disintegrate cellulose fibers into cellulose nanofibers (CNFs). Turbak et al. (1983) were the pioneers in isolating cellulose CNF with lateral diameters in the nanometer range from pulp made from softwood using a high-pressure homogenizer. Turbak's CNFs exhibited a highly entangled network of crystalline and amorphous structures that get consolidated in water (Nechyporchuk et al., 2016). Mechanical operations include homogenization, grinding, refining, extrusion, rapid mixing, ultrasonic treatment, cryocrushing, thermal exploding, and ball milling (Li et al., 2015). Among these, homogenization, purification, and crushing are identified as extremely effective (Khalil et al., 2016).

7.2.2.2 Acid Hydrolysis Procedure

Acid hydrolysis is widely utilized to isolate cellulose nanocrystals (CNCs) (Liu et al., 2016). CNCs were initially isolated from sulfuric acid by Rånby (1951). Other acids like hydrochloric, phosphoric, bromic, and maleic acids were later utilized (Lu and Hsieh, 2010). Sulfuric acid is still favored because it stabilizes CNC suspensions by electrostatic repulsion (Jiang et al., 2013). Acids in the form of hydrochloric and hydrobromic can hydrolyze cellulose without surface esterification, but CNCs isolated by them have an agglomeration tendency due to the lack of surface charge (Yu et al., 2017). Acid hydrolysis de-amorphinates cellulose, isolating needle-shaped CNCs with high crystallinity and smaller aspect ratios in comparison with CNCs from mechanical or TEMPO-mediated methods (Sacui et al., 2014).

7.3 FUNCTIONAL PROPERTIES OF NANO-SPHERICAL CELLULOSE IN FOOD SYSTEMS

Natural plant components like lignin have been utilized to confer steric hindrance for enhancing nanocellulose redispersibility. Lignin plays an adhesive/filler role in plant cell walls (Wang et al., 2015), while its phenolic functionalities allow for interactions with cellulose and other polymers. CNFs that contain lignin, which were prepared by a heat-intensified disk nano-grinder, can be redispersed in alcohol solutions (Sirviö and Visanko, 2017). Such CNFs that have been subjected to drying from nanopaper that can likewise be redispersed in water. Kraft lignin likewise enhances CNF redispersibility by steric hindrance and lowered hydrogen bonding. Likewise, alkaline lignin prevents CNF aggregation upon oven-drying, enhancing redispersion in addition to stability in dispersion (Kim et al., 2022). Polysaccharide degradation products like glucose have been observed to increase redispersibility. Adsorbed glucose on CNC surfaces in solvent exchange with ethanol hinders interfibrillar bonding (Zhang et al., 2016). Polysaccharide degradation products have been found to exist commonly in suspensions of nanocelluloses (Peri et al., 2012) and can be adsorbed onto CNCs to improve dispersibility.

7.4 APPLICATIONS OF NANOCELLULOSE-METAL OR METAL OXIDE HYBRID NANOPARTICLES

As the focus has shifted to sustainability, renewable materials have become the preferred choice for plastics, as they can be recycled. Cellulose is also used more widely due to its biodegradability, cost-effectiveness, barrier properties, and non-toxicity (Khalil et al., 2016). Cellulose fibers have been used on a large commercial scale for packaging materials such as paper, paperboard, corrugated fiberboard, and pulp

TABLE 7.1
The sources, pre-treatment methods, and food advantages of nonspherical cellulose

Source Material	Pre-treatment Method	Key Advantage for Food Application	References
Wood	Acid hydrolysis	Biodegradable	Habibi et al., 2010; Klemm et al., 2011
Cotton linters	Alkaline+ bleaching	Emulsifier stabilizer in food emulsions	Bondeson et al., 2006; Lin and Dufresne, 2014
Agricultural waste	Steam explosion	Eco-friendly and cost-effective source	Trache et al., 2020; Phanthong et al., 2018
Bacterial cellulose	Mild acid treatment	High purity	Jorfi and Foster, 2015; Ul-Islam et al., 2012
Fruit peels	Microwave-assisted treatment	Natural antioxidant	Trache et al., 2020; Azeredo et al., 2019

containers. In addition to their conventional application in packaging materials such as cellulose acetate and regenerated cellulose (cellophane), cellulose is used more widely for packaging materials such as polyethylene-coated paper for packaging foods such as milk and orange juice. Antimicrobial packaging materials are developed by adding agents such as Ag, ZnO, or CuO, NPs to polymers such as PLA, PBAT, LDPE, PHBV, or PVC, alginate, gelatin, and carbohydrate-based biopolymers (Shankar and Rhim, 2018). However, metal or metal oxide NPs decrease film physical properties. Therefore, in addition to nanocelluloses, inorganic NPs in combination with nanocelluloses act as multifunctional nanofillers to improve film physical properties as well as antibacterial properties (Khalil et al., 2016).

Fu et al. (2017) prepared ZnO-Ag hybrid NP film using regenerated cellulose without harmful chemicals. This hybrid showed high antimicrobial performance and that is suitable for large-scale packaging for food. Magnetic cellulose beads were prepared by mixing cellulose/NaOH/urea solution with magnetic NPs (Fe_2O_3) in a submerged circulative reactor by Luo and Zhang (2009). The beads were responsive to magnetic fields and showed efficacy in adsorbing organic dyes in wastewater. Magnetic chitosan-cellulose-coated hydrogel microspheres in which cellulose added strength while chitosan-stabilized Fe_3O_4 had high capacities for adsorbing heavy metals (Cu^{2+}, Fe^{2+}, Pb^{2+}) and could be reused. These materials have potential in food packaging, biomedicine, healthcare, and catalysis (Liu et al., 2012).

7.5 APPLICATIONS FOR FOOD PACKAGING AND COATINGS

Food packaging is an integral part of the food industry, shielding food from physical harm and chemical dangers (Han et al., 2018). Plastics from synthetic polymers, paperboard, paper (Rovera et al., 2020), and metal comprise most common packaging materials, depending on the type. Approximately 40% of global plastic production is used for packaging applications. About 5 billion polyethylene bags are used every year across the world. After consumption, plastic packaging is commonly incinerated or landfilled, polluting the environment (Suleman et al., 2022). Today, development in terms of "sustainable development" has stimulated the development of biodegradable, sustainable, and renewable food packaging materials (Table 7.2).

Currently, cellulose fibers are processed into nanoscale particles by using methodologies such as acid hydrolysis (Spiliopoulos et al., 2021), homogenization (Serial et al., 2021), grinding (Shamshina et al., 2022), and blending (Qian et al., 2021), which have been grouped as nanocellulose (Phanthong et al., 2018). Though the terminology for nanocellulose is not standardized, it is reported to occur in three groups depending on the production process and morphological properties: CNC, CNF, and BC (Blanco et al., 2018). The coating process involves using consumable, environmental friendly natural polymers that can be immersed or applied directly to the food surface. These coating materials can often include nanocellulose-based materials as well as other naturally derived biologically active chemicals. According to studies, such chemical composite films have been reported to possess good antibacterial properties as well as higher efficacy in terms of moisture maintenance and controlling food oxidation rate. The use of biopolymers such as nanocellulose as covering materials may effectively regulate the nutrient loss in the ecosystem. The results demonstrated that such covering may effectively decrease microbial growth and increase prolonged storage (Quintana et al., 2021).

Nanocellulose is widely employed as a filler in paper and packaging films, where it enhances mechanical strength, barrier properties, and overall material performance

TABLE 7.2
Applications of nano-spherical cellulose in the food industry

Application Area	Description	Benefits	References
Barrier films	Used as a barrier layer in food packaging	Reduce permeability to moisture, oxygen, and oils	Azeredo., et al 2017; Blanco et al., 2018
Biodegradable packaging	Incorporated into biopolymer matrixes to replace plastic	Environmentally friendly	Dufresne, A., 2013; Blanco et al., 2018
Antimicrobial coatings	Functionalized with antimicrobial agents	Enhance food safety by inhibiting microbial growth	Vilela et al., 2018; Azeredo et al., 2017
Edible coatings	Applied directly on food surfaces	Enhance shelf life and maintain food freshness	Lin and Dufresne, 2014
Smart packaging	Integrated with sensors and indicators	Enable real-time monitoring of food condition	Vilela et al., 2018
Active packaging system	Used to deliver active compounds like antioxidants or preservatives	Maintain food quality and extend shelf life	Vilela et al., 2018; Dufresne, A., 2013

(Sharma and Bhardwaj, 2020). Since NC is a natural polyose, it has gained attention in the manufacturing of foods making it an excellent stabilizing agent in products such as cream, sauce, additives, icing, foams, salad dressing, soups, puddings, desserts, dips, texture additives, frozen dairy products, and thickeners (Zanchetta et al., 2017). NC has been an exceptional hydrocolloid food additive (da Gama and Dourado, 2017). NC serves as an option for less-calorie additives to foods, quality additives, stabilizing additives, and thickening agents. The features of NC are excellent and unquestionable, thus making them an exceptional choice to use in foods (Perumal et al., 2019).

Using coaxial 3D printing, cellulose nanofiber (CNF)-based dual-functional labels were developed that preserve freshness in fruits and allow f monitoring of product quality during storage (Ezati et al., 2022). Cellulose nanofibrils-based ink that contained anthocyanin, which was used to form the shell of fibers, exhibiting formability, printing accuracy, and a responsive color-based visual freshness indicator (Nascimento et al., 2018). 1-methylcyclopropene (1-MCP)-infused chitosan was introduced to the resonating channels in fibers, where 1-MCP was detected by means of the electrostatic association between cellulose nanofibers (CNF) and chitosan. The 3D printed labels extended six days in terms of freshness retention of litchis and effectively highlighted freshness variations, as revealed by Headspace Gas Chromatography Ion Spectrometer (Zhou et al., 2021).

Nano cellulose (NC) can act as an emulsifying agent as well as stabilizing such as hydrophilic polysaccharides from various materials, including seaweed, microorganisms, plant seeds, and insoluble polysaccharides (Yano et al., 2012). Cellulose had organoleptic properties, including texture, fat-based, cream-based fillings, which it is substituting. CNC can act as an oil–water stabilizer mainly for Pickering emulsion formulation (Shapira-Zaltsberg et al., 2018). CNC was utilized in preparing polystyrene beads encapsulating monodispersed water microdroplets through the water-in-oil Pickering emulsion process followed by suspension polymerization (Ajelou et al., 2019).

NC is used as a food stabilizer to increase shelf life and maintain frozen dessert's structure. Increasing the concentration of food stabilizers can improve shape retention; however, excessive amounts may adversely affect product quality and flavor addition of vegetable nanocellulose prolonged the duration of structural retention without affecting other properties of the dessert (Winuprasith and Suphantharika, 2015). Corral et al. (2017) examined the effect on baking quality that results from adding BC to wheat flour by changing the viscoelastic properties of biomass. Adding an appropriate quantity of BC increases moisture retention and decreases the value of the browning index.

Currently, more interest in food-grade Pickering mixtures is being generated. Owing to their excellent strength and rheological characteristics, these mixtures have useful applications in dairy products, including substituting butter in baking cakes, cream, ice cream, and frozen yogurt. Velásquez-Cock et al. (2019) studied the use of cellulose nanofibers in ice cream to explore the impact of cellulose nanofibers on the structure and functionality of ice over products. According to research, CNF's impact is dependent on CNF's adsorption on the lipid matrix. Despite heat treated specimens, the addition of CNF enhanced sensory properties in samples that were not high in fat. The hardness, rheology at temperatures below −2°C, and formation of ice crystals in ice cream were not affected by the addition of CNF. Esparza et al. (2020) have developed milled formulations through spray wiping for Pickering blends stabilized by hempseed oil with cellulose nanocrystals. From this study, it is evident that the formulation with hempseed oil in addition to other constituents can be stabilized and microencapsulated with cellulose nanocrystals. Pickering suspensions offer enhanced strength as well as being eco-sociable compared with conservative emulsions (Yadav et al., 2013). Further study is needed to enhance Pickering emulsions in foods. The applications of nano-spherical cellulose in the food industry are given in Table 7.2.

7.6 ROLE OF NANO-SPHERICAL CELLULOSE IN ENCAPSULATION AND DELIVERY SYSTEMS

Several cellulose nanoparticle variants have been reported to act as drug nanocarriers. Polat et al. (2019) synthesized pH-responsive carboxymethyl cellulose-coated magnetite nanoparticles for delivery of doxorubicin. Oxidized cellulose was also prepared by 2,2,6,6-Tetramethylpiperidine-1-oxyl (TEMPO) oxidation and applied as a doxorubicin nanocarrier for cancer treatment at 7.4 pH (Habibi et al., 2006; Kumari et al., 2018). The carboxyl groups of nanoparticles were found to physically bind to the amino groups in drugs through electrostatic forces and amide linkages, allowing for pH-responsive controlled release at targeted locations (Li et al., 2015). Carboxylic groups in nanoparticles have the capability to form hydrogen bonds with the drug. Also, the carbonyl group in an ester could potentially be utilized to join in physical bonding to the drug through hydrophobic interactions (Chin et al., 2019).

Most of the studies with cellulose-based spherical particles focus on developing new methods of drug delivery, primarily in terms of nano-sized particles prepared by nanoprecipitation and emulsification processes. Examples include modified celluloses such as oxidized cellulose (Kumari et al., 2018) and their analogs such as hydroxypropyl methyl cellulose and hydroxyethyl cellulose (Yang et al., 2020) for encapsulation of anticancer drugs such as doxorubicin (Elumalai et al., 2015), 5-fluorouracil, and paclitaxel (Lu et al., 2020). Encapsulation is not limited to anticancer drugs. Antiretroviral drugs (e.g., zidovudine [Joshy et al., 2017]), antibiotics (e.g., penicillin G [Ho et al., 2020] and tetracycline [Wang et al., 2015]), anti-inflammatory drugs (e.g., diclofenac [Nethaji et al., 2016], ibuprofen [Pooresmaeil et al., 2020], ketoprofen lysinate [Guarino et al., 2019], naproxen [Yeap et al., 2019]), vaccines (e.g., foot-and-mouth disease virus [FMDV]) subunit inoculation (Liu et al., 2019), antiseptics (Vidal-Romero et al., 2020), and pesticides (Huang et al., 2018) avermectin (Lee et al., 2018), other bioactive molecules (Döge et al., 2016) and curcumin (Dewangan et al., 2017) have been encapsulated in cellulose-based particles for treating other diseases. El-Habashy et al. (2016) demonstrated the preparation of EC particles to alter piroxicam's release profile in order to prevent ulcerogenicity from this anti-inflammatory drug upon oral administration. Encapsulation greatly inhibited piroxicam's gastrointestinal ulceration potential in rats by providing 66% reduction in the mean ulcer index.

Ethyl cellulose has been utilized as an exemplary matrix material in microsphere preparation for prolonged delivery of drugs because it is chemically stable, insoluble in water, elastic, and inexpensive (Nath et al., 2010). Among the compounds encapsulated in such delivery systems sodium bicarbonate has attracted attention due to its ability to neutralized intratumoral lactic acid, thereby promoting improved anti-tumor effects since intratumoral lactic acid plays an essential role in causing resistance in cancer cells to glucose starvation-mediated cell death. However, the formulation of sodium bicarbonate is challenging because it is highly water-soluble (9.6 g/100 mL at 21°C), exhibits very limited solubility in organic solvents, and readily decomposes upon heating. Sodium bicarbonate is thus used as an example for thermally unstable and highly soluble in water medication with low molecular weight (Chao et al., 2016).

7.7 SAFETY, REGULATORY ASPECTS, AND CONSUMER ACCEPTANCE

As consumers and manufacturers are increasingly shifting towards sustainable and renewable materials, cellulose nitrate (CN) has emerged as a promising alternative to fossil fuel-based plastics. Manufacturers can employ CN's unique properties, including color, tensile strength, transparency, lightness, and biodegradability, for both new and traditional purposes. CN has attracted considerable interest for a broad range of applications, including packaging, paper products, polymer composites, insulation materials, protective coatings, aerogels, cosmetics, air and water filtration systems, and recyclable electronic devices (Wei et al., 2014; Shatkin et al., 2014). Other potential applications, such as sensors and scaffolds for tissue engineering, are being developed in the biomedicine field (Jorfi et al., 2015).

The studies of CN safety have mainly concentrated on occupational and inhalation exposure and short-term bioassays (Roman, 2015). Cellulose nanocrystals manufactured via a proprietary method to reduce sulfate content have been extensively evaluated for their potential human and environmental toxicological effects.

Occupational cellulose nanofibrils exposure was investigated in a conductive simulation and targeted *in vivo* and *in vitro* investigations. Collectively this study and a few additional publications related to CN toxicity, provide an insubstantial foundation to evaluate possible health and safety issues in many commercial applications. Existing uncertainty regarding nanomaterial's safety restricts their use in commercial processes. A systematic and extensive risk assessment was made to help develop an environmental health and safety roadmap for CN because it is being investigated for numerous applications with the potential for large-scale production and widespread use across multiple sectors of the economy (Vartiainen et al., 2011).

The two cellulose nanomaterial variants, cellulose nanofiber (CNF) and cellulose nanocrystal (CNC), have distinctive physical and chemical attributes that make them different from conventional cellulose as well as from each other. CNC displays more crystalline structure compared with CNF (Sacui et al., 2014; Moon et al., 2011). CNC particles even tend to exist in rod- or whisker-shaped objects that may have sizes 50–501 nm in length by 3–5 nm in width. CNF particles tend to be more fibrous with some amorphous segments; they still exist as a nanoscale material, but not solely in three-dimensional terms. The physical storage in powder form for dry CN is one aspect in maintaining material security in an occupational setting (Yano et al., 2012).

Standard cellulose dust burns at 240°C , while an experimental deflagration index (KSt) value of 229 confirms that tested CN is a powerful explosive according to the UN Globally Harmonized System of Classification and Labeling of Chemicals (GHS) (NFPA, 2007). Owing to their high surface areas and conceivably higher reactivity compared with typical cellulose, CNs could pose an increased safety threat, as indicated by comparative igniting experiments with micro- and nanoparticulates of metal oxides. Standard cellulose is very combustible. CNF, CNC-sulfated, and CNC-neutralized variants have been tested for flammability, the investigation showed that, CNF possessed maximum thermal stability (320°C onset temperature) and maximum flammability parameters (Vartiainen et al., 2011).

Fox et al. (2012) reported that cytotoxicity and inflammatory reaction in human and mouse macrophages were not produced by cellulose microfibrils (CMF). Pitkänen et al. (2010) examined two nanocelluloses, cellulose nanofibrils (CNF) and whisker-type ultrafine cellulose (UFC), for *in vitro* cytotoxicity and genotoxicity. Short-term cytotoxicity could not be detected in terms of determination of maximum tolerated dose and observing change in total protein content, sub-lethality, or genotoxicity by the standard Ames' test. Rouhiainen et al. (2012) made an occupational risk assessment for carbon nanofibers (CNFs), including relevant tests for toxicity and hazard assessment. *In vitro* exposures to CNF samples indicated that CNF samples were non-cytotoxic, yet some effects on immunotoxicity were observed. Besides, mild genotoxic effects were observed in one of two tests done, but another test indicated an adverse finding. In an *in vivo* pharyngeal aspiration test done in mice, CNF showed an inflammatory reaction. However, we cannot conclude here whether CNF or bacterial contamination in CNF or their combined effects induced an inflamed reaction, thus, we need to have further studies on this question. Bhattacharya et al. (2012) have also indicated a lack of cytotoxicity and determined the intrinsic biocompatibility of natural cellulose nanofibril hydrogels as hepatic cell culture substrates. The cytotoxicity was examined in two indigenous nanofibril-based constructions, one being dense but thin, while the other was porous but open. Neither construct caused acute cytotoxicity in 3T3 fibroblast cells based on cell membrane changes, mitochondrial function, and DNA replication by indirect cell assays and direct cell contact assays.

7.8 FUTURE PROSPECTS AND CHALLENGES

Due to the high interest in sustainability, finding renewable and abundantly available natural raw material such as cellulose is appreciated in all fields of contemporary science and technology. Cellulose is inherently an adaptable natural macromolecule that is easy to functionalize and mix with other molecules and macromolecules to present numerous particle-engineering possibilities. Its inability to dissolve in water as well as most common solvents is mainly to blame for making cellulose hard to process. But new solvents such as ionic liquids, deep eutectic solvents, switchable solvents, and organic electrolyte solutions have been introduced to replace limitations with respect to environmental friendly dissolution as well as chemical transformation of cellulose. Derivatives such as carboxymethylcellulose, cellulose acetate, and ethyl cellulose are dominating cellulose-based spherical particle research due to their ease in processibility (Kostag et al., 2018).

Cellulose microparticles can be used in a number of niche markets, indicating their commercial scalability. Examples include Cellufine™ chromatography media used for protein purification, enzymes and other biomolecules, Macroporous Bead Cellulose MT used in biomolecule separation by gel filtration and Cytopore macroporous microcarriers used for culturing cells, producing recombinant proteins, and immobilizing insect cells, plant cells, recombinant organisms and bacteria in stirred suspension culture systems. Conversely, numerous formidable impediments have been linked to cellulose (or their derivative) reduction to cell-like particles by dissolution followed by coagulation, alongside insufficient regulation for their safe application and disposal, which is hindering commercial deployment in most of the reviewed spherical cellulose-based nanomaterials (Klemm et al., 2018). Promises in material science and nanotechnology will most certainly shape the future use of cellulose in biomedicine. Combining cellulose with advanced technologies such as 3D bioprinting, smart hydrogels, and bioelectronics could transform regenerative medicine, tissue regeneration, and individualized healthcare. In addition, with environmental sustainability gaining more focus in medical material creation, cellulose's renewability will increase its value as an eco-friendly material to replace traditional synthetic medical materials (Shah et al., 2021).

7.9 CONCLUSION

In conclusion, nano-spherical cellulose displays a wide range of applications in the food sector. Its structural features and inherent traits provide notable benefits to food texture, stability, and packaging, and also translate to new concepts in delivering bioactive compounds. The development of more sustainable production technologies further reinforces its viability as a sustainable and eco-friendly ingredient. As the research field continues to proceed at an accelerated pace, a profound understanding of the regulatory policies and extensive safety evaluations are essential to ensure the successful transfer of such applications to commercially acceptable food products. Further exploration of new functionalities and optimization of current applications of nano-spherical cellulose are avenues that will unlock its maximum potential to develop healthier, safer, and more sustainable food systems.

REFERENCES

Abou El-Nour, K.M., Eftaiha, A.A., Al-Warthan, A. and Ammar, R.A., 2010. Synthesis and applications of silver nanoparticles. *Arabian Journal of Chemistry*, *3*(3), pp. 135–140.

Ajelou, Z., Nikfarjam, N., Deng, Y. and Taheri-Qazvini, N., 2019. Expanded polystyrene via stabilized water droplet by in-situ modified starch nanocrystals. *Colloids and Surfaces A: Physicochemical and Engineering Aspects*, *582*, p.123863.

Azeredo, H. M. C., Rosa, M. F. and Mattoso, L. H. C., 2017. Nanocellulose in bio-based food packaging applications. *Industrial Crops and Products*, *97*, 664–671.

Bettaieb, F., Khiari, R., Hassan, M.L., Belgacem, M.N., Bras, J., Dufresne, A. and Mhenni, M.F., 2015. Preparation and characterization of new cellulose nanocrystals from marine biomass Posidonia oceanica. *Industrial Crops and Products*, *72*, pp.175–182.

Bhattacharya, M., Malinen, M.M., Lauren, P., Lou, Y.R., Kuisma, S.W., Kanninen, L., Lille, M., Corlu, A., GuGuen-Guillouzo, C., Ikkala, O. and Laukkanen, A., 2012. Nanofibrillar cellulose hydrogel promotes three-dimensional liver cell culture. *Journal of Controlled Release*, *164*(3), pp.291–298.

Blanco, A., Monte, M.C., Campano, C., Balea, A., Merayo, N. and Negro, C., 2018. Nanocellulose for industrial use: Cellulose nanofibers (CNF), cellulose nanocrystals (CNC), and bacterial cellulose (BC). In C. M. Hussain (Ed.), *Handbook of Nanomaterials for Industrial Applications* (pp. 74–126). Elsevier.

Bondeson, D., Mathew, A. and Oksman, K., 2006. Optimization of the isolation of nanocrystals from microcrystalline cellulose by acid hydrolysis. *Cellulose*, *13*(2), pp.171–180.

Brogniart, A., Pelonze, A.B. and Dumas, R., 1839. Report on a memoir of M. Payen, on the composition of the woody nature. *Comptes Rendus*, *8*(51–53), p.7.

Chandrasekharan, S., Chinnasamy, G. and Bhatnagar, S., 2022. Sustainable phyto-fabrication of silver nanoparticles using *Gmelina arborea* exhibit antimicrobial and biofilm inhibition activity. *Scientific Reports*, *12*(1), p.156.

Chao, M., Wu, H., Jin, K., Li, B., Wu, J., Zhang, G., Yang, G. and Hu, X., 2016. A nonrandomized cohort and a randomized study of local control of large hepatocarcinoma by targeting intratumoral lactic acidosis. *Elife*, *5*, p.e15691.

Chin, S.F., Romainor, A.N., Pang, S.C., Lee, B.K. and Hwang, S.S., 2019. pH-responsive starch-citrate nanoparticles for controlled release of paracetamol. *Starch-Stärke*, *71* (9–10), p.1800336.

Corral, M.L., Cerrutti, P., Vázquez, A. and Califano, A., 2017. Bacterial nanocellulose as a potential additive for wheat bread. *Food Hydrocolloids*, *67*, pp.189–196.

da Gama, F.M.P. and Dourado, F., 2017. Bacterial NanoCellulose: What future?. *BioImpacts: BI*, *8*(1), p.1.

Dewangan, A.K., Perumal, Y., Pavurala, N., Chopra, K. and Mazumder, S., 2017. Preparation, characterization and anti-inflammatory effects of curcumin loaded carboxymethyl cellulose acetate butyrate nanoparticles on adjuvant induced arthritis in rats. *Journal of Drug Delivery Science and Technology*, *41*, pp.269–279.

Döge, N., Hönzke, S., Schumacher, F., Balzus, B., Colombo, M., Hadam, S., Rancan, F., Blume-Peytavi, U., Schäfer-Korting, M., Schindler, A. and Rühl, E., 2016. Ethyl cellulose nanocarriers and nanocrystals differentially deliver dexamethasone into intact, tape-stripped or sodium lauryl sulfate-exposed ex vivo human skin-assessment by intradermal microdialysis and extraction from the different skin layers. *Journal of Controlled Release*, *242*, pp.25–34.

Dufresne, A., 2013. Nanocellulose: A new ageless bionanomaterial. *Materials Today*, *16*(6), pp.220–227.

El-Habashy, S.E., Allam, A.N. and El-Kamel, A.H., 2016. Ethyl cellulose nanoparticles as a platform to decrease ulcerogenic potential of piroxicam: Formulation and in vitro/in vivo evaluation. *International Journal of Nanomedicine*, *11*, pp.2369–2380.

Elumalai, R., Patil, S., Maliyakkal, N., Rangarajan, A., Kondaiah, P. and Raichur, A.M., 2015. Protamine-carboxymethyl cellulose magnetic nanocapsules for enhanced delivery of anticancer drugs against drug resistant cancers. *Nanomedicine: Nanotechnology, Biology and Medicine*, *11*(4), pp.969–981.

Esparza, Y., Ngo, T.D. and Boluk, Y., 2020. Preparation of powdered oil particles by spray drying of cellulose nanocrystals stabilized pickering hempseed oil emulsions. *Colloids and Surfaces A: Physicochemical and Engineering Aspects*, *598*, p.124823.

Ezati, P., Rhim, J.W., Molaei, R., Priyadarshi, R. and Han, S., 2022. Cellulose nanofiber-based coating film integrated with nitrogen-functionalized carbon dots for active packaging applications of fresh fruit. *Postharvest Biology and Technology*, *186*, p.111845.

Fernandes, I.J., Aroche, A.F., Schuck, A., Lamberty, P., Peter, C.R., Hasenkamp, W. and Rocha, T.L., 2020. Silver nanoparticle conductive inks: Synthesis, characterization, and fabrication of inkjet-printed flexible electrodes. *Scientific reports*, *10*(1), p.8878.

Ferrer, A., Pal, L. and Hubbe, M., 2017. Nanocellulose in packaging: Advances in barrier layer technologies. *Industrial Crops and Products*, *95*, pp.574–582.

Fox, D.M., Lee, J., Zammarano, M., Katsoulis, D., Eldred, D.V., Haverhals, L.M., Trulove, P.C., De Long, H.C. and Gilman, J.W. 2012. Char-forming behavior of nanofibrillated cellulose treated with glycidyl phenyl POSS. *Carbohydrate Polymers, 88*(3), 847–858.

Fu, T., Moon, R.J., Zavattieri, P., Youngblood, J. and Weiss, W.J., 2017. Cellulose nanomaterials as additives for cementitious materials. In *Cellulose-reinforced nanofibre composites* (pp. 455–482). Woodhead Publishing.

Ghramh, H.A., Ibrahim, E.H. and Kilany, M., 2020. Study of anticancer, antimicrobial, immunomodulatory, and silver nanoparticles production by Sidr honey from three different sources. *Food Science & Nutrition*, *8*(1), pp.445–455.

González, E.A., Leiva, N., Vejar, N., Sancy, M., Gulppi, M., Azócar, M.I., Gomez, G., Tamayo, L., Zhou, X., Thompson, G.E. and Páez, M.A., 2019. Sol–gel coatings doped with encapsulated silver nanoparticles: Inhibition of biocorrosion on 2024-T3 aluminum alloy promoted by pseudomonas aeruginosa. *Journal of Materials Research and Technology*, *8*(2), pp.1809–1818.

Guarino, V., Altobelli, R., Caputo, T., Ambrosio, L., Caserta, S., Calcagnile, P. and Demitri, C., 2019. Mono-and bi-phasic cellulose acetate micro-vectors for anti-inflammatory drug delivery. *Pharmaceutics*, *11*(2), p.87.

Habibi, Y., Chanzy, H. and Vignon, M.R., 2006. TEMPO-mediated surface oxidation of cellulose whiskers. *Cellulose*, *13*, pp.679–687.

Habibi, Y., Lucia, L.A. and Rojas, O.J., 2010. Cellulose nanocrystals: chemistry, self-assembly, and applications. *Chemical Reviews*, *110*(6), pp.3479–3500.

Han, J.W., Ruiz-Garcia, L., Qian, J.P. and Yang, X.T., 2018. Food packaging: A comprehensive review and future trends. *Comprehensive Reviews in Food Science and Food Safety*, *17*(4), pp.860–877.

Hiasa, S., Iwamoto, S., Endo, T. and Edashige, Y., 2014. Isolation of cellulose nanofibrils from mandarin (Citrus unshiu) peel waste. *Industrial Crops and Products*, *62*, pp.280–285.

Ho, B.K., Chin, S.F. and Pang, S.C., 2020. pH-responsive carboxylic cellulose acetate nanoparticles for controlled release of penicillin G. *Journal of Science: Advanced Materials and Devices*, *5*(2), pp.224–232.

Hua, K., Carlsson, D.O., Ålander, E., Lindström, T., Strømme, M., Mihranyan, A. and Ferraz, N., 2014. Translational study between structure and biological response of nanocellulose from wood and green algae. *RSC Advances*, *4*(6), pp.2892–2903.

Huang, A., Li, X., Liang, X., Zhang, Y., Hu, H., Yin, Y. and Huang, Z., 2018. Solid-phase synthesis of cellulose acetate butyrate as microsphere wall materials for sustained release of emamectin benzoate. *Polymers*, *10*(12), p.1381.

Ingle, A.P., Duran, N. and Rai, M., 2014. Bioactivity, mechanism of action, and cytotoxicity of copper-based nanoparticles: A review. *Applied Microbiology and Biotechnology*, *98*, pp.1001–1009.

Jiang, F., Han, S. and Hsieh, Y.L., 2013. Controlled defibrillation of rice straw cellulose and self-assembly of cellulose nanofibrils into highly crystalline fibrous materials. *RSC Advances*, *3*(30), pp.12366–12375.

Jorfi, M. and Foster, E.J., 2015. Recent advances in nanocellulose for biomedical applications. *Journal of Applied Polymer Science*, *132*(14) 41719.

Joshy, K.S., Snigdha, S., George, A., Kalarikkal, N., Pothen, L.A. and Thomas, S., 2017. Core–shell nanoparticles of carboxy methyl cellulose and compritol-PEG for antiretroviral drug delivery. *Cellulose*, *24*, pp.4759–4771.

Khalil, H.A., Davoudpour, Y., Saurabh, C.K., Hossain, M.S., Adnan, A.S., Dungani, R., Paridah, M.T., Sarker, M.Z.I., Fazita, M.N., Syakir, M.I. and Haafiz, M.K.M., 2016. A review on nanocellulosic fibres as new material for sustainable packaging: Process and applications. *Renewable and Sustainable Energy Reviews*, *64*, pp.823–836.

Khan, A., Huq, T., Khan, R.A., Riedl, B. and Lacroix, M., 2014. Nanocellulose-based composites and bioactive agents for food packaging. *Critical Reviews in Food Science and Nutrition*, *54*(2), pp.163–174.

Kim, Y., Park, J., Bang, J., Kim, J., Jin, H.J. and Kwak, H.W., 2022. Highly efficient Cr (VI) remediation by cationic functionalized nanocellulose beads. *Journal of Hazardous Materials*, *426*, p.128078.

Klemm, D., Cranston, E.D., Fischer, D., Gama, M., Kedzior, S.A., Kralisch, D., Kramer, F., Kondo, T., Lindström, T., Nietzsche, S. and Petzold-Welcke, K., 2018. Nanocellulose as a natural source for groundbreaking applications in materials science. Today's state. *Materials Today*, *21*(7), pp.720–748.

Klemm, D., Kramer, F., Moritz, S., Lindström, T., Ankerfors, M., Gray, D. and Dorris, A., 2011. Nanocelluloses: A new family of nature-based materials. *Angewandte Chemie International Edition*, *50*(24), pp.5438–5466.

Kostag, M., Jedvert, K., Achtel, C., Heinze, T. and El Seoud, O.A., 2018. Recent advances in solvents for the dissolution, shaping and derivatization of cellulose: Quaternary ammonium electrolytes and their solutions in water and molecular solvents. *Molecules*, *23*(3), p.511.

Kumari, S., Ram, B., Kumar, D., Ranote, S. and Chauhan, G.S., 2018. Nanoparticles of oxidized-cellulose synthesized by green method. *Materials Science for Energy Technologies*, *1*(1), pp.22–28.

Lee, H.B., Yoon, S.Y., Singh, B., Oh, S.H., Cui, L., Yan, C., Kang, S.K., Choi, Y.J. and Cho, C.S., 2018. Oral immunization of FMDV vaccine using pH-sensitive and mucoadhesive thiolated cellulose acetate phthalate microparticles. *Tissue Engineering and Regenerative Medicine*, *15*, pp.1–11.

Li, Y.M., Jiang, T., Lv, Y., Wu, Y., He, F. and Zhuo, R.X., 2015. Amphiphilic copolymers with pendent carboxyl groups for high-efficiency loading and controlled release of doxorubicin. *Colloids and Surfaces B: Biointerfaces*, *132*, pp.54–61.

Lin, N. and Dufresne, A., 2014. Nanocellulose in biomedicine: Current status and future prospect. *European Polymer Journal*, *59*, 302–325.

Liu, D., Dong, Y. and Sui, G., 2016. Isolation of cellulose nanowhiskers and their nano composites. In: Stoyko Fakirov (ed) *Nano-Size Polymers: Preparation, Properties, Applications* (pp. 155–177). Springer Cham, Switzerland.

Liu, G., Lin, G., Lin, X., Zhou, H., Chen, H., Hao, L. and Zhou, X., 2019. Enzyme and pH dual-responsive avermectin nano-microcapsules for improving its efficacy. *Environmental Science and Pollution Research*, *26*, pp.25107–25116.

Liu, Y., Ahmed, S., Sameen, D.E., Wang, Y., Lu, R., Dai, J., Li, S. and Qin, W., 2021. A review of cellulose and its derivatives in biopolymer-based for food packaging application. *Trends in Food Science & Technology*, *112*, pp.532–546.

Liu, Z., Jiao, Y., Wang, Y., Zhou, C. and Zhang, Z., 2008. Polysaccharides-based nanoparticles as drug delivery systems. *Advanced Drug Delivery Reviews*, *60*(15), pp.1650–1662.

Liu, D., Yuan, X. and Bhattacharyya, D., 2012. The effects of cellulose nanowhiskers on electrospun poly (lactic acid) nanofibres. *Journal of Materials Science*, *47*(7), pp.3159–3165.

Lu, A., Petit, E., Jelonek, K., Orchel, A., Kasperczyk, J., Wang, Y., Su, F. and Li, S., 2020. Self-assembled micelles prepared from bio-based hydroxypropyl methyl cellulose and polylactide amphiphilic block copolymers for anti-tumor drug release. *International Journal of Biological Macromolecules*, *154*, pp.39–47.

Lu, P. and Hsieh, Y.L., 2010. Preparation and properties of cellulose nanocrystals: rods, spheres, and network. *Carbohydrate polymers, 82*(2), pp. 329–336.

Luo, X. and Zhang, L., 2009. High effective adsorption of organic dyes on magnetic cellulose beads entrapping activated carbon. *Journal of Hazardous Materials*, *171*(1–3), pp.340–347.

Moon, R.J., Martini, A., Nairn, J., Simonsen, J. and Youngblood, J., 2011. Cellulose nanomaterials review: Structure, properties and nanocomposites. *Chemical Society Reviews*, *40*(7), pp.3941–3994.

Moriana, R., Vilaplana, F. and Ek, M., 2016. Cellulose nanocrystals from forest residues as reinforcing agents for composites: A study from macro-to nano-dimensions. *Carbohydrate Polymers*, *139*, pp.139–149.
Nascimento, D.M., Nunes, Y.L., Figueirêdo, M.C., de Azeredo, H.M., Aouada, F.A., Feitosa, J.P., Rosa, M.F. and Dufresne, A., 2018. Nanocellulose nanocomposite hydrogels: Technological and environmental issues. *Green Chemistry*, *20*(11), pp.2428–2448.
Nath, B., Nath, L.K., Mazumder, B., Kumar, P., Sharma, N. and Sahu, B.P., 2010. Preparation and characterization of salbutamol sulphate loaded ethyl cellulose microspheres using water-in-oil-oil emulsion technique. *Iranian Journal of Pharmaceutical research: IJPR*, *9*(2), p.97.
National Fire Protection Association (NFPA), 2007. *NFPA 68: Standard on Explosion Protection by Deflagration Venting*. NationalFireProtectionAssoc.
Nechyporchuk, O., Belgacem, M.N. and Bras, J., 2016. Production of cellulose nanofibrils: A review of recent advances. *Industrial Crops and Products*, *93*, pp.2–25.
Nethaji, R., Shanub, S., Manikandan, P., Surendiran, N.S. and Babu, G., 2016. Preparation and characterization of ethyl cellulose microspheres containing diclofenac sodium. *International Journal of Research in Pharmaceutical and Nano Sciences*, *5*, pp.224–234.
Oun, A.A. and Rhim, J.W., 2016. Characterization of nanocelluloses isolated from Ushar (Calotropis procera) seed fiber: Effect of isolation method. *Materials Letters*, *168*, pp.146–150.
Payen, A., 1838. Mémoire sur la composition du tissu propre des plantes et du ligneux. *Comptes Rendus de l'Académie des Sciences*, *7*, 1052–1056.
Peri, S., Muthukumar, L., Nazmul Karim, M. and Khare, R., 2012. Dynamics of cello-oligosaccharides on a cellulose crystal surface. *Cellulose*, *19*(6), pp.1791–1806.
Perumal, A.B., Nambiar, R.B., Sellamuthu, P.S. and Sadiku, E.R., 2019. Application of biosynthesized nanoparticles in food, food packaging and dairy industries. In *Biological Synthesis of Nanoparticles and Their Applications* (pp. 145–158). CRC Press.
Phanthong, P., Reubroycharoen, P., Hao, X., Xu, G., Abudula, A., Guan, G., 2018. Nanocellulose: Extraction and application. *Carbon Resources Conversion*, *1*, 32–43.
Pickering, K.L., Efendy, M.A. and Le, T.M., 2016. A review of recent developments in natural fibre composites and their mechanical performance. *Composites Part A: Applied Science and Manufacturing*, *83*, pp.98–112.
Pitkänen, M., Honkalampi, U., Von Wright, A., Sneck, A., Hentze, H.P., Sievänen, J., Hiltunen, J. and Hellen, E., 2010. Nanofibrillar cellulose: In vitro study of cytotoxic and genotoxic properties. In *2010 TAPPI International Conference on Nanotechnology for the Forest Product Industry* (pp. 246–261). TAPPI Press.
Polat, T.G. and Topel, S.D., 2019. pH-responsive carboxymethyl cellulose conjugated superparamagnetic iron oxide nanocarriers. *Journal of Scientific Perspectives*, *3*(2), pp.99–110.
Pooresmaeil, M., Javanbakht, S., Nia, S.B. and Namazi, H., 2020. Carboxymethyl cellulose/mesoporous magnetic graphene oxide as a safe and sustained ibuprofen delivery bio-system: Synthesis, characterization, and study of drug release kinetic. *Colloids and Surfaces A: Physicochemical and Engineering Aspects*, *594*, p.124662.
Qian, M., Lei, H., Villota, E., Zhao, Y., Wang, C., Huo, E., Zhang, Q., Mateo, W. and Lin, X., 2021. High yield production of nanocrystalline cellulose by microwave-assisted dilute-acid pretreatment combined with enzymatic hydrolysis. *Chemical Engineering and Processing-Process Intensification*, *160*, p.108292.
Quintana, S.E., Llalla, O., García-Risco, M.R. and Fornari, T., 2021. Comparison between essential oils and supercritical extracts into chitosan-based edible coatings on strawberry quality during cold storage. *The Journal of Supercritical Fluids*, *171*, p.105198.

Rafique, M., Sadaf, I., Rafique, M.S. and Tahir, M.B., 2017. A review on green synthesis of silver nanoparticles and their applications. *Artificial Cells, Nanomedicine, and Biotechnology*, *45*(7), pp.1272–1291.
Rånby, B.G., 1951. Fibrous macromolecular systems. Cellulose and muscle. The colloidal properties of cellulose micelles. *Discussions of the Faraday Society*, *11*, pp.158–164.
Reiniati, I., Hrymak, A.N. and Margaritis, A., 2017. Recent developments in the production and applications of bacterial cellulose fibers and nanocrystals. *Critical Reviews in Biotechnology*, *37*(4), pp.510–524.
Roman, M., 2015. Toxicity of cellulose nanocrystals. A review. *Industrial Biotechnology*, *11*(1), pp.25–33.
Rouhiainen, J., Väänänen, V., Tsitko, I. and Kautto, J., 2012, June. Risk assessment of nanofibrillated cellulose in occupational settings. In *SUNPAP Final Conference*. Milan.
Rovera, C., Türe, H., Hedenqvist, M.S. and Farris, S., 2020. Water vapor barrier properties of wheat gluten/silica hybrid coatings on paperboard for food packaging applications. *Food Packaging and Shelf Life*, *26*, p.100561.
Sacui, I.A., Nieuwendaal, R.C., Burnett, D.J., Stranick, S.J., Jorfi, M., Weder, C., Foster, E.J., Olsson, R.T. and Gilman, J.W., 2014. Comparison of the properties of cellulose nanocrystals and cellulose nanofibrils isolated from bacteria, tunicate, and wood processed using acid, enzymatic, mechanical, and oxidative methods. *ACS Applied Materials & Interfaces*, *6*(9), pp.6127–6138.
Salar, R.K., Sharma, P. and Kumar, N., 2015. Enhanced antibacterial activity of streptomycin against some human pathogens using green synthesized silver nanoparticles. *Resource-Efficient Technologies*, *1*(2), pp.106–115.
Serial, M.R., Velichko, E., Nikolaeva, T., den Adel, R., Terenzi, C., Bouwman, W.G. and van Duynhoven, J.P., 2021. High-pressure homogenized citrus fiber cellulose dispersions: Structural characterization and flow behavior. *Food Structure*, *30*, p.100237.
Shah, S.S., Shaikh, M.N., Khan, M.Y., Alfasane, M.A., Rahman, M.M. and Aziz, M.A., 2021. Present status and future prospects of jute in nanotechnology: A review. *The Chemical Record*, *21*(7), pp.1631–1665.
Shamshina, J.L., Acharya, S., Rumi, S.S., Liyanage, S., Parajuli, P. and Abidi, N., 2022. Cryogenic grinding of cotton fiber cellulose: The effect on physicochemical properties. *Carbohydrate Polymers*, *289*, p.119408.
Shapira-Zaltsberg, G., Grynspan, D., Quintana, M.V., Dominguez, P.C., Reddy, D., Davila, J.H. and Miller, E., 2018. MRI features of the placenta in fetuses with and without CNS abnormalities. *Clinical Radiology*, *73*(9), pp.836–e9.
Shankar, S. and Rhim, J.W., 2018. Antimicrobial wrapping paper coated with a ternary blend of carbohydrates (alginate, carboxymethyl cellulose, carrageenan) and grapefruit seed extract. *Carbohydrate Polymers*, *196*, pp.92–101.
Sharma, C. and Bhardwaj, N.K., 2020. Fabrication of natural-origin antibacterial nanocellulose films using bio-extracts for potential use in biomedical industry. *International Journal of Biological Macromolecules*, *145*, pp.914–925.
Shatkin, J.A., T.H. Wegner, E.M. (Ted), 2014. Bilek and J. Cowie, *Tappi Journal*, *13*, 9–16.
Sirviö, J.A. and Visanko, M., 2017. Anionic wood nanofibers produced from unbleached mechanical pulp by highly efficient chemical modification. *Journal of Materials Chemistry A*, *5*(41), pp.21828–21835.
Spiliopoulos, P., Spirk, S., Pääkkönen, T., Viljanen, M., Svedström, K., Pitkänen, L., Awais, M. and Kontturi, E., 2021. Visualizing degradation of cellulose nanofibers by acid hydrolysis. *Biomacromolecules*, *22*(4), pp.1399–1405.
Suleman, R., Amjad, A., Ismail, A., Javed, S., Ghafoor, U. and Fahad, S., 2022. Impact of plastic bags usage in food commodities: An irreversible loss to environment. *Environmental Science and Pollution Research*, *29*(33), pp.49483–49489.

Syafiuddin, A., 2019. Toward a comprehensive understanding of textiles functionalized with silver nanoparticles. *Journal of the Chinese Chemical Society*, *66*(8), pp.793–814.

Trache, D., Tarchoun, A.F., Derradji, M., Hamidon, T.S., Masruchin, N., Brosse, N. and Hussin, M.H., 2020. Nanocellulose: from fundamentals to advanced applications. *Frontiers in Chemistry*, *8*, p.392.

Turbak, A.F., Snyder, F.W. and Sandberg, K.R., 1983, January. Microfibrillated cellulose, a new cellulose product: Properties, uses, and commercial potential. *Journal of Applied Polymer Science Applied Polymer Symposium*, 37 (9), pp. 815–827.

Vartiainen, J., Pöhler, T., Sirola, K., Pylkkänen, L., Alenius, H., Hokkinen, J., Tapper, U., Lahtinen, P., Kapanen, A., Putkisto, K. and Hiekkataipale, P., 2011. Health and environmental safety aspects of friction grinding and spray drying of microfibrillated cellulose. *Cellulose*, *18*, pp.775–786.

Velásquez-Cock, J., Serpa, A., Vélez, L., Gañán, P., Hoyos, C.G., Castro, C., Duizer, L., Goff, H.D. and Zuluaga, R., 2019. Influence of cellulose nanofibrils on the structural elements of ice cream. *Food Hydrocolloids*, *87*, pp.204–213.

Velmurugan, P., Sivakumar, S., Young-Chae, S., Seong-Ho, J., Pyoung-In, Y., Jeong-Min, S. and Sung-Chul, H., 2015. Synthesis and characterization comparison of peanut shell extract silver nanoparticles with commercial silver nanoparticles and their antifungal activity. *Journal of Industrial and Engineering Chemistry*, *31*, pp.51–54.

Vidal-Romero, G., Zambrano-Zaragoza, M.L., Martínez-Acevedo, L., Leyva-Gómez, G., Mendoza-Elvira, S.E. and Quintanar-Guerrero, D., 2020. Design and evaluation of pH-dependent nanosystems based on cellulose acetate phthalate, nanoparticles loaded with chlorhexidine for periodontal treatment. *Pharmaceutics*, *11*(11), p.604.

Vilela, C., Kurek, M., Hayouka, Z., Röcker, B., Yildirim, S., Antunes, M. D. C., Nilsen-Nygaard, J., Pettersen, M. K. and Freire, C. S. R. (2018). A concise guide to active agents for active food packaging. *Trends in Food Science & Technology*, *80*, 212–222.

Wang, F.J., Lu, F.S., Cui, M. and Shao, Z.Q., 2015. Biocompatible microcapsule of carboxymethyl cellulose/chitosan as drug carrier. *Advanced Materials Research*, *1118*, pp.227–236.

Wei, H., Rodriguez, K., Renneckar, S. and Vikesland, P.J., 2014. Environmental science and engineering applications of nanocellulose-based nanocomposites. *Environmental Science: Nano*, *1*(4), pp.302–316.

Winuprasith, T. and Suphantharika, M., 2015. Properties and stability of oil-in-water emulsions stabilized by microfibrillated cellulose from mangosteen rind. *Food Hydrocolloids*, *43*, pp.690–699.

Wsoo, M.A., Shahir, S., Bohari, S.P.M., Nayan, N.H.M. and Abd Razak, S.I., 2020. A review on the properties of electrospun cellulose acetate and its application in drug delivery systems: A new perspective. *Carbohydrate Research*, *491*, p.107978.

Yadav, M., Rhee, K.Y., Jung, I.H. and Park, S.J., 2013. Eco-friendly synthesis, characterization and properties of a sodium carboxymethyl cellulose/graphene oxide nanocomposite film. *Cellulose*, *20*, pp.687–698.

Yang, X., Jiang, X., Yang, H., Bian, L., Chang, C. and Zhang, L., 2020. Biocompatible cellulose-based supramolecular nanoparticles driven by host–guest interactions for drug delivery. *Carbohydrate Polymers*, *237*, p.116114.

Yano, H., Abe, K., Kase, Y., Kikkawa, S. and Onishi, Y., 2012. US 20140342075 A1. *Washington, DC: US Patent Application No*, *14*(344,158).

Yeap, E.W., Acevedo, A.J. and Khan, S.A., 2019. Microfluidic extractive crystallization for spherical drug/drug-excipient microparticle production. *Organic Process Research & Development*, *23*(3), pp.375–381.

Yu, Z., Alsammarraie, F.K., Nayigiziki, F.X., Wang, W., Vardhanabhuti, B., Mustapha, A. and Lin, M., 2017. Effect and mechanism of cellulose nanofibrils on the active functions of biopolymer-based nanocomposite films. *Food Research International*, *99*, pp.166–172.

Zanchetta, G., Rocchi, E. and Piazza, L., 2017. Seeing is believing: Coupling between liquid crystalline ordering and rheological behavior in cellulose nanocrystals suspensions. *Chemical Engineering Transactions*, *57*, pp.1933–1938.

Zhang, F., Wu, W., Zhang, X., Meng, X., Tong, G. and Deng, Y., 2016. Temperature-sensitive poly-NIPAm modified cellulose nanofibril cryogel microspheres for controlled drug release. *Cellulose*, *23*, pp.415–425.

Zhou, W., Wu, Z., Xie, F., Tang, S., Fang, J. and Wang, X., 2021. 3D printed nanocellulose-based label for fruit freshness keeping and visual monitoring. *Carbohydrate Polymers*, *273*, p.118545.

8 Food Applications of Nano-Spherical Cellulose

Kalai Rajan Poomalaisundar, Sridevi Sankar, Ahamed Ibrahim Syed Maghdoom, Kavitha Duraisamy, Nithishkumar Sengottaiyan, and Ruby Celsia Arul Selvaraj

8.1 INTRODUCTION TO NANO-SPHERICAL CELLULOSE

As a result of its role in propelling the creation of high-tech materials with improved characteristics, nanotechnology has recently arisen as a revolutionary force across several sectors. Nanocellulose stands out among the others because of its many uses, low environmental impact, and adaptability. In the field of nanocellulose, nano-spherical cellulose is rising to the top as an innovative material with unique physical and chemical characteristics and a spherical form. To meet the growing need for environmentally benign and highly functional materials, nano-spherical cellulose offers a novel alternative to conventional cellulose fibers. Because of its non-toxicity and biocompatibility, nano-spherical cellulose is highly suitable for use in biomedical fields. Targeted drug-delivery systems benefit from its spherical shape since it permits homogeneous drug loading and controlled release. Its versatility in tissue engineering, wound healing, and diagnostic platforms is further enhanced by its large surface area and simplicity of surface modification, which allow the attachment of bioactive compounds [1].

The ability to chemically functionalize nano-spherical cellulose and its large surface area make it an excellent adsorbent for heavy metals, dyes, and other contaminants. Because of this, it shows promise as a sustainable and environmentally benign substitute for conventional adsorbents in wastewater treatment and environmental remediation. There is hope for nano-spherical cellulose as a novel food packaging material in the food sector. Because it acts as a barrier, keeping out oxygen and moisture, it can make food last longer on store shelves. It also helps keep perishable foods from going bad, which means less food waste, thanks to its antibacterial characteristics. There are new possibilities for flexible electronics made possible by nano-spherical cellulose's exceptional crystallinity and mechanical strength. Its incorporation into sensors, conductive films, and other electrical devices can improve performance while reducing environmental impact because it is both lightweight and renewable. The biodegradable and biocompatible properties of nano-spherical cellulose make it a versatile ingredient in personal care products, serving as a thickener, stabilizer, or carrier. It satisfies the growing demand for environmentally friendly products by providing a safe and natural substitute for synthetic materials [2].

DOI: 10.1201/9781003596127-9

The long-term viability of nano-spherical cellulose is one of its main selling points. It lessens reliance on materials derived from petroleum because it is a renewable resource and a derivative of natural cellulose. Its ability to break down naturally means it won't harm the environment like other plastics. In addition, its adaptability to various functions makes it an attractive option for several uses in fields as varied as biology, packaging, electronics, and water purification.

In the long run, nano-spherical cellulose has a lot of potential. Scientists are looking for new synthesis methods to improve its size, shape, and surface characteristics, which will increase its use. Also, improvements in mass production will probably bring prices down and make it easier for businesses to use it. With the increasing concern for the environment, nano-spherical cellulose has the potential to play a pivotal role in creating sustainable, high-performance materials. Nano-spherical cellulose is a cutting-edge nanomaterial that is extremely functional, adaptable, and environmentally benign. Numerous prospects arise in fields including medicine, ecology, technology, and packaging as a result of its singular spherical shape and the intrinsic characteristics of cellulose. The significance of nano-spherical cellulose in the quest for creative and sustainable solutions is being further underscored by the expanding scope of possible new applications and commercial feasibility as a result of ongoing research and development [3].

The use of nano-spherical cellulose represents a leap toward a greener, more technologically sophisticated future at a time when both renewable and high-performance materials are in great demand. Nanomaterials derived from biomacromolecules found in nature have recently garnered a lot of attention. The tremendous structural alteration capability of cellulose makes it an ideal material for use in nanomaterial technologies. Cellulose nanofibrils (CNFs), cellulose nanocrystals (CNCs), and cellulose nanospheres (CNSs) are the three main types of nanocellulose that may be distinguished by their characteristics, size, and aspect ratio (length/diameter). Mechanical delamination from native cellulose is the usual method for producing CNFs, which generally have lengths of 500–2000 nm and widths of 20–50 nm. Because of their environmentally friendly preparation, CNFs produced by ultrasonication, milling, microfluidization, high-pressure homogenization, and freeze-disintegration often have desirable qualities, including biocompatibility and biodegradability. CNCs are structures that resemble needles and have geometrical dimensions that vary between 100 and 500 nm in length and 5–50 nm in diameter. They are created by removing amorphous regions from native cellulose through strong acid hydrolysis or an oxidant, leading to the formation of extremely rigid and pure crystalline structures. The spherical morphologies of CNSs may be achieved using either the bottom-up approach of cellulose dissolution and regeneration, which reveals the crystalline form of cellulose II, or the top-down methods of chemical treatment, mechanical disintegration, and enzymatic hydrolysis [4].

The qualities and uses of nanocelluloses might vary greatly due to their various dimensions and morphologies. In the paper and polymer industries, CNFs are extensively employed to create gas barriers; in the biomedical and packaging sectors, CNCs are reinforcing agents; and in the magnetoelectric and Pickering emulsion stabilization fields, CNSs are prevalent. Various morphologies of nanocelluloses have recently been the subject of papers detailing their synthesis, characterization, and potential applications. Nanocelluloses of different morphologies produced from the same native cellulose materials have, however, received little attention in the literature. Furthermore, there has been a dearth of research comparing their morphologies and exploring the possible applications of each. There have been several technological advancements in the past few years that aim to substitute saturated and trans fats in meals. Because of its possible uses in the food and pharmaceutical sectors, the development of oleogel is one such technology that garners a lot of interest. Oleogels are semi-solid materials that maintain the physicochemical properties of the trapped oil by enclosing it in a three-dimensional network. Materials that form network structures to retain oils are known as oleogelators. Common compounds utilized as oleogelators range from those with a high molecular weight, such as proteins and polysaccharides, to those with a low molecular weight, including waxes, monoglycerides, phytosterols, ceramides, and lecithin. Depending on the characteristics of the oleogelators, oleogels can be made either directly or indirectly using vegetable oil and an oleogelator of choice. The choice of oleogelator is dependent on the desired oleogels' characteristics. Approaches that include solvent exchange, emulsion-template, and foam-template are examples of indirect approaches [5].

One indirect approach that shows promise for the creation of oleogels employing food-approved polymers, such as polysaccharides and proteins, is the

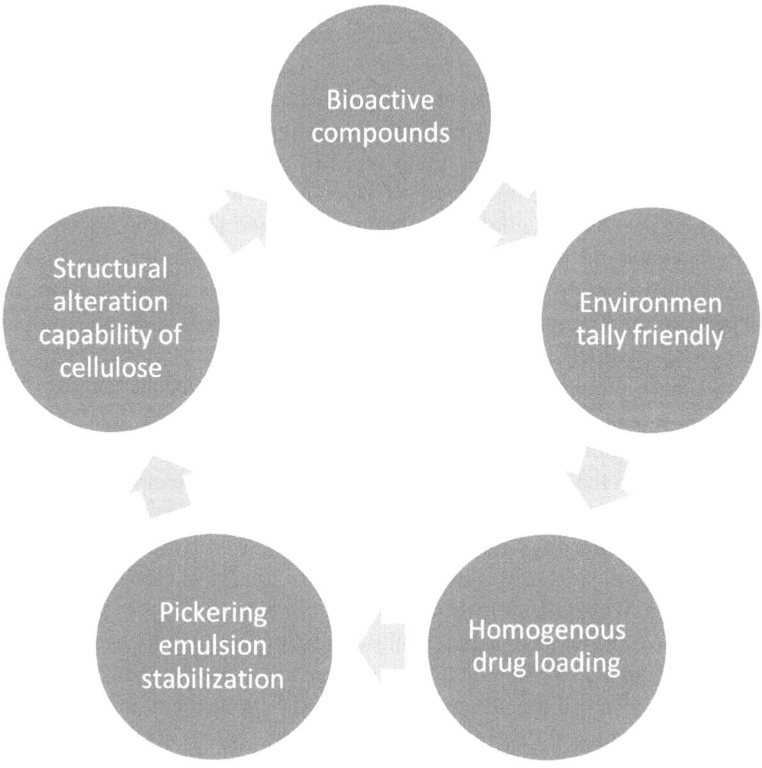

FIGURE 8.1 Introduction to nano-spherical cellulose.

emulsion-templated technique. But because of their hydrophilic nature, these polymers can't be distributed in oil to create the oleogelation structure that's required. With this method, you start by making an oil-in-water emulsion with the right polymer (Figure 8.1).

8.2 PHYSICOCHEMICAL PROPERTIES OF NANO-SPHERICAL CELLULOSE

One interesting type of nanocellulose material is nano-spherical cellulose, which has a distinctive spherical shape. This new kind of cellulose nanoparticle is spherical in shape and has nanoscale dimensions; therefore, it has improved physicochemical characteristics while retaining many of cellulose's fundamental attributes. Included in these characteristics are a large surface area, crystallinity, mechanical strength, and the capacity to undergo chemical modification for specific uses. One defining feature of nano-spherical cellulose is its consistently round form. The normal range for particle size, depending on the production technique, is a few hundred nanometers to a few thousand nanometers. In drug delivery or adsorption procedures, for example, a high surface-to-volume ratio made possible by this morphology is essential for effective interaction with other materials. In addition to facilitating effective usage and minimizing agglomeration, the spherical form improves dispersion stability in both water- and non-water-based environments. Its enhanced surface area is one of nano-spherical cellulose's most striking features [6].

The spherical nanoparticles have a much larger surface area because of their geometric form and decreased size compared with microcrystalline cellulose or conventional cellulose fibers. Particle reactivity, surface functionalization, and interaction with other substances are all improved by this increased surface area. You may manage the synthesis process to vary the porosity of nano-spherical cellulose, which affects its permeability and adsorption capability in applications like catalysis and filtering. Excellent mechanical strength and stiffness are displayed by nano-spherical cellulose particles, despite their small size. This strong mechanical feature is a result of cellulose's great crystallinity and strong intermolecular hydrogen bonding. Applications requiring structural integrity and stability, including the manufacturing of composites or reinforcing materials, rely heavily on this strength [7].

Additionally, by incorporating polymers or other reinforcing agents into nano-spherical cellulose, the mechanical characteristics can be enhanced even further. The reactivity and adaptability of nano-spherical cellulose are determined in large part by its surface chemistry. Hydroxyl groups abound on the nanoparticles' surfaces, making them ideal building blocks for various chemical reactions. Hydrophobicity, hydrophilicity, and binding affinity are just a few of the qualities that may be improved or added to by functionalizing these hydroxyl groups. From medication delivery systems and biosensors to environmental cleanup, nano-spherical cellulose may be customized for specific purposes by adjusting the surface chemistry. Because it is derived from natural cellulose, nano-spherical cellulose is biocompatible and biodegradable. Because of these qualities, it is a sustainable material that may be used in biomedicine, packaging, and other green fields. Its lack of toxicity makes it suitable for use in biomedical devices such as scaffolds for tissue engineering and medication delivery systems. Biomedicine, environmental science, packaging, and innovative materials are just a few of the many areas that can benefit from nano-spherical cellulose's physicochemical features, which include a uniform morphology, high crystallinity, big surface area, and variable surface chemistry [8].

A further selling point as a sustainable material across sectors is its intrinsic biodegradability and biocompatibility. Because of these characteristics, nano-spherical cellulose is a material that researchers and businesses are very interested in using to build new, environmentally friendly products. Academic and industry scientists are showing a growing interest in electrospinning since it is a novel technique for separating membranes. The electro spun membrane separation materials have a micro-nano structure, high porosity, specific surface area, and flexibility. Moreover, by controlling the parameters of the spinning process, one may create micro-nano materials with a variety of architectures, such as fibers with imitation lotus leaves, spider-like, pea-like, porous, tube, necklace-like, and other shapes. The outstanding hydrophobicity, resistance to abrasion, chemicals, and fouling of polyvinylidene fluorides frequently utilized raw material for membrane separation has attracted attention. Nevertheless, issues persist, including inadequate mechanical qualities and poor separation performance in real-world scenarios. Recent studies have mostly concentrated on methods to modify membrane surface hydrophobicity, such as adding hydrophobic inorganic nanoparticles to spinning solutions or varying the solution's polymer content and solvent ratio [9].

The water crisis is one of the most critical global issues of our day, and nanotechnology, a new and promising area of study, has the ability to solve it. Conventional approaches to water purification are becoming increasingly ineffective as the global population continues to rise and the effects of climate change become more apparent. Advanced water treatment methods utilizing functionalized nanomaterials have been investigated, as have filtration setups based on nanoparticles for the removal of various contaminants from water. Additionally, there has been talk on how to guarantee the delivery of clean water by using nanomaterials with photocatalytic or antimicrobial capabilities to neutralize pollutants. In this chapter, nano water discusses the problem of freshwater scarcity and offers suggestions for desalination methods that are safe and won't harm the environment. Nano water highlights how nanotechnology has the ability to transform water treatment by making it more eco-friendly, efficient, and effective. To help academics and businesses find a solution to the world water problem, this chapter also covers the ethical considerations and problems that can arise from using nanotechnology in water treatment on a large scale [10] (Figure 8.2).

8.3 ROLE OF NANO-SPHERICAL CELLULOSE AS A FOOD STABILIZER

Sustainable and environmentally friendly food additives are all the rage; therefore, scientists are looking for new ways to use ingredients made from renewable sources. An intriguing example of this type of material is nano-spherical cellulose, a nanostructured version of cellulose with exceptional physicochemical qualities that make it an excellent food stabilizer. Meeting consumer expectations for clean-label

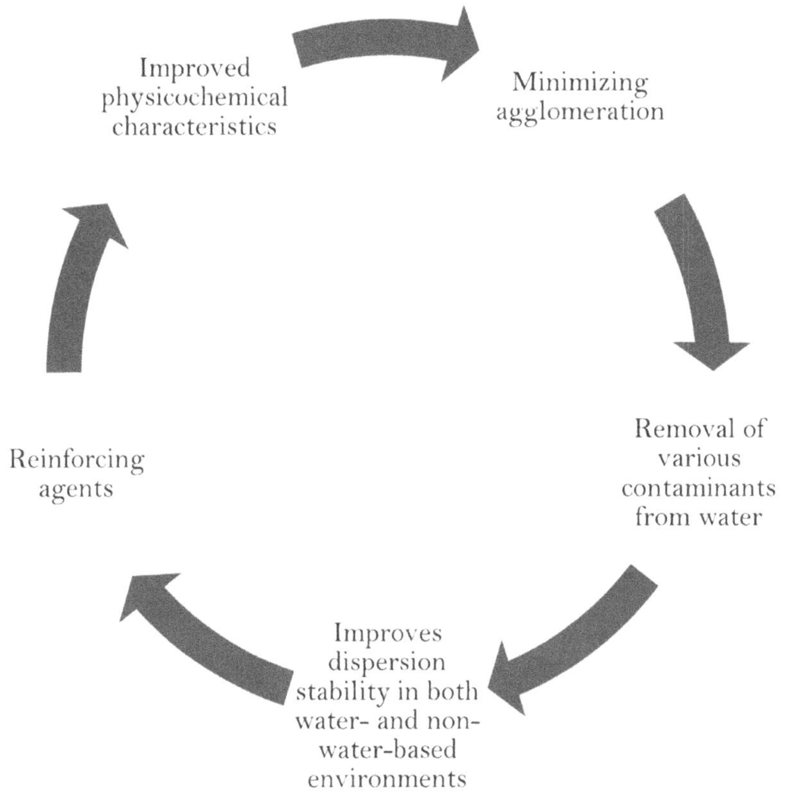

FIGURE 8.2 Physicochemical properties of nano-spherical cellulose.

ingredients, nano-spherical cellulose, a naturally occurring polymer, offers a number of benefits to food items, such as increased stability, texture, and shelf life. The gelling ability, large surface area, and exceptional water retention capacity of nano-spherical cellulose make it an ideal stabilizer for use in food items. Particles of a spherical shape are less likely to aggregate and experience phase separation in water, making them more stable and easier to disperse [11].

A three-dimensional network is formed when water molecules mix with nano-spherical cellulose. Desserts, soups, sauces, and dressings all benefit from the enhanced viscosity and consistency that results from this interaction, which forms a gel-like structure. Its capacity to establish a solid network improves the control over food flow and texture, which in turn improves the product's mouthfeel and overall quality. Mayonnaise, salad dressings, and dairy substitutes are emulsion-based foods that effectively employ nano-spherical cellulose as a stabilizer. Adsorption at the oil–water interface is possible due to its large surface area, which lowers interfacial tension and stops oil droplets from coalescing. The end product has a longer shelf life and looks better because of the stable emulsion that is created. Some foods, such as fruit pulp in juices or spices in sauces, include insoluble particles that tend to settle over time. Distributing these particles uniformly over the product matrix, nano-spherical cellulose aids in their suspension. This quality is due to its thixotropic nature and gel-forming ability, which keep the suspended particles evenly dispersed even when stored. With its many practical uses in texture, emulsion stability, water retention, and fat reduction, nano-spherical cellulose is quickly becoming a popular food stabilizer [12].

It is a sustainable and natural option for food makers that want to satisfy health-conscious customers because of its unique physicochemical qualities, which include a large surface area, great water-binding capacity, and biocompatibility. As a stabilizing and structuring agent, nano-spherical cellulose shows promise for broad use in the food sector due to its ability to improve product quality, extend shelf life, and conform to clean-label trends. On a global scale, nanotechnology has become an important and effective scientific tool. Numerous fields have been profoundly impacted by the advancements in materials science, including bioscience, food, medicinal, agricultural, cosmetics, and environmental processes. Understanding the variations in the chemo-physical characteristics of nano-sized composites is greatly aided by this approach. Silver nanoparticles stand out among other metal nanoparticles for a number of intriguing reasons, such as their antibacterial and antioxidant activities, high stability, low cost, large specific surface area, impressive anti-cancer efficacy, and catalytic activity [13].

Metal poisoning is a condition characterized by the buildup of harmful heavy metals in living organisms' cells; the symptoms of this condition are associated with specific metals. At parts per billion levels, the body can't operate properly without several heavy metals. But these poisons can cause immediate harm to living cells if they reach toxic levels. In terms of the harmful impact on human health, the most prevalent heavy metals are lead, mercury, arsenic, and cadmium. One nonbiodegradable and carcinogenic hazard is from mercury ions, which are extremely poisonous and found all throughout the environment. Improved solubility, increased in vivo circulation time of free drugs, crossing biological barriers, high disease target site selectivity, space/time control release, combination therapies that overcome drug resistance, and novel imaging techniques that benefit from disease and material property-oriented design and developments of materials are just a few of the remarkable and inspiring properties of nano-therapy that have contributed to its recent success in disease treatment [14]. Over time, biodegradable polymers have piqued the interest of researchers in their pursuit of important nano-devices for biomedical applications and the eventual triumph over the myriad formidable obstacles that have long stood in the way of effective disease therapies. Chitosan is an essential component of the natural polysaccharide macromolecule, which is a popular material for nanomedicine. Because of its controllable and active amino groups, chitosan is a naturally occurring cationic polysaccharide. It is produced by deacetylating chitin, which is mainly found in crab and insect shells and in the cell walls of certain fungi. The natural world is teeming with chitosan, a byproduct of chitin breakdown by acid hydrolysis, enzymatic degradation, or both. The biodegradable structure and inherent safety of chitosan make it an advantageous polymer material. The U.S. Food and Drug Administration has approved chitosan as a food additive, and the fact that it is a natural product makes it even safer than many other polymer materials [15] (Figure 8.3).

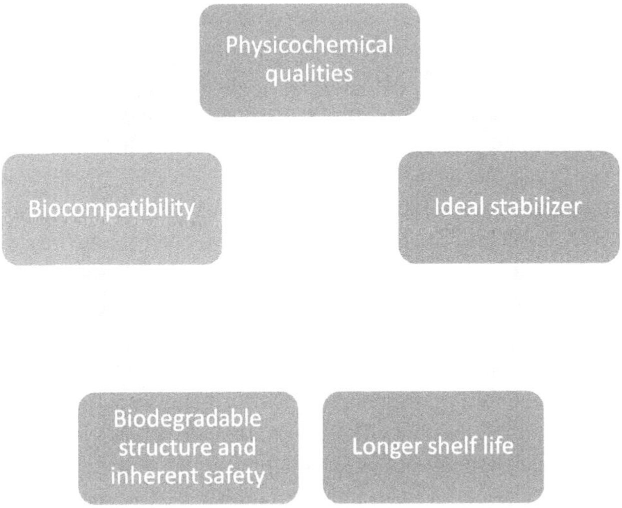

FIGURE 8.3 Role of nano-spherical cellulose as a food stabilizer.

8.4 EMULSIFICATION PROPERTIES OF NANO-SPHERICAL CELLULOSE

Because of its exceptional emulsifying properties, nano-spherical cellulose has recently gained attention as a potentially useful natural material. A lot of people are interested in how nano-spherical cellulose stabilizes emulsions because of the trend toward sustainable and environmentally friendly chemicals in industry. Oil and water are two examples of immiscible liquids that can be stabilized by the process of emulsification. For a long time, emulsifiers have been used to keep two liquids from separating by lowering the interfacial tension. Nano-spherical cellulose has the ability to revolutionize the process of creating stable and effective emulsions due to its unique nanoscale size and spherical shape, which provide several benefits over traditional emulsifiers [16].

A key component of nano-spherical cellulose's emulsification capabilities is its distinctive shape. Emulsion stabilization is improved by its enormous surface area for interaction, given by its nanoscale size and spherical shape. A physical barrier is formed at the oil–water interface by the efficient adsorption of nano-spherical cellulose, thanks to its large surface area, which hinders the coalescence of scattered oil droplets. Because nano-spherical cellulose's surface characteristics may be chemically altered to increase or decrease hydrophobicity, it is capable of driving this interfacial adsorption. Due to these alterations, nano-spherical cellulose may be used successfully in several emulsion systems; thus, it finds use in many culinary, cosmetic, and medicinal products. The creation of a three-dimensional network surrounding oil droplets is one of the main ways in which nano-spherical cellulose stabilizes emulsions. Emulsion stability is improved because this network forms a protective layer that physically prevents oil droplets from coalescing. Its stabilizing impact is amplified by the structural integrity of the network, which is enhanced by the strong hydrogen bonding connections between cellulose molecules [17].

Thus, nano-spherical cellulose can be employed as a Pickering emulsifier, a technique that substitutes conventional surfactants with solid particles to stabilize emulsions. The utilization of nano-spherical cellulose as a Pickering emulsifier has several benefits, including the possibility of producing extremely stable emulsions regardless of environmental changes, enhanced biocompatibility, and a decreased dependence on synthetic surfactants. The capacity of nano-spherical cellulose to alter the rheological characteristics of emulsions is another remarkable quality of this material. The viscosity of cosmetic formulations and the texture and mouthfeel of food items are both influenced by rheology, which is the study of how materials behave when subjected to flow and deformation. Because of its strong ability to bind water and the interactions between its particles, nano-spherical cellulose, when distributed in water, may create a gel-like structure. Emulsions benefit from increased stability and textural quality brought about by this gelation action, which gives them a desired viscosity and consistency. Variables, including particle concentration, particle size, and surface changes, allow for the fine-tuning of the rheological characteristics of nano-spherical cellulose-stabilized emulsions. Because of its malleability, nano-spherical cellulose may be tuned to meet the specific rheological requirements of a wide range of industrial applications [18].

It would be irresponsible to ignore the health and environmental benefits of employing nano-spherical cellulose as an emulsifier. The increasing demand from consumers for products without synthetic ingredients, known as "clean-label," is perfectly met by nano-spherical cellulose, a naturally occurring and biodegradable substance. Because of its biocompatibility, cellulose is safe to use in food, medicine, and personal care items, whether taken orally or applied topically. In addition, cellulose's renewable status as a product of plant biomass helps make the emulsification process more eco-friendly by decreasing the need for synthetic and petroleum-based emulsifiers. One of the main reasons why nano-spherical cellulose is becoming more and more popular in business is its environmentally favorable quality. The emulsifying properties of nano-spherical cellulose extend beyond the stability of basic oil-in-water and water-in-oil emulsions. According to the research, nano-spherical cellulose may also stabilize systems that are more complicated, including microemulsions or multiple emulsions. For pharmaceutical and cosmetic applications that need a controlled release of active ingredients or encapsulation of bioactive substances, multiple emulsions, such as water-in-oil-in-water emulsions, are very desirable. The enormous surface area and varied surface chemistry of nano-spherical cellulose make it an attractive choice for stabilizing these complex emulsions, opening the door to the development of innovative formulations with improved functions [19].

To enhance its emulsification capability, nano-spherical cellulose's surface chemistry can be further adjusted. Stabilizing emulsions with high oil content or forming stable nanoemulsions are both made possible by functionalizing the surface hydroxyl groups of nano-spherical cellulose with hydrophobic or amphiphilic moieties, which enhances its emulsifying potential. The optical clarity, stability, and bioavailability of encapsulated chemicals are all improved using nanoemulsions, which are emulsions with droplet sizes in the nanoscale range. New uses for nanoemulsions stabilized by nano-spherical cellulose may emerge in industries, including cosmetics, food, and pharmaceuticals that value stable formulations with

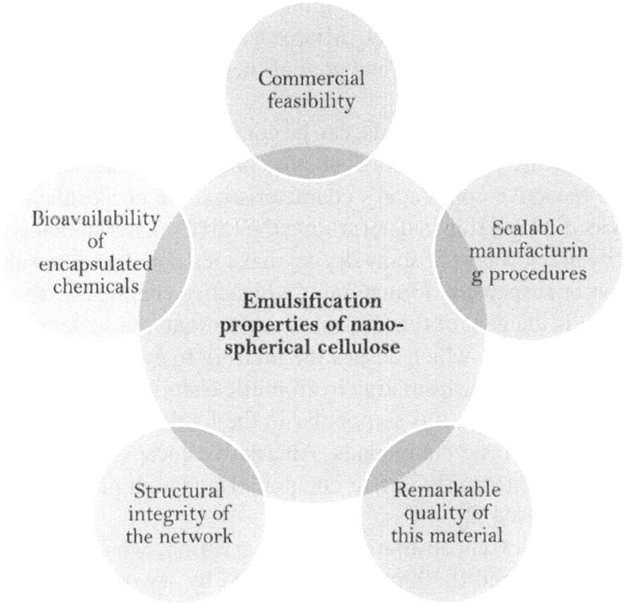

FIGURE 8.4 Emulsification properties of nano-spherical cellulose.

excellent clarity. Although nano-spherical cellulose has many benefits, there are still several obstacles that must be overcome before its emulsification potential can be completely realized. Finding the sweet spot in manufacturing for consistent particle size and surface properties is a major obstacle. The emulsifying ability and other characteristics of nano-spherical cellulose can change drastically depending on the cellulose source, synthesis technique, and processing conditions. In order to guarantee repeatability and commercial feasibility, it is important to design standardized and scalable manufacturing procedures. To build optimum formulations that may suit specific industrial needs, it is vital to understand the interaction processes between nano-spherical cellulose and other emulsion components, including proteins or polysaccharides [20] (Figure 8.4).

8.5 ENCAPSULATION OF BIOACTIVE COMPOUNDS

Research and development in the food, pharmaceutical, and nutraceutical sectors has shifted its focus to encapsulation of bioactive substances. Bioactive substances, including vitamins, minerals, antioxidants, enzymes, probiotics, and essential oils, can be stabilized, made more bioavailable, and released under regulated conditions by using this method. Although bioactive chemicals have health benefits, they can lose some of those benefits when exposed to certain environmental conditions, such as light, oxygen, temperature, and pH. By encasing them, these chemicals may be kept safe, transported more efficiently, and kept functioning until they reach their intended location. There are several methods for encapsulation, and each one is adapted to the specifics of the bioactive chemical and its intended use. Some of the most common ways include emulsification, nanoprecipitation, spray drying, coacervation, and liposome entrapment. Microcapsules and nanocapsules, encased in various carrier materials, are the end goal of these methods for producing bioactive chemicals [21].

The matrixes, or carrier materials, can be composed of a variety of substances that offer regulated release and protection, such as polymers, proteins, lipids, or polysaccharides. The bioactive compound's characteristics, the encapsulation method, and the application all play a role in determining the carrier material that is most suitable. The scalability and low cost of spray drying make it a popular encapsulation technology. Emulsion or suspension formation of a bioactive chemical with an appropriate carrier material is the goal of this procedure. After that, the solvent is sprayed into a chamber with heated air, which causes the mixture to evaporate quickly and forms encapsulated particles. Encapsulating hydrophilic chemicals is a typical use case for spray drying, which is why it is so popular in the food business for things like vitamins, probiotics, and taste compounds. Alternative encapsulating methods may be necessary for heat-sensitive bioactive compounds, as high processing temperatures might restrict their use [22].

Another method of encapsulating is coacervation, which involves creating a coacervate phase around the bioactive molecule by separating colloidal particles in a solution. At this stage, a polymeric substance encases the bioactive center, creating a protective covering. Using one or two polymers determines the complexity of the coacervation process. In complex coacervation, a protective matrix is formed by the interaction of two polymers with opposing charges. With its many benefits, including controlled release of the active agent and great encapsulation effectiveness, this approach is perfect for encasing hydrophobic bioactive chemicals. The process of liposome entrapment involves enclosing bioactive chemicals in lipid bilayers to form vesicles called liposomes. Liposomes are perfectly suited to transport hydrophilic and lipophilic bioactive substances due to their spherical shape, hydrophilic interior, and hydrophobic bilayer. Encapsulated chemicals are protected from degradation and may be delivered to specific locations, thanks to the lipid bilayer. Liposomes have the potential to increase the therapeutic effectiveness of medications, are biocompatible, and have increased bioavailability, making them ideal for pharmaceutical applications. In addition, the active substances can be controllably released by liposome-based encapsulation, resulting in long-lasting benefits [23].

One more flexible method that works well for encasing bioactive chemicals that are oil-soluble is emulsification. To use this technique, an emulsion of oil and water or water and oil is prepared, with the bioactive ingredient spread throughout the oil. By preventing the droplets from coalescing and increasing the stability of the encapsulating substance, emulsifiers and surfactants stabilize emulsions. Nanoemulsions, which have droplet sizes in the nanoscale range, have more surface area to interact with biological membranes, are more stable, and have better bioavailability. It is possible to create sustainable, clean-label goods with improved functioning by combining bioactive chemicals with natural emulsifiers like proteins or polysaccharides. Another common method for encasing bioactive substances in polymeric nanoparticles is nanoprecipitation. The technique involves slowly adding a polymer solution containing the bioactive chemical to a non-solvent, resulting in the precipitation of nanoparticles. Due to the formation of nanoparticles with a hydrophobic core, this method is ideal for encasing hydrophobic chemicals [24].

Among the many benefits of nanoprecipitation are its ease of use, minimal energy needs, and capacity to create uniformly sized nanoparticles. Nanoparticles made in this way have potential uses in the pharmaceutical and nutraceutical industries due to their malleability in terms of targeting, controlled release, and improved cellular absorption. Enhancing the bioavailability of bioactive substances is a key benefit of encapsulation. Because of their instability, fast metabolism, or poor solubility, many bioactive compounds have low bioavailability. To overcome these obstacles, encapsulation is used to create carriers that protect the molecule from the gut's hostile environment, allowing it to be absorbed and retained in the body for longer [25].

Nanoemulsions increase the surface area and solubilization of lipophilic bioactives, which improves their absorption in the digestive fluids, while liposomal encapsulation increases the bioavailability of hydrophilic and lipophilic substances. Another important advantage of encapsulating bioactive substances is controlled release. The continuous or targeted release of active substances is made possible by encapsulation, which is crucial for preserving the therapeutic efficacy over time. This is accomplished by utilizing encapsulation methods and carrier materials that are sensitive to environmental factors like temperature, pH, or enzyme activity. Bioactive chemicals can be protected from breakdown in the acidic stomach by having enteric coatings manufactured from pH-sensitive polymers. This allows for their release in the intestines instead. Minimizing adverse effects, increasing patient compliance, and decreasing dosage frequency are all benefits of controlled release of bioactive substances. Encapsulation not only prevents light, heat, oxygen, and moisture from degrading bioactive substances, but it also increases their stability. These conditions greatly diminish the functional characteristics of

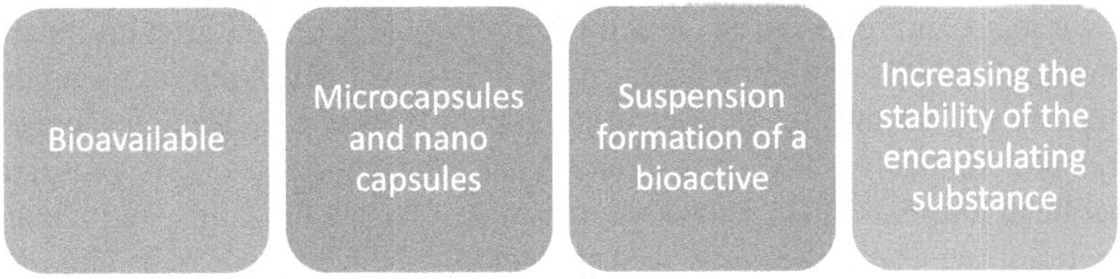

FIGURE 8.5 Encapsulation of bioactive compounds.

several bioactive molecules, including vitamins, polyphenols, and omega-3 fatty acids [26] (Figure 8.5).

8.6 NANO-SPHERICAL CELLULOSE IN FOOD PACKAGING

Because of its exceptional mechanical, barrier, and structural qualities, nano-spherical cellulose (NSC) is becoming more popular for use in food packaging. This nanomaterial has the potential to revolutionize sustainable packaging. It is generated from cellulose using chemical or mechanical methods and has several advantages, such as being non-toxic, renewable, and biodegradable. Because of its spherical shape, NSC is more useful; it has better mechanical strength, thermal stability, and barrier qualities, all of which are necessary for maintaining food quality and increasing its shelf life. The outstanding mechanical qualities of NSC-based food packaging are one of the main reasons for its growing popularity. The strong inter-fibrillar connections and large surface area of NSC provide its high modulus and tensile strength. When it comes to food packaging, these qualities are crucial, as they ensure that the product will last, remain flexible, and not be damaged easily. The mechanical performance of bio-nanocomposite films made by combining NSC with polymer matrixes like chitosan, polylactic acid, or polyvinyl alcohol is superior to that of conventional packaging materials. Not only does NSC increase mechanical strength, it also helps food packaging have better barrier qualities. Nanostructured spherical particles obstruct the passage of gases, moisture, and oxygen by means of a circuitous pathway they fashion [27].

Because food spoils and loses nutritional value when exposed to too much oxygen and moisture, this is vital for food preservation. Films treated with NSC have a much-reduced capacity to let oxygen and water vapor through, extending the time that perishable items may be stored. For example, NSC helps keep perishable goods fresh and avoids rancidity when used in packaging for items that are oxidation-sensitive. Among the many critical areas where NSC is crucial is thermal stability. Crystalline NSC increases the heat resistance of packing materials, making them more suitable for processing or storing foods at higher temperatures. Materials used for packaging that will be subjected to sterilization, pasteurization, or microwave heating will benefit greatly from this quality. With enhanced thermal stability, NSC-based films may be used in more contexts and the packaging can withstand different temperatures without compromising its structural integrity. There is a growing interest in the potential of NSC as a component in the creation of active packaging materials. To enhance the advantages of packaging materials, active packaging incorporates functional components. These components can manage the release of preservatives or operate as antimicrobials. Bioactive substances like essential oils, antioxidants, or antibacterial agents may be easily incorporated into NSC because of its large surface area and functional groups [28].

As a consequence, we have packaging materials that may actively prevent microbial development or scavenge free radicals, increasing the shelf life of food. Intelligent packaging solutions may also be designed using films based on NSC. In order to keep tabs on the state of the food within, "intelligent packaging" incorporates sensors or indications within the material. Functionalization of NSC with pH-sensitive dyes, oxygen indicators, or temperature-responsive compounds is possible because of its strong reactivity. By providing up-to-the-minute data on food quality, these systems can increase security and decrease food waste. One example is a film made of natural semiconducting compounds that changes color to alert customers to potential spoiling or contamination by microorganisms. As a biodegradable substitute for conventional plastic packaging, NSC is good for the environment [29].

Packaging materials made from NSCs are safe for the environment since cellulose, one of the most prevalent natural polymers, is biodegradable by nature. Sustainable packaging solutions that promote a circular economy and lessen plastic pollution are in high demand, and this fits right in with that trend. To further improve resource efficiency and reduce dependence on non-renewable resources, NSC may be made from renewable sources such as agricultural waste or wood pulp. The practical use of NSC in food packaging still faces obstacles, despite its many advantages. It is essential to achieve NSC homogeneous dispersion in polymer matrixes in order to fully use it. The films' mechanical and barrier qualities can be severely compromised due to agglomeration, which can be caused by poor dispersion. Creating efficient methods of surface modification or utilizing appropriate polymer mixes are necessary to tackle this difficulty. For NSC to be a practical choice for food packaging on a grand scale, its cost-effectiveness and scalability need to be determined. Researchers are

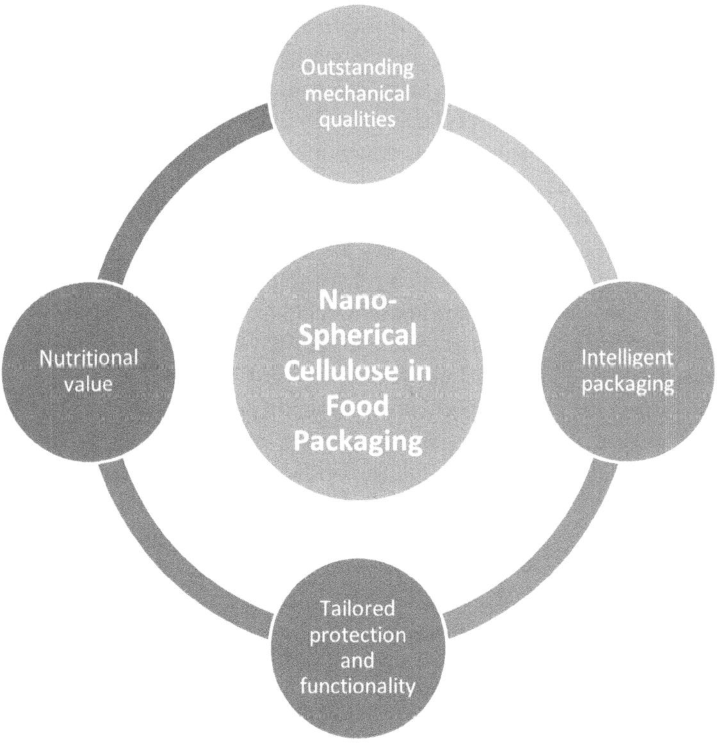

FIGURE 8.6 Nano-spherical cellulose in food packaging.

also looking at ways to improve packaging materials' overall performance by incorporating NSC into multi-layered systems. It is feasible to develop product-specific packaging solutions by integrating NSC-based films with other functional layers, such as layers that absorb moisture or seal in heat. In order to keep food safe and of high quality all the way through the supply chain, these multi-layered solutions provide tailored protection and functionality [30] (Figure 8.6).

8.7 HEALTH AND SAFETY CONSIDERATIONS

Numerous domains place a premium on health and safety issues, such as public health, food production, industrial environments, and consumer safety. If we care about people, places, and things, we must take care of these details. Various practices, legislation, and initiatives are covered by this topic with the goal of reducing risks and preventing hazards that might cause harm to people, the environment, or both. The creation of safety standards and regulations is an important part of health and safety. Industries, groups, and people are all subject to the rules and regulations imposed by various governing authorities and regulatory agencies. Ensuring safe working conditions, product quality and safety, and environmental protection are the goals of these standards. In the United States, for instance, institutions are subject to inspections by the Occupational Safety and Health Administration to ensure that they are in conformity with safety regulations for the workplace. Similarly, ISO 45001, which details best practices for health and safety management systems in the workplace, is one of the standards developed by the International Organization for Standardization (ISO). In addition to reducing the likelihood of accidents, these rules and regulations help make workplaces safer for everyone. Because of the prevalence of potentially dangerous substances, heavy machinery, and procedures in the industrial sector, health and safety concerns take on an even greater significance. To better understand, evaluate, and control potential dangers, several businesses have instituted all-encompassing health and safety management systems. An essential part of this process is risk assessment, which entails finding possible dangers and figuring out how likely and damaging they are. Physical hazards include things like tripping, falls, and slips; chemical hazards include things like exposure to harmful chemicals. Personal protective equipment, safety training programs, and emergency protocols can all be prioritized after careful consideration of risk assessments [31].

Construction, manufacturing, and mining are examples of high-risk occupations where worker health is an additional important factor to consider. Respiratory illnesses, hearing loss, and musculoskeletal difficulties are among the chronic health problems that can develop from prolonged exposure to chemicals, dust, radiation, and noise. This highlights the critical need of providing employees with health monitoring, frequent medical exams, and safety training. Designing jobs, equipment, and workstations to decrease physical strain and avoid injuries is another important aspect of ergonomics that protects workers' health. In addition to protecting individuals, a proactive approach to worker health increases productivity and decreases absenteeism. The preservation of both people and the environment is an integral part of public health's focus on health and safety. Ensuring the safety of food and water sources, controlling the spread of illnesses, and preventing outbreaks of diseases are all goals of public health laws and regulations. Important parts of public health plans include things like immunization programs, cleanliness standards, and laws governing food safety [32].

Hygiene techniques, social distance, and protective gear were all emphasized during the COVID-19 pandemic as ways to keep the people healthy and stop the spread of infectious illnesses. Considerations of health and safety are also critical in the food industry. In order to prevent food poisoning, it is crucial that all food items be safe from the moment they are harvested until they reach the customer. The Hazard Analysis and Critical Control Points system is one example of a food safety standard that is put into place to find possible dangers at every point in the food supply chain and to put controls in place to avoid contamination. Strict hygiene procedures, frequent inspections, and traceability systems are necessary for food processing plants to detect and resolve safety concerns promptly. Furthermore, it is critical to utilize appropriate packing materials and store food properly to ensure its quality and safety. Creating and regulating products in a way that keeps consumers safe is also part of consumer safety. Making sure things like toys, cosmetics, medications, and electrical equipment are safe falls under this category [33].

Manufacturers are obligated by product safety standards to carry out safety testing, adhere to labeling requirements, and offer clear directions for usage. In the United States, for example, the Consumer Product Safety Commission is responsible for investigating customer complaints, enforcing product recalls, and monitoring product safety. Not only does this keep buyers safe, but it also makes them more confident in businesses and products. Sustainable development and prevention of harmful activities are the goals of environmental health and safety concerns. Pollution control, waste management, and the promotion of environmentally friendly goods and activities are the main areas of environmental regulation. When it comes to things like emissions management, hazardous waste disposal, and water and air quality, industries must follow the rules. In order to make well-informed decisions and put mitigation strategies into action, it is essential to conduct environmental impact assessments to determine the possible consequences of projects and activities on the environment. Achieving a balance between economic growth, environmental preservation, and social well-being is a key component of sustainable development policies that prioritize health and safety [34].

As digitalization, automation, and AI continue to grow in prominence, new health and safety concerns are emerging within the framework of technological progress. Opportunities to enhance safety do arise with technological advancements, but new difficulties often arise. For instance, although manufacturing robots lessen repetitive stress injuries, they also pose new dangers in the form of human–robot interactions. Consequently, in order to deal with these new dangers, safety regulations and standards will need to change as technology does. Another important aspect of health and safety is being prepared for emergencies. All levels of government, as well as local towns and organizations, should have a plan in place to deal with catastrophic events like fires, chemical spills, and pandemics. Evaluations of potential danger, methods of contact, protocols for leaving the area, and supplies for first aid should all be part of any comprehensive emergency plan. People must participate in training and exercises on a regular basis to guarantee

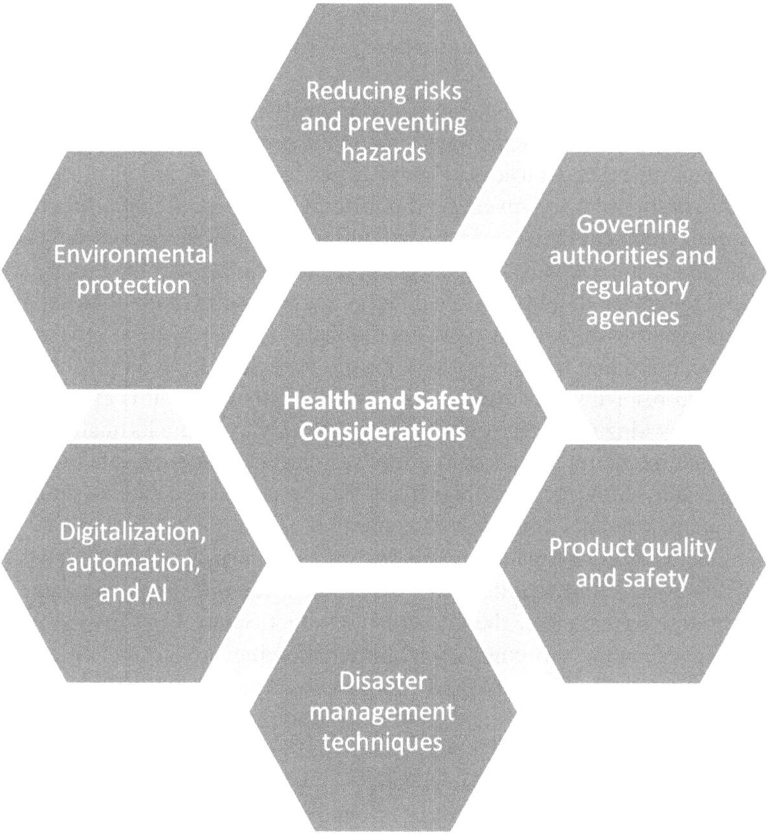

FIGURE 8.7 Health and safety considerations of nano-spherical cellulose.

they know what to do in an emergency. Minimizing the effect of calamities, protecting lives, and ensuring a rapid recovery are the goals of disaster management techniques [35] (Figure 8.7).

8.8 CHALLENGES IN THE APPLICATION OF NSC

There has been a lot of buzz about NSC's potential uses because of its remarkable mechanical, structural, and functional qualities. Nevertheless, there are a number of obstacles that prevent NSC from being widely used, even though it has great promise in many sectors, including food, medicines, packaging, and biomedicine. The manufacture, characterization, functionalization, integration, and long-term stability of the material are all aspects that provide a problem. In order to make full use of NSC and enable its effective commercial application, it is essential to address these obstacles [36].

Achieving consistent and scalable manufacturing is one of the main problems in utilizing NSC. Temperature, concentration, and reaction time are only a few of the critical variables that must be precisely controlled during NSC synthesis. To create

NSC, a number of synthetic techniques are used, including as acid hydrolysis, enzymatic treatment, and mechanical processing. But there are restrictions on purity, particle size distribution, and yield with each approach. One such process, acid hydrolysis, uses powerful acids that can be harmful to the environment and can produce particles with unusual shapes. However, mechanical processing methods might result in a wider range of particle sizes and need a lot of energy. Reproducibility and large-scale manufacturing are hindered by the absence of established production processes and the difficulty in attaining homogeneous particle sizes and morphologies [37].

Another set of difficulties arises when NSC is functionalized. Surface modification is frequently required to improve NSC compatibility with various polymer matrixes or to include new capabilities. Chemically modifying NSC without altering its beneficial characteristics or structural integrity is a challenging process. Polymer chain grafting onto cellulose surfaces, reactive group introduction by chemical processes, and nanomaterial coating are common functionalization approaches. Because agglomeration and the degradation of desirable qualities like mechanical strength and thermal stability might result from inconsistencies, it is critical to attain consistent and repeatable functionalization. Furthermore, the use of harmful chemicals in various functionalization procedures raises questions regarding the product's safety and the effect on the environment [38].

Problems also develop when trying to include NSC into composite materials. Bio-nanocomposites with improved mechanical, barrier, or thermal characteristics are resultant from the incorporation of NSC into polymers. Because of its hydrophilic nature, NSC can be difficult to homogeneously disperse inside a polymer matrix. The mechanical and functional characteristics of the composite might be negatively impacted by agglomeration and weak interfaces caused by poor dispersion. To enhance dispersion, surface modification or compatibilizers can be used; however, these extra procedures might make the production process more difficult and expensive. Overdosing on NSC can cause brittleness or poor film-forming capabilities; therefore, it's important to find the sweet spot between mechanical strength, flexibility, and processability when adding it to composites. Another important issue is the stability of NSC when it is stored and used. The reactive nature and large surface area of NSC make it potentially susceptible to deterioration and property changes with time. Its stability can be impacted by environmental conditions as humidity, temperature, and light exposure [39].

Hydrolysis, which can alter the size and crystallinity of NSC particles in environments with high humidity, is one example. The processing or storage conditions of NSC can also be affected by its interactions with other formulation components, such as additives, bioactive chemicals, or polymers. The performance and safety of products based on NSC must be guaranteed by ensuring their stability over the long term. Aside from these technological hurdles, economic and regulatory obstacles also prevent NSC from being widely used. The requirement for certain machinery, energy sources, and purification procedures might make NSC manufacture expensive. This means that NSC-based goods could not be as cost-effective as those made from more traditional materials. Furthermore, producers and end-users are left with a sense of ambiguity due to the absence of clear norms and standards for the manufacturing, characterization, and application of nanomaterials. Market adoption of nanocellulose-based products may be delayed as regulatory bodies work on standards to evaluate their safety, environmental effect, and biodegradability. The possible hazards to human and environmental health and the environment posed by NSC should also be carefully examined from a safety standpoint. Despite cellulose's reputation as a biocompatible and non-toxic material, NSC particles, due to their nanoscale size and strong reactivity, may provide unique dangers. For instance, NSC particles may cause respiratory problems if inhaled, and unexpected toxicological consequences may occur if applied to the skin or consumed. Hence, in order to determine if NSC is safe for use in different contexts, especially in food and biomedical goods, extensive toxicity studies and risk evaluations are required. Transparency in risk communication and adherence to safety regulations are essential for ensuring consumer safety and winning public approval [40].

To sum up, NSC's unique features make it highly promising for a wide range of applications. However, there are a few obstacles that need to be overcome before it can be widely used. Production consistency and scalability, functionalization, dispersion in composites, and storage stability are all obstacles. To top it all off, getting NSC-based goods into the hands of consumers requires meticulous consideration of economic considerations, regulatory uncertainty, and safety concerns. Progress in synthesis methodologies, standardization of characterization procedures, development of novel functionalization techniques, and joint efforts by academics,

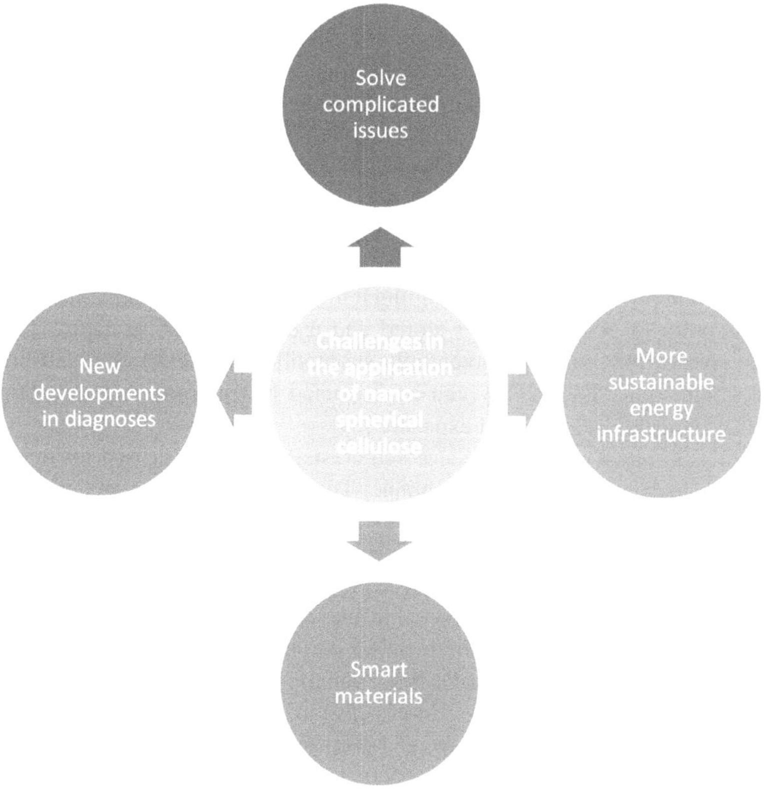

FIGURE 8.8 Challenges in the application of nano-spherical cellulose.

businesses, and government organizations are all necessary to overcome these obstacles. Solving these problems will allow NSC to reach its full potential in many different industries, which will boost innovation and sustainability [41]. Figure 8.8 show the challenges in the application of NSC.

8.9 FUTURE PERSPECTIVES AND INNOVATIONS

Potentially transforming industries, improving human well-being, and addressing global concerns are the future views and breakthroughs connected to improved materials, new technologies, and expanding scientific paradigms. To adapt to a dynamic and unpredictable environment, the next wave of innovations will leverage cutting-edge materials and technology while prioritizing efficiency, integration, and sustainability. The emergence of functional nanomaterials is expected to bring about substantial advancements in the field of advanced materials. Engineered at the molecular and atomic levels, these materials have special qualities that make them superior to more common substances in carrying out certain tasks [42].

There is a great deal of interest in the potential uses of nanomaterials in fields as diverse as energy storage, medicine, electronics, environmental remediation, and carbon nanotubes, graphene, and metallic nanoparticles. For instance, there has been a lot of buzz about NSC due to its biodegradability, low weight, and ability to improve composites' mechanical strength and thermal stability. Researchers hope to one day make these nanomaterials widely available and affordable by perfecting their synthesis and functionalization procedures and increasing manufacturing volumes. Emerging technologies like the Internet of Things, artificial intelligence, and robots are altering existing sectors, just as material innovations. New research methods and faster discoveries are on the horizon because of the convergence of artificial intelligence, nanotechnology, and materials science. Artificial intelligence systems can foretell how materials will behave, enhance synthesis procedures, and discover novel compounds with desirable features. AI-powered prediction models used to material design can cut down on experimental time and money. Additionally, automation and robots in material synthesis, quality control, and production will increase efficiency and precision, allowing for the mass manufacture of complex materials with little waste [43].

In the realm of innovation, the creation of environmentally beneficial and long-lasting technology is quickly rising to the forefront. Researchers are actively seeking environmentally friendly and energy-efficient materials and methods due to the growing concern about climate change, resource depletion, and environmental contamination. For instance, scientists are working to find alternatives to traditional plastics that end up in landfills and the ocean, such as biodegradable polymers and nanomaterials derived from plants. Green material solutions are on the horizon, thanks to innovations in biopolymer synthesis, waste valorization, and the concepts of the circular economy. More sustainable energy infrastructure is being pursued through the development of renewable energy materials like perovskite solar cells and energy-efficient storage systems. The domain of biomedical applications is another important area of innovation. Researchers in the fields of biotechnology and materials science are aiming to create smart medication delivery systems, tissue engineering, and individualized medicine. Innovative new ways to cure illnesses are being developed through the invention of nanomaterials that are more biocompatible, have the ability to be targeted, and have controlled release mechanisms [44].

By reacting to changes in pH or temperature, for instance, nano-carriers can release therapeutic chemicals at the exact location where they are needed, enhancing treatment efficacy while decreasing adverse effects. New developments in diagnoses, treatments, and regenerative medicine are predicted to emerge in the biomedical field as a whole as a result of the merging of nanotechnology, genomics, and artificial intelligence. There is hope in the field of smart materials and gadgets as well. Smart materials are well-suited for use in adaptive systems, actuators, and sensors due to their capacity to detect and react to environmental stimuli. Some examples of materials that are finding applications in robotics, aircraft, and consumer electronics include shape-memory alloys, piezoelectric materials, and self-healing polymers. Future developments in smart materials will aim to make them more responsive, reliable, and versatile. Smart environments and autonomous systems that can interact with their environs smoothly will be possible with the integration of these materials with IoT-based technologies [2].

Technological breakthroughs in areas like neuromorphic engineering, quantum computing, and flexible electronics are anticipated to propel IT and electronics into the future. With their lightweight, adjustable gadgets that incorporate sensing and display features, flexible and wearable electronics are set to transform healthcare monitoring, entertainment, and communication. New opportunities in cryptography, optimization, and large-scale simulations will be unlocked as a result of the exponential improvement in processing capacity that is promised by the development of quantum computing. The field of neuromorphic engineering, which draws inspiration from the human brain, is working toward the goal of developing computational systems that are both energy-efficient and possess superior cognitive capacities. These systems have the potential to bring about a revolution in robotics, AI, and brain–machine interfaces. It is critical to acknowledge the difficulties and ethical concerns that come with these advancements, even while the future is full of possibilities. Concerns regarding the safety and environmental effect of nanomaterials, for example, arise when their manufacture and application are done on a big scale [45].

One distinguishing aspect of forthcoming developments will be the merging of transdisciplinary domains. Scientists from several fields will work together to solve complicated issues; this includes material science, computer science, biology, chemistry, engineering, and physics. Innovations in precision agriculture, food safety, and

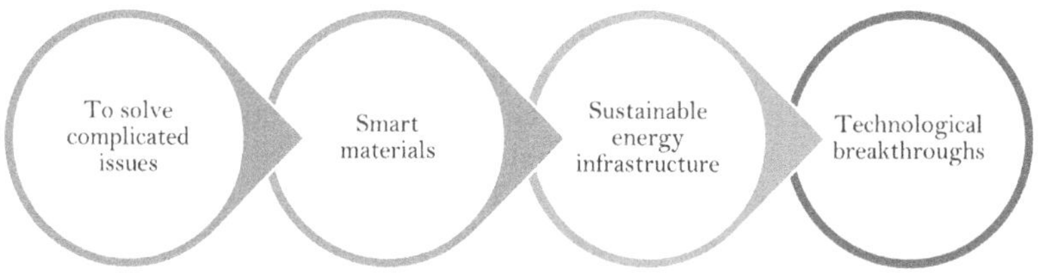

FIGURE 8.9 Future perspectives and innovations of nano-spherical cellulose in food applications.

environmental monitoring are just a few examples of how nanotechnology and biotechnology may work together. Whole solutions that tackle several problems at once will be the hallmark of future innovations that make use of multidisciplinary techniques (Figure 8.9).

8.10 CONCLUSION

The adaptable and promising NSC has tremendous promise for use in a wide range of culinary applications. Its unique physicochemical qualities, such as a large surface area, biocompatibility, mechanical strength, and stability, facilitate the improvement of food quality, safety, and sustainability. NSC's qualities as a stabilizer, emulsifier, and carrier for bioactive chemicals enhance the nutritional profile, consistency, and texture of food items. It also helps food retain its freshness for longer and has less of an effect on the environment when used as a packaging material instead of more traditional options. Maximizing the use of NSC in the food business requires addressing various hurdles, notwithstanding its benefits. Making sure everything is safe, following all regulations, and improving production procedures to maintain high quality are all part of this. As research and development progress, incorporating NSC into food systems can greatly enhance new solutions that meet the changing needs of the food industry in terms of sustainability, health, and functionality. In general, NSC has tremendous promise for use in the food industry and may significantly alter the formulation, preservation, and packaging of food goods in the future. Better food quality, happier customers, and less impact on the environment can result from the food industry's adoption of more sustainable and efficient procedures, made possible by expanding knowledge of NSC.

REFERENCES

1. Zhu Y, Wei Z, Jiang F, Hu W, Yu X, Du SK. Comparative analysis of millet bran nanocelluloses with various morphologies: Revealing differences in the formation mechanism and structure characteristics. *Carbohydrate Polymers*. 2024 Jun 21;342:122419.
2. Abou-Elsoud M, Salama M, Li Z, Wang S, Cai Z, Ahn DU, Huang X. Structuring low density lipoprotein-based oleogels with pectin via an emulsion-templated approach: Formation and characterization. *Journal of Food Engineering*. 2024 Sep 27;387:112340.
3. Yang W, Zhang S, Feng A, Li Y, Wu P, Li H, Ai S. Water-insoluble tea polyphenol nanoparticles as fillers and bioactive agents for pectin films to prepare active packaging for fruit preservation. *Food Hydrocolloids*. 2024 Nov 1;156:110364.
4. Contreras-López G, Carrillo-López LM, Vargas-Bello-Pérez E, García-Galicia IA. Microencapsulation of feed additives with potential in livestock and poultry production: A systematic review. *Chilean Journal of Agricultural & Animal Sciences*. 2024 Apr;40(1):229–249.
5. Al-Harbi N, Abd-Elrahman NK. Physical methods for preparation of nanomaterials, their characterization and applications: A review. *Journal of Umm Al-Qura University for Applied Sciences*. 2024 May 30:1–22.
6. Wang L, Liu L, Chen R, Jiao Y, Zhao K, Liu Y, Zhu G. Carbonized polymer dots-based molecular imprinting: An adsorbent with enhanced selectivity for highly efficient recognition and removal of ceftiofur sodium from complex samples. *Journal of Hazardous Materials*. 2024 Jul 15;473:134637.

7. Liu S, Liu Y, Li Q, Song Y, Zhang L, Peng F, Ma C. Oleanolic acid nanoparticles-stabilized W/O pickering emulsions: Fabrication, characterization, and delivery application. *Food Chemistry*. 2024 Jun 30;444:138598.
8. Yan J, Huang J, Pei F, Bai T, Chen Y, Wang D, Cheng W, Han G. Biomimetic dendritic-networks-inspired oil-water separation fiber membrane based on TBAC and CNC/PVDF porous nanostructures. *Chemical Engineering Journal*. 2024 Jul 15;492:152307.
9. Alshammari AS. Impact of lignocellulosic biomass-derived graphene-titanium oxide nanocomposite as an electrode for sustainable performance in microbial fuel cell. *International Journal of Environmental Science and Technology*. 2024 Mar;21(5):5185–202.
10. Tawiah B, Abraham AM, Kappen JJ, Raju NM, Rajan CS, George SC. Nanotechnology and water: Emerging solutions for a global crisis. In *Nanotechnology in Societal Development* (Eds: Soney C. George, Benjamin Tawiah). 2024 Sep 18 (pp. 333–374). Singapore: Springer Nature Singapore.
11. Sahu M, Singh S, Prajapati S, Verma DK, Shin DK. From green chemistry to biomedicine: The sustainable symphony of cobalt oxide nanoparticles. *RSC Advances*. 2024;14(45):32733–32758.
12. Parveen M, Mahar IA, Tahira A, Thebo GM, Hulio AA, Bhatti MA, Naz I, Naqvi AZ, Shah AA, Dawi E, Jakhrani MA. Nickel oxide (NiO) nanostructures modified by rice straw extract phytochemicals for efficient urea oxidation in alkaline media using non-enzymatic sensing. *Microchemical Journal*. 2024 Sep 1;204:111188.
13. Chen H, Tian C, Zhang L. Design and synthesis of dexmedetomidine capped silver nanoparticles: Investigation of its catalytic application for reduction of nitrobenzenes and its effects on anesthetic, anti-inflammatory, antioxidant and analgesic applications. *Journal of Organometallic Chemistry*. 2024 Sep 1;1017:123262.
14. El-Sewify IM, Shenashen MA, ElAgamy RF, Selim MS, Khairy M, Shahat A, Selim MM, Elmarakbi A, Ebara M, El-Safty SA. Visual detection and tracking of toxic mercury in biological cells using biocompatible mesoporous silica nanospheres. *Sensors and Actuators B: Chemical*. 2024 Oct 15;417:136166.
15. Wang Z, Yu F, Hu F. Functional chitosan and its derivative-related drug delivery systems for nano-therapy: Recent advances. *Pharmaceutics*. 2024 Feb 28;16(3):337.
16. Wang L, Liu L, Chen R, Jiao Y, Zhao K, Liu Y, Zhu G. High selectivity and anti-interference molecularly imprinted polymer based on spheroidal carbonized polymer dots for effective sulfadiazine sodium removal. *Separation and Purification Technology*. 2024 Dec 18;350:127954.
17. Kim H, Kim M, Oh S, Hwang Y, Choe JS. Optimization of enzyme-assisted extraction from ginger (Zingiber officinale) leaf and its immune-stimulating effects on macrophages. *Preventive Nutrition and Food Science*. 2024 Jun 6;29(2):228.
18. Idris DS, Roy A. Antioxidant and dye degradation activity of green synthesized silver-iron oxide (Ag-Fe2O3) bimetallic nanoparticles. *Nano-Structures & Nano-Objects*. 2024 May 1;38:101142.
19. Li H, Lee LM, Yu D, Chan SH, Li A. An optimized multi-technique based analytical platform for identification, characterization and quantification of nanoplastics in water. *Talanta*. 2024 May 15;272:125800.
20. Elbatanony MM, Safwat EM, El-Sherif S, Hassan ML. Resin-based dental pulp capping restoration enclosing silica and portlandite nanoparticles from natural resources. *Scientific Reports*. 2024 Jul 17;14(1):16554.
21. Ali SQ, Aziz WJ, Kadhim DA, Abid MA, Sabry RS. Corn plant leaf extract inhibits microbial growth with the photocatalytic activities of MB, MO, and RdB dye-mediated nano-Cr2O3. *Nano-Structures & Nano-Objects*. 2024 Sep 1;39:101257.
22. Song Y, Hou Y, Ren S, Qu C, Li M, Tu Y, Wang R. Pickering emulsion stabilized using modified gluten protein. *Journal of Food Engineering*. 2024 Jun 1;371:111985.

23. Guo S, Cui H, Agarwal T, Zhang LG. Nanomaterials in 4D printing: Expanding the frontiers of advanced manufacturing. *Small.* 2024 Mar 3;20(30):2307750.
24. Saleh AK, Hussein AS, Ray JB, Elzaref AS. Comparative versatility and diverse biological applications of eco-friendly zinc oxide and cobalt oxide nanoparticles using Punica granatum L. peel extract. *Nano-Structures & Nano-Objects.* 2024 May 1;38:101174.
25. Mondal M, Banerjee S, Banerjee S, Naik SN, Ali SW. Biocompatibility and toxicity of nanoparticles in textile finishing. In *Nanotechnology in Textile Finishing: Advancements and Applications* (Eds: Satyaranjan Bairagi, Shakeel Ahmed, S. Wazed Ali). 2024 Sep 25 (pp. 461–486). Singapore: Springer Nature Singapore.
26. Mulindwa P, Kasule JS, Nantaba F, Wasswa J, Expósito AJ. Bioadsorbents for removal of microplastics from water ecosystems: A review. *International Journal of Sustainable Engineering.* 2024 Dec 31;17(1):1–8.
27. He X, Deng G, Zhang Z, Mao H, Cai L. Enhanced coloration and functionality of wool fabric by Hydroxypropyl-β-cyclodextrin coated magnetic nanoparticles. *Arabian Journal of Chemistry.* 2024 Sep 1;17(9):105923.
28. Alshammari AS. Role of graphene-silver nanocomposite as anode to boost single-chamber microbial fuel cell performance. *Chemistry Africa.* 2024 Jul 17;7(8):1–3.
29. Timothy UJ, Mamudu U, Solomon MM, Umoren PS, Igwe IO, Anyanwu PI, Aharanwa BC, Lim RC, Uchechukwu TO, Umoren SA. In-situ biosynthesized plant exudate gums-silver nanocomposites as corrosion inhibitors for mild steel in hydrochloric acid medium. *International Journal of Biological Macromolecules.* 2024 Jun 1;269:132065.
30. Chelimela N, Alavala RR, Satla SR. Curcumin–bioavailability enhancement by prodrug approach and novel formulations. *Chemistry & Biodiversity.* 2024 May;21(5): e202302030.
31. Saini N, Awasthi A, Pandey K, Awasthi K. Engineering of nanochannels in polymer membranes for energy and biological applications. *The European Physical Journal Special Topics.* 2024 Jun 4;233(21):1–9.
32. Abraham DA, Li AD, Sanmugam A, Wadaan MA, Baabbad A, Kanagaraj K, Karuppasamy K, Maiyalagan T, Kim HS, Vikraman D. Highly sensitive and selective detection of dopamine using atomic layer deposited HfO2 ultra-thin films. *Electrochimica Acta.* 2024 Jun 10;497:144574.
33. Mansour AM, Abou Hammad AB, El Nahrawy AM. Synthesis and optical properties of erbium-doped sodium silicate in sol-gel matrix. *Silicon.* 2024 Mar 22;16(9):1–9.
34. Chan YB, Aminuzzaman M, Rahman MK, Win YF, Sultana S, Cheah SY, Watanabe A, Wong LS, Guha SK, Djearamane S, Rajendran V. Green synthesis of ZnO nanoparticles using the mangosteen (Garcinia mangostana L.) leaf extract: Comparative preliminary in vitro antibacterial study. *Green Processing and Synthesis.* 2024 Apr 16;13(1):20230251.
35. Luong HV, Le TP, Le TL, Dang HG, Tran TB. A graphene oxide based composite granule for methylene blue separation from aqueous solution: Adsorption, kinetics and thermodynamic studies. *Heliyon.* 2024 Apr 15;10(7):e28648.
36. Le CV, Yoon H. Advances in the use of conducting polymers for healthcare monitoring. *International Journal of Molecular Sciences.* 2024 Jan 26;25(3):1564.
37. Gontarek-Castro E, Castro-Muñoz R. How to make membrane distillation greener: A review of environmentally friendly and sustainable aspects. *Green Chemistry.* 2024;26(1):164–185.
38. Balasubramanian A, Rajesh V, Bajda T, Tseng WL, Kumar AS, Rajesh N. Enhancing the adsorptive recovery of europium from real-world solid samples through the synergistic impact of lysozyme fibril and melamine phytate. *Journal of Molecular Liquids.* 2024 Jan 1;393:123529.
39. Feitosa AP, Rodrigues K, Martins WE, Rodrigues SM, Pereira L, Silva GM. Enhancing greywater treatment: High-efficiency constructed wetlands with seashell and ceramic brick substrates. *Applied Sciences.* 2024 Oct 6;14(19):9011.

40. Shen L, Ye T, Chen Y, Chu B, Chen H, Hu J, Yu Y. Facile synthesis of a novel AgIO3/CTF heterojunction and its adsorption–photocatalytic performance with organic pollutants. *Toxics*. 2024 Feb 6;12(2):133.
41. Sabbagh M, Hussaini M, Ismail U, Ahmed HR, Al-Suwaiyan M, Vohra M. Novel nafion-palygorskite composite for Pb/Lead treatment. *International Journal of Environmental Science and Technology*. 2024 May 11;22(2):1–6.
42. Zhang S, Xu X, Ye Z, Liu Y, Wang Q, Chen Q, Jiang Y, Qi J, Tian D, Xu J, Wu F. A large-nanosphere/small-nanosphere (cellulose/silver) antibacterial composite with prominent catalytic properties for the degradation of p-nitrophenol. *Applied Surface Science*. 2023 Jan 15;608:155192.
43. Ghaffar A, Salman M, Yameen M, Iqbal SZ, Altaf S, Munir B. Sustainable biomedical applications of cellulose. In *Regenerated Cellulose and Composites: Morphology-Property Relationship* (Ed: Mohd Shabbir). 2023 Apr 26 (pp. 347–379). Singapore: Springer Nature Singapore.
44. Singh BP, Rohit, Manju KM, Sharma R, Bhushan B, Ghosh S, Goel G. Nano-conjugated food-derived antimicrobial peptides as natural biopreservatives: A review of technology and applications. *Antibiotics*. 2023 Jan 25;12(2):244.
45. Rajpoot SR, Ahmad K, Asif HM, Madni MA, Tasleem MW, Zafar F, Majeed H, Khalid A, Hashmi HA, Aslam MR, Hayee A. Study of anti-inflammatory and anti-arthritic potential of curcumin-loaded Eudragit L100 and hydroxy propyl methyl cellulose (HPMC) microparticles. *Polymer Bulletin*. 2024 Apr;81(5):4335–4350.

9 Harnessing Spherical Nanocellulose for Next-Generation Energy Storage Systems

Rajni Verma, Mahapara Abbass, Shalom Akhai, Amandeep Singh Wadhwa, and Amar Nath Sharma

9.1 INTRODUCTION

Energy storage systems are crucial for balancing supply and demand in energy production, particularly in integrating renewable energy sources into the grid [1]. Spherical nanocellulose (SNC) has emerged as a favorable sustainable material, offering advantages such as availability in nature in abundance, biodegradability, and superior mechanical characteristics [2]. It has shown potential in improving the performance of various energy storage technologies, such as in supercapacitors, lithium-ion batteries (LIBs), sodium-ion batteries, and fuel cells [3]. Despite its potential, several challenges remain in scaling up SNC for commercial use. These include addressing issues related to scalability, standardization, recycling, and sustainability [4]. Scalability involves developing cost-effective production methods for large-scale manufacturing [5]. Standardization is required to ensure uniform performance and quality in various applications [6]. Recycling concerns focus on confirming that SNC-based materials can be efficiently reused post-use [7]. Sustainability considerations include minimizing environmental impacts during production while optimizing the material's lifecycle [8]. Further research is crucial for addressing such challenges. Integrating SNC with other nanomaterials could further improve materials' performance with respect to energy storage systems [9]. Additionally, developing new characterization techniques will help us to better understand SNC's interactions within energy storage devices paves the way for novel applications. Thus, the successful integration of SNC into advanced energy storage solutions can contribute significantly to the energy landscape. Overall, SNC serves as a flexible and sustainable material for next-generation energy storage systems [10]. Overcoming existing challenges through further research and development is a must for exploring its full potential in the drive for sustainable energy solutions.

DOI: 10.1201/9781003596127-10

9.2 FUNDAMENTALS OF SPHERICAL NANOCELLULOSE

SNC material is a cellulose particle with a sphere-shaped morphology, with enhanced properties for several applications, particularly in energy storage systems [11]. SNC has an amorphous cellulose coating around a crystalline core, which provides it with unique mechanical and thermal characteristics [12]. The crystallinity of SNC is found to vary drastically depending on the source and extraction methods used, typically ranging from 40% to 70% [13]. Key characteristics of SNC involve a high surface area, mechanical strength, thermal stability, biodegradability and good resistance to chemical activity [14]. Its spherical shape, combined with its large surface area, makes it a suitable reinforcing material [15]. The structural stability of SNC is maintained because of its thermal stability and its ability to withstand exposure to various chemicals, thereby making it resistant to extreme environments [16]. Furthermore, SNC is nontoxic due to its biodegradable nature [17]. The production of SNC can be done using several methods, primarily through the hydrolysis of cellulose sources. Common methods include acid hydrolysis, ultrasound-assisted disintegration, and regeneration from cellulose solutions [18]. These techniques are fundamental in determining the final properties of SNC, such as size distribution, crystallinity, and functional characteristics [19]. Ongoing advances in these methods are enhancing scalability and reducing production costs without compromising on the quality of SNC [20].

9.3 SNC IN ENERGY STORAGE APPLICATIONS

The potential of SNCs is revolutionizing energy storage applications by improving performance, safety, and sustainability [21]. Nevertheless, some challenges related to scalable synthesis, interfacial engineering, and standardization are still to be met. Table 9.1 discusses the developments and challenges in sphere-nanocellulose-based energy storage technologies, including its role in supercapacitors, batteries, and fuel cells, its potential in next-generation energy storage systems, and its applications, advantages, and challenges.

9.3.1 Supercapacitors

SNC has been considered an ideal electrode material for supercapacitors because of its large surface area and good mechanical properties [22]. The spherical morphology of SNC increases its ability to store charges, which enhances the electrochemical performance of supercapacitors [23]. It has been observed that the energy density and cycling stability can be improved remarkably when SNC is introduced into composite materials with conductive polymers or metal oxides [9]. Furthermore, the lightweight nature of SNC contributes to the overall efficiency of supercapacitors, which are appropriate for applications in portable and flexible electronic devices [24].

9.3.2 Lithium-Ion Batteries

SNC is an effective anode material for LIBs with higher specific capacities than graphite; hence, it can be considered one of the most promising anodes for the next

TABLE 9.1
SNC applications in next-generation energy storage systems

Application	Key Advantages	Challenges	Performance and Potential
SNC in energy storage applications [32]	• Improves performance, safety, and sustainability in energy storage • Enhances efficiency and stability of energy storage systems • Lightweight and cost-effective material for energy storage solutions	• Scalable synthesis difficulties • Interfacial engineering challenges • Need for standardization in manufacturing processes	Revolutionizing energy storage systems; improving performance and sustainability
Supercapacitors [22,33]	• High surface area and favorable mechanical properties • Spherical morphology enhances charge storage capacity • Increases energy density and cycling stability when combined with conductive polymers or metal oxides • Lightweight nature boosts efficiency	• Need for better conductivity and stability of composites • Challenges in large-scale production	Ideal for portable electronics and flexible energy storage applications
Lithium-ion batteries [34]	• Higher specific capacities than graphite • Spherical structure accommodates volume changes during cycles, enhancing stability • Improved rate capability and energy density when combined with other nanomaterials • Efficient for lightweight energy storage	• Limited lithium availability • High production costs • Need for enhanced life cycle and long-term stability	Next-generation energy storage solutions for electric vehicles and portable devices
Sodium-ion batteries [29,35]	• Higher specific capacities and better cycling stability than conventional materials • Spherical morphology aids in uniform ion distribution • Sodium is abundant and inexpensive, making it suitable for large-scale storage	• Sodium-ion conductivity issues • Optimization needed for large-scale commercial use	Promising for large-scale energy storage and grid applications
Fuel cells [30]	• Biodegradable and excellent proton conductivity • Strong structural integrity and thermal stability enhance performance • Aligns with demand for sustainable energy sources	• Fine-tuning required for specific applications • Need for optimizing SNC properties and manufacturing processes	Potential for next-generation, sustainable energy storage and fuel-cell technologies

generation of LIB applications [25]. Spherical SNC can better accommodate the stresses associated with volume changes during charge–discharge cycles, which improves its cycle stability and life [26]. By compositing with other nanomaterials, SNC can further increase its electrochemical performance due to its enhanced rate capability and general energy density [27]. These improvements together put SNC at the forefront of next-generation LIB development, especially in applications that require lightweight and efficient-energy storage solutions.

9.3.3 Sodium-Ion Batteries

With its good electrochemical performance, SNC is expected to be one of the best candidates for NIB applications [28]. SNC-based anodes in previous studies demonstrated better specific capacities and longer cycling stability compared with conventional materials. The uniform spherical morphology of SNC might promote a homogeneous distribution of ions along the surface and therefore further enhance the insertion and extraction kinetics of sodium ions [29]. Thus, SNC should be considered one of the most promising materials for NIBs, with sodium being abundant and cheap relative to lithium, thus possibly serving as a precursor for large-scale energy storage.

9.3.4 Fuel Cells

There is growing research interest in using SNC as a membrane material in fuel cells because of its biodegradable nature and excellent proton conductivity. Its structural integrity and thermal stability are believed to further enhance performance under operating conditions [30]. Researchers are working to fine-tune SNC's properties for specific fuel-cell applications. SNC's distinctive structural features are in perfect harmony with the rapidly growing demand for sustainable sources of energy, and research being conducted to fine-tune its properties and manufacturing processes will accelerate its adoption in next-generation energy storage technologies [31].

9.4 ELECTROCHEMICAL PERFORMANCE OF SNC-BASED ELECTRODES

SNC-based electrodes have been found to be promising candidates for applications due to their high specific capacitance and excellent cycle stability. Their fast charge/discharge rates combined with the environmental benefits make these materials a strong candidate for sustainable energy storage solutions [36]. Table 9.2 examines the electrochemical performance of SNC-based electrodes, their prospects for sustainable energy storage, comparison with graphite electrodes, and their potential for next-generation energy storage.

9.4.1 Specific Capacitance and Cycle Stability

Super capacitance is one of the most important parameters in determining the energy storage capacity of supercapacitors and batteries. SNC-based electrodes have shown excellent specific capacitance. The surface-rich, highly porous structure of

TABLE 9.2
Advancements in SNC-based electrodes for sustainable energy storage solutions

Application	Technology	Key Advantages	Challenges	Performance and Potential
Electrochemical performance of SNC-based electrodes	All technologies	• High specific capacitance • Excellent cycle stability • Rapid charge/discharge rates • Environmentally friendly and biodegradable	• Scaling production for mass adoption • Ensuring long-term stability in large-scale applications	Strong candidate for sustainable energy storage, providing high performance and environmental benefits
Specific capacitance and cycle stability	Supercapacitors and batteries	• Highly porous structure enhances ion accessibility • Excellent cycle stability with minimal degradation • High charge storage capacity	• Need for continuous optimization for long-term stability in diverse environments	Vital for practical applications requiring long-lasting energy storage
Charge/discharge rates	Supercapacitors and batteries	• High charge/discharge rates due to efficient ion transport • Spherical morphology reduces diffusion distances • Ideal for applications requiring quick energy delivery	• Optimization needed to maintain high rates over many cycles • Challenges in scalability for high-power applications	Ideal for electric vehicles, portable electronics, and other fast-response applications
Comparison with graphite-based electrodes	Lithium-ion batteries and supercapacitors	• Higher specific capacities and better cycle stability than graphite • Ability to accommodate volume changes during cycling • Biodegradable and environmentally friendly	• Production costs may be higher compared with traditional graphite • Need for further improvement in commercial scalability	Promising alternative to graphite-based electrodes, particularly for next-generation energy storage technologies

SNC enhances ion accessibility, thereby improving charge storage capabilities [37]. Moreover, SNC has excellent cycle stability, with its performance retained even after several charge and discharge cycles with minimal degradation. This stability is very important for practical applications to ensure long-lasting energy storage solutions.

9.4.2 Charge/Discharge Rates

The charge and discharge rates of SNC-based electrodes are significantly high and thus highly suitable for applications requiring quick energy delivery [38]. The spherical morphology of SNC offers the advantage of efficient transport of ions and reduces the distances for diffusion within the electrode material [39]. This beneficial electrochemical kinetics allow rapid charging and discharging, which is desirable in applications like electric vehicles and portable electronics, in which quick access to energy is required [40,41].

9.4.3 Comparison with Graphite-Based Electrodes

Compared with traditional graphite-based electrodes, SNC-based electrodes have shown several advantages. Graphite has been extensively used in LIBs as the anode due to the excellent conductivity and demonstrated performance as an anode material; however, SNC exhibits more specific capacities along with remarkable cycle stability. The ease of accommodating the volume variation in cycling increases its lifespan by a considerable extent, as opposed to that of the graphite structure degrading after a period of time. Further, SNC is biodegradable—a very important environmental benefit, which graphite cannot offer and its demand is on rise in energy storage technologies owing to the increasing concern over green materials. Overall, the electrochemical performance properties of SNC-based electrodes make them superior alternatives to traditional materials such as graphite. With high specific capacitance, excellent cycle stability, fast charge/discharge rates, and environmental sustainability, SNC is a promising candidate for next-generation energy storage solutions [42].

9.5 INNOVATIVE USES OF SNC IN ENERGY STORAGE

The increasing application of SNC as an emerging material for energy storage systems and anode materials in batteries and fuel-cell membranes underscores its unique properties [43]. Further improvement in performance and achievement toward sustainability goals in the energy sector is expected through additional research on the development of energy storage technologies.

9.5.1 SNC as an Anode Material

SNC has gained much attention as an anode material in a wide range of energy storage devices, such as lithium-ion and sodium-ion batteries (NIBs) [44]. The spherical geometry of SNC enhances electrochemical performance by providing more space for lithium or sodium ions to intercalate, increasing the surface area and improving ion accessibility for charge storage [45]. Another flexibility in SNC to absorb the volume changes during cycling adds better cycle stability and makes it a suitable substitute for graphite-based anode materials. Also, the biodegradable property of this material aligns it with the rising demands of renewable energy storage [46,47].

9.5.2 SNC in NIBs

SNC presents a lot of benefits over conventional materials in NIBs. The spherical morphology of SNC provides uniform distribution and transport of ions, which is important for high specific capacities and rate capabilities. Research studies have shown that SNC-based anodes can display superior performance metrics, such as reversible capacity and excellent cycling stability [48]. This means that SNC not only improves performance but also diminishes reliance on expensive and relatively scarce materials such as lithium. Thus, SNC is more sustainable for large-scale energy storage applications [49].

9.5.3 SNC for Fuel-Cell Membranes

SNC demonstrates excellent proton conductivity combined with mechanical strength, thus representing a promising material for membrane applications in fuel cells [50]. Its biodegradable properties enhance its eco-friendliness compared with traditional membrane materials. The unique structural properties of SNC allow it to transport ions efficiently while maintaining structural integrity under operating conditions [51]. Ongoing research should be directed toward optimizing performance through chemical modification and composite formulations of SNC-based membranes that may form the basis for more efficient and sustainable fuel-cell technologies.

9.6 MODIFICATION TECHNIQUES TO ENHANCE SNC PERFORMANCE

Modification techniques like chemical functionalization, composite creation, and porous structure engineering greatly improve the performance of SNC for energy storage applications. These modification techniques improve electrochemical properties, including charge storage capacity, ion transport, and cycling stability. These modifications have now extended the applicability of SNC by tuning its surface chemistry, hybridization with other materials, and optimization of pore networks for use in the most advanced energy systems such as batteries, supercapacitors, and fuel cells. These developments render SNC as a more efficient and sustainable material for the next-generation energy storage technologies. Table 9.3 discusses the innovative applications and advancements of SNC in energy storage systems, the benefits and challenges it holds, and its utilization in SNC-based electrodes and membranes with a focus on the enhancement of performance and sustainability for energy storage solutions.

9.6.1 Chemical Functionalization

Functionalization of chemical nature: the surface chemistry of SNC can be modified to enhance its functionality and compatibility with other material properties [52].

TABLE 9.3
Innovative applications and advancements of SNC in energy storage

Application	Technology	Key Advantages	Challenges	Performance and Potential
Innovative uses of SNC in energy storage	All technologies	• Biodegradable • High surface area • Enhanced electrochemical performance • Contributes to sustainability goals	• Scalability of synthesis processes • Ensuring consistency in performance	Potential for advancing energy storage systems with improved performance and sustainability
SNC as an anode material	Lithium-ion and sodium-ion batteries	• Increased surface area for ion intercalation • Enhanced ion accessibility • Better cycle stability due to flexibility. • Sustainable alternative to graphite	• Need for optimized synthesis for large-scale production • Balancing flexibility with mechanical strength	Viable alternative to graphite, offering better performance and sustainability for energy storage devices
SNC in sodium-ion batteries	Sodium-ion batteries	• Uniform ion distribution. • High reversible capacity • Excellent cycling stability • Reduced reliance on expensive lithium	• Challenges in scaling production for large-scale applications • Further optimization of rate capability	Promising material for sustainable and high-performance large-scale energy storage
SNC for fuel-cell membranes	Fuel cells	• Excellent proton conductivity • High mechanical strength • Biodegradable and eco-friendly • Efficient ion transport	• Performance optimization through chemical modifications • Enhancing membrane stability under operating conditions	Potential for efficient, sustainable fuel cells with reduced environmental impact

(*Continued*)

TABLE 9.3 (*Continued*)
Innovative applications and advancements of SNC in energy storage

Application	Technology	Key Advantages	Challenges	Performance and Potential
Modification techniques to enhance SNC performance	All technologies	• Improved electrochemical properties • Enhanced charge storage capacity • Better ion transport • Customizable surface properties	• Balancing modification with scalability. • Ensuring long-term stability of modifications.	Expanded use of SNC in advanced energy storage systems, improving overall efficiency and sustainability
Chemical functionalization	All technologies	• Improved compatibility with electrolytes • Enhanced hydrophilicity and ion transport • Provides bonding sites for polymers	• Need for precise control over functional group addition • Ensuring uniform performance in large-scale applications	Significant improvement in the electrochemical performance of SNC for batteries and supercapacitors
Composite preparation	Lithium-ion, sodium-ion batteries	• Enhanced electrical conductivity • Improved mechanical strength • Synergy with other materials for better performance	• Difficulty in optimizing composites for specific applications • Maintaining material stability over time	Improves rate capabilities, energy densities, and overall performance of energy storage systems
Porous structure engineering	Supercapacitors, batteries	• Increased surface area for ion exchange • Improved electrolyte infiltration • Faster ion transport • Lightweight and efficient	• Challenges in achieving consistent pore structures • Balancing porosity with structural integrity	Optimized porous structure leads to efficient-energy storage and faster charge/discharge rates

This technique allows various functional groups to be present on the surface of the SNC, thus promoting interaction with electrolytes or other components in energy storage systems. For instance, this modification increases the hydrophilicity, the transport of ions, and offers anchoring sites for bonding with polymers or conductive materials [53]. These modifications significantly enhance the electrochemical performance of SNC in applications ranging from batteries to supercapacitors by improving charge storage and cycling stability [33].

9.6.2 Composite Preparation

Composite preparation is the mixing of SNC with other materials to combine their strengths and hence improve overall performance. In particular, SNC can be mixed with conductive polymers, metal oxides, or carbon-based materials for improved electrical conductivity, mechanical strength, and thermal stability [54,55]. The composites can be tailored for special energy storage applications, such as lithium-ion or sodium-ion batteries, where they show better rate capabilities and energy densities [56,57]. Generally, this hybrid exhibits better electrochemical activity than the use of pure SNC and other materials.

9.6.3 Porous Structure Engineering

Porous structure engineering involves the design of a pore network within SNC to optimize its characteristics. This can be achieved through templating, freeze-drying, and controlled degradation processes [58]. Introducing porosity increases the surface area for ion exchange and enhances the infiltration of electrolytes into the pores, which is important for energy storage applications [53]. A well-engineered porous structure leads to faster ion transport, enhancing the charge/discharge rate of supercapacitors and batteries. Further, the reduction of weight without sacrificing strength by porosity makes it feasible to design lightweight efficient-energy storage systems [59,42].

9.7 CHALLENGES AND LIMITATIONS

The SNC might bring about a revolution in energy storage technologies through high capacitance, excellent cycle stability, and environmental sustainability [60]. Challenges, however, include scalability, standardization, compatibility, recyclability, and sustainability [61,62]. All these will be improved with further work on manufacturing processes, protocols for testing, and recycling techniques to ensure the integration of SNC in large energy storage systems. SNC will, in due time, play a huge role in the future of energy storage.

9.7.1 Scalability and Cost-Effectiveness

Some significant hurdles in the large-scale deployment of SNC for energy storage applications relate to scaling up production processes [63]. Despite the existence of various manufacturing methods, scaling these techniques to meet industrial demands while maintaining consistent quality remains challenging. Additionally, the cost-effectiveness of producing SNC at a competitive price, comparable with traditional materials like graphite or metal oxides, is a major concern. High production costs hinder its commercialization, highlighting the need for more efficient and economical manufacturing methods to make SNC-based technologies viable at large scales.

9.7.2 Standardization and Interfacial Compatibility

Standardization of the properties of SNC is crucial for ensuring consistent performance across various applications. Variability in the source material, production methods, and post-processing techniques can result in discrepancies in the electrochemical performance of SNC [64]. To address this, standardized protocols for characterizing and testing SNC must be developed to facilitate its seamless integration into energy storage systems. Additionally, interfacial compatibility between SNC and other materials, such as those used in composites or electrodes, is vital for optimizing performance [65]. Poor compatibility between the electrodes and electrolyte layers can lead to issues like delamination or reduced conductivity, ultimately affecting the overall performance of energy storage devices [53].

9.7.3 Recyclability and Sustainability

SNC derived from biodegradable biomass, holds significant promise for providing sustainable energy storage solutions [66]. However, recyclability remains a critical aspect of its overall sustainability [67]. To fully realize its potential, effective recycling processes must be developed to enable SNC to be recovered and reused after its life cycle in energy storage applications. This approach would minimize the environmental impact throughout the lifecycle of SNC. As demand for eco-friendly materials continues to grow, innovations in recycling technologies and the adoption of sustainable practices in SNC manufacturing will enhance its role in next-generation energy storage systems [42]. These advancements will not only make SNC a more attractive alternative to traditional materials but also support the development of a circular economy within the energy storage sector. This will ultimately contribute to the broader goal of reducing the environmental footprint of energy storage technologies.

9.8 FUTURE DIRECTIONS IN SNC-BASED ENERGY STORAGE

The future directions for SNC-based energy storage encompass a range of innovative research areas, including novel applications in flexible devices, integration with other nanomaterials to boost performance, development of robust characterization methodologies, and modeling techniques to predict performance outcomes. These efforts will be vital for unlocking the full potential of SNC in next-generation energy storage technologies [68].

9.8.1 Research Areas and Novel Applications

Future research on SNC is poised to explore a variety of innovative applications in energy storage. Key areas of focus include the development of flexible energy storage devices suitable for wearable electronics, such as smartwatches and health monitoring systems [69]. Additionally, there is significant potential for SNC in advanced battery technologies, including lithium-sulfur and zinc-ion batteries, which require materials capable of accommodating substantial volume changes during operation. Researchers are also investigating the use of SNC in hybrid energy storage systems that combine the advantages of supercapacitors and batteries, aiming to enhance overall performance metrics like energy density and power output [26].

9.8.2 Integration of SNC with Other Nanomaterials

Integrating SNC with other nanomaterials is a strategic approach to enhance its electrochemical performance. This integration can involve creating composites with conductive polymers, metal oxides, or carbon-based materials to improve electrical conductivity and mechanical stability [54,55]. Such composites can be tailored for specific applications in energy storage systems, resulting in improved charge storage capabilities and cycling stability. For instance, combining SNC with graphene or carbon nanotubes can facilitate better charge transport within electrodes, thereby boosting the overall efficiency of batteries and supercapacitors [70].

9.8.3 Development of Characterization Methodologies

Advancing characterization methodologies is crucial for understanding the properties and performance of SNC in energy storage applications. Developing standardized protocols for evaluating electrochemical behavior, mechanical properties, and thermal stability will enable more consistent comparisons across different studies and applications. Techniques such as scanning electron microscopy, transmission electron microscopy, and electrochemical impedance spectroscopy can provide valuable insights into the structure–property relationships of SNC. Enhanced characterization methods will guide future modifications and optimizations of SNC for specific energy storage applications.

9.8.4 Modeling of SNC-Based Energy Storage Processes

Modeling the processes can significantly improve our understanding of their performance dynamics of the systems [71,72]. Computational models that simulate ion transport, charge/discharge kinetics, and thermal behavior could provide valuable predictive insights into how SNC performs under various conditions [73,74]. This predictive capability will inform the design of new materials and structures ultimately, leading to more efficient-energy storage devices [75,76]. Additionally, ML algorithms can be employed to identify key parameters influencing material behavior, allowing for an optimized material synthesis and design [77–80]. The application of AI, ML, and internet of things (IoT) can also facilitate the exploration of novel materials and structures [81–83]. Collaborative efforts between experimentalists and computational scientists will be essential in advancing this area.

9.9 CONCLUSIONS AND FUTURE SCOPE

- SNC is a promising material for future energy storage systems because of its unique structure and characteristics. SNC, a biodegradable and renewable biomass-based material, has been shown to significantly enhance the energy storage capacity of a wide range of supercapacitors, LIBs, NIBs, and fuel cells. The electrochemical performance of SNC electrodes exhibits higher capacitance, excellent cycle stability, and faster charge/discharge rates. In addition, SNC has been used as an alternative to traditional graphite-based anodes and has further improved reversible capacity, rate capability, and cycle stability.
- NIBs with SNC have demonstrated superior performance in terms of higher specific capacities, cycle stabilities, and rate capability. It is being explored as an environmentally acceptable and nontoxic substitute for fuel-cell membranes and exhibits excellent proton conductivity, mechanical strength, and thermal stability. A great number of modification processes, including chemical functionalization, creation of composites, and design of porous structures, have been employed for enhanced performance; however, many challenges persist, such as scalability and cost-effectiveness, standardization, interface compatibility, recyclability, and sustainability.
- SNC has the potential to transform energy storage applications through improved performance, safety, and sustainability. However, challenges that need to be addressed include scalable synthesis, interfacial engineering, and standardization.
- NIBs have good electrochemical properties, such as higher specific capacities and good cycling stability. Due to its spherical shape, SNC provides better and more uniform ion distribution, making it more suitable for use as a precursor in large-scale energy storage. Due to its biodegradable nature and excellent proton conductivity, SNC holds great promise for sustainable energy storage solutions.
- A comparison between graphite-based electrodes and SNC-based electrodes reveals many advantages of the latter. Though graphite has proven to be a valid anode material in LIBs, providing satisfactory conductivity and performance, SNC offers greater specific capacities and better cycle stability. The accommodation of volume changes during cycling prolongs the life of SNC, whereas the structure of graphite degrades gradually over time. The electrochemical performance properties of SNC-based electrodes position them as superior alternatives to conventional materials like graphite. High specific capacitance, excellent cycle stability, fast charge/discharge rates, and environmental sustainability make SNC a promising candidate for next-generation energy storage solutions.
- SNC is gaining popularity as a material for energy storage systems, anode materials in batteries, and fuel-cell membranes. Its unique properties enhance performance and contribute to sustainability goals in the energy sector. SNC's spherical geometry promotes electrochemical performance by providing additional space for lithium or sodium ions to intercalate,

increasing surface area and enhancing ion accessibility for charge storage. Its flexibility to accommodate volume changes during cycling is beneficial for cycle stability, making it a potential alternative to graphite-based anode materials.

- In NIBs, SNC has major advantages over its classical counterparts due to its spherical morphology; this favors the uniform transportation and distribution of ions, leading to high specific capacities and high-rate capability. SNC-based anodes have outperformed traditional materials, indicating higher reversible capacity with superior cycling stability. The adoption of SNC in NIBs improves performance while reducing dependency on expensive and scarce materials, as in the case of lithium, making it better suited for large-scale energy storage applications.
- Another promising application of SNC is in fuel-cell membranes, as it has excellent proton conductivity and good mechanical strength. Its biodegradable nature also contributes to its eco-friendliness compared with standard membrane materials. Further research should focus on optimizing its performance through chemical modifications and composite formulations of SNC-based membranes, which are anticipated to lead to more efficient and sustainable fuel-cell technologies.
- Techniques including chemical functionalization, composite fabrication, and structuring of porous features enhance the performance of SNC in energy applications. Enhancements of this nature make SNC a more efficient and sustainable material for next-generation energy storage technology applications. It is critical to design the optimal pore network within SNC through porous structure engineering, thus increasing the surface area available for ion exchange and improving electrolyte infiltration. A well-engineered porous structure leads to faster ion transport, enhancing the charge/discharge rate of supercapacitors and batteries.
- Despite its potential in energy storage, SNC faces challenges such as scalability, standardization, compatibility, recyclability, and sustainability. Further work on manufacturing processes, testing protocols, and recycling techniques will ensure SNC's integration into large-scale energy storage systems.

REFERENCES

1. Meena, S. B., Patil, P. R., Kandharkar, S. R., Hemalatha, N., Khade, A., Dixit, K. K., & Chinthamu, N. (2024). The evolution of smart grid technologies: Integrating renewable energy sources, energy storage, and demand response systems for efficient energy distribution. *Nanotechnology Perceptions*, *20*(S11), 1098–1109.
2. Srivastava, S. (2024). Advancements in nanocellulose derived from plant waste: Strategies for sustainable innovations and applications. *The Journal of Solid Waste Technology and Management*, *50*(3), 543–563.
3. Lan, G. X., Liu, Y., Zhou, N., Guo, D. Q., & Ma, M. G. (2023). Multifunctional nanocellulose-based composites for potential environmental applications. *Cellulose*, *30*(1), 39–60.

4. Gu, Y., Yao, F., Gong, R., Di, Y., Srinivasan, V., Hu, X., ... & Li, Y. (2024). A full green, sustainable paper-based packaging material with high-strength, water resistance, and thermal insulation. *Polymers*, *17*(1), 6.
5. Akhai, S., & Khang, A. (2024). Innovations in Medical Manufacturing: A Review of 3D Printing, Robotics, and Internet of Things (IoT). In A. Khang & K. C. Rath (Eds.), *The Quantum Evolution: Application of AI and Robotics in the Future of Quantum Technology* (pp. 226–241). CRC Press.
6. Mekhzoum, M. E. M., El Bourakadi, K., & Bouhfid, R. (2021). Cellulose nanocrystal/nanoparticles hybrid nanocomposites: From preparation to applications. In *Cellulose Nanocrystal/Nanoparticles Hybrid Nanocomposites* (pp. 1–25). Woodhead Publishing. https://doi.org/10.1016/B978-0-12-822906-4.00008-6
7. Szafraniec, M., Grabias-Blicharz, E., Barnat-Hunek, D., & Landis, E. N. (2022). A critical review on modification methods of cement composites with nanocellulose and reaction conditions during nanocellulose production. *Materials*, *15*(21), 7706.
8. Abbass, M., & Akhai, S. (2025). Exploring the sustainability potential of geopolymer concrete with coal bottom ash and basalt fibres. *Multiscale and Multidisciplinary Modeling, Experiments and Design*, *8*(3), 199.
9. Kumar, S., Ngasainao, M. R., Sharma, D., Sengar, M., Gahlot, A. P. S., Shukla, S., & Kumari, P. (2022). Contemporary nanocellulose-composites: A new paradigm for sensing applications. *Carbohydrate Polymers*, *298*, 120052.
10. Jayanthi, B., Vinoth, S., Hariharan, M., Raja, R. K., Kamaraj, C., & Narayanan, M. (2024). Valorization of agro-industry wastes for nanocellulose fabrication and its multifunctional applications. *Biocatalysis and Agricultural Biotechnology*, *57*, 103124.
11. Mokhena, T. C., Mochane, M. J., Mtibe, A., Sigonya, S., Ntsendwana, B., Masibi, E. G., ... & Motsoeneng, T. S. (2024). Recent advances on nanocellulose-graphene oxide composites: A review. *Cellulose*, *31*(12), 7207–7249.
12. Dias, I. K., Lacerda, B. K., & Arantes, V. (2023). High-yield production of rod-like and spherical nanocellulose by controlled enzymatic hydrolysis of mechanically pretreated cellulose. *International Journal of Biological Macromolecules*, *242*, 125053.
13. Gobi, M., Kumar, A., Singh, J., Singh, S., & Ramamurthy, P. C. (2024). Nanocellulose-Based Adsorption for the removal of heavy metal from wastewater—A review. *Water Conservation Science and Engineering*, *9*(1), 24.
14. Manimaran, M., Norizan, M. N., Kassim, M. H. M., Adam, M. R., Norrrahim, M. N. F., & Knight, V. F. (2024). Critical assessment of the thermal stability and degradation of chemically functionalized nanocellulose-based polymer nanocomposites. *Nanotechnology Reviews*, *13*(1), 20240005.
15. Chen, Y., Zhang, L., Yang, Y., Pang, B., Xu, W., Duan, G., ... & Zhang, K. (2021). Recent progress on nanocellulose aerogels: Preparation, modification, composite fabrication, applications. *Advanced Materials*, *33*(11), 2005569.
16. Chen, S., Li, X., Zhang, C., Long, L., Zhao, X., Zhu, C., ... & Yi, X. (2024). Superhydrophobic nanocellulose aerogel for thermal insulation and thermal resistance up to 1300°C. *Chemical Engineering Journal*, *499*, 156043.
17. Dhali, K., Ghasemlou, M., Daver, F., Cass, P., & Adhikari, B. (2021). A review of nanocellulose as a new material towards environmental sustainability. *Science of the Total Environment*, *775*, 145871.
18. Liu, B., Li, Y., Yuan, Y., Zheng, B., Liu, C., Zhou, L., & Zhang, J. (2023). Controllable self-assembly of cellulose nanospheres through phosphoric acid triggered dissolution-regeneration and degradation. *International Journal of Biological Macromolecules*, *243*, 125119.
19. Yang, Y., Fu, J., Duan, Q., Xie, H., Dong, X., & Yu, L. (2024). Strategies and methodologies for improving toughness of starch films. *Foods*, *13*(24), 4036.

20. Kareem, F., Chandrawati, R., & Ahmed, M. U. (2024). Tailoring nanocellulose: A comprehensive exploration of functionalisation with small, macro, and inorganic molecules for enhanced sensing and dual-mode biosensing. *Measurement, 239*, 115499. https://doi.org/10.1016/j.measurement.2024.115499
21. Almeida, R. L. J., Rios, N. S., & dos Santos, E. S. (2022). Modification of red rice starch by a combination of hydrothermal pretreatments and α-amylase hydrolysis. *Carbohydrate Polymers, 296*, Article 119963. https://doi.org/10.1016/j.carbpol.2022.119963
22. Sezali, N. A. A., Ong, H. L., Jullok, N., Villagracia, A. R., & Doong, R. A. (2021). A review on nanocellulose and its application in supercapacitors. *Macromolecular Materials and Engineering, 306*(12), 2100556.
23. Chen, J., Chen, H., Chen, M., Zhou, W., Tian, Q., & Wong, C. P. (2022). Nacre-inspired surface-engineered MXene/nanocellulose composite film for high-performance supercapacitors and zinc-ion capacitors. *Chemical Engineering Journal, 428*, 131380.
24. Rana, A. K., Hart, P., & Thakur, V. K. (2024). Nanocellulose/zero, one-and two-dimensional inorganic additive based electrodes for advanced supercapacitors. *Materials Today Electronics, 8*, 100103.
25. Chen, W., Yu, H., Lee, S. Y., Wei, T., Li, J., & Fan, Z. (2018). Nanocellulose: A promising nanomaterial for advanced electrochemical energy storage. *Chemical Society Reviews, 47*(8), 2837–2872.
26. Xu, T., Du, H., Liu, H., Liu, W., Zhang, X., Si, C., ... & Zhang, K. (2021). Advanced nanocellulose-based composites for flexible functional energy storage devices. *Advanced Materials, 33*(48), 2101368.
27. Naderi, L., & Shahrokhian, S. (2024). Ni, S co-doped Cu dendrites decorated with core–shell architecture assisted by MOF and Fe 0.92 Co 0.08 S nanoflakes on nanocellulose/graphene fibers for fabrication of flexible wire-type micro-supercapacitor. *Nanoscale, 16*(43), 20260–20279.
28. Mittal, U., Djuandhi, L., Sharma, N., & Andersen, H. L. (2022). Structure and function of hard carbon negative electrodes for sodium-ion batteries. *Journal of Physics: Energy, 4*(4), 042001.
29. Fan, X., Kong, X., Zhang, P., & Wang, J. (2024). Research progress on hard carbon materials in advanced sodium-ion batteries. *Energy Storage Materials, 69*, 103386.
30. Vilela, C., Silvestre, A. J., Figueiredo, F. M., & Freire, C. S. (2019). Nanocellulose-based materials as components of polymer electrolyte fuel cells. *Journal of Materials Chemistry A, 7*(35), 20045–20074.
31. Osman, A. I., Chen, Z., Elgarahy, A. M., Farghali, M., Mohamed, I. M., Priya, A. K., ... & Yap, P. S. (2024). Membrane technology for energy saving: Principles, techniques, applications, challenges, and prospects. *Advanced Energy and Sustainability Research, 5*(5), 2400011.
32. Siddiqa, A., Yhobu, Z., Nagaraju, D. H., Padaki, M., Budagumpi, S., Pasupuleti, V. R., & Lim, J. W. (2023). Review and perspectives of sustainable lignin, cellulose, and lignocellulosic carbon special structures for energy storage. *Energy & Fuels, 37*(4), 2498–2519.
33. Tafete, G. A., Abera, M. K., & Thothadri, G. (2022). Review on nanocellulose-based materials for supercapacitors applications. *Journal of Energy Storage, 48*, 103938.
34. Lin, W., Zhao, S., Lu, B., Jiang, F., Lu, Z., & Xu, Z. (2024). Structures, performances and applications of green biomass derived carbon in lithium-ion batteries. *Energy Materials, 4*(5), N–A.
35. Wang, K., Li, M., Zhu, Z., Ai, W., Wu, H., Wang, B., ... & Huang, W. (2024). Hard carbon with embedded graphitic nanofibers for fast-charge sodium-ion batteries. *Nano Energy, 124*, 109459.
36. Hsu, H. H., & Zhong, W. (2019). Nanocellulose-based conductive membranes for free-standing supercapacitors: A review. *Membranes, 9*(6), 74.

37. Durairaj, A., Maruthapandi, M., Saravanan, A., Luong, J. H., & Gedanken, A. (2022). Cellulose nanocrystals (CNC)-based functional materials for supercapacitor applications. *Nanomaterials*, *12*(11), 1828.
38. Lasrado, D., Ahankari, S., & Kar, K. (2020). Nanocellulose-based polymer composites for energy applications—a review. *Journal of Applied Polymer Science*, *137*(27), 48959.
39. Xie, L. J., Tang, C., Song, M. X., Guo, X. Q., Li, X. M., Li, J. X., ... & Chen, C. M. (2022). Molecular-scale controllable conversion of biopolymers into hard carbons towards lithium and sodium ion batteries: A review. *Journal of Energy Chemistry*, *72*, 554–569.
40. Kang, W., Zeng, L., Liu, X., He, H., Li, X., Zhang, W., ... & Zhang, C. (2022). Insight into cellulose nanosizing for advanced electrochemical energy storage and conversion: A review. *Electrochemical Energy Reviews*, *5*(3), 8.
41. Kim, J. H., Lee, D., Lee, Y. H., Chen, W., & Lee, S. Y. (2019). Nanocellulose for energy storage systems: Beyond the limits of synthetic materials. *Advanced Materials*, *31*(20), 1804826.
42. Maity, C. K., De, S., Verma, K., Moniruzzaman, M., & Sahoo, S. (2023). Nanocellulose: A versatile nanostructure for energy storage applications. *Industrial Crops and Products*, *204*, 117218.
43. Bishnoi, P., Siwal, S. S., Kumar, V., & Thakur, V. K. (2024). Cellulose-based smart materials: Novel synthesis techniques, properties, and applications in energy storage and conversion devices. *Electron*, *2*(2), e42.
44. Azman, N. A. N. M. N., Krishnasamy, K., Asmadi, M., Adrus, N., See, W. Q., Ayub, M., ... & Amin, N. A. S. (2023). Cellulose morphologies for energy applications. In: M. Shabbir (Ed.), *Regenerated Cellulose and Composites. Engineering Materials* (pp. 197–221). Springer Nature Singapore. https://doi.org/10.1007/978-981-99-1655-9_8
45. Ahmed, B., Hossain, M. J., Al Parvez, A., Talukder, A., Al-Amin, M., Al Mahmud, M. A., & Islam, T. (2024). Recent advancements of MXene/Nanocellulose-based hydrogel and aerogel: A review. *Advanced Energy and Sustainability Research*, *5*(3), 2300231.
46. Molaiyan, P., Bhattacharyya, S., dos Reis, G. S., Sliz, R., Paolella, A., & Lassi, U. (2024). Towards greener batteries: Sustainable components and materials for next-generation batteries. *Green Chemistry*, *26*(13), 7508–7531.
47. Norizan, M. N., Shazleen, S. S., Alias, A. H., Sabaruddin, F. A., Asyraf, M. R. M., Zainudin, E. S., ... & Norrrahim, M. N. F. (2022). Nanocellulose-based nanocomposites for sustainable applications: A review. *Nanomaterials*, *12*(19), 3483.
48. Sekar, S., Aqueel Ahmed, A. T., Kim, D. Y., & Lee, S. (2020). One-pot synthesized biomass C Si nanocomposites as an anodic material for high-performance sodium-ion battery. *Nanomaterials*, *10*(9), 1728.
49. Peter, S., Lyczko, N., Gopakumar, D., Maria, H. J., Nzihou, A., & Thomas, S. (2022). Nanocellulose and its derivative materials for energy and environmental applications. *Journal of Materials Science*, *57*(13), 6835–6880.
50. Chibac-Scutaru, A. L., & Coseri, S. (2023). Advances in the use of cellulose-based proton exchange membranes in fuel cell technology: A review. *International Journal of Biological Macromolecules*, *247*, 125810.
51. Chen, C., & Hu, L. (2018). Nanocellulose toward advanced energy storage devices: Structure and electrochemistry. *Accounts of Chemical Research*, *51*(12), 3154–3165.
52. Tortorella, S., Vetri Buratti, V., Maturi, M., Sambri, L., Comes Franchini, M., & Locatelli, E. (2020). Surface-modified nanocellulose for application in biomedical engineering and nanomedicine: A review. *International Journal of Nanomedicine*, *15*, 9909–9937. https://doi.org/10.2147/IJN.S266103

53. Guo, R., Zhang, L., Lu, Y., Zhang, X., & Yang, D. (2020). Research progress of nanocellulose for electrochemical energy storage: A review. *Journal of Energy Chemistry, 51*, 342–361.
54. Cui, S., Wei, P., & Li, L. (2018). Preparation of poly (propylene carbonate)/graphite nanoplates-spherical nanocrystal cellulose composite with improved glass transition temperature and electrical conductivity. *Composites science and technology, 168*, 63–73.
55. Krystyjan, M., Khachatryan, G., Khachatryan, K., Krzan, M., Ciesielski, W., Żarska, S., & Szczepankowska, J. (2022). Polysaccharides composite materials as carbon nanoparticles carrier. *Polymers, 14*(5), 948.
56. Kim, J. H., Lee, D., Lee, Y. H., Chen, W., & Lee, S. Y. (2019). Nanocellulose for energy storage systems: Beyond the limits of synthetic materials. *Advanced Materials, 31*(20), 1804826.
57. Sirichaibhinyo, T., & Ummartyotin, S. (2024). *Development of cellulose-based composite for energy storage applications* (Doctoral dissertation, Thammasat University).
58. Lorevice, M. V., Claro, P. I., Aleixo, N. A., Martins, L. S., Maia, M. T., Oliveira, A. P., ... & Gouveia, R. F. (2023). Designing 3D fractal morphology of eco-friendly nanocellulose-based composite aerogels for water remediation. *Chemical Engineering Journal, 462*, 142166.
59. Wang, L., Han, J., Kong, D., Tao, Y., & Yang, Q. H. (2019). Enhanced roles of carbon architectures in high-performance lithium-ion batteries. *Nano-Micro Letters, 11*, 1–23.
60. Jose, J., Thomas, V., Vinod, V., Abraham, R., & Abraham, S. (2019). Nanocellulose based functional materials for supercapacitor applications. *Journal of Science: Advanced Materials and Devices, 4*(3), 333–340.
61. Balea Martín, A., Blanco Suárez, M. Á., Delgado-Aguilar, M., Monte Lara, M. C., Tarrés, Q., Mutjé, P., & Negro Álvarez, C. M. (2021). Nanocellulose characterization challenges.
62. Clauser, N. M., Felissia, F. F., Area, M. C., & Vallejos, M. E. (2022). Technological and economic barriers of industrial-scale production of nanocellulose. In V. K. Thakur & M. K. Thakur (Eds.), *Green nanomaterials for industrial applications. Micro and Nano Technologies* (pp. 21–39). Elsevier. https://doi.org/10.1016/B978-0-12-823296-5.00015-0
63. Kargupta, W., Stevenson, T., Sharman, S., Tanner, J., & Batchelor, W. (2023). Sustainable production of nanocellulose: Technoeconomic assessment, energy savings and scalability. *Journal of Cleaner Production, 425*, 138748.
64. Chesley, M. P. (2019). *Characterization of nano-cellulose based composites for biomedical applications.* The University of Maine.
65. Han, J., Wang, S., Zhu, S., Huang, C., Yue, Y., Mei, C., ... & Xia, C. (2019). Electrospun core–shell nanofibrous membranes with nanocellulose-stabilized carbon nanotubes for use as high-performance flexible supercapacitor electrodes with enhanced water resistance, thermal stability, and mechanical toughness. *ACS Applied Materials & Interfaces, 11*(47), 44624–44635.
66. Lv, P., Lu, X., Wang, L., & Feng, W. (2021). Nanocellulose-based functional materials: From chiral photonics to soft actuator and energy storage. *Advanced Functional Materials, 31*(45), 2104991.
67. Haron, G. A. S., Mahmood, H., Noh, M. H., Alam, M. Z., & Moniruzzaman, M. (2021). Ionic liquids as a sustainable platform for nanocellulose processing from bioresources: Overview and current status. *ACS Sustainable Chemistry & Engineering, 9*(3), 1008–1034.
68. Mahajan, P., & Sharma, M. (2024). A comprehensive review on developments and future perspectives of biopolymer-based materials for energy storage. *Energy Storage, 6*(4), e634.
69. Horta-Velázquez, A., & Morales-Narváez, E. (2022). Nanocellulose in wearable sensors. *Green Analytical Chemistry, 1*, 100009.

70. Zhang, S., Li, L., Liu, Y., & Li, Q. (2024). Nanocellulose/carbon nanotube/manganese dioxide composite electrodes with high mass loadings for flexible supercapacitors. *Carbohydrate Polymers, 326*, 121661.
71. Wadhwa, Singh, A., Mahapara, A., Akhai, S., Kumar, D., & Kumar, P. (2024). Integrating taguchi optimization for multi-criteria decision making in engineering applications. In S. Aouadni & I. Aouadni (Eds.), *Recent Theories and Applications for Multi-Criteria Decision-Making* (Vol. 1, No. 1, pp. 125–150). IGI Global. https://doi.org/10.4018/979-8-3693-6502-1.ch005
72. Kumar, H., Wadhwa, A. S., Akhai, S., & Kaushik, A. (2024). Parametric analysis, modeling and optimization of the process parameters in electric discharge machining of aluminium metal matrix composite. *Engineering Research Express, 6*(2), 025542.
73. Khang, A., & Akhai, S. (2025). E-waste and lithium-ion battery recycling insights for sustainable transportation. In A. Khang (Ed.), *Driving Green Transportation System Through Artificial Intelligence and Automation: Approaches, Technologies and Applications. Lecture Notes in Intelligent Transportation and Infrastructure* (pp. 203–230). Springer Nature Switzerland. https://doi.org/10.1007/978-3-031-72617-0_11
74. Sharma, V. K., Kumar, P., Akhai, S., Kumar, V., & Joshi, R. S. (2024). Corrosion inhibition analysis on cerium induced hydrophobic surface of Al-6061/SiC/Al2O3 hybrid composites. *Proceedings of the Institution of Mechanical Engineers, Part B: Journal of Engineering Manufacture, 238*(14), 2126–2138.
75. Akhai, S., Walia, J., Bhullar, B. S., & Wadhwa, A. S. (2024). Advances in bandgap and conductivity engineering of graphene-based semiconductors. In N. Nirmala, J. Arun, & S. S. Dawn (Eds.), *Carbon-Based Materials and Environmental Remediation: Graphene, Biochar, and More* (pp. 211–228). IGI Global. https://doi.org/10.4018/979-8-3693-8257-8.ch007
76. Akhai, S., Singh, S., Kaur, S., & Kaur, K. (2024). Advancing aerospace inspection with terahertz technology. In S. Khalid & N. N. Siddiqui (Eds.), *New Innovations in AI, Aviation, and Air Traffic Technology* (pp. 195–211). IGI Global. https://doi.org/10.4018/979-8-3693-1954-3.ch010
77. Abbass, M., Abbas, U., Jafri, R., Arif, S. M., & Akhai, S. (2025). AI and machine learning applications in sustainable smart cities. In S. N. S. Al-Humairi, A. I. Hajamydeen, & A. Mahfoudh (Eds.), *Sustainable Smart Cities and the Future of Urban Development* (pp. 1–32). IGI Global Scientific Publishing. https://doi.org/10.4018/979-8-3693-6740-7.ch001
78. Akhai, S. (2024). Towards trustworthy and reliable AI the next frontier. In U. Kose, N. Sengoz, X. Chen, & J. A. M. Saucedo (Eds.), *Explainable Artificial Intelligence (XAI) in Healthcare* (Vol. 1, pp. 89–99). CRC Press, Taylor & Francis Group. https://doi.org/10.1201/9781003426073-7
79. Kanda, R., Akhai, S., & Bansal, R. (2013). Analysis of MOST technique for elimination of ideal time by synchronization of different lines. *International Journal of Research in Advent Technology, 1*(4), 151–158.
80. Motia, K., Kumar, R., & Akhai, S. (2024). AI and smart manufacturing: Building industry 4.0. In K. Jermsittiparsert, N. Phongkraphan, & N. Lekhavichit (Eds.), *Modern Management Science Practices in the Age of AI* (pp. 1–28). IGI Global. https://doi.org/10.4018/979-8-3693-6720-9.ch001
81. Akhai, S., Abbass, M., Kaur, P., & Kaur, T. (2025). Digital transformation across generations: robotics and AI in action. In M. Anshari, M. N. Almunawar, & P. Ordóñez de Pablos (Eds.), *Impacts of Digital Technologies Across Generations* (pp. 23–40). IGI Global Scientific Publishing. https://doi.org/10.4018/979-8-3693-6366-9.ch002

82. Akhai, S. (2024). Smart home environment for society: An overview of IoT-enabled air conditioners. In S. P. Singh & A. Solanki (Eds.), *Modelling of Virtual Worlds Using the Internet of Things* (1st ed., pp. 74–87). CRC Press. https://doi.org/10.1201/9781003480181-3
83. Abbas, M., Akhai, S., Abbas, U., Jafri, R., & Arif, S. M. (2025). AI-enabled sustainable urban planning and management. In S. Mallik, S. K. Mathivanan, S. K. B. Sangeetha, & B. O. Soufiene (Eds.), *Real-World Applications of AI Innovation* (pp. 233–260). IGI Global Scientific Publishing. https://doi.org/10.4018/979-8-3693-4252-7.ch012

Section 2

Cellulose Microspheres

10 Regenerated Cellulose Microspheres
Preparation and Characterization

Oluwatosin David Abodunrin, Ebenezer Oluwatosin Atoyebi, Samson Kolawole Fasogbon, Ayodele James Oyejide, Tolulope Motunrayo Omotayo, and Nnamdi Chukwunenye Igwe

10.1 INTRODUCTION

Cellulose is one of the most abundant biopolymers in nature, a basic structural component of cell walls in green plants, algae, and some bacteria. Cellulose consists of chains or strings made of units of β-D-glucose, connected to one another through β-1,4-glycosidic bonds, with unique properties such as high tensile strength, biodegradability, and non-toxicity. It has increasingly become a subject of attention among many applications from textiles to paper products, food processing, and pharmaceuticals. Cellulose has a very strong and versatile nature, and this has impelled research and development of cellulose for advanced materials and composites [1,2]. Sustainability and environmental friendliness necessitate the increasing demand for cellulose as a biopolymer that is becoming increasingly important in environmental applications, fostering innovations consistent with principles of green chemistry and sustainable development. Given cellulose prominence, it needs to be differentiated as either native cellulose or regenerated cellulose, with native cellulose primarily depicting the structural integrity of plant matter whereas, regenerated cellulose is derived from the processing of native cellulose aimed at altering its physical, chemical, or functional properties. Besides, native cellulose and regenerated cellulose share chemical similarities in their common backbone derived from β(1→4)-linked glucose units but differ in structures and properties as a result of processing methods applied during their production. Native cellulose, found primarily as cellulose I, has a crystalline structure (70%–90%) with parallel chains highly stabilized by hydrogen bonds, resulting in its considerable strength and limited solubility. Regenerated cellulose, which is cellulose II, undergoes a process (see Figure 10.1), which leads to structural alteration changing the chains to an antiparallel arrangement, reduced crystallinity (40%–60%), and a different set of mechanical properties [3,4]. Native cellulose appears to have a degree of polymerization that ranges from 300 to over

DOI: 10.1201/9781003596127-12

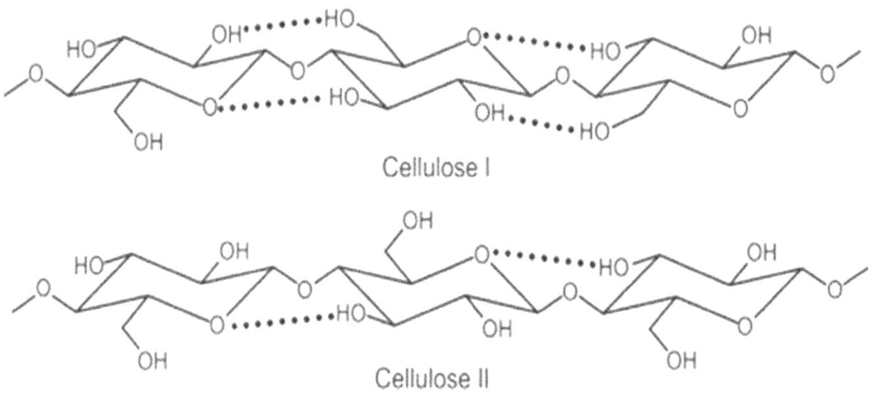

FIGURE 10.1 Chemical structure of cellulose I (native cellulose) and cellulose II (regenerated cellulose), reproduced with permission [7].

10,000 [5,6], but it loses this property considerably during regeneration due to processing conditions. The native cellulose hydroxyl group interactivity ensures structural intramolecular integrity in native cellulose but limits modification, while different bonding orientations in regenerated cellulose contribute to its high solubility and reactivity. Morphologically, the native cellulose possesses hierarchical fibrillar structures, while most times, regenerated cellulose obtains less well-organized forms. Hence, these differences make them different in their special physical properties and applications; therefore, they need to know about the comparison for optimization of cellulose materials in sustainable technologies.

Such transformation from native to regenerated cellulose usually includes dissolving cellulose into solvent-based processes and then regenerating it, thereby leading to materials of different characteristics that can be processed for specific applications. Regenerated cellulose is thus of dual significance: one, it adapts to different requirements of material, and secondly, it offers a possibility to serve as a sustainable substitution for synthetic polymers by permitting the reuse of renewable resources. Besides, this transformation process has progressed through significant changes brought about over more than a century by innovations in material science. In 1838, cellulose itself was identified for the first time by Anselme Payen, defining the future of cellulose for possible industrial applications. By 1855, George Audemars had produced the first regenerated nitrocellulose fiber, but its flames made the material unsuitable for use. From this time forward, a series of significant developments: 1890 cuprammonium and 1891 viscose could revolutionize their applications in artificial silk manufacturing. Viscose became popularly known after being patented by Courtaulds in 1904, pushing up production to reach the highest global output of 3.86 million tons of fiber in 1973. Thus, the innovations created include cellophane, which was invented in 1905, contributing further to applications in cellulose [4,8,9].

Typically, three steps encompass the fabrication of cellulose microspheres: the dissolution of cellulose into a specific solvent, molding it into spheres and revitalizing it or solidifying it. This cellulosic restructuring will change its physical structure and therefore improve its properties and widen its prospects for application.

Regenerated cellulose has several advantages over conventional cellulose, especially in terms of mechanical properties, solubility, and processing capabilities. It is most suitable for biomedical applications and environmental remediation. Regeneration cellulose microsphere (RCM), in particular, has great promise for overcoming the challenges in drug delivery systems, biocatalysis, and advanced adsorption processes. The unique characteristics of these microspheres, such as their spherical shape, low density, and high surface-to-volume ratio, augment their loading capacity and release kinetics in drug delivery. Moreover, the surface hydroxyl groups that can be modified favorably enhance interactions with biological molecules making the microspheres suitable for targeted and controlled release of therapeutics. This feature is very important in the specific customization of functionalities in personalized medicine [10–12]. In addition to applications in biomedicine, regenerated cellulose microspheres display significant potential in environmental applications such as waste-water treatment and pollution management [13]. Their specialized adsorptive properties enable them to adsorb a wide range of pollutants, making these materials conceivable in the filtering and purification of contaminated water. The development of regenerated cellulose microspheres as adsorbents represents both their dual functional applications and provides considerable visibility for the cellulose attributes as a renewable resource, hence contributing toward the sustainability discourse [14,15]. The purpose of this chapter is to describe and review the synthesis methods, affecting parameters, and characterization techniques regarding regenerated cellulose microspheres, thus emphasizing their suitability for various applications and promoting further innovation in their development.

10.2 PREPARATION OF REGENERATED CELLULOSE MICROSPHERES

10.2.1 Raw Materials and Sources

As the main raw material for regenerated cellulose microspheres, cellulose is mainly derived from plants and can be obtained from two of the most common feedstocks: wood pulp and cotton. Because of its high cellulose content and affordability, wood pulp, which is mostly derived from coniferous and deciduous trees is a useful starting material for extensive industrial uses. It goes through a rigorous processing cycle to eliminate lignin and hemicellulose, resulting in extremely pure cellulose. The choice between different types of wood pulp can have a substantial impact on the final microspheres' shape, surface area, and porosity. In contrast, cotton gives a purified cellulose (90%–95% content) and it is preferred for uses that call for exceptional purity and performance, but its availability and higher manufacturing costs limit its use. Because the procurement of the raw material must be considered as having environmental repercussions, the choice ultimately influences not only the sustainability of the regenerated microspheres but also their properties [2,16]. As the demand for regenerated cellulose microspheres rises, it leads to research on alternative sources of cellulose; therefore, the use of agricultural residues, such as bagasse from sugarcane, straw from rice, and straw from wheat, has been highlighted for the potential use as feedstock sources. They are waste materials, contributing to the lesser environmental footprints associated with sources of cellulose and complementing a circular economy. Additionally, improvements in biotechnological techniques for cellulose extraction from non-woody plants like flax and hemp open the door to more environmentally friendly means of producing microspheres. By utilizing the distinctive qualities of various cellulose sources, this diversification of cellulose sources not only takes ecological concerns into account but also helps to improve the microspheres' functional qualities [17–19].

10.2.2 Solvents and Additives Used in Regeneration

Regeneration of cellulose microspheres is carried out with various solvents and additives that play crucial roles in the dissolution and subsequent forming of these materials. Alkali-based solvents, non-derivatizing solvents, and additives are combined in most cases to optimize the properties of regenerated cellulose microspheres. Not only does this optimization increase the efficiency of the regeneration process, but it also makes the final product suitable for specific application in areas such as agriculture or biomedical engineering. Table 10.1 summarizes the solvents and additives used in the regeneration steps.

10.2.3 Methods of Preparation

There are different methods to fabricate regenerated cellulose microspheres as highlighted in Table 10.2, each having its associated advantages and disadvantages. However, the method chosen is usually influenced by the target properties of the microspheres, such as their size, shape, porosity, and surface characteristics.

10.2.3.1 Solvent-Evaporation/Extraction Method

This technique as shown in Figure 10.2 is a versatile and an efficient fabrication method for RCMs. The method consists of three steps: dissolution, emulsification, and regeneration. This involves creating micro-scale droplets of cellulose solution in a continuous phase and subsequently regenerating the cellulose into solid particles. They allow controlled size, shape, and composition of microspheres. Initially, cellulose is dissolved in a solvent system (for instance, NaOH/urea, ionic liquids, TEAOH/urea/H_2O, or LiCl/DMAC) to develop a homogeneous solution. This solution is subsequently emulsified in a hydrophobic oil phase, generally stabilized using emulsifiers or surfactants like polysorbate 80, through means such as mechanical stirring, ultrasonication, or high-pressure homogenization to produce a 'water'-in-oil emulsion. Then, the cellulose is regenerated and solidified into microspheres upon introduction of an anti-solvent (e.g., water or ethanol, or an acid), after which washing is performed followed by drying to eliminate any residual solvents and oil. The process can easily be fine-tuned with modifications such as stirring speed and emulsifier concentration to produce microspheres having controlled size, shape, and encapsulation efficiencies for different applications [20,21].

The emulsification-based method of producing RCMs has several advantages but also has some limitations. Some of its advantages include precise control of parameters such as particle size, morphology, and uniformity that are critical to the

TABLE 10.1
Solvents and additives commonly used for cellulose dissolution and regeneration

Category	Solvent/Additive	Function	Disadvantages
Solvents	Ionic Liquids (e.g., 1-ethl-3-methylimidazolium acetate ([Emim][OAc]), and 1-butyl-3-methylimidazolium chloride ([Bmim]Cl)	Dissolve cellulose effectively at lower temperatures, Efficient dissolution, reversible process	Can be expensive, with potential toxicity concerns
	Sodium hydroxide (NaOH)/ lithium hydroxide (LiOH) with Thiourea Aqueous Solution	Effective solvent for cellulose, relatively inexpensive and environmentally friendly	Requires higher temperatures and longer dissolution times
	N-Methylmorpholine-*N*-oxide (NMMO)	Dissolves cellulose effectively, relatively non-toxic and recyclable	More expensive than NaOH/thiourea
	Lithium chloride (LiCl)/ Dimethylacetamide (DMAc)	Dissolves cellulose effectively, relatively stable and efficient	Can be corrosive
	Tetraethylammonium Hydroxide (TEAOH)/Urea System	Effective solvent system for dissolving cellulose	Potential toxicity issues
Co-solvent	Dimethylformamide (DMF), Dimethyl sulfoxide (DMSO)	Enhance cellulose dissolution within Ionic liquids	Both DMF and DMSO are toxic with possible negative effects on the environment
	Ethanol and Methanol	Used to adjust viscosity and precipitation rates during regeneration	Limited solubility
Additives	Surfactants (e.g., Span 80, Tween 80)	Stabilize emulsions, and control microsphere size and morphology, Improve microsphere uniformity	May affect microsphere properties
	Cross-linking Agents (e.g., epichlorohydrin, glutaraldehyde)	Enhance mechanical strength and stability	Can affect biocompatibility
	Porogen (e.g., sodium chloride, sugar, silica particles)	Create pores in microspheres, enhance surface area and drug loading	May affect microsphere morphology
	Plasticizers (e.g., glycerol, polyethylene glycol, Polyvinyl alcohol)	Increase flexibility and processability	May affect microsphere stability
	Functionalization Agents (e.g Carboxylic acids or anhydrides, Amines)	Provides carboxylic groups that improve adsorption capabilities and amines impart amino functionality for bioconjugation of ion exchange	Limited compatibility

TABLE 10.2
Techniques used in production of regenerated cellulose microspheres

Primary Groups	Method	Key Features	Size Range
Solvent-based techniques	Solvent-evaporation/extraction	Uses immiscible liquids; surfactants stabilize droplets; scalable and versatile	100 μm – 1 mm
	Phase separation (coacervation)	Induces separation via polymer incompatibility or solvent shift; high control over morphology	10–420 μm
	Precipitation techniques	Rapid mixing of cellulose solutions with non-solvent; simple and fast	30–240 μm
Droplet-based techniques	Spray drying	Rapid drying of atomized droplets; scalable and efficient	1–100 μm
	Electrospraying	Uses electric field to form fine droplets; produces small, uniform particles	50 μm
Thermal-, chemical-, or physical-based techniques	Thermal gelation	Relies on temperature-induced phase changes for gel formation	10 μm – 1 mm
	Mechanical milling	Physically grinds cellulose into microspheres; low-cost but less precise	100 nm – 10 μm
	Chemical cross-linking	Uses cross-linking agents to stabilize structure; tunable mechanical properties	10 μm – 1 mm
Specialized/advanced techniques	Template-assisted synthesis	Relies on predefined templates (e.g., molds or beads) for uniform size and shape	10–500 μm
	Freeze-drying (lyophilization)	Freezes solution and removes solvent via sublimation; porous structure	1–100 μm
	Microfluidic techniques	Precise control of droplet size; highly uniform particles; high-cost equipment	10–100 μm
	Layer-by-layer	Potential for creating microspheres with intricate architecture and multiple layers of functionality	1 μm – 1 mm

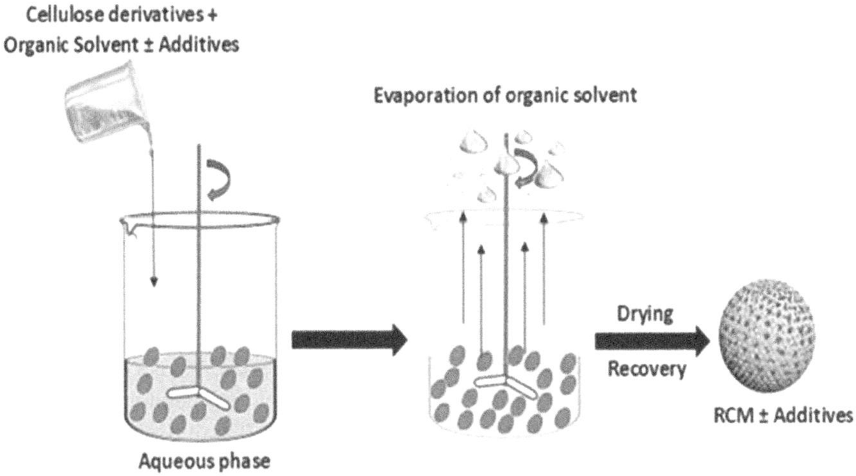

FIGURE 10.2 Schematic representation of solvent-evaporation technique.

applications requiring it to perform consistently. Furthermore, the method is highly versatile because it is amenable to almost all solvents of cellulose and regenerating agents, making it scalable to industrial production and economically viable. Hence, porous microspheres can be prepared for sustained drug release applications, and thermally sensitive compounds can be efficiently encapsulated due to moderate processing temperatures [22,23]. Limitations, though, are associated with this method. Waste management becomes indispensable because it raises environmental and safety issues with organic solvents as well as surfactant use. It might also take quite some time to optimize some process parameters such as stirring speed and emulsifier concentration to achieve stable emulsions with a uniform size distribution [21–23]. The final properties and biocompatibility of the microspheres can also be affected by contamination of the final product with residual solvents, aggregation of microspheres during drying, and the environmental profile of the solvent and surfactant being used.

There are a lot of parameters that affect the quality as well as the characteristics of emulsion-based RCMs. Among these are the concentration and type of cellulose solvent that determine solubility and regeneration. The oil-to-water phase ratio varies, and so does the type and concentration of surfactants affect emulsion stability and particle size distribution. When it comes to stirring speed, the emulsified droplets are typically smaller with faster-stirring rates. Moreover, all these micrometer dimensions and morphology have widely been influenced by the polymerization degree of cellulose and porosity parameters, which include the viscosity of the forming solution. The methods of coagulation, including the type of coagulant and its concentration, significantly affect the porosity, mechanical behavior, and encapsulation efficiency of the microspheres. Similarly, optimum emulsification temperatures have a role in affecting the evaporation rates, consequentially influencing the size and morphology of microspheres [24].

10.2.3.2 Phase Separation (Coacervation)

Coacervation as indicated in Figure 10.3 is one of the popular techniques employed in the manufacture of regenerated cellulose microspheres. The technique is based on the separation of a polymer solution into two distinct phases. One phase, which is the coacervate phase, is rich in polymer, while the latter phase is a continuous phase containing the solvent. Coacervation can be classified as either simple or complex coacervation. Simple coacervation occurs when a single polymer goes through phase separation. Complex coacervation is, however, the result of interaction between two oppositely charged polymers, usually proteins and polysaccharides [25].

Preparation of RCM through coacervation involves some critical steps. First, cellulose is dissolved in solvents, such as NMMO, or ionic liquids. Then a phase separation occurs when a non-solvent (water or ethanol) is added or the conditions, such as temperature or pH, are changed to promote coacervation. These droplets are stabilized through the aggregation of hydrocolloids or polymers surrounding them, which is activated by surface tension differences. In addition to this, cross-linking or gelling agents are included for added stability. Then, microspheres are collected through filtration or centrifugation and washed and dried to the preferred size, morphology, and water insolubility [26,27]. They are capable of encapsulating an extremely wide variety of materials, ranging from hydrophilic to hydrophobic compounds aimed at uses in pharmaceuticals and food industries. All these enable the controlled release of encapsulated materials by the composition and structure of microspheres. More importantly, regenerated cellulose is biocompatible and biodegradable, thereby increasing its use for different applications. The efficiency and the results of the coacervation process are affected by some parameters. The adjustment of pH could further significantly influence solubility and charge interaction between the polymers and, thus, phase separation. Temperature changes may affect whether or not gelation occurs or may alter viscosity and thereby result in changes in droplet formation and stability during coacervation. The polymer concentration in solution determines viscosity and interfacial tension, both of which are necessary for successful microsphere formation [28,29].

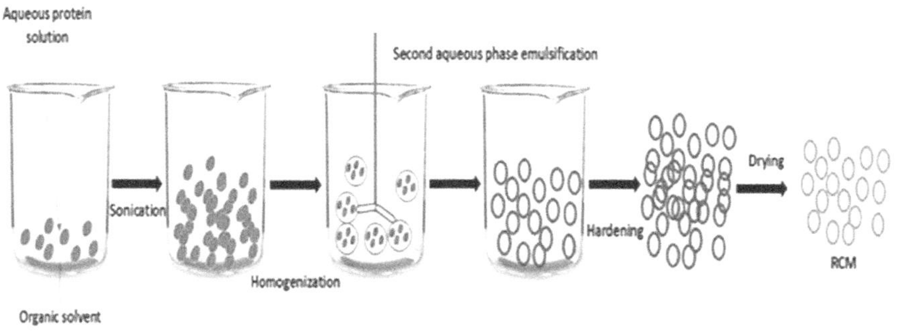

FIGURE 10.3 Phase separation technique.

10.2.3.3 Precipitation Technique

The preparation of RCMs that involve the precipitation process follows two major steps, which are dissolution and precipitation. The cellulose is dissolved in a solvent system, like ([Emim][OAc]), NaOH/urea, or (TEAOH)/urea, which breaks the hydrogen bonding network within it. By emulsification or dripping, this homogeneous solution is then transferred into a non-solvent phase, where it forms droplets. The non-solvent, usually oil-containing surfactants such as polysorbate 80, facilitate the gradual precipitation and regeneration of cellulose microspheres according to the evaporation of the solvent and regenerating intermolecular hydrogen bonds. Further hardening of the microspheres may occur by coagulation in acids such as acetic acid. The resultant particles will be subjected to washing for residual solvent and impurity removal before they will be dried or functionalized further as required. Some of the most important parameters that affect the property of the produced microspheres include the solvent system, composition of non-solvent, surfactants' concentrations, stirring speed, and coagulation time, all of which can then be optimized for particular properties [21,30,31].

10.2.3.4 Spray Drying

This is a recognized technique for preparing RCMs, which are efficient, scalable, and can produce uniform particles. Preparation of cellulose solution is the first step involved in spraying drying as indicated in Figure 10.4, usually carried out by dissolving cellulose in a solvent system like NMMO or in ionic liquids that almost completely break down the cellulose into homogeneous solutions. Cross-linking agents or functional modifiers can be added to improve microsphere properties for specific applications. In the atomization process stage, the cellulose solution enters the spray dryer fitted with an atomizer like a rotary disk, pressure nozzle, or ultrasonic device. Different atomizers influence droplet size, thereby affecting the produced microsphere particle's final size [32,33]. The atomized droplets are discharged into the

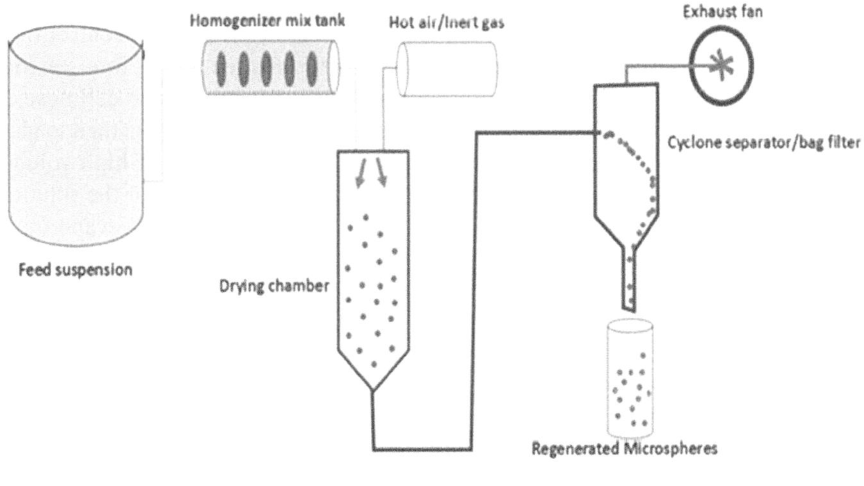

FIGURE 10.4 Schematics showing spray drying technique.

drying chamber where hot air or inert gas is passed through. The temperature and flow rates of the drying medium are strictly monitored to promote rapid solvent evaporation without resulting in any thermal degradation of the cellulose. Microspheres will form from the droplets as soon as the solvent evaporates. Drying parameters such as inlet and outlet temperatures directly relate to the morphology, porosity, and mechanical properties of the prepared microspheres. By adjusting inlet and outlet temperatures, either hollow or dense microspheres can be obtained. Once dried, the microspheres are separated using a cyclone separator or a bag filter, with minimal losses and contaminations. Regenerated cellulose microspheres might need further processing, depending on the desired application. For example, cross-linking treatment may increase chemical and thermal stability, while surface functionalization imparts specific properties such as hydrophobicity, ion exchange capacity, or bioactivity. Such post-processing amendments make the use of the microspheres in various fields more extended [34–36].

The concentration of cellulose solution in feed affects the microspheres significantly in size and morphology. Higher concentrations give larger particles usually, which influence the smoothness of surfaces. The type and configuration of the atomizer, combined with the air pressure used, play a vital part in determining the size of the initial droplets produced. All of these parameters on atomization will influence microsphere sizes directly. The atomization effect can be optimized concerning particular applications, thus adapting the properties of microspheres. Temperature during drying is another parameter that is critical in its control. The inlet and outlet temperatures of the spray dryer have to quickly evaporate solvent, while also being thermally stable for the cellulose. Too high a temperature may lead to cellulose degradation, while low temperatures may lead to incomplete drying of the cellulose, thereby compromising the integrity of the particle [21,35,37,38].

10.2.3.5 Electrospraying

The electrospraying method shown in Figure 10.5 is a kind of droplet-based method versatile for producing nano-micro particle materials offering good control over particle size and morphology. The cellulose solution is prepared before starting the electrospraying process. This solution usually contains dissolved cellulose in the relevant solvent system: NMMO, ionic liquids, etc. The solution is then loaded in a syringe device with a metallic needle. When the application of high voltage is made between the needle and collector plate, the surface tension of the solution gets overcome. A fine jet of liquid is ejected, which undergoes rapid evaporation of the solvent or coagulation in a non-solvent bath, creating solid microspheres, depending on the setup. The microspheres are collected on the grounded plate or within the coagulation medium for further processing and analysis [39,40]. The electrospraying process and properties of regenerated cellulose microspheres are highly influenced by some parameters. Among these is the voltage; the higher the voltage, the smaller the droplets that are produced because of the increased electrostatic repulsion; flow rate, which influences droplet formation, at slower rates, larger particles are formed; distance between the needle and the collector affects time of flight of droplets and the extent of evaporation; concentration of the

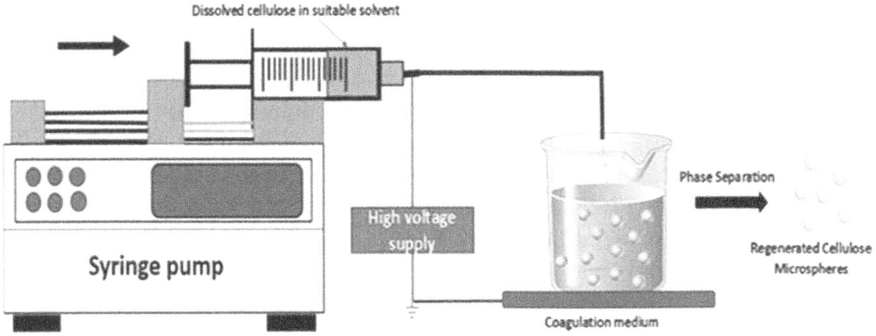

FIGURE 10.5 Electrospraying technique for cellulose regeneration.

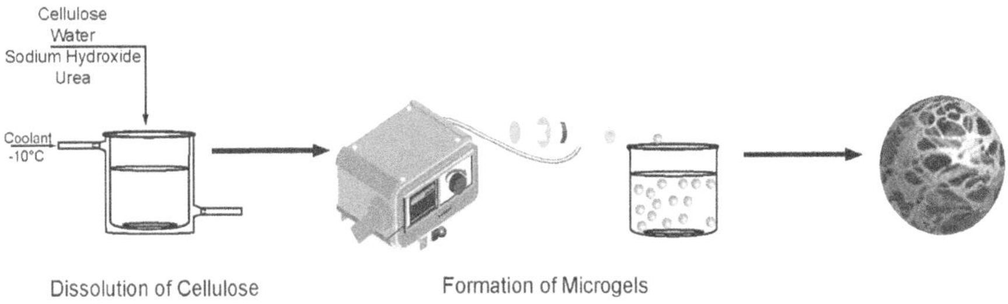

FIGURE 10.6 Thermal gelation technique for cellulose regeneration [47].

solution in higher concentrations increases the viscosity affecting the formation of droplets; as well-environmental conditions like humidity and temperature can influence solvent-evaporation rates, thus affecting the resulting particle morphology [41,42].

10.2.3.6 Thermal Gelation

Thermal gelation as indicated in Figure 10.6 is an important method for the production of RCMs in which microspheres of controlled dimensions and morphologies are formed by thermally induced phase transitions. The process typically commences, like all the other methods, with the dissolution of cellulose in an appropriate solvent system, such as NaOH/urea or NaOH/thiourea aqueous solutions, at low temperatures to have complete dissolution. The cellulose solution is heated to undergo sol–gel transitions that induce gelation via physical cross-linking of cellulose chains. Factors such as temperature, cellulose concentration, and time can influence the transition. Thus, a rheological evaluation of the gelation process could reveal that the storage modulus increases as gelation progresses. After gelation, the cellulose solution is emulsified into microdroplets within a continuous phase, stabilized with surfactants or polymers. These microdroplets are then subjected to controlled heating to complete the gelation process. The gel microspheres are then washed to remove residual solvents and impurities and either freeze-dried or air-dried to give stable, porous particles that can be applied [43–46].

10.2.3.7 Mechanical Milling

The mechanical milling process involves several stages. First, in the preparation stage, cellulose is dissolved in a suitable solvent, such as DMAc/LiCl, to create a homogeneous solution. Next, during milling, the cellulose solution is subjected to mechanical forces using various types of mills, such as hammer, ball, or jet mills, which apply impact, shear, and attrition to reduce particle size. Mechanical milling, therefore, refers to the application of a variety of mechanical forces to break down the cellulosic material into micro-scale particulate streams. The next stage is regeneration, in which an anti-solvent for ending precipitating cellulose microspheres from the solution is introduced. The most commonly utilized anti-solvents for this regeneration procedure are water and alcohol. The step is very important since it will have a great effect on the crystalline pattern and morphology of the final product. The last stage is the harvesting where the microspheres formed are collected, washed to remove excess solvents, and dried for either storage or further applications. This process allows for control of the microsphere sizes and characteristics, from a few micrometers to several hundred meters [48–51].

10.2.3.8 Chemical Cross-Linking

This can be accomplished by either physical or chemical methods; while physical cross-linking consists essentially of hydrogen binding or ionic interaction, chemical cross-linking utilizes such agents as epichlorohydrin or glutaraldehyde to make covalent bonds between cellulose chains. The process enhances the microsphere's structural integrity. With methods like emulsification or through the incorporation of dissolved cellulose dropwise into a non-solvent, microsphere cellulose is formed. Moreover, emulsification helps in creating the spherical shape via cellulose solution

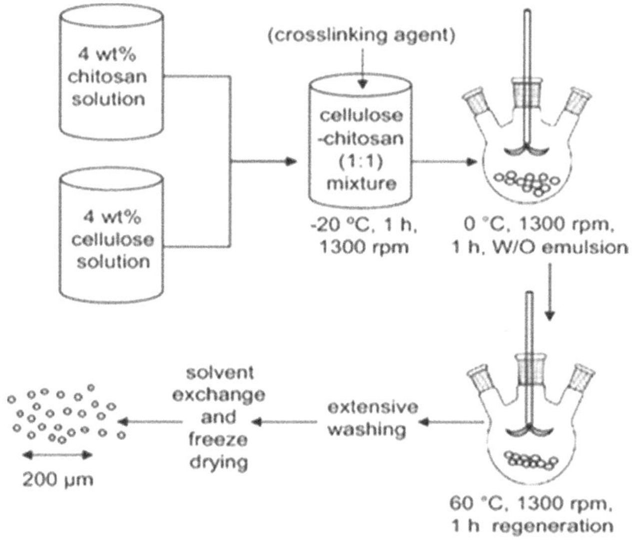

FIGURE 10.7 Schematic showing the cross-linking method of preparing RCMs [56].

combination with an oil phase while stabilizing it using surfactants. After shaping, the microspheres go through a regeneration process in which they are hardened using an acid solution or solvent exchange, causing the cellulose to precipitate out and form a structure resembling gel. Cross-linking agents as indicated in Figure 10.7 can be introduced during or after the regeneration process to increase the stability and usefulness of these microspheres. Suspension cross-linking is a popular technique for cross-linking in which small droplets of a polymer solution are created in a liquid that does not mix with it, and these droplets are subsequently solidified by covalent bonding. Additional biopolymers, such as chitosan or agarose, can be added to the process to improve it and produce composite microspheres with even higher qualities [52–55].

10.2.3.9 Specialized/Advanced Techniques

Template-assisted synthesis is a technique that has been used for making microspheres of well-defined shapes, sizes, and porosities. Sacrificial templates like porous molds, beads, or regenerated cellulose sponges are used to mold precursors such as cellulose acetate into shapes by filling a template with the precursor material and then regenerating it using such solvents like NaOH or ethanol. Once the cellulose structure is formed, it is typical for the template to be removed by dissolution or thermal degradation, thus yielding the intended configurations of microspheres. This enables controlled features of the microspheres and makes possible surface modifications via the introduction of functional groups to further enhance application suitability [57–59]. Freeze-drying (also called lyophilization) is a method of preservation that is particularly useful for heat-sensitive materials, as it helps them retain their original structure and biological activity. Firstly, the cellulose solution gets frozen under low temperatures (around −40°C). From there, the pressure is lowered to create a vacuum environment, highly suitable for sublimation. This technique has been used to prepare RCMs for enhanced adsorption capacity [60,61]. The manipulation of tiny fluid volumes through the use of microchannels, or microfluidics, enables exact control over the synthesis conditions. In this process, continual flow mixing and reaction within limited regions create homogeneous microspheres. The production of RCMs with uniform size and shape is made possible by some parameters, such as flow rates and channel geometry, in microfluidic devices. Additionally, microfluidics has benefits, including improved reproducibility, shorter processing times, and less reagent usage. Additionally, adding sensors to the microfluidic system enables real-time reaction monitoring, which improves control over the characteristics of the microsphere [62]. Layer-by-layer methods rely on subsequent alternate deposition of oppositely charged polymers on the template to form the electrostatically assembled multilayered cellulose microspheres. The layers can be designed with specific functionalities, such as mechanical strength or controlled release properties. For example, polyelectrolytes could be coated on a cellulose core to produce magnetic cellulose beads that can be directly applied for enzyme immobilization. This method offers the possibility of fabricating microspheres with complex structures and multilayered functions. Once the desired thickness has been obtained, the template is removed, leaving behind the hollow or solid cellulose spheres. However, this requires a long time to prepare and also seems to need advanced equipment for layer deposition and monitoring uniformity [63,64].

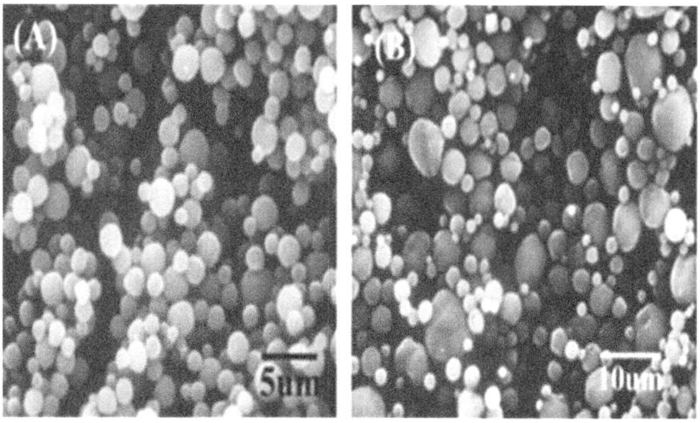

FIGURE 10.8 Sample SEM image of RCMs reproduced with permission [33].

10.3 CHARACTERIZATION OF REGENERATED CELLULOSE MICROSPHERES

10.3.1 Morphological Analysis

The surface and internal morphologies of RCMs are critical for understanding their structure–function properties, which can be efficiently investigated by Scanning Electron Microscopy (SEM) and Transmission Electron Microscopy (TEM). SEM is designed to provide image details of surface morphology, allowing users to observe exactly how spherical the materials are, alongside their variable diameter, surface roughness, and porosity (see Figure 10.8). Moreover, SEM, along with methods of particle size determination is useful in establishing the uniformity and dispersion of the particles on microsphere surfaces that can further affect their performance in target applications. For instance, Wu et al. discovered that the microspheres produced had a 3D porous structure and an increased surface area. It could therefore be said that the images also showed some small and big holes on the surface that were important in applications where there is a need for a high adsorption capacity [40]. In several other studies, SEM images revealed a highly homogeneous spherically shaped structure with a porous framework within cellulose microspheres [22,41,45,65,66].

However, the TEM analysis supplements SEM findings with internal features at a nanoscale level as shown in Figure 10.9, which reveals the crystalline and amorphous regions within RCM materials that influence the mechanical property and degradation rates. It can also be used to determine the size and distribution of nanoparticles embedded into the cellulose matrix, which will provide insight into how these nanoparticles interact with cellulose structure. Hence, SEM and TEM analysis of RCM are critical in clearly understanding the morphological and structural characteristics of such materials concerning various applications.

10.3.2 Particle Size Distribution

Particle size distribution is important in establishing the functional features of RCMs. Typically, RCM particle sizes range from 1 μm to 1 mm, with numerous factors

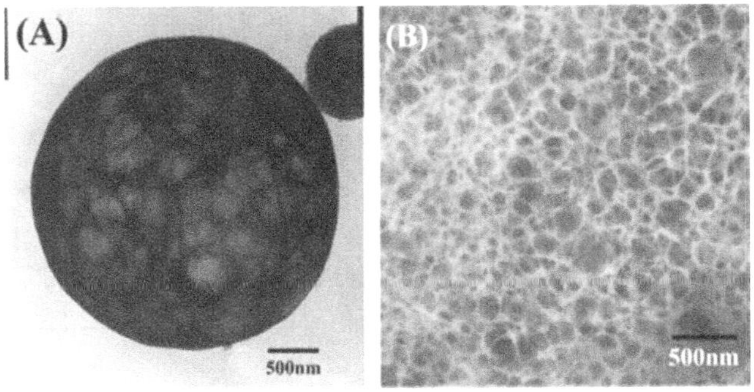

FIGURE 10.9 TEM image showing internal morphology of a RCM, reproduced with permission [33].

impacting the size dispersion [21,67]. For example, high cellulose concentrations produce large microspheres. Other factors that influence size distributions include solvent composition, stirring speed, and preparation processes. Furthermore, uniform particle size has been demonstrated to maintain consistency in drug delivery applications, where size has a direct impact on bioavailability. Small particle sizes have been proven in catalysis to alter the specific surface area to volume ratio, which increases reaction speeds. To assure the PSD assessment reliability, several characterization techniques, in addition to SEM and TEM, could be used to accurately estimate particle size distribution. Examples include dynamic light scattering [68], laser diffraction, particle size analyzer (PSA) [69], and optical microscope [70]. For example, Dong et al. [71] used PSA to determine the size of their RCM, which was around 200 microns. Meryem et al. [70] determined PSD using an optical microscope with a magnification range of 37–779 microns. Furthermore, as shown in Figure 10.10, Ranjv et al. [72] created cellulose beads with a size distribution in the millimeter range. These procedures ensure that RCM shape and size parameters are fully characterized.

10.3.3 Surface Area Measurement

The surface area property is one of the principal parameters required for evaluating the performance of RCMs. This is especially important for application areas such as adsorption and catalysis. The most available and regarded as the most used surface area method for RCMs is the Brunauer–Emmett–Teller (BET) method. Briefly, it involves nitrogen adsorption using cryogenic temperatures to develop an isotherm, from which values for the specific surface area can be calculated. However, it is commonly used as it offers accurate measurements for surface area, which is important when considering these materials in terms of adsorption capacity and kinetics. Rina et al.'s [73] study, for instance, indicated that RCMs possess a large BET surface area of 108 m^2/g as highlighted by their potential use as effective adsorbents or carriers. Moreover, it was highlighted in the study of Donia et al. [74] that upon modification, cellulose samples had their surface areas significantly increased with values reaching up to 706 m^2/g, thereby enhancing their affinities for adsorption in heavy metals. The amounts of adsorbents per cellulosic microsphere surface directly relate them to

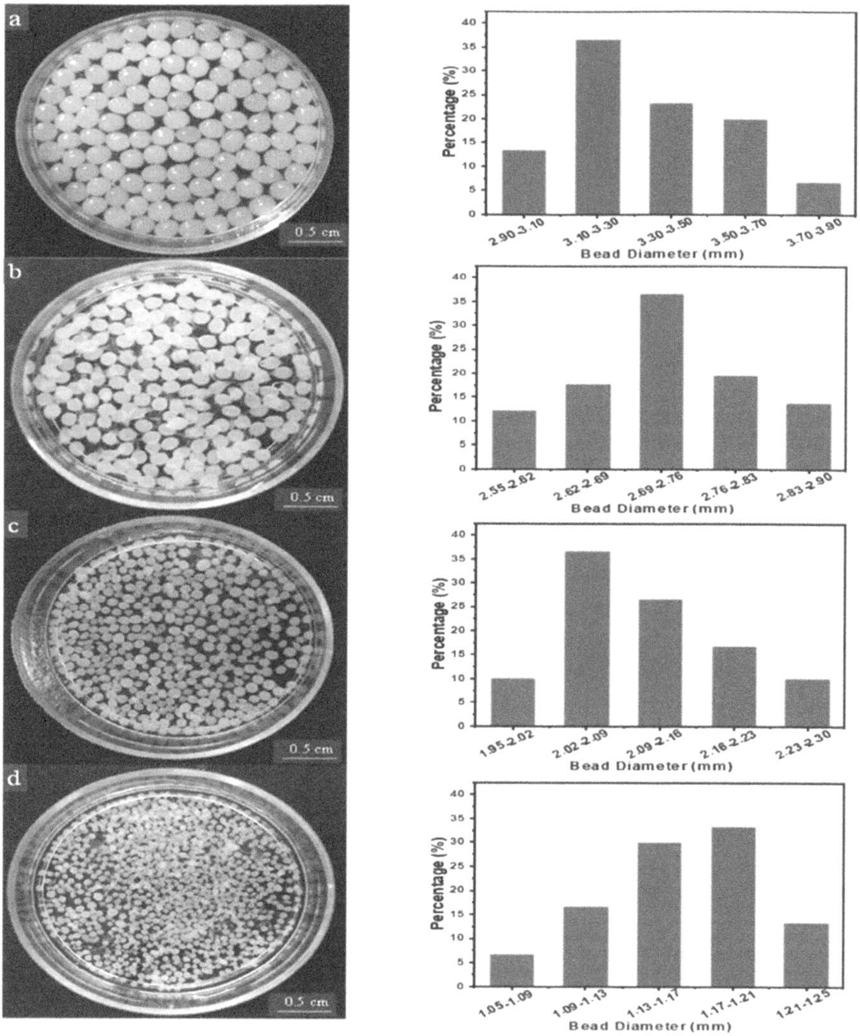

FIGURE 10.10 Particle size distribution of RCMs [72].

their capacity and efficiency. Consequently, an increase in surface area means having more sites available for adsorbing gases. This is especially important when it comes to applications like pollutant removal and enzyme immobilization. For example, Shimaa et al. in their study highlighted an improved adsorption performance for Congo red dye, probably caused by increasing BET values related to cellulose modification [75]. It is also vital to highlight that surface area analysis is just as relevant as alternative forms of characterization for comprehending these microspheres' textural and structural features, which can then be used to optimize how they perform for various uses.

10.3.4 Functional Group Analysis

The major functional groups present in the RCMs include hydroxyl groups (–OH), which are essential for hydrogen bonding, contributing to its hydrophilic property and solubility; ether linkages (–O–), which indicate integrity in cellulosic parts; and carboxylate groups (–COO–), which are usually found in modified cellulose forms for their enhanced affinity or interaction with other molecules. The chemical structure of RCM comprises linear chains of glucose units joined through β(1→4) glycosidic linkages. The regeneration process alters the arrangement of crystalline cellulose from cellulose I to cellulose II. These variations in the crystalline structure significantly impact several of the physical properties of the regenerated cellulose, including porosity and the adsorption capacity. The qualitative presence of these functional groups is crucial to their performance in several applications. Fourier Transform Infrared Spectroscopy (FTIR) is one of the major methods used for characterizing the functional groups and molecular structure of RCMs. It tells us about the chemical bond types and molecular interactions under infrared light with certain wavelengths of absorption. These spectra usually show several typical absorption bands as shown in Figure 10.11. For instance, a broad O–H stretching band at around 3400 cm^{-1} indicates the presence of hydroxyl groups; C–H stretching vibration at 2800–3000 cm^{-1} is associated with aliphatic structures. Whereas peaks near 1730 cm^{-1} would signify the presence of carbonyl in modified cellulose, C–O–C stretching at around 1160 cm^{-1} is associated with the glycosidic linkages; Moreso C–O stretching vibrations found at around 1100 cm^{-1} are important in identifying

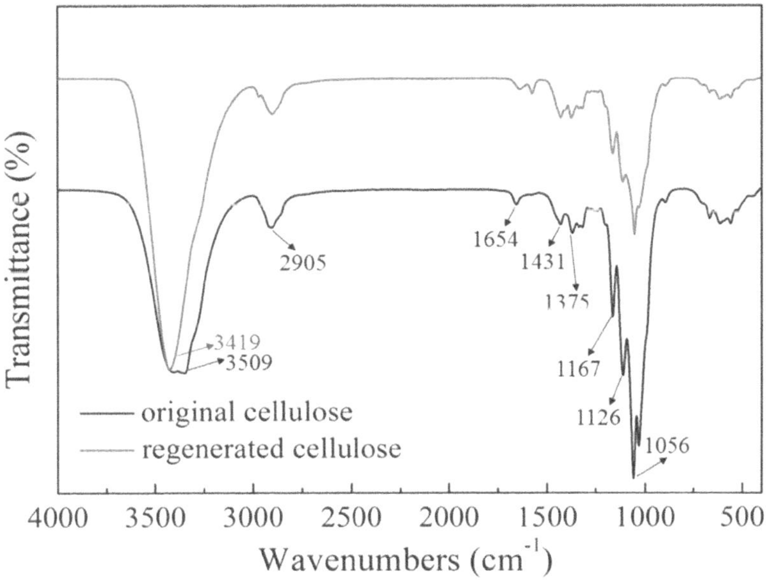

FIGURE 10.11 FTIR image showing changes in functional group before and after regeneration [77].

ether linkages. These bands are very important for the confirmation of the chemical identity of RCM and hence the understanding of their interactions with solvents or any other additives/compounds. FTIR analysis is important for some reasons. It makes it possible to identify functional groups and gives an accurate assessment of the chemical structure of compounds. Keeping an eye on modifications in the chemical structure also guarantees uniformity in production procedures, which is essential for quality assurance. Moreover, FTIR makes it easier to investigate novel uses of RCM across a range of domains [11,21,31,45,76,77].

10.3.5 Phase-Change Analysis

X-ray diffraction (XRD) is one of the critical characterization methods applied to investigate the phase structure of RCMs; it gives insight into the crystalline and amorphous domains, crystal defects, and average grain size present in the material. XRD involves measuring the intensity of the diffracted X-rays as a function of the 2-theta scattering angle, which then allows for the determination of the crystallinity index (CI) by calculating the percentage of area occupied by crystalline peaks to the area occupied by all peaks [78]. Scans are performed on the samples that usually extend from 2° to 40°. The spectra as indicated in Figure 10.12 display characteristic peaks that correspond to certain atomic planes. For cellulose, strong peaks usually occur at about 2-theta angles of 12.3°, 14–18°, 20.4°, and 22.3°, denoting the various crystalline forms [79]. One major distinguishing feature of the XRD patterns of regenerated cellulose is that in general, they show significant peaks due to the different crystalline structures. Peaks at approximately 15.3°, 16.1°, and 22.7° are indications of cellulose form I, whereas peaks at about 12.3° and 20.7° indicate form II from regenerated cellulose microspheres, the latter being the stable form that resulted from the regeneration process. A lack of or peaks broadening can

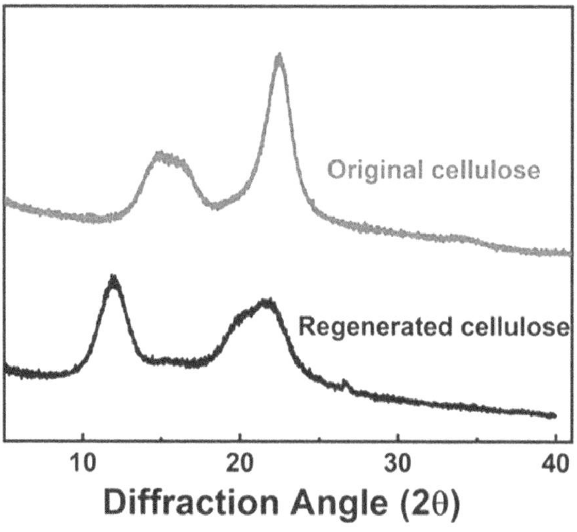

FIGURE 10.12 XRD image showing changes in phase before and after regeneration [79].

indicate an increase in amorphous content, something that is typical of regenerated cellulose, especially because the dissolution and regeneration processes alter hydrogen bonds.

Literature has established the great importance that crystallinity possesses for regenerated cellulose. For example, Gentile et al. [80] demonstrated that wide-angle x-ray scattering highlights the distinction between crystalline and amorphous phase contributions that enhances the accuracy of crystallinity determinations in the regenerated cellulose material. Besides, investigation on the use of various solvents indicates that the mode of dissolution greatly affected the degree of crystallinity of the regenerated cellulose with some solvent leading to lower CI values that were due to higher amorphous content [23]. In another study by Wu et al. [73], they noted that their regenerated cellulose microparticles showed a relatively lower degree of crystallinity in comparison with starting materials, which points to the fact that regeneration affects structural features. There are several reasons why the crystallinity of the regenerated cellulose microspheres is important, for instance, materials with higher crystallinity are likely to have greater strength and stability. Also, the ratio of the crystalline and amorphous parts has a bearing on the water absorption and swelling characteristics that are required in hydrogels and other water-sensitive applications.

10.3.6 Thermal Characterization

For RCMs to be effectively utilized, their thermal properties must be predicted and analyzed. Therefore, two common techniques are usually employed to study RCMs' thermal properties, thermogravimetric analysis (TGA) and differential scanning calorimetry (DSC) as indicated in Figure 10.13a and b [81]. TGA analyzes material weight change during heating, providing insights into thermal stability and degradation behavior. Moisture evaporation typically accounts for some mass loss at less than 150°C while cellulose decomposition follows usually at an initial degradation around 250°C –300°C. Whereas, significant mass loss occurs at around 300°C –400°C,

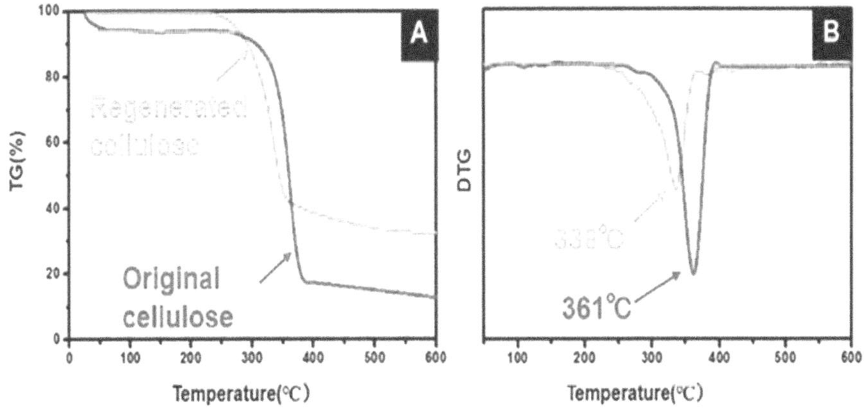

FIGURE 10.13 Thermal characterization monitoring of RCMs [79].

which can occur through mechanisms, including dehydration, depolymerization, and char formation [82]. On the other hand, DSC thermally analyzes the flow of heat in relation to phase transitions, which encompasses melting temperatures (T_m), crystallization temperatures (T_c), and glass transition temperatures (T_g).

For RCMs, a range of 50°C–120°C is where T_g is expected to be observed, meaning that there has been a transition from a rubber state to a glassy state. T_m, which is the state around where the material turns fluid, also is estimated to be 320°C –350°C. Grasping these thermal properties is vital since the thermal integrity makes it possible for these materials to perform under harsh ambient conditions, and also knowing the transition temperatures helps to achieve the processing condition for drying and for molding optimally. It is noteworthy that RCMS are typically less thermally stable than native cellulose due to changes in structural features and composition throughout regeneration. Loss of crystalline order, which occurs when native cellulose is converted to RCM, causes thermal stability in RCM to significantly decrease. Native cellulose has a very high degree of crystallinity, which prevents thermal degradation; however, regeneration disturbs these ordered regions to cause a more amorphous structure that degrades at lower temperatures. Chemical treatments during regeneration are also part of the modes of introducing structural defects. This could lead to ruptured hydrogen bonds, which in turn reduce molecular weight, thus further making the material more susceptible to thermal degradation. Residual solvents or additives left after the regeneration process, such as NaOH or ionic liquids, could also contribute to lower thermal stability [79,81,83].

10.4 CONCLUSION AND PERSPECTIVES

In this chapter, we have detailed the properties of RCMs, their common synthesis strategies as well as their characterization techniques. While RCMs hold great promise in various applications such as in catalysis, environmental remediation, and bioengineering because of their tunable properties. Issues regarding reproducibility exist, whereby differences in synthesis methods can lead to inconsistency in RCMs' properties. Also, in terms of scalability, translating laboratory-scale techniques to effective industrial production is an issue that needs to be addressed. We have also discussed the impact of raw materials selection, emphasizing how the choice of material type could influence the final RCM properties, besides, the use of agricultural residues is encouraged as a sustainable and environmentally friendly means of RCM production. Moreover, we discussed various solvents and additives that enhance cellulose dissolution before regeneration. Furthermore, future research should be devoted to up-scaling green production techniques and developing harmonized synthesis protocols for improving reproducibility, such that they could be applied most universally in a variety of novel applications.

REFERENCES

1. S. Bhaladhare, D. Das, Cellulose: A fascinating biopolymer for hydrogel synthesis, *Journals of Materials Chemistry B* 10 (2022) 1923–1945. https://doi.org/10.1039/D1TB02848K.

2. S. Varnaitė-Žuravliova, J. Baltušnikaitė-Guzaitienė, Properties, production, and recycling of regenerated cellulose fibers: Special medical applications, *Journal of Functional Biomaterials* 15 (2024) 348. https://doi.org/10.3390/jfb15110348.
3. D.V. Zlenko, D.N. Vtyurina, S.V. Usachev, A.A. Skoblin, M.G. Mikhaleva, G.G. Politenkova, S.N. Nikolsky, S.V. Stovbun, On the orientation of the chains in the mercerized cellulose, *Science Reports* 11 (2021) 8765. https://doi.org/10.1038/s41598-021-88040-x.
4. N.S.N. Mazlan, S. Zakaria, S. Gan, C.C. Hua, K.W. Baharin, Comparison of regenerated cellulose membrane coagulated in sulphate based coagulant, *Cerne* 25 (2019) 18–24. https://doi.org/10.1590/01047760201925012586.
5. H.U. Usmanov, Physical chemistry of native cellulose, *Journal of Polymer Science* 23 (1957) 831–838. https://doi.org/10.1002/pol.1957.1202310426.
6. C. Holt, W. Mackie, D.B. Sellen, Degree of polymerization and polydis-persity of native cellulose, *Journal of Polymer Science: Polymer Symposia* 42 (1973) 1505–1512. https://doi.org/10.1002/polc.5070420353.
7. A.P. Manian, T. Pham, T. Bechtold, Regenerated cellulosic fibers, in: *Handbook of Properties of Textile and Technical Fibres*, Woodhead Publishing, 2018: pp. 329–343. https://doi.org/10.1016/B978-0-08-101272-7.00010-9.
8. C. Woodings, 1- A brief history of regenerated cellulosic fibres, in: C. Woodings (Ed.), *Regenerated Cellulose Fibres*, Woodhead Publishing, 2001: pp. 1–21. https://doi.org/10.1533/9781855737587.1.
9. C. Woodings, Cellulose fibers, regenerated, in: *Encyclopedia of Polymer Science and Technology*, Wiley Portico, 2003: pp. 532–569. https://doi.org/10.1002/0471440264.pst046.
10. M. Badshah, H. Ullah, F. He, F. Wahid, U. Farooq, M. Andersson, T. Khan, Development and evaluation of drug loaded regenerated bacterial cellulose-based matrices as a potential dosage form, *Frontiers in Bioengineering and Biotechnology* 8 (2020) 579404. https://doi.org/10.3389/fbioe.2020.579404.
11. Y. Gong, S. Mohd, S. Wu, S. Liu, Y. Pei, X. Luo, pH-responsive cellulose-based microspheres designed as an effective oral delivery system for insulin, *ACS Omega* 6 (2021) 2734–2741. https://doi.org/10.1021/acsomega.0c04946.
12. L.K. Voon, S.C. Pang, S.F. Chin, Porous cellulose beads fabricated from regenerated cellulose as potential drug delivery carriers, *Journal of Chemistry* 2017 (2017) 1–11. https://doi.org/10.1155/2017/1943432.
13. T.T. Dele-Afolabi, M.A.A. Hanim, O.J. Ojo-Kupoluyi, E.O. Atoyebi, Biobased materials for wastewater treatment, in: Ahmed & Annu (Eds.), *Advanced Applications of Biobased Materials*, Elsevier, 2023: pp. 593–624. https://doi.org/10.1016/B978-0-323-91677-6.00025-8.
14. K. Mishra, S.S. Siwal, T. Sithole, N. Singh, P. Hart, V.K. Thakur, Biorenewable materials for water remediation: The central role of cellulose in achieving sustainability, *Journal of Bioresources and Bioproducts* 9 (2024) 253–282. https://doi.org/10.1016/j.jobab.2023.12.002.
15. S.S. Jaffar, S. Saallah, M. Misson, S. Siddiquee, J. Roslan, S. Saalah, W. Lenggoro, Recent development and environmental applications of nanocellulose-based membranes, *Membranes* 12 (2022) 287. https://doi.org/10.3390/membranes12030287.
16. I. Ghosh, Q. Haider, C. Sharma, "Department of Paper Technology, IIT Roorkee, Saharanpur Campus, Saharanpur-247001, Uttar-Pradesh", Characterization of regenerated cellulose prepared from different pulp grades using a green solvent, *Cellulose Chemistry and Technology* 53 (2019) 643–653. https://doi.org/10.35812/CelluloseChemTechnol.2019.53.63.
17. N. Sharma, B.J. Allardyce, R. Rajkhowa, R. Agrawal, Rice straw-derived cellulose: A comparative study of various pre-treatment technologies and its conversion to nanofibres, *Scientific Reports* 13 (2023) 16327. https://doi.org/10.1038/s41598-023-43535-7.

18. S. Janaswamy, M.P. Yadav, M. Hoque, S. Bhattarai, S. Ahmed, Cellulosic fraction from agricultural biomass as a viable alternative for plastics and plastic products, *Industrial Crops and Products* 179 (2022) 114692. https://doi.org/10.1016/j.indcrop.2022.114692.
19. S.M. Costa, P.G. Mazzola, J.C.A.R. Silva, R. Pahl, A. Pessoa, S.A. Costa, Use of sugar cane straw as a source of cellulose for textile fiber production, *Industrial Crops and Products* 42 (2013) 189–194. https://doi.org/10.1016/j.indcrop.2012.05.028.
20. F. Xu, B.-U. Cho, Preparation of porous regenerated cellulose microstructures via emulsion-coagulation technique, *Cellulose* 29 (2022) 1527–1542. https://doi.org/10.1007/s10570-022-04428-6.
21. F. Xu, B.-U. Cho, Preparation and optimization of porous regenerated cellulose microspheres from cellulose tetraethyl-ammonium/urea solution for adsorption of cationic methylene blue dye, *BioResources* 18 (2022) 748–766. https://doi.org/10.15376/biores.18.1.748-766.
22. J.-H. Wu, X.-J. Wang, S.-J. Li, X.-Y. Ying, J.-B. Hu, X.-L. Xu, X.-Q. Kang, J. You, Y.-Z. Du, Preparation of ethyl cellulose microspheres for sustained release of sodium bicarbonate, *IJPR* 18 (2019) 556–568. https://doi.org/10.22037/ijpr.2019.1100651.
23. S. Park, Y. Oh, D. Jung, S.H. Lee, Effect of cellulose solvents on the characteristics of cellulose/Fe2O3 hydrogel microspheres as enzyme supports, *Polymers* 12 (2020) 1869. https://doi.org/10.3390/polym12091869.
24. K.S. Soppimath, A.R. Kulkarni, T.M., Cellulose acetate microspheres prepared by o/w emulsification and solvent evaporation method, *Journal of Microencapsulation* 18 (2001) 811–817. https://doi.org/10.1080/02652040110065486.
25. A. Napiórkowska, M. Kurek, Coacervation as a novel method of microencapsulation of essential oils—A review, *Molecules* 27 (2022) 5142. https://doi.org/10.3390/molecules27165142.
26. J. Xia, A.W.T. King, I. Kilpeläinen, V. Aseyev, Phase-separation of cellulose from ionic liquid upon cooling: Preparation of microsized particles, *Cellulose* 28 (2021) 10921–10938. https://doi.org/10.1007/s10570-021-04230-w.
27. S.O. Ilyin, V.V. Makarova, T.S. Anokhina, V.Y. Ignatenko, T.V. Brantseva, A.V. Volkov, S.V. Antonov, Diffusion and phase separation at the morphology formation of cellulose membranes by regeneration from N-methylmorpholine N-oxide solutions, *Cellulose* 25 (2018) 2515–2530. https://doi.org/10.1007/s10570-018-1756-9.
28. B. Wang, T.O. Akanbi, D. Agyei, B.J. Holland, C.J. Barrow, Coacervation technique as an encapsulation and delivery tool for hydrophobic biofunctional compounds, in: *Role of Materials Science in Food Bioengineering*, Academic Press, 2018: pp. 235–261. https://doi.org/10.1016/B978-0-12-811448-3.00007-3.
29. Y. Wei, J. Yu, B. Xiong, H. Chen, Z. Song, W. Zhou, Investigating key factors in the phase separation step of microspheres fabrication via coacervation, *International Journal of Pharmaceutics* 666 (2024) 124781. https://doi.org/10.1016/j.ijpharm.2024.124781.
30. X. Yan, J. Bernard, F. Ganachaud, Nanoprecipitation as a simple and straightforward process to create complex polymeric colloidal morphologies, *Advances in Colloid and Interface Science* 294 (2021) 102474. https://doi.org/10.1016/j.cis.2021.102474.
31. J.P.F. Carvalho, A.C.Q. Silva, A.J.D. Silvestre, C.S.R. Freire, C. Vilela, Spherical cellulose micro and nanoparticles: A review of recent developments and applications, *Nanomaterials* 11 (2021) 2744. https://doi.org/10.3390/nano11102744.
32. A. Ziaee, A.B. Albadarin, L. Padrela, T. Femmer, E. O'Reilly, G. Walker, Spray drying of pharmaceuticals and biopharmaceuticals: Critical parameters and experimental process optimization approaches, *European Journal of Pharmaceutical Sciences* 127 (2019) 300–318. https://doi.org/10.1016/j.ejps.2018.10.026.
33. Q. Wang, A. Fu, H. Li, J. Liu, P. Guo, X.S. Zhao, L.H. Xia, Preparation of cellulose based microspheres by combining spray coagulating with spray drying, *Carbohydrate Polymers* 111 (2014) 393–399. https://doi.org/10.1016/j.carbpol.2014.05.002.

34. A. Sosnik, K.P. Seremeta, Advantages and challenges of the spray-drying technology for the production of pure drug particles and drug-loaded polymeric carriers, *Advances in Colloid and Interface Science* 223 (2015) 40–54. https://doi.org/10.1016/j.cis.2015.05.003.
35. L.P. Pui, A.K.S. Lejaniya, Effects of spray-drying parameters on physicochemical properties of powdered fruits, *Foods and Raw Materials* 10(2) (2022) 235–251. https://doi.org/10.21603/2308-4057-2022-2-533.
36. Y. Peng, D.J. Gardner, Y. Han, Drying cellulose nanofibrils: In search of a suitable method, *Cellulose* 19 (2012) 91–102. https://doi.org/10.1007/s10570-011-9630-z.
37. R. Vehring, Pharmaceutical particle engineering via spray drying, *Pharmaceutical Research* 25 (2008) 999–1022. https://doi.org/10.1007/s11095-007-9475-1.
38. R. Nardi, M.H.L. Silveira, G. Siqueira, Understanding important aspects of spray drying microfibrillated cellulose through statistical analysis, *Cellulose* 27 (2020) 10707–10717. https://doi.org/10.1007/s10570-020-03392-3.
39. Ku, Min Kyung, Park, Hyunkyu, Kim, Hyungsup, Preparation of cellulose microparticles by electrospraying, *Textile Science and Engineering* 50 (2013) 187–192. https://doi.org/10.12772/TSE.2013.50.187.
40. W. Wu, J. Gu, G. Zhou, L. Zhang, M. Gong, H. Dai, Fabrication of natural cellulose microspheres via electrospraying from NaOH/Urea aqueous system, *Journal of Applied Polymer Science* 131 (2014) app.40656. https://doi.org/10.1002/app.40656.
41. N. Kadivar, H. Tavanai, A. Allafchian, Fabrication of cellulose nanoparticles through electrospraying, *IET Nanobiotechnology* 12 (2018) 807–813. https://doi.org/10.1049/iet-nbt.2018.0044.
42. Q. Zhao, M. Wang, Manipulating the release of growth factors from biodegradable microspheres for potentially different therapeutic effects by using two different electrospray techniques for microsphere fabrication, *Polymer Degradation and Stability* 162 (2019) 169–179. https://doi.org/10.1016/j.polymdegradstab.2019.02.009.
43. Y. Yu, W. Wang, Y. Liu, Enhancing the mechanical properties of regenerated cellulose through high-temperature pre-gelation, *Materials* 17 (2024) 4886. https://doi.org/10.3390/ma17194886.
44. H. Hu, J. Pereira, L. Xing, G. Zhou, W. Zhang, Thermal gelation and microstructural properties of myofibrillar protein gel with the incorporation of regenerated cellulose, *LWT* 86 (2017) 14–19. https://doi.org/10.1016/j.lwt.2017.07.015.
45. X. Luo, L. Zhang, Creation of regenerated cellulose microspheres with diameter ranging from micron to millimeter for chromatography applications, *Journal of Chromatography A* 1217 (2010) 5922–5929. https://doi.org/10.1016/j.chroma.2010.07.026.
46. L. Weng, L. Zhang, D. Ruan, L. Shi, J. Xu, Thermal gelation of cellulose in a NaOH/thiourea aqueous solution, *Langmuir* 20 (2004) 2086–2093. https://doi.org/10.1021/la035995o.
47. M.T. Garnett, S.A. Seyed Esfahani, A.P. Yingst, L.T. May, S.L.M. Alexander, Achieving 3-D structural uniformity in cellulose gel beads via salt screening, *Polymers* 16 (2024) 3519. https://doi.org/10.3390/polym16243519.
48. L. Lan, H. Chen, D. Lee, S. Xu, N. Skillen, A. Tedstone, P. Robertson, A. Garforth, H. Daly, C. Hardacre, X. Fan, Effect of ball-milling pretreatment of cellulose on its photoreforming for H_2 production, *ACS Sustainable Chemistry & Engineering* 10 (2022) 4862–4871. https://doi.org/10.1021/acssuschemeng.1c07301.
49. L. Zhang, T. Tsuzuki, X. Wang, Preparation of cellulose nanofiber from softwood pulp by ball milling, *Cellulose* 22 (2015) 1729–1741. https://doi.org/10.1007/s10570-015-0582-6.
50. H. Liimatainen, J. Sirviö, A. Haapala, O. Hormi, J. Niinimäki, Characterization of highly accessible cellulose microfibers generated by wet stirred media milling, *Carbohydrate Polymers* 83 (2011) 2005–2010. https://doi.org/10.1016/j.carbpol.2010.11.007.

51. L. Huang, Q. Wu, Q. Wang, M. Wolcott, Mechanical activation and characterization of micronized cellulose particles from pulp fiber, *Industrial Crops and Products* 141 (2019) 111750. https://doi.org/10.1016/j.indcrop.2019.111750.
52. R. Arshady, Microspheres and microcapsules: A survey of manufacturing techniques. Part 1: Suspension cross-linking, *Polymer Engineering & Science* 29 (1989) 1746–1758. https://doi.org/10.1002/pen.760292404.
53. R. Nicu, G. Lisa, R.N. Darie-Nita, M.I. Avadanei, A. Bargan, D. Rusu, D.E. Ciolacu, Tailoring the structure and physico-chemical features of cellulose-based hydrogels using multi-epoxy crosslinking agents, *Gels* 10 (2024) 523. https://doi.org/10.3390/gels10080523.
54. M. Gao, X. Xie, T. Huang, N. Zhang, Y. Wang, Glutaraldehyde-assisted crosslinking in regenerated cellulose films toward high dielectric and mechanical properties, *Cellulose* 29 (2022) 8177–8194. https://doi.org/10.1007/s10570-022-04785-2.
55. T.T. Dele-Afolabi, A.H. Mohamed Ariff, O.J. Ojo-Kupoluyi, E.O. Atoyebi, Chitosan nanocomposites as wound healing materials: Advances in processing techniques and mechanical properties, *JST* 31 (2022) 543–575. https://doi.org/10.47836/pjst.31.1.32.
56. J. Yang, J. Duan, L. Zhang, B. Lindman, H. Edlund, M. Norgren, Spherical nanocomposite particles prepared from mixed cellulose–chitosan solutions, *Cellulose* 23 (2016) 3105–3115. https://doi.org/10.1007/s10570-016-1029-4.
57. A.M. Claro, C.C. Alves, K.S. Dos Santos, E.G. Da Rocha, M. De Lima Fontes, G.C. Monteiro, G.S.G. De Carvalho, J.M.A. Caiut, A. Moroz, S.J.L. Ribeiro, H.S. Barud, Regenerated cellulose sponge as sacrificial template for the synthesis of three-dimensional porous alumina-silica scaffold for tissue engineering, *Journal of Sol-Gel Science and Technology* 107 (2023) 83–95. https://doi.org/10.1007/s10971-022-05990-y.
58. Y. Tanaka, Q. Li, M. Hatakeyama, T. Kitaoka, Synthesis and structural design of microspheres comprising cellulose nanofibers and artificial lignin polymer by enzyme-mediated Pickering emulsion templating, *RSC Sustainability* 2 (2024) 1580–1589. https://doi.org/10.1039/D4SU00067F.
59. N. Gangwar, C. Gangwar, J. Sarkar, A review on template-assisted approaches & self assembly of nanomaterials at liquid/liquid interface, *Heliyon* 10 (2024) e36810. https://doi.org/10.1016/j.heliyon.2024.e36810.
60. N. Kommineni, A. Butreddy, V.G.S. Sainaga Jyothi, P. Angsantikul, Freeze-drying for the preservation of immunoengineering products, *iScience* 25 (2022) 105127. https://doi.org/10.1016/j.isci.2022.105127.
61. E. Jakubowska, J. Lulek, The application of freeze-drying as a production method of drug nanocrystals and solid dispersions – A review, *Journal of Drug Delivery Science and Technology* 62 (2021) 102357. https://doi.org/10.1016/j.jddst.2021.102357.
62. J. Yu, T. Huang, Z.H. Lim, R. Luo, R.R. Pasula, L. Liao, S. Lim, C. Chen, Production of hollow bacterial cellulose microspheres using microfluidics to form an injectable porous scaffold for wound healing, *Advanced Healthcare Materials* 5 (2016) 2983–2992. https://doi.org/10.1002/adhm.201600898.
63. D. Rawtani, Y.K. Agrawal, Emerging strategies and applications of layer-by-layer self-assembly, *Nanobiomedicine (Rij)* 1 (2014) 8. https://doi.org/10.5772/60009.
64. J. Borges, J. Zeng, X.Q. Liu, H. Chang, C. Monge, C. Garot, K. Ren, P. Machillot, N.E. Vrana, P. Lavalle, T. Akagi, M. Matsusaki, J. Ji, M. Akashi, J.F. Mano, V. Gribova, C. Picart, Recent developments in layer-by-layer assembly for drug delivery and tissue engineering applications, *Advanced Healthcare Materials* 13 (2024) 2302713. https://doi.org/10.1002/adhm.202302713.
65. S. Tanisood, Y. Baimark, P. Srihanam, Preparation and characterization of cellulose/silk fibroin composite microparticles for drug-controlled release applications, *Polymers* 16 (2024) 3020. https://doi.org/10.3390/polym16213020.

66. W. Tang, Z. Yuan, B. Sun, K. Li, Y. Li, J. Zhong, Facile and scalable fabrication of regenerated cellulose microspheres with high strength and porosity as a potential matrix for hemoperfusion, *Journal of Macromolecular Science, Part B* 63 (2024) 588–603. https://doi.org/10.1080/00222348.2024.2314413.
67. K. Wei Wei Tay, S. Fun Chin, M.E. Wasli, Z. Musa, Response surface methodology: A versatile tool for the optimization of particle sizes of cellulose beads, *JST* 31 (2023) 2805–2822. https://doi.org/10.47836/pjst.31.6.10.
68. J. Stetefeld, S.A. McKenna, T.R. Patel, Dynamic light scattering: A practical guide and applications in biomedical sciences, *Biophysical Reviews* 8 (2016) 409–427. https://doi.org/10.1007/s12551-016-0218-6.
69. K. Thümmler, S. Fischer, A. Feldner, V. Weber, M. Ettenauer, F. Loth, D. Falkenhagen, Preparation and characterization of cellulose microspheres, *Cellulose* 18 (2011) 135–142. https://doi.org/10.1007/s10570-010-9465-z.
70. M. Mouffok, I. Abdelmalek, A. Mesli, A.A. Moulay, investigation of factors affecting particle size distribution and sustained release of a water-soluble drug from cellulose derivatives microspheres, *Chemistry Africa* 6 (2023) 163–173. https://doi.org/10.1007/s42250-022-00323-6.
71. D.-Y. Zhang, R.-Z. Hu, X.-Q. Zhang, Y. Wan, X.-H. Yao, T. Chen, Y.-J. Fu, Z.-Y. Liu, S. Hua, Preparation of modified cellulose microspheres and application of fast enrichment of chlorogenic acid from mulberry leaf extract, *Journal of Materials Research and Technology* 15 (2021) 911–923. https://doi.org/10.1016/j.jmrt.2021.08.066.
72. R. Meng, L. Liu, X. Su, W. Gong, X. Luo, H. Gao, Facile preparation of cellulose beads with tunable graded pores and high mechanical strength, *Polymers* 16 (2024) 725. https://doi.org/10.3390/polym16060725.
73. R. Wu, P. Huang, B. He, Preparation and characterization of regenerated cellulose microspheres and the adsorption of pectinase, *BioResources* 11 (2016) 4146–4158. https://doi.org/10.15376/biores.11.2.4146-4158.
74. A.M. Donia, A.M. Yousif, A.A. Atia, H.M. Abd El-Latif, Preparation and characterization of modified cellulose adsorbents with high surface area and high adsorption affinity for Hg(II), *Journal of Dispersion Science and Technology* 35 (2014) 380–389. https://doi.org/10.1080/01932691.2013.791626.
75. S.M. Ali, B. Al-Oufi, Synergistic sorption performance of cellulose-modified La0.9Sr0.1FeO3 for organic pollutants, *Cellulose* 27 (2020) 429–440. https://doi.org/10.1007/s10570-019-02814-1.
76. Y.P. Yang, Y. Zhang, Y.X. Lang, M.H. Yu, Structural ATR-IR analysis of cellulose fibers prepared from a NaOH complex aqueous solution, *IOP Conference Series: Materials Science and Engineering* 213 (2017) 012039. https://doi.org/10.1088/1757-899X/213/1/012039.
77. H. Zhang, Y. Xu, Y. Li, Z. Lu, S. Cao, M. Fan, L. Huang, L. Chen, Facile cellulose dissolution and characterization in the newly synthesized 1,3-Diallyl-2-ethylimidazolium acetate ionic liquid, *Polymers* 9 (2017) 526. https://doi.org/10.3390/polym9100526.
78. K.S. Salem, N.K. Kasera, Md.A. Rahman, H. Jameel, Y. Habibi, S.J. Eichhorn, A.D. French, L. Pal, L.A. Lucia, Comparison and assessment of methods for cellulose crystallinity determination, *Chemical Society Reviews* 52 (2023) 6417–6446. https://doi.org/10.1039/D2CS00569G.
79. J. Cui, P. Lu, Y. Li, K. Xu, Y. Li, H. Shen, C. Liu, T. Zhang, D. Liu, The flexible and transparent film heaters based on regenerated cellulose and carbon nanotubes, *Frontiers in Energy Research* 10 (2022) 879257. https://doi.org/10.3389/fenrg.2022.879257.
80. L. Gentile, H. Sixta, C. Giannini, U. Olsson, A novel X-ray diffraction approach to assess the crystallinity of regenerated cellulose fibers, *IUCrJ* 9 (2022) 492–496. https://doi.org/10.1107/S205225252200570X.

81. L.C. Yeng, M.U. Wahit, N. Othman, Thermal and flexural properties of regenerated cellulose(rc)/poly(3-hydroxybutyrate)(phb)biocomposites, *Jurnal Teknologi* 75 (2015) 11. https://doi.org/10.11113/jt.v75.5338.
82. N.M. Nurazzi, M.R.M. Asyraf, M. Rayung, M.N.F. Norrrahim, S.S. Shazleen, M.S.A. Rani, A.R. Shafi, H.A. Aisyah, M.H.M. Radzi, F.A. Sabaruddin, R.A. Ilyas, E.S. Zainudin, K. Abdan, Thermogravimetric analysis properties of cellulosic natural fiber polymer composites: A review on influence of chemical treatments, *Polymers* 13 (2021) 2710. https://doi.org/10.3390/polym13162710.
83. K. Li, H. Yang, L. Jiang, X. Liu, P. Lang, B. Deng, N. Li, W. Xu, Glycerin/NaOH aqueous solution as a green solvent system for dissolution of cellulose, *Polymers* 12 (2020) 1735. https://doi.org/10.3390/polym12081735.

11 Degradation of Pharmaceutical Pollutants in Wastewater via Cellulose Microsphere

Zarqua Hashmi, Abu Taha, and Athar Adil Hashmi

11.1 INTRODUCTION

With the exponential growth in human population, the industrial revolution, and urbanization, the need for a better life is also increasing. Various methods have come to meet the requirements of human beings, but with their useful applications, there are many kinds of problems, including the emergence of waste materials [1]. With good living standards and changing production and consumption methods, they are leading to the development of new molecules and compounds. Out of these compounds, medicines such as antibiotics, analgesics, and non-steroidal anti-inflammatories are mainly used in hospitals and health sectors. These drugs are synthetic or natural organic molecules that have a particular mode of action in human and animal bodies. Advancement in medical technology has further helped in the development of various synthetic therapeutic materials [2]. The development of various characterization techniques from the past 25 years has paved the way for the detection of pharmaceutical wastes in the water bodies. In early times, the discharges from manufacturing were considered the prime source of active pharmaceutical ingredients (APIs) in the environment [3,4], but at this time, these studies received little attention. The detection of estrogen in wastewater effluents as a cause of the feminization of fish in the late 1990s sparked a sudden increased interest in research on the impact of pharmaceutical waste on the environment [5,6]. Metropolitan wastewater contaminated with drugs and medical substances has been found to be the primary result of pharmaceutical waste pollution [7], emerging from medicine industries and the throwaway of unused drug from hospitals and households. Martín et al. [8] also documented that human excreta also contains defecated pharmaceuticals, which circulate through the sewage system and end up in wastewater treatment plants.

Discernment, identification, and investigation of drugs at a very minute level in wastewater is also a great challenge [9]. A number of drugs such as diclofenac, sulfamethoxazole, metoprolol, and carbamazepine were detected in the surface water of Slovakia. At a very low concentration in the water, some medicines can be harmful to the environment and ecosystem [10]. Lesser and team carried out a study on the most extensive untreated wastewater irrigation system of Mexico and observed more than

DOI: 10.1201/9781003596127-13

two hundred organic pollutants, including pharmaceutical products and hormones in the wastewater [11]. Numerous remediation methods, such as adsorption, biotic treatment, laminate-related processes, and advanced oxidation activity, are being explored by environmentalists for pharmaceutical wastewater treatment. Many scientists have published their work on the remediation of pharmaceutical pollutants from the sewage water treatment industry [12] and municipal wastewater, while other scientists have focused on single methods such as membrane bioreactor [13], constructed wetlands [14], adsorption [15], or hybrid/integrated technology like Fenton-aerobic [16], and biological membrane ultrasound [17]. Only some researchers have summarized the development of pharmaceutical pollutants removal by standard and advanced technologies [18].

Cellulose-base polymers are widely used in drugs and generally considered as biocompatible. Cellulose bead was first described as cellulose pellets in US patent [19]. Being a natural polymer, cellulose is available on earth in large amounts and can be used in a plethora of applications due to its chemical structure, which can be modified easily. Cellulose can be modified into CMSs, which are natural in origin, nonpoisonous, and decomposable [20]. Due to the presence of hydroxyl groups on their surface [21], CMSs have a wide range of applications, such as separator, carrier structure, and adsorbent in blood purification [22,23]. Use of cellulose is definitely beneficial for environmental protection and sustainable development due to its biocompatibility, biodegradability, and renewable resource [24]. CMSs, when mixed with some metal nanoparticles, act as good catalysts as they provide a large surface area due to their very highly porous nature [25]. There are various methods to prepare CMSs using cellulose or cellulose derivatives, as depicted in Figure 11.1 [26]. A large number of

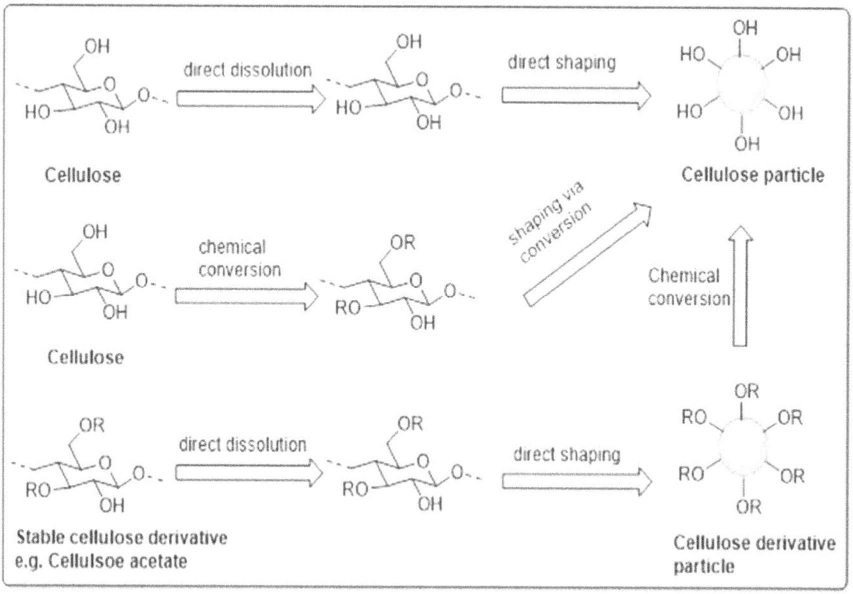

FIGURE 11.1 Preparation of CMSs via different route.

catalysts have been developed using CMSs with different kinds of metal nanoparticles for the degradation of pharmaceutical waste. In this chapter, we will explore various cellulose microsphere-based catalysts for the degradation of pharmaceutical pollutants.

11.2 DEGRADATION OF ANTIBIOTICS USING CMSS

Antibiotics are chemicals that have the potential to kill or inhibit the growth of microorganism, such as bacteria. These are classified on the basis of their mode of action or chemical structure. The parent compound of antibiotics can be very stable and bypass the water treatment process, leading to environmental pollution [27]. Even very minute concentrations of antibiotics in the environment can have harmful effects on human health and ecosystems [28]. The principal sources of antibiotics in the environment are wastewater treatment plants, pharmaceutical plants, and aquaculture [29]. Some of the antibiotics, such as ampicillin, amoxicillin, and ciprofloxacin, are very harmful to aquatic ecosystems even at very low concentrations [30–32]. Due to these reasons, we need to monitor the concentration of antibiotics in the environment and degrade them using suitable techniques. Figure 11.2 shows the structures of some antibiotics.

X. Luo and coworkers synthesized magnesium/aluminum LDH embedded on phosphorylated cellulose microspheres (Mg/Al LDH@PCMs) for tetracycline hydrochloride removal. Mg/Al LDH@PCMs were analyzed using various techniques such as scanning electron microscopy, specific surface area test, Fourier transform infrared spectroscopy, X-ray diffraction, ζ potential, and X-ray electron energy. The adsorption value of tetracycline was found to be 25.107 mg/g and the adsorption was due to the electrostatic attraction between LDH and the anionic species. Figure 11.3 shows anionic attraction of tetracycline toward Mg/Al LDH@PCMs. The kinetic pattern of adsorption of tetracycline hydrochloride was determined to be of pseudo second order, while the isotherm fitted with the Freundlich model [33].

In another similar work, X. Luo and coworkers synthesized a cellulose-based composite bead for photocatalytic degradation of tylosin. Tylosin is a hydrophobic antibiotic, which makes it difficult to get dissolved in water, and hence their removal becomes very difficult. The carboxylated cellulose beads were made with TiO_2-doped gold nanoparticles. The synthesized cellulose bead catalyst was well characterized using SEM images, energy dispersive X-ray spectrometry, infrared spectroscopy, X-ray diffraction, X-ray photoelectron spectroscopy, UV–visible diffused reflectance spectra, and electron spin resonance. The cellulose bead was efficient in removing 92% of tylosin in 180 minutes. The photocatalyst activity was ascribed to the separation of the electron-hole pair, and the redox mechanism of photocatalysis is shown in Figure 11.4 [34].

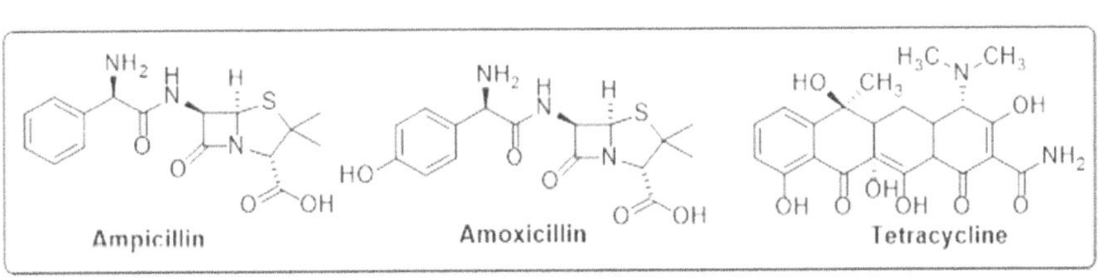

FIGURE 11.2 Structure of ampicillin, amoxicillin, and tetracycline.

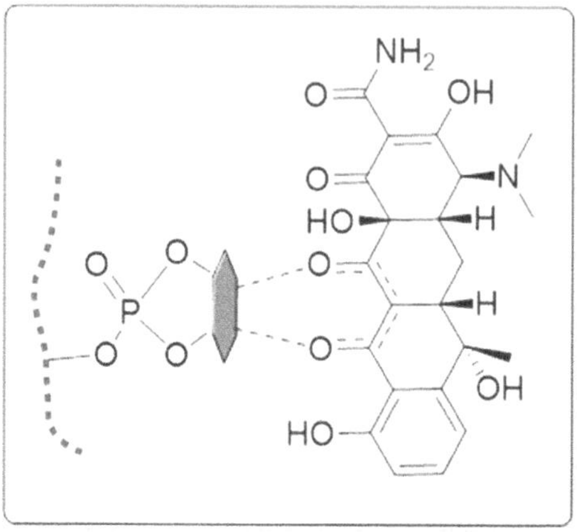

FIGURE 11.3 Adsorption of tetracycline on Mg/Al LDH@PCMs.

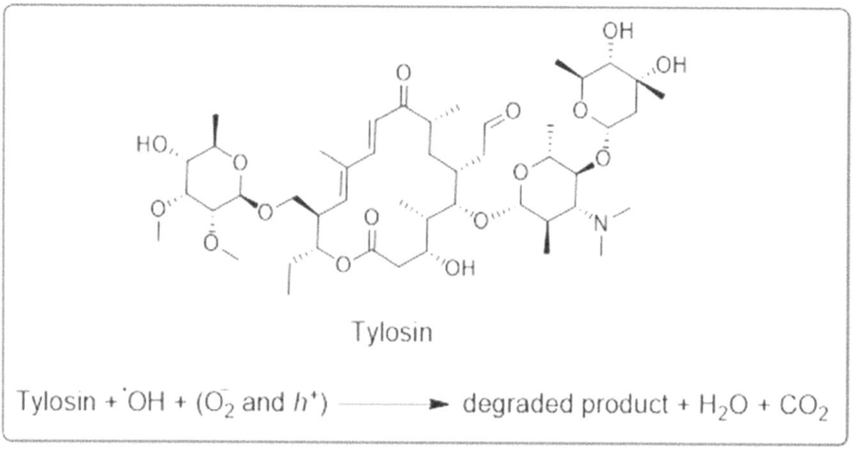

FIGURE 11.4 Structure of tylosin and its degradation by hydroxyl radicals.

A metal organic framework (MOF) embedded with cellulose acetate beads (UiO-66/ZIF-8/PDA@CA) was developed for the adsorptive removal of tetracycline by A. M. Omar and coworkers. 66/ZIF-8/PDA@CA beads were characterized using different analytical techniques such as FTIR, XRD, SEM, and BET. Under optimum condition, 66/ZIF-8/PDA@CA beads exhibit a large surface area having good adsorption and short equilibrium time (60 min). Maximum adsorption capacity of 66/ZIF-8/PDA@CA beads was found to be 290.69 mg/g at 25°C where kinetics was governed by pseudo second order. Thermodynamic studies revealed endothermic and spontaneous adsorption for eight consecutive cycles [35]. A low-concentration removal of amoxicillin was reported by X. Luo and team by micro-/nano-porous cellulose beads. Quatarnized cellulose beads (QCB) were synthesized from a homogeneous mixture in a one-step cellulose moderation and reconstruction mechanism.

Morphology and structure of QCB were evaluated using SEM images, infrared spectroscopy, and XPS. The QCB was tested for the adsorption of amoxicillin, where it was ascertained to be multilayer and controlled by physical adsorption as well as chemical adsorption. Removal efficiency of amoxicillin was found to be 60 mg/L, and adsorption capacity reached 30.23 mg/g. A recycle test was performed three times without any loss of binding efficiency; Figure 11.5 shows the adsorption of amoxicillin on QCB [36].

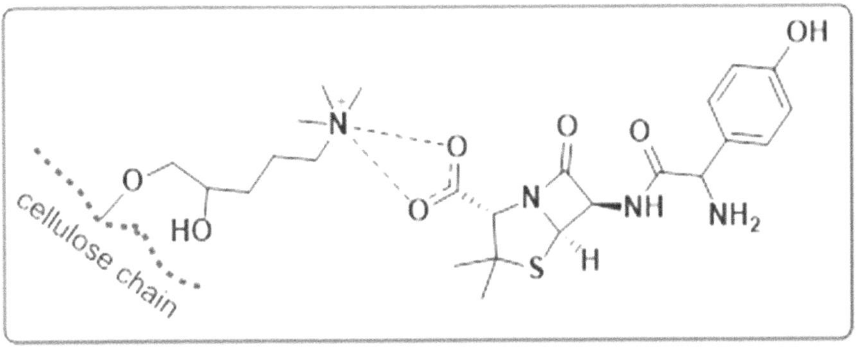

FIGURE 11.5 Adsorption of amoxicillin on QCB.

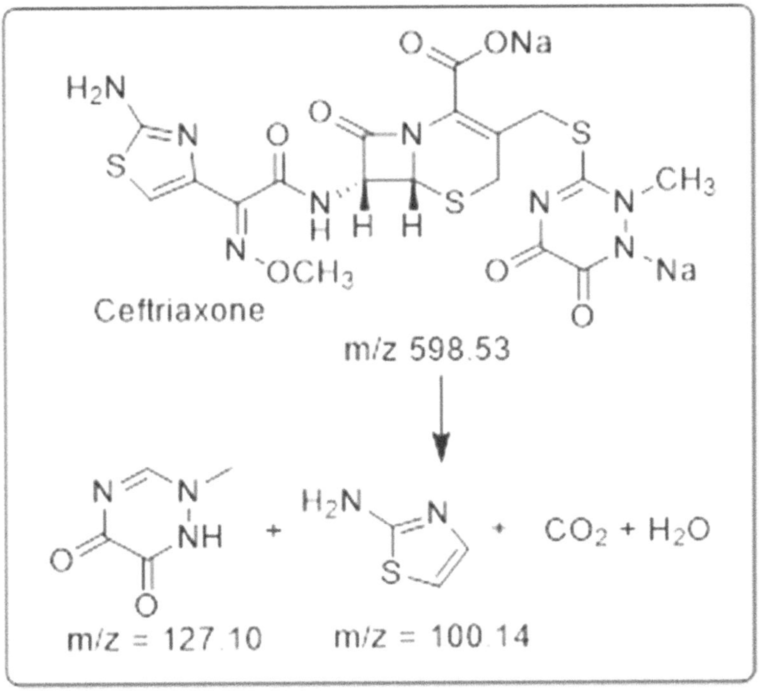

FIGURE 11.6 Degradation pathway of ceftriaxone.

In another work, silver-doped TiO_2 cellulose carbon beads were synthesized by J. Yang and X. Luo for the degradation of ceftriaxone sodium. About 0.5% Ag was the best catalyst for degradation under simulated sunlight. The degradation follows first-order kinetics and highest apparent rate constant (K_{app}) of 0.0091/min. The degradation result shows that ceftriaxone was broken into smaller particles and finally into CO_2 and water. A short degradation pathway for ceftriaxone sodium is shown in Figure 11.6 [37].

T. Liu and coworkers synthesized a recyclable boronic acid modified bacterial cellulose microsphere via emulsion for tetracycline removal from hoggery sewer. The CMSs were 415 nm in size and analyzed by SEM images, surface analysis, infrared, and thermal analysis. Tetracycline shows reversible interaction between the diol groups in tetracycline and the boronic acid ligand. Adsorption shows that tetracycline follows pseudo second order, Elovich kinetic, Sips, and Redlich-Peterson isothermal models. The adsorption capacity of cellulose microsphere for tetracycline was 614.2 mg/g with 18 recycle times and 84.5% adsorption recovery [38]. Sulfamethoxazole is found to remain persistent in wastewater and drinking water in the range of 100–2500 ng/L and 12 ng/L respectively. It is found that sulfamethoxazole shows low biodegradation and cause harmful effect on human and animal. For this, W. Wang and team developed an aminated cellulose nanocomposite microsphere (ACNM) for low-concentration removal of sulfamethoxazole from water. The ACNM were characterized by different analytical techniques such as SEM, FTIR, XRD, XPS, and BET. A 10 mg/L of sulfamethoxazole solution was made to simulate low-concentration adsorption. The results show that sulfamethoxazole reached equilibrium in 30 minutes and the removal efficiency of sulfamethoxazole was 99.18% and adsorption capacity was found to be 13.59 mg/g. The adsorption follows pseudo-second-order kinetics and Langmuir isotherm model in which monolayer chemical adsorption was dominant. Electrostatic interaction between the anion of sulfamethoxazole and amine group of ACNM was responsible for the adsorption of sulfamethoxazole on ACNM as shown in Figure 11.7 [39].

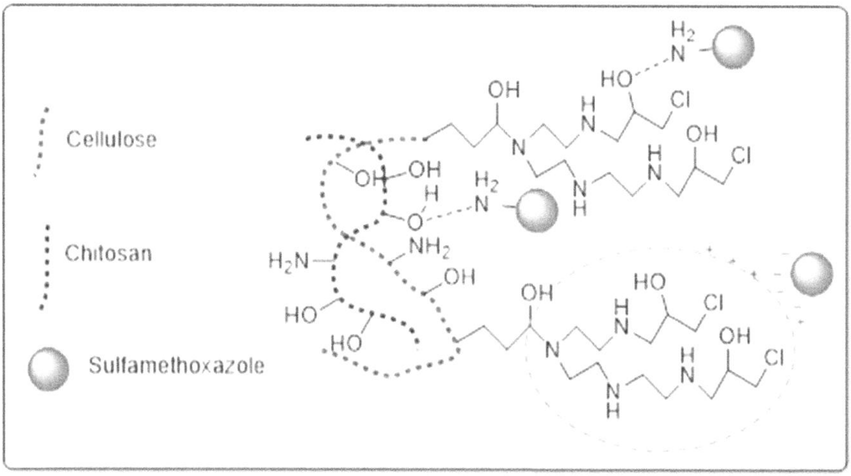

FIGURE 11.7 Adsorption of sulfamethoxazole on ACNM.

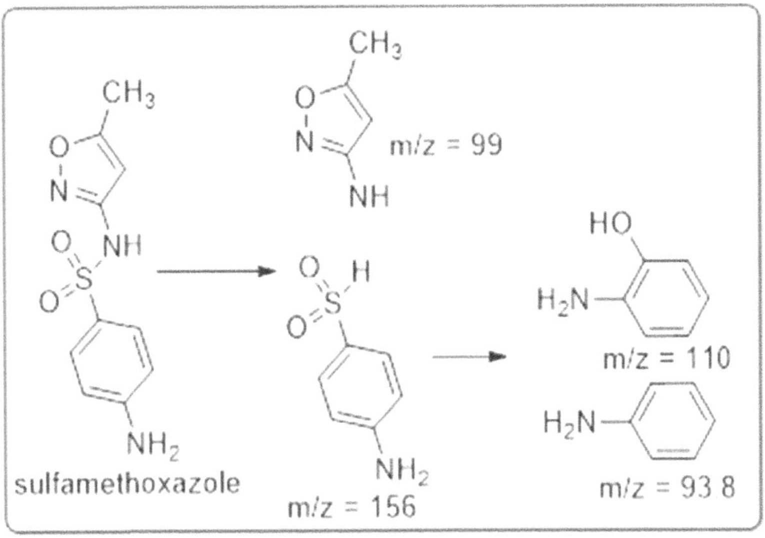

FIGURE 11.8 Degradation of sulfamethoxazole by cellulose magnetic bead.

In another similar work, cellulose-embedded Fe_3O_4/Fe^{3+} composite beads were fabricated by P. S. Kumar and team for the degradation of sulfamethoxazole from aqueous solution. Morphology and functional group present on the magnetic beads were confirmed using SEM and FTIR. Degradation of sulfamethoxazole was investigated, including the pH of the reaction mixture, dosage of catalyst, and initial sulfamethoxazole concentration. Under optimal conditions, 81.89% of sulfamethoxazole was degraded in 40 min using H_2O_2. The degradation follows first-order kinetics where an •OH radical was generated due to the Fe particle. The degradation pathway of sulfamethoxazole using magnetic beads is shown in Figure 11.8 [40].

11.3 DEGRADATION OF ANALGESICS USING CMSS

Analgesics are the most used class of medicines, having very complex structures and not getting degraded easily when exposed to the environment. These drugs also posed a threat to non-targeted species and have the potential to damage aquatic environments [41]. These medicines have been designed to relieve pain and include subcategories such as non-steroidal anti-inflammatory drugs (NSAIDs), opioids, and nonnarcotic medications [42]. NSAIDs have demonstrated mitochondrial dysfunction, reproductive dysfunction, endocrine disorder, and nucleic acid damage [43]. Nonnarcotic analgesics such as paracetamol and tramadol are harmful to cyanobacteria and their pigments [44]. These harmful effects of analgesics have made researchers develop methods for their removal from aquatic environments. Structures of some analgesics are shown in Figure 11.9.

S. H. Lee and coworkers prepared a reswellable alginate-activated carbon carboxymethyl cellulose-based hydrogel (Alg/AC/CMC) beads for the effective adsorption of ibuprofen. The dried Alg/AC/CMC shows 86% recovery with 34.0 mg/g

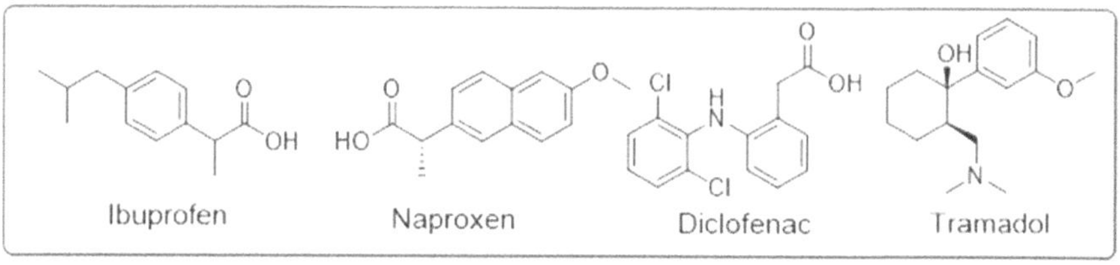

FIGURE 11.9 Structure of some analgesics.

adsorption of ibuprofen, while the initial hydrogel shows an adsorption capacity of 39.6 mg/g of ibuprofen. The hydrogel bead follows pseudo-second-order kinetics and the Langmuir isotherm for the adsorption of ibuprofen. The adsorption capacity of hydrogel beads was found to be more than 93% even after 10 cycles [45]. Diclofenac degradation was performed using peroxydisulfate (PDS) and 2,2′-azino-bis (3-et hylbenzothiazoline-6-sulfonate) (ABTS) encapsulated into cellulose beads. Both ABTS@Ce beads and PDS/ABTS@Ce beads show excellent degradation of diclofenac. The degradation efficiency was achieved to 100% in 60 minutes after 6 cycles. The rate of diclofenac degradation decreases with the consecutive use of ABTS@ Ce, which indicates a release of ABTS or $ABTS^{\bullet+}$ from ABTS@Ce during diclofenac removal [46]. J. S. Gonzalez and workers prepared magnetic carboxymethyl cellulose beads for effective diclofenac adsorption. The TEM images show the size of magnetic beads around 8.3 nm. FTIR confirm interaction of carboxymethyl cellulose with iron oxide nanoparticles. The adsorption study show that 50% adsorption reached in 500 min in the lowest concentration (10 mg/L), while it shows only 30% adsorption in the highest concentration (2000 mg/L). Overall, the magnetic gel bead shows 666.7 mg/g of diclofenac adsorption with reusability due to the magnetic nature of the catalyst [47].

11.4 DEGRADATION OF PESTICIDES USING CMSS

Pesticides are the chemicals that are used in gardens, agricultural fields, and other areas to kill unwanted organisms. These chemicals run to water bodies and affect them adversely. The World Health Organization data shows that about 1,000,000 humans are being affected by acute poisoning due to pesticide contamination [48,49]. Owing to these problems, cellulose-base catalysts were synthesized for the degradation of these harmful chemicals from aqueous media. D. C. Whitehead and coworkers synthesized an amine-functionalized cellulose-based microcrystal (CNC-PEI) for the degradation of melathion, deltamethrin, and permethrin, showing 100%, 95%, and 78% degradation, respectively. The CNC-PEI shows good durability for the degradation of melathion for two wash-and-reuse cycles [50]. The degradation pathway for melathion and deltamethrin by CNC-PEI is shown in Figure 11.10.

11.5 CONCLUSION

This chapter highlights the effective use of CMSs in the degradation of pharmaceutical waste during wastewater treatment. It is established that CMSs serve as a

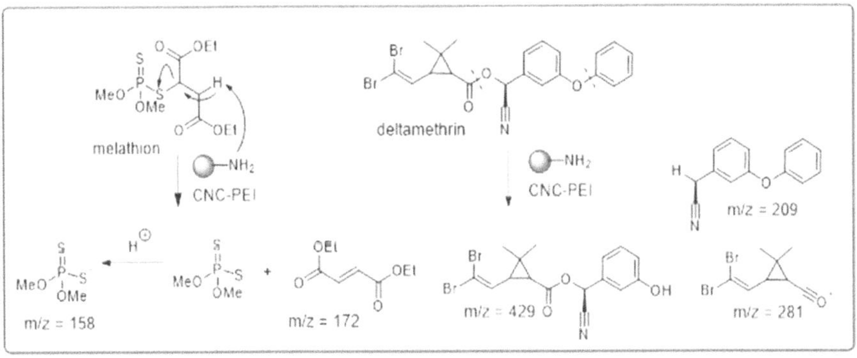

FIGURE 11.10 Degradation of melathion and deltamethrin by CNC-PEI.

solid foundation for the development of various photocatalysts that facilitate the degradation of contaminants like antibiotics, analgesics, and pesticides. The discussion includes insights into the mechanisms underlying the degradation of these organic pollutants. Furthermore, the review of literature indicates a significant gap in research concerning the advancement of CMSs aimed at removing other pharmaceutical waste materials such as personal care products, analgesics, pesticides, and hormones from wastewater. Addressing this research need is crucial for enhancing wastewater remediation strategies.

REFERENCES

1. X. Chen, S. Xue, M. Lv, R. Wang, Pharmaceutical industry in China: Policy, market and IP, in: K.-C. Liu, U.S. Racherla (Eds.), *Innovation, Economic Development, and Intellectual Property in India and China, Springer Nature Singapore*, Singapore, 2019: pp. 215–250. https://doi.org/10.1007/978-981-13-8102-7_10.
2. M.D'Alessio, S. Onanong, D.D. Snow, C. Ray, Occurrence and removal of pharmaceutical compounds and steroids at four wastewater treatment plants in Hawai'i and their environmental fate, *Science of the Total Environment* 631–632 (2018) 1360–1370. https://doi.org/10.1016/j.scitotenv.2018.03.100.
3. E.M. Materon, G. Ibáñez-Redín, N, Joshi, D. Gonçalves, O.N. Oliveira, R.C. Faria, Analytical detection of pesticides, pollutants, and pharmaceutical waste in the environment, in: S. Kumar Tuteja, D. Arora, N. Dilbaghi, E. Lichtfouse (Eds.), *Nanosensors for Environmental Applications, Springer International Publishing*, Cham, 2020: pp. 87–129. https://doi.org/10.1007/978-3-030-38101-1_3.
4. A.A. Renita, P.S. Kumar, S. Srinivas, S. Priyadharshini, M. Karthika, A review on analytical methods and treatment techniques of pharmaceutical wastewater, *Desalination and Water Treatment* 87 (2017) 160–178. https://doi.org/10.5004/dwt.2017.21311.
5. L. Racz, R.K. Goel, Fate and removal of estrogens in municipal wastewater, *Journal of Environmental Monitoring* 12 (2010) 58–70. https://doi.org/10.1039/B917298J.
6. D.G.J. Larsson, Pollution from drug manufacturing: Review and perspectives, *Philosophical Transaction of the Royal Society of London* B 369 (2014) 20130571. https://doi.org/10.1098/rstb.2013.0571.
7. O.A.H. Jones, N. Voulvoulis, J.N. Lester, Human pharmaceuticals in wastewater treatment processes, *Critical Reviews in Environmental Science and Technology* 35 (2005) 401–427. https://doi.org/10.1080/10643380590956966.

8. J. Martín, D. Camacho-Muñoz, J.L. Santos, I. Aparicio, E. Alonso, Occurrence of pharmaceutical compounds in wastewater and sludge from wastewater treatment plants: Removal and ecotoxicological impact of wastewater discharges and sludge disposal, *Journal of Hazardous Materials* 239–240 (2012) 40–47. https://doi.org/10.1016/j.jhazmat.2012.04.068.
9. J.O. Eniola, R. Kumar, M.A. Barakat, Adsorptive removal of antibiotics from water over natural and modified adsorbents, *Environmental Science and Pollution Research* 26 (2019) 34775–34788. https://doi.org/10.1007/s11356-019-06641-6.
10. T. Mackuľak, S. Černanský, M. Fehér, L. Birošová, M. Gál, Pharmaceuticals, drugs, and resistant microorganisms — environmental impact on population health, *Current Opinion in Environmental Science & Health* 9 (2019) 40–48. https://doi.org/10.1016/j.coesh.2019.04.002.
11. L.E. Lesser, A. Mora, C. Moreau, J. Mahlknecht, A. Hernández-Antonio, A.I. Ramírez, H. Barrios-Piña, Survey of 218 organic contaminants in groundwater derived from the world's largest untreated wastewater irrigation system: Mezquital Valley, Mexico, *Chemosphere* 198 (2018) 510–521. https://doi.org/10.1016/j.chemosphere.2018.01.154.
12. J.L. Tambosi, L.Y. Yamanaka, H.J. José, R.D.F.P.M. Moreira, H.F. Schröder, Recent research data on the removal of pharmaceuticals from sewage treatment plants (STP), *Química Nova* 33 (2010) 411–420. https://doi.org/10.1590/S0100-40422010000200032.
13. M. Kim, P. Guerra, A. Shah, M. Parsa, M. Alaee, S.A. Smyth, Removal of pharmaceuticals and personal care products in a membrane bioreactor wastewater treatment plant, *Water Science and Technology* 69 (2014) 2221–2229. https://doi.org/10.2166/wst.2014.145.
14. Y. Li, G. Zhu, W.J. Ng, S.K. Tan, A review on removing pharmaceutical contaminants from wastewater by constructed wetlands: Design, performance and mechanism, *Science of the Total Environment* 468–469 (2014) 908–932. https://doi.org/10.1016/j.scitotenv.2013.09.018.
15. A.M. Awad, S.M.R. Shaikh, R. Jalab, M.H. Gulied, M.S. Nasser, A. Benamor, S. Adham, Adsorption of organic pollutants by natural and modified clays: A comprehensive review, *Separation and Purification Technology* 228 (2019) 115719. https://doi.org/10.1016/j.seppur.2019.115719.
16. R. Changotra, H. Rajput, A. Dhir, Treatment of real pharmaceutical wastewater using combined approach of Fenton applications and aerobic biological treatment, *Journal of Photochemistry and Photobiology A: Chemistry* 376 (2019) 175–184. https://doi.org/10.1016/j.jphotochem.2019.02.029.
17. P. Alfonso-Muniozguren, E.A. Serna-Galvis, M. Bussemaker, R.A. Torres-Palma, J. Lee, A review on pharmaceuticals removal from waters by single and combined biological, membrane filtration and ultrasound systems, *Ultrasonics Sonochemistry* 76 (2021) 105656. https://doi.org/10.1016/j.ultsonch.2021.105656.
18. Y. Xu, T. Liu, Y. Zhang, F. Ge, R.M. Steel, L. Sun, Advances in technologies for pharmaceuticals and personal care products removal, *Journal of Materials Chemistry A* 5 (2017) 12001–12014. https://doi.org/10.1039/C7TA03698A.
19. K. Thümmler, S. Fischer, A. Feldner, V. Weber, M. Ettenauer, F. Loth, D. Falkenhagen, Preparation and characterization of cellulose microspheres, *Cellulose* 18 (2011) 135–142. https://doi.org/10.1007/s10570-010-9465-z.
20. I. Chen, C. Xu, J. Peng, D. Han, S. Liu, M. Zhai, Novel functionalized cellulose microspheres for efficient separation of lithium ion and its isotopes: Synthesis and adsorption performance, *Molecules* 24 (2019) 2762. https://doi.org/10.3390/molecules24152762.
21. L. Wu, X. Lin, X. Du, X. Luo, Biosorption of uranium(VI) from aqueous solution using microsphere adsorbents of carboxymethyl cellulose loaded with aluminum(III), *Journal of Radioanalitical and Nuclear Chemistry* 310 (2016) 611–622. https://doi.org/10.1007/s10967-016-4859-5.

22. J.A. Kaster, W. De Oliveira, W.G. Glasser, W.H. Velander, Optimization of pressure-flow limits, strength, intraparticle transport and dynamic capacity by hydrogel solids content and bead size in cellulose immunosorbents, *Journal of Chromatography A* 648 (1993) 79–90. https://doi.org/10.1016/0021-9673(93)83289-5.
23. B. Wolf, W. Schmitz, H. Schneider, Composites of bead cellulose and hydrophilic solubilizers, *International Journal of Pharmaceutics* 139 (1996) 87–94. https://doi.org/10.1016/0378-5173(96)04600-5.
24. R.P. Swatloski, S.K. Spear, J.D. Holbrey, R.D. Rogers, Dissolution of cellose with ionic liquids, *Journal of the American Chemical Society* 124 (2002) 4974–4975. https://doi.org/10.1021/ja025790m.
25. J. Wu, N. Zhao, X. Zhang, J. Xu, Cellulose/silver nanoparticles composite microspheres: Eco-friendly synthesis and catalytic application, *Cellulose* 19 (2012) 1239–1249. https://doi.org/10.1007/s10570-012-9731-3.
26. J.P.F. Carvalho, A.C.Q. Silva, A.J.D. Silvestre, C.S.R. Freire, C. Vilela, Spherical cellulose micro and nanoparticles: A review of recent developments and applications, *Nanomaterials* 11 (2021) 2744. https://doi.org/10.3390/nano11102744.
27. K. Kümmerer, A. Henninger, Promoting resistance by the emission of antibiotics from hospitals and households into effluent, *Clinical Microbiology and Infection* 9 (2003) 1203–1214. https://doi.org/10.1111/j.1469-0691.2003.00739.x.
28. P.K. Jjemba, Excretion and ecotoxicity of pharmaceutical and personal care products in the environment, *Ecotoxicology and Environmental Safety* 63 (2006) 113–130. https://doi.org/10.1016/j.ecoenv.2004.11.011.
29. A.C. Singer, H. Shaw, V. Rhodes, A. Hart, Review of antimicrobial resistance in the environment and its relevance to environmental regulators, *Frontiers in Microbiology* 7 (2016). https://doi.org/10.3389/fmicb.2016.01728.
30. T. Ando, H. Nagase, K. Eguchi, T. Hirooka, T. Nakamura, K. Miyamoto, K. Hirata, A novel method using cyanobacteria for ecotoxicity test of veterinary antimicrobial agents, *Environmental Toxicology and Chemistry* 26 (2007) 601–606. https://doi.org/10.1897/06-195R.1.
31. R. Andreozzi, V. Caprio, C. Ciniglia, M. De Champdoré, R. Lo Giudice, R. Marotta, E. Zuccato, Antibiotics in the environment: Occurrence in Italian STPs, fate, and preliminary assessment on algal toxicity of amoxicillin, *Environmental Science and Technology* 38 (2004) 6832–6838. https://doi.org/10.1021/es049509a.
32. B. Halling-Sorensen, Environmental risk assessment of antibiotics: Comparison of mecillinam, trimethoprim and ciprofloxacin, *Journal of Antimicrobial Chemotherapy* 46 (2000) 53–58. https://doi.org/10.1093/jac/46.suppl_1.53.
33. C. Yang, J. Yuan, Y. Guo, X. Luo, In situ nano-assembly of Mg/Al LDH embedded on phosphorylated cellulose microspheres for tetracycline hydrochloride removal, *Cellulose* 28 (2021) 301–316. https://doi.org/10.1007/s10570-020-03533-8.
34. P. Zhao, Y. Yang, Y. Pei, X. Luo, TEMPO-oxidized cellulose beads embedded with Au-doped TiO2 nanoparticles for photocatalytic degradation of Tylosin, *Cellulose* 30 (2023) 1133–1147. https://doi.org/10.1007/s10570-022-04935-6.
35. E.M. Abd El-Monaem, A.M. Omer, R.E. Khalifa, A.S. Eltaweil, Floatable cellulose acetate beads embedded with flower-like zwitterionic binary MOF/PDA for efficient removal of tetracycline, *Journal of Colloid and Interface Science* 620 (2022) 333–345. https://doi.org/10.1016/j.jcis.2022.04.010.
36. Y. Gong, J. Yuan, Y. Pei, S. Liu, X. Luo, One-step quaternization and macromolecular reconstruction to prepare micro-/nano-porous cellulose beads from homogeneous solution for low-concentration amoxicillin removal, *Carbohydrate Polymers* 315 (2023) 120985. https://doi.org/10.1016/j.carbpol.2023.120985.
37. J. Yang, X. Luo, Ag-doped TiO2 immobilized cellulose-derived carbon beads: One-Pot preparation, photocatalytic degradation performance and mechanism of ceftriaxone sodium, *Applied Surface Science* 542 (2021) 148724. https://doi.org/10.1016/j.apsusc.2020.148724.

38. G. Zhang, L. Li, G. Zhou, Z. Lin, J. Wang, G. Wang, F. Ling, T. Liu, Recyclable aminophenylboronic acid modified bacterial cellulose microspheres for tetracycline removal: Kinetic, equilibrium and adsorption performance studies for hoggery sewer, *Environmental Pollution* 307 (2022) 119544. https://doi.org/10.1016/j.envpol.2022.119544.
39. Y. Gong, L. Liu, F. Wang, Y. Pei, S. Liu, R. Lyu, X. Luo, Aminated chitosan/cellulose nanocomposite microspheres designed for efficient removal of low-concentration sulfamethoxazole from water, *Journal of Molecular Liquids* 339 (2021) 116407. https://doi.org/10.1016/j.molliq.2021.116407.
40. A.K. Alanazi, P.S. Kumar, M. Shanmugapriya, G. Prasannamedha, H.M. Abo-Dief, Two-step fabrication of cellulose embedded Fe3O4/Fe3+ composite beads as catalyst in degradation of sulfamethoxazole in floating bed reactor, *Chemosphere* 335 (2023) 139158. https://doi.org/10.1016/j.chemosphere.2023.139158.
41. S. Singh, A. Majumder, P. Mandal, M.K. Yadav, Analgesics in wastewater matrix: A comprehensive review on occurrence, toxicity, and sustainability assessment of biological, tertiary, and hybrid treatment processes, *Environmental Nanotechnology, Monitoring & Management* 23 (2025) 101039. https://doi.org/10.1016/j.enmm.2024.101039.
42. R.H. Perry, Theoretical study of the adsorption of analgesic environmental pollutants on pristine and nitrogen-doped graphene nanosheets, *Physical Chemistry Chemical Physics* 23 (2021) 1221–1233. https://doi.org/10.1039/D0CP05543C.
43. E.A. Tyumina, G.A. Bazhutin, A.D.P. Cartagena Gómez, I.B. Ivshina, Nonsteroidal anti-inflammatory drugs as emerging contaminants, *Microbiology* 89 (2020) 148–163. https://doi.org/10.1134/S0026261720020125.
44. S. Fatima, N. Asif, R. Ahmad, T. Fatma, Toxicity of NSAID drug (paracetamol) to non-target organism—Nostoc muscorum, *Environmental Science and Pollution Research* 27 (2020) 35208–35216. https://doi.org/10.1007/s11356-020-09802-0.
45. J.W. Lee, J. Han, Y.-K. Choi, S. Park, S.H. Lee, Reswellable alginate/activated carbon/carboxymethyl cellulose hydrogel beads for ibuprofen adsorption from aqueous solutions, *International Journal of Biological Macromolecules* 249 (2023) 126053. https://doi.org/10.1016/j.ijbiomac.2023.126053.
46. Y. Huang, J. Zou, J. Lin, H. Yang, M. Wang, J. Li, W. Cao, B. Yuan, J. Ma, ABTS as both activator and electron shuttle to activate persulfate for diclofenac degradation: Formation and contributions of $ABTS^{•+}$, $SO_4^{•-}$, and •OH, *Environmental Science & Technolnology* 57 (2023) 18420–18432. https://doi.org/10.1021/acs.est.2c04318.
47. Y.P. Barragan Medina, V.A. Alvarez, P. Mendoza Zélis, J.S. Gonzalez, Ecofriendly magnetic gels beads based on carboxymethylcellulose and iron oxides for diclofenac adsorption, *Discover Chemical Engineering* 4 (2024) 22. https://doi.org/10.1007/s43938-024-00060-x.
48. M. Eddleston, Poisoning by pesticides, *Medicine* 48 (2020) 214–217. https://doi.org/10.1016/j.mpmed.2019.12.019.
49. Z.-Q. Jia, Y.-C. Zhang, Q.-T. Huang, A.K. Jones, Z.-J. Han, C.-Q. Zhao, Acute toxicity, bioconcentration, elimination, action mode and detoxification metabolism of broflanilide in zebrafish, Danio rerio, *Journal of Hazardous Materials* 394 (2020) 122521. https://doi.org/10.1016/j.jhazmat.2020.122521.
50. M.I. Swasy, B.R. Brummel, C. Narangoda, M.F. Attia, J.M. Hawk, F. Alexis, D.C. Whitehead, Degradation of pesticides using amine-functionalized cellulose nanocrystals, *RSC Advances* 10 (2020) 44312–44322. https://doi.org/10.1039/D0RA08308A.

12 Degradation of Synthetic Dyes

Pollutants in Wastewaters via Cellulose Microspheres

Josue Abraham Lara Zavala, Silva Beltrán Eduardo, and Naveen Kumar Reddy Bogireddy

12.1 INTRODUCTION

Synthetic dyes, widely used in various industries such as textiles (1), food, cosmetics, paints, and pharmaceuticals (2) for their vibrant and stable colors, have emerged as a significant environmental concern. It has been reported that during the dyeing process, a significant 10%–15% of the dyes are released into the water (3). These dyes, categorized based on their water solubility, chromophore nature, and application (4), are complex organic compounds that are challenging to degrade by conventional methods, which can negatively impact aquatic ecosystems (5) and human health.

When synthetic dyes are discharged into water bodies without proper treatment, they can alter water quality, affect aquatic biodiversity, and pose risks to organisms that depend on these resources, including humans (6). These compounds' effects on human health vary, ranging from skin irritations to more serious effects such as genetic mutations and carcinogenicity, especially in chronically exposed individuals (7). It has been reported in the scientific literature that synthetic dyes can affect different organs and systems of the human body, such as the nervous, enzymatic, and reproductive systems (8), in addition to the imminent carcinogenic risk of these coloring substances (9).

On the other hand, aquatic ecosystems can suffer significant alterations in their biodiversity due to the toxicity of these chemicals, interfering with the natural balance of aquatic species (10). At the regulatory level, several international organizations have established permissible concentration limits for synthetic dyes in wastewater to mitigate their harmful effects (11–14). However, the effective implementation of these regulations remains challenging due to the difficulty of treating water contaminated with these compounds (15). In this context, there is a need to explore innovative solutions that allow an efficient and sustainable degradation of synthetic dyes in contaminated water. A promising alternative is cellulose microspheres (16), biodegradable materials with a high potential to adsorb and degrade toxic compounds (17).

This chapter presents the application of cellulose microspheres and their combination with other materials (composite microspheres) as an emerging strategy for

DOI: 10.1201/9781003596127-14

removing synthetic dyes in wastewater. This approach offers an innovative solution and responds to the growing demand for environmentally responsible technologies that effectively remove contaminants.

12.2 SYNTHETIC DYES WASTEWATERS TREATMENT PROCESSES

Cellulose microspheres, with their spherical shape and three-dimensional structure derived from cellulose, a natural polymer in plant cell walls, are remarkably adaptable for various environmental applications. Their biocompatibility, biodegradability, and ease of modification make them an ideal candidate for treating contaminated water (3,18–20). The porous network of cellulose microspheres, with its large specific surface area, is particularly promising for the adsorption of contaminants, including organic substances such as synthetic dyes (21). The high abundance of hydroxyl (–OH) groups on the surface of cellulose-based nanomaterials, coupled with their high reactivity, gives them excellent potential for the sorption of various contaminants (22). Moreover, their structure can be easily modified to enhance their adsorption or degradation capacity, making them a versatile and inspiring tool for removing toxic compounds in wastewater.

Modifying the surface chemistry of cellulose is a versatile tool that can enhance its interaction with other matrices (23). These modifications, which can alter the material's hydrophobic or hydrophilic nature by grafting suitable molecules, can increase its water absorption capacity, ionic exchangeability, and elasticity. The main methods for this versatile modification include etherification, halogenation, oxidation, and alkalization (24). Numerous studies have proven these methods effective, demonstrating the versatility of modified cellulosic materials in adsorbing synthetic dyes (25).

The mechanism by which this process occurs is relatively simple to explain; once the contaminated water encounters the surface of the cellulose-based adsorbent, the charge of the surface functional groups plays a significant role in the easy attraction of oppositely charged dye molecules. This underscores the crucial role of the cellulose-based adsorbent in the process. However, the degradation processes of these dyes are not so simple since synthetic dyes are characterized by being chemically stable. These substances generally have complex aromatic structures; their toxicity depends on the dye charge. Dyes with cationic charges are usually more toxic than anionic ones (21).

12.3 TREATMENT MECHANISMS

12.3.1 Adsorption Mechanisms

Cellulose microspheres can adsorb dyes due to the physical and chemical interactions between the cellulose structures and the functional groups of the dyes (see Figure 12.1) (26). This adsorption phenomenon occurs when the dye molecules adhere to the surface of the microspheres due to physical forces such as electrostatic attraction, π–π interactions, ion exchange, and complexation. Notably, Van der Waals interactions (16) play a significant role in the process.

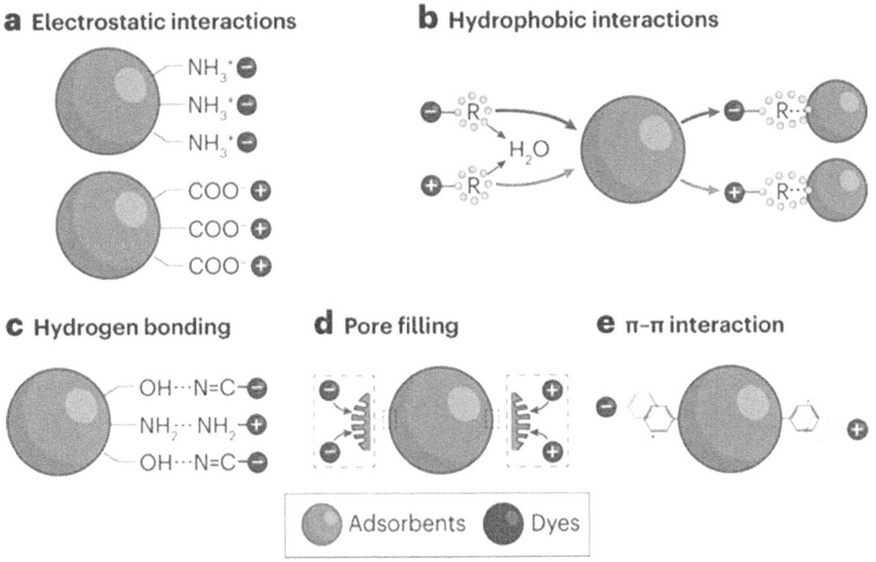

FIGURE 12.1 (a) Electrostatic, (b) hydrophobic, (c) hydrogen bonding, (d) pore filling, and (e) π–π interaction between dyes into the porous structure of adsorbents (6).

These attractions and interactions occur especially for aromatic adsorbents since these also contribute to the adsorbent capacity of the material (27). By nature, cellulose has hydroxyl groups that can form bonds with dye molecules. This facilitates the uptake of cationic dyes through electrostatic attraction, allowing their capture and subsequent removal. Meanwhile, anionic dyes can interact through hydrogen bonds or chemical modifications on the cellulose surface (see Figure 12.2) (28).

Furthermore, the pH can influence adsorption, the solution's ionic strength, and the microspheres' functionalization, optimizing their capacity for removing contaminants in aqueous media (29). The main factor that allows dyes to be adsorbed is their hydrophobicity; this property is related to the polarity and size of the dyes, as well as the protonation state of the ionizable molecules. Understanding the protonation state of these molecules is of utmost importance as it directly influences the adsorption process. This property is especially useful when synthetic, negatively charged dyes adsorb microspheres modified with positive functional groups (30). Dye degradation occurs when microspheres, especially those modified with additional compounds such as metallic nanoparticles or catalysts, facilitate the chemical decomposition of the dyes into less toxic compounds. Depending on the materials used to modify the microspheres, these processes can include reduction, oxidation, or photocatalysis reactions (31).

Metal nanoparticles, with their ability to act as active sites for redox reactions, and catalysts, which can accelerate reactions without being consumed, are a combination that significantly improves process efficiency. Furthermore, photocatalysis, in combination with UV or visible light, can generate reactive oxygen species that decompose dyes into less harmful substances. Dye degradation in contaminated water has been studied as an alternative to adsorption, and cellulose microspheres could play a key role in this process due to their high specific surface area and ability to be easily

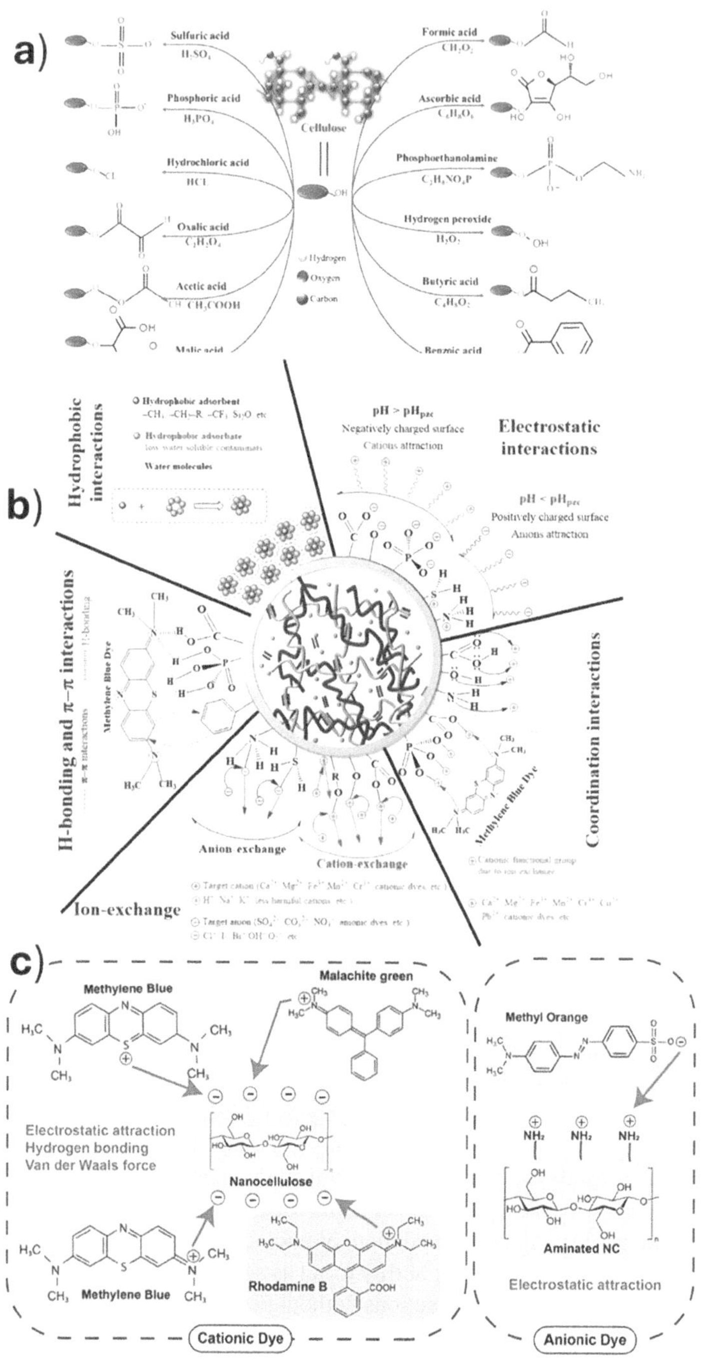

FIGURE 12.2 (a) Cellulose modifications methods (34), (b) cellulose–dye degradation interaction mechanisms (26), and (c) cationic and anionic dye molecule-based mitigation mechanism through cellulose as an adsorbent (21).

functionalized with various active agents (32). Unlike other traditional methods, such as adsorption, which can be less efficient in removing high-concentration dyes, cellulose microspheres offer a more sustainable and potentially more effective option for wastewater treatment by taking advantage of both their adsorption capacity and their catalytic and photocatalytic properties (33).

12.3.2 Degradation Mechanisms

Dealing with synthetic dyes by cellulose microspheres is a complex process involving chemical and biological elements. These microspheres facilitate adsorption, followed by catalytic degradation or biodegradation (35,36). Cellulose, a biopolymer, can support catalysts, such as metallic nanoparticles or oxidative enzymes, which facilitate the chemical degradation of dyes to less toxic products (37). However, the biodegradation process, which involves the action of microorganisms capable of decomposing the azo or aromatic bonds in the dyes, is equally crucial. It adds a layer of complexity and appreciation for the natural systems involved (38). Factors such as pH, temperature, and the presence of oxidizing agents can influence the efficiency of the process, allowing the development of sustainable systems for the treatment of contaminated water (39). Cellulose microspheres can be modified with photocatalysts such as titanium dioxide (TiO_2) (40,41) or silver nanoparticles, which, when activated by ultraviolet (UV) light, can induce redox reactions that decompose the dyes (42). This process has been investigated for many dyes, including methyl orange, Rhodamine B, Methylene blue, and Congo red (43), demonstrating its broad applicability.

As an organic material, cellulose can be a substrate for biodegrading microorganisms in bioremediation processes (44–46). Although these degradation processes are not as commonly used as adsorption due to their complexity and the need for controlled conditions, there is intriguing potential in modifying cellulose microspheres. These modifications can facilitate adsorption and degradation simultaneously, offering a new avenue to improve the efficiency of removing dyes from wastewater (47).

Complete degradation of dyes can eliminate toxicity, unlike adsorption, which only removes the dye from the medium without necessarily destroying it. Biodegradation allows the recycling of the degraded dye into less harmful compounds. On the other hand, degradation can be slower and requires specific conditions (pH, temperature, and the presence of suitable catalysts or microorganisms). Some dyes, especially those with complex structures or high stability, may be more difficult to degrade by biological or chemical pathways.

12.4 SYNTHETIC DYES

Synthetic dyes represent one of the main sources of contamination in water bodies due to their high chemical stability and difficulty in being degraded naturally (47). In this context, cellulose microspheres, in their pure and modified forms, have emerged as an efficient strategy for removing these contaminants through adsorption and degradation processes. In general terms, synthetic dyes can be classified in two ways: (a) according to their chemical structure and (b) according to their solubility in water. Understanding the basis of these classifications is essential to analyze the degradation mechanisms of dyes when cellulose microspheres are used in their removal, which are briefly described here.

12.4.1 Chemical Structure

This classification, based on the composition of the dyes and their chromophore grouping, is a powerful tool. It enables us to predict the chemical reactions between the dye and various reducing agents, oxidizing agents, or other chemical compounds. For instance, synthetic dyes can be classified under names such as azo (monoazo, diazo, and triazo), which represent 70% of world production and are characterized by having azo groups (–N=N–) coupled to auxochromes (OH, NH_2) (7, 48). Following the azo, the anthraquinone group, derived from anthracene, has hydroxyl or amino groups on a quinone nucleus. This unique chemical characteristic gives dyes in this category an intriguingly high light resistance (48, 49).

On the other hand, indigo dyes include homologs of selenium, sulfur, and oxygen, ranging in color from orange to indigo blue (50). There are also xanthenic dyes (51), such as fluorescein, which are not just dyes but also serve practical purposes as tracers due to their fluorescent properties. Other types of synthetic dyes, such as phthalocyanine dyes (52), obtained by the interaction of dicyanobenzene with metals (Cu, Ni, Co, Pt), stand out for their chemical stability and light resistance, with copper phthalocyanine being the most commonly used. Some others, like nitrated and nitrosated dyes, are inexpensive and contain hydroxyl or amino groups in an ortho position to a nitro group (–NO_2). Triphenylmethane dyes, such as fuchsin (53) and malachite green (54), can be sulfonated to produce acid dyes, which find practical applications in various industries. The last of these groups is polymethine (cyanines), which contains a polymethine chain linked to heterocycles and is used in optical applications (55). This is how synthetic dyes, with their diverse properties and practical applications, are usually classified based on their chromophore structure.

In addition to this classification, another one is based on the solubility of these dyes in water. In the literature, you will find that in addition to these classifications, they can also be divided by application or material, as is the case with dyes that have textile applications, which are divided according to the nature of the fiber dyed (34).

12.4.2 Solubility in Water

Synthetic dyes can also be classified based on their water solubility. In this sense, some are more soluble in water (water-soluble), which are commonly used in industries such as textiles, cosmetics, and pharmaceuticals. These are direct, acidic (anionic), basic (cationic), metalliferous, and reactive dyes. On the other hand, some are insoluble in water, such as vat dyes, sulfur dyes, and dispersed dyes (56). These classifications are of utmost importance since they allow us to determine their application in different materials and their environmental impact, especially in wastewater treatment.

This chapter focuses on classifying synthetic dyes based on their water solubility rather than their chemical structure to explain the application of cellulose microspheres in the degradation of dyes in water. This approach provides a clear and comprehensive understanding of this important application.

12.5 TREATMENT OF SYNTHETIC DYES IN WASTEWATER

As mentioned above, this section presents the main categories of synthetic dyes based on their water solubility. Their applications, potential effects on human health and the environment, and the use of cellulose microspheres for treating water contaminated with these compounds will be briefly described. Notably, not all dyes have been evaluated with a material as specific as cellulose microspheres; therefore, the examples mentioned in each category correspond to the most recent research in which they have been applied. To ensure that this chapter reflects the state of the art, only the most relevant studies, which we consider the most significant, and those published in the last five years, i.e., between 2020 and Spring 2025, were considered. This currency of our research ensures that the information presented is highly relevant and applicable to current environmental issues.

12.5.1 Acidic Dyes

Acidic dyes are a type of water-soluble dyes that present acidic functional groups, such as sulfonic ($-SO_3H$) or carboxylic ($-COOH$), which confer a negative charge in aqueous solution (57). Due to this characteristic, they have a high affinity for positively charged substrates. These dyes are usually widely used in textile industries, particularly in the dyeing of natural fibers such as wool or silk (58). They are called acid dyes precisely because of the characteristic functional groups they contain and because, in aqueous solution, they tend to ionize and confer a negative charge to the dye molecule (59). Furthermore, these dyes are usually applied in acidic media, where protein fibers, such as wool and silk, acquire a positive charge, facilitating the electrostatic interaction between the dye and the material to be dyed (60). In this sense, "acid" does not mean that they are strongly acidic substances but are designed to be used in acidic conditions and have acidic functional groups in their chemical structure.

Acid or anionic dyes, a diverse group of compounds, present a variety of chromophoric systems. They exhibit characteristic structural differences but share a common trait—the presence of ionic substituents. These substituents are crucial in rendering the dyes water-soluble (61). These dyes owe water solubility to a chromophore group or one or more sulfonate groups (62). In general, these types of dyes are triaryl methanes, anthraquinones, xanthenes, or azo dyes (7), each with its unique properties, as briefly described below:

With their $-N{=}N-$group, azo dyes play a significant role in various fields. For instance, acid red 88 (63) is a key component in textiles and inks, acid yellow 17 (64) in paper and fiber coloring, and Congo red (65) is a crucial pH indicator and amyloid detection tool. Similarly, acid dyes based on triphenylmethane, such as acid green 3 (66) in cosmetics and food and aniline blue (67) for cell staining, are important in their respective fields.

Further expanding the spectrum of dyes, we have anthraquinone derivatives, known for their robust resistance and stability under light. Acid Blue 25 (68) is a prime example, widely used in the textile industry and printing inks, while acid red 52 (69) is a popular choice in cosmetics and food. A subcategory of acid dyes, xanthene dyes, is also worth noting for their unique chemical structure. Examples include eosin Y (70), a crucial dye in tissue staining, and fluorescein (71), a significant dye in fluorescent markers and various medical tests, each playing a vital role in their respective fields.

Acid dyes have been reported to cause various health effects in humans, and prolonged exposure to them can cause allergic reactions, skin irritations, respiratory problems, and damage to the nervous system (72). This same water-soluble property of these dyes can promote easy contamination of water bodies such as rivers and lakes, affecting aquatic life since they can be toxic to fish and other aquatic organisms (73–75). Some of the most recent case studies are presented below to address how cellulose microspheres have been used to treat dye-contaminated water.

Recently, an innovative development in wastewater treatment has been reported: the amphoteric cellulose microspheres (ACMs). These microspheres have shown a remarkable capacity for removing anionic and cationic dyes, particularly in the adsorption of methyl orange, a widely used anionic dye in the textile industry. The ACMs, with their high concentration of carboxyl (1.52 mmol/g) and amino (1.51 mmol/g) functional groups, enable electrostatic interaction with dyes of opposite charges. Their hierarchically porous structure and high specific surface area (123.97 m^2/g) further enhance their effectiveness, optimizing the capture of contaminants. This research reported that these materials had an adsorption capacity of up to 345.32 mg/g, demonstrating the considerable potential of ACMs in wastewater treatment.

Furthermore, the mechanical strength and reusability of the microspheres were remarkable, as they retained more than 90% of their performance after eight adsorption–desorption cycles. Additionally, tests with real effluents from the printing and dyeing industry confirmed that the ACMs could significantly reduce the pollutant load, decreasing nitrogen and phosphorus concentrations in the treated water by more than 50%. These findings underscore the potential of ACMs as an efficient and sustainable alternative for remediating contaminated water and highlight their versatility, as they show promise for treating a variety of pollutants, not just methyl orange, but also other acidic and basic industrial dyes.

Another interesting case study recently published deals with the synthesis and application of cellulose microspheres with a porous three-dimensional structure modified with different groups of ionic liquids for the adsorption of the Acid Orange 7 (AO7) dye (76). This work indicated that cellulose microspheres modified with 4-methylimidazole (CMs4) presented the highest adsorption capacity, reaching a value of 218.6 mg/g. In addition, they showed exceptionally fast adsorption kinetics, achieving an adsorption of 100 mg/g in just ten minutes. The study also identified the adsorption mechanism, which primarily involves the electrostatic attraction, hydrogen bonding, and π–π interactions between the ionic liquid groups and the dye molecules, underscoring the crucial role of these groups in the process.

Similarly, CMs4's pH-dependent selective adsorption was demonstrated, allowing the separation of cationic and anionic dyes by adjusting the pH of the solution. This feature was further validated in a small-scale water purification device, where CMs4's efficiency in contaminated water treatment was highlighted. Moreover, case studies have shown the synthesis and application of cellulose microspheres supported with palladium nanoparticles (PdNPs) as catalysts for degrading synthetic dyes, particularly for the Congo red dye (37). The microspheres, fabricated by microwave heating, resulted in better dispersion and smaller size of the palladium nanoparticles (3.1 ± 1.4 nm), enhancing their catalytic activity. In degradation tests, the microspheres could remove up to 99% of Congo red in nine minutes, with a rate constant of 0.3886 per min, further underlining their potential for large-scale environmental remediation.

Furthermore, these catalysts displayed excellent recyclability, maintaining over 90% efficiency after five cycles. The research provided a simple, sustainable, and efficient method for fabricating PdNP-based catalysts, highlighting the potential of cellulose as a reducing agent and support for applications in pollutant remediation. While these are just a few research cases, they highlight the effectiveness of cellulose microspheres as adsorbent and catalytic materials in the removal and degradation of synthetic dyes. Thanks to their structural properties, surface modulation capacity, and functionalization with active groups, cellulose microspheres offer a promising and inspiring solution for wastewater decontamination. Their reusability and low cost make them a sustainable option for treating dye-contaminated water, although performance may vary depending on the dye characteristics and the application environment.

12.5.2 Reactive Dyes

Reactive dyes are synthetic dyes characterized by having functional groups capable of forming covalent bonds with the fibers or surfaces to which they adhere, generally in dyeing or dyeing processes (77). They are called "reactive" because these dyes contain reactively active functional groups capable of forming covalent bonds with the molecules of the fibers being dyed. The chemical reaction between the dye and the fiber distinguishes reactive dyes from other dyes that do not form this permanent chemical bond. This makes the dye adhere more strongly and stably to the fiber, giving greater resistance to washing, friction, and fading, characteristics highly appreciated in the textile industry (78). The term "reactive" refers to this capacity for chemical reaction, which is the basis of its fixation mechanism. These dyes contain reactive groups, such as amino, epoxide, or isothiocyanate, which react with functional groups in the fibers, such as hydroxyl or amino, to form stable bonds (77,79). This type of dye is widely used in the textile industry, particularly for dyeing cellulosic fibers such as cotton (80). Some common products or fabrics dyed with reactive dyes include cotton t-shirts, bed linens, and towels.

The main advantage of reactive dyes is their high stability and wash resistance, which allows the colors to remain on garments longer. In addition, they are appreciated for their ability to generate bright and vibrant colors (81). Reactive dyes usually include chromophore groups derived from the azo, anthraquinone, and phthalocyanine families. These dyes have greater water solubility, especially in cotton dyeing processes and potentially in the coloring of wool and polyamides (77). The residues of these dyes come mainly from the textile and paper industries and are released when the dye does not properly adhere to or bond with the fibers, a process known as 'fixing'.

The most used reactive dyes in the textile industry, such as Reactive Red 120, Reactive Blue 19, Reactive Yellow 145, Reactive Black 5, and Reactive Green 5, are known for fixing natural fibers such as cotton (82). However, some of these dyes, especially those from the azo dye family (like Reactive Blue 19 and Reactive Red 120), have been studied for their potential toxicity. They can release toxic compounds such as anilines or heavy metals, harming the environment and aquatic organisms (83). These risks have led to a call for more environmentally friendly alternatives and, importantly, the need for stricter regulations on industrial waste management to reduce their environmental impact.

Reactive dyes can have negative effects on both human health and the environment. In health, they can cause skin and eye irritation, allergies, or sensitization reactions. Some, especially in the azo family, are associated with carcinogenic, mutagenic, and possible effects on the reproductive system (84). In the environment, residues of these dyes can be toxic to aquatic organisms, altering the fauna and flora of aquatic ecosystems. They are also persistent and bioaccumulative in living beings (85). They can also contaminate the soil and affect terrestrial biodiversity. These risks have driven the search for more sustainable alternatives and implementing regulations to minimize their impact. An example of these alternatives for their control is the use of cellulose microspheres for their removal from aqueous media (86,87). Below are some case studies that show the application of these new materials for treating these contaminants.

The synthesis and characterization of porous cellulose acetate (CA) microspheres functionalized with 3-aminopropyltriethoxysilane (APTES-CA) for the removal of the anionic dye Reactive blue 19 (RB19) from wastewater was recently published (88). The porous CA microspheres were synthesized using the double emulsion (W/O/W) technique and subsequently functionalized with APTES, which introduced amine and silane groups on the surface, significantly improving their adsorption capacity compared with unmodified microspheres. The characterization confirmed the successful modification of the material, evidencing a greater interaction with RB19. The optimal pH for dye adsorption was determined to be two, and the process followed a chemisorption mechanism, indicating a strong interaction between RB19 and the functional groups of the microspheres.

Furthermore, isotherm studies revealed the existence of heterogeneous binding sites, suggesting the involvement of multiple adsorption mechanisms. APTES-CA microspheres demonstrated high removal efficiency for RB19, positioning themselves as a promising material for treating water contaminated with anionic dyes. Another relevant study investigated chitosan/cellulose microspheres (CCM) to remove the synthetic dye Reactive Black 5 (29). The microspheres were obtained by emulsification in an alkali/urea solvent, which prevented agglomeration of oppositely charged polymers and improved the homogeneity of the mixture. Furthermore, the CCMs presented higher mechanical strength and anti-swelling properties than pure chitosan microspheres.

The adsorption mechanism mainly involved electrostatic interactions between the amine groups of chitosan and the sulfonate groups of the dye, as well as hydrogen bonding and π–π interactions. The CCMs demonstrated a 214.36 mg/g adsorption capacity and rapid kinetics, reaching 76% adsorption in just 20 minutes. Furthermore,

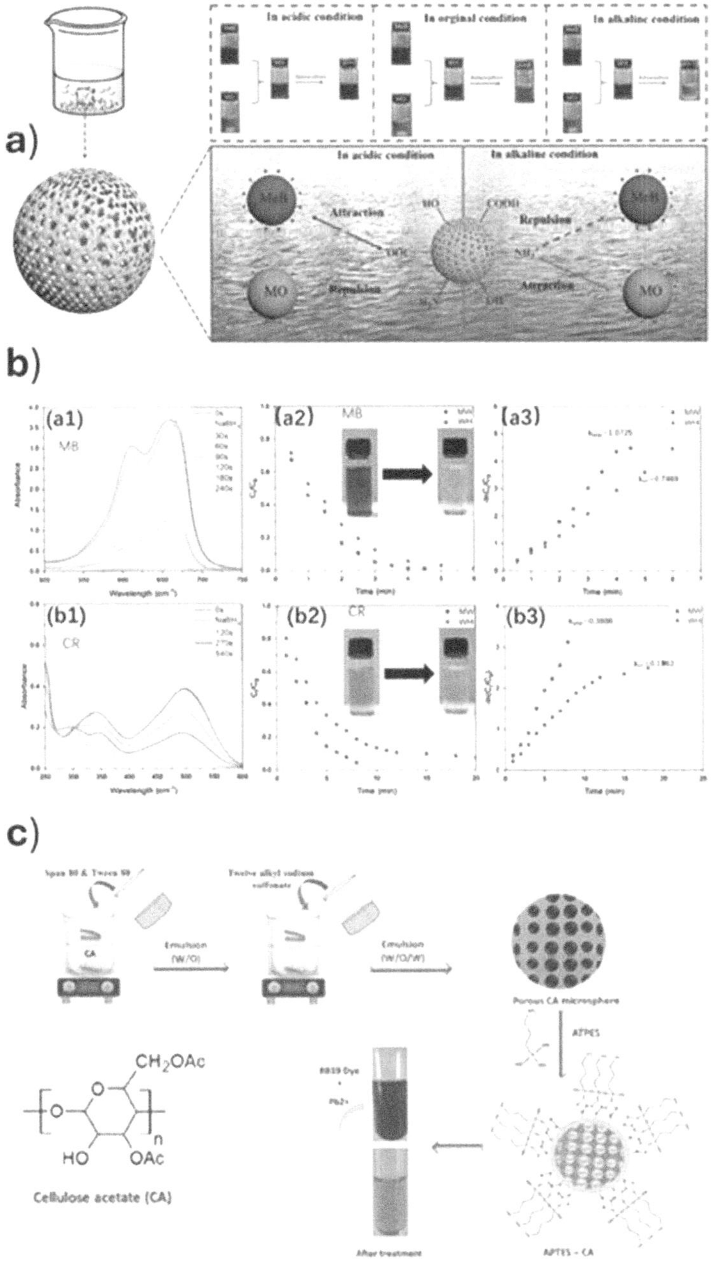

FIGURE 12.3 (a) Amphoteric cellulose microspheres as an anionic/cationic dye removal catalyst (20), (b) their corresponding example absorbance spectra with (a1) methylene blue (MB) and (b1) Congo red (CR). The kinetic study (a2, b2, a3, and b3) with dye concentration *vs.* reaction time during the degradation of MB and CR dye reaction solutions, respectively (37). (c) Schematic representation of APTES-modified CA, RB19 degradation, and Pb^{2+} adsorption (88).

they showed excellent reusability, maintaining 75% efficiency after five adsorption/desorption cycles. These results demonstrate the effectiveness of CCM microspheres in removing synthetic dyes from contaminated water and suggest their promising potential for wider application in water treatment. Other relevant case studies of synthetic dye removal using cellulose microspheres are shown in Figure 12.3.

12.5.3 Disperse Dyes

Disperse dyes, also known as soluble dyes, are applied as fine powders dispersed in the dye bath and are highly insoluble in water. They are used at high temperatures, a crucial factor, to allow the dye to penetrate the synthetic fibers before setting, maintaining their stability. They are especially effective in coloring synthetic fibers, such as polyester and polyamide, in the presence of a dispersing agent that ensures uniform dye distribution (89). Disperse dyes are used in the textile industry to dye synthetic fibers, and their residues can be released into the environment during textile washing or production. Like other dyes, these pose risks to health and the environment since some are linked to mutagenicity and carcinogenicity, such as Disperse Red 1 (90) and Disperse Blue 79 (91).

A recent study reported preparing and applying cellulose triacetate (CTA) microspheres as highly efficient adsorbents for removing dispersed dyes from water (92). CTA was synthesized from ramie fiber, a natural resource abundant in China, via an acetylation reaction. CTA microspheres, with a honeycomb-like structure, were prepared using the solvent evaporation method. These microspheres exhibited an average diameter of 14 μm, with a size distribution ranging from 10 to 16 μm, and a high-area rough surface, enhancing their adsorption capacity. The characterization of the microspheres was performed using FTIR, NMR, and XRD techniques, which confirmed the successful conversion of cellulose to CTA with a high degree of substitution (DS = 2.93), ensuring that the synthetic material met the requirements for efficient adsorbent applications. The adsorption performance of the CTA microspheres was evaluated relative to the disperse dye Disperse Red S-R, showing a removal capacity of up to 97% at 50°C and pH 7. Adsorption was favored at higher temperatures and neutral pH conditions, which improved dye diffusion and microsphere stability.

Moreover, the unique properties of CTA microspheres, such as their structure and interactions with dyes, made them more efficient in adsorbing dispersed dyes than reactive dyes. This study showcases the successful creation of cellulose triacetate microspheres from ramie fiber and their effective use in eliminating dispersed dyes from wastewater under optimized conditions. This method underscores the potential of biopolymers as eco-friendly adsorbent materials for purifying waters tainted with synthetic dispersed dyes.

12.5.4 Basic Colorants

Basic dyes are organic compounds that can dye fibers with anionic sites when combined with metal salts. They are common in the textile industry and cosmetic products, but their release into wastewater during dyeing and washing processes can be problematic. These dyes are known to cause skin irritations, respiratory problems, and neurotoxic effects; some are even classified as carcinogenic. In the environment, their presence is highly toxic to aquatic ecosystems, altering the biological balance and reducing biodiversity (93). To mitigate these effects, a promising option is using cellulose microspheres, which, due to their high surface area and adsorption capacity, could help remove these dyes from wastewater more efficiently. Below, we will look at some case studies in which cellulose microspheres have been applied to address this problem, demonstrating their effectiveness in removing basic dyes from contaminated water.

Recently, a significant breakthrough has been made in environmental chemistry with the discovery of the Pd@SA/Na-CMC/GA catalyst. This catalyst has been found to efficiently degrade rhodamine B (RhB) (94), a highly toxic dye commonly used in the textile and paper industries. The study reported that the Pd@SA/Na-CMC/GA catalyst reduced RhB in just 35 seconds under mild reaction conditions, demonstrating its high efficiency. Furthermore, a rate constant of 0.058 per second for the reduction of RhB was calculated, indicating favorable kinetics for the elimination of this compound. The catalyst support, composed of sodium alginate (SA), sodium carboxymethylcellulose (Na-CMC), and gum arabic (GA), exhibited key functional groups such as –OH (hydroxyl), COO– (carboxylate), and C–O–C (ether), which played a crucial role in stabilizing and dispersing the Pd nanoparticles on the material surface. These findings underscore the potential of the Pd@SA/Na-CMC/GA catalyst for efficiently remedying rhodamine B-contaminated water, offering a promising alternative for treating industrial effluents under favorable operating conditions. In another case study, a green and facile method was carried out to efficiently prepare highly porous CNC/MnO_2/SA microspheres to remove dyes from wastewater (95). The porous structure was formed using an air bubble mold and ionic crosslinking technology.

The microspheres, with an expansion ratio of 2, presented a honeycomb-like structure with high porosity (98.23%) and low density (0.027 g/cm^3). The CNC/MnO2/SA microspheres demonstrated excellent dye removal capacity, achieving a decolorization rate of 95.4% for methylene blue in ten minutes and an equilibrium decolorization capacity of 114.5 mg/g. The high porosity and specific surface area (13 m^2/g) allowed both adsorption and oxidative degradation of the dyes. The dye removal mechanism was based on electrostatic adsorption on negatively charged carboxyl groups and oxidative degradation by the MnO_2 component. The synergistic effect between adsorption and degradation led to high dye removal efficiency. Furthermore, the CNC/MnO_2/SA microspheres showed excellent recyclability performance, maintaining their efficiency for ten cycles.

Overall, this study demonstrates the potential of developed porous microspheres for the sustainable and efficient treatment of dye-contaminated wastewater. Similarly, a case study on hydrangea-like nanocellulose microspheres was conducted, specifically focusing on their application for dye adsorption using TEMPO-oxidized cellulose nanofibers (TOCN) (96). The emulsion method prepared TOCN microspheres with different sizes (3 μm, 105 μm, and 544 μm). These microspheres exhibited a stable, porous, three-dimensional structure with a high specific surface area (up to 172 m^2/g) and mesoporous pore sizes. TOCN microspheres demonstrated excellent adsorption capacity for the cationic dye methylene blue (MB, up to 412.1 mg/g) and the anionic dye methyl orange (MO, up to 286.5 mg/g). Electrostatic interactions for MB and hydrogen bonding for MO drove the adsorption. Furthermore, TOCN microspheres showed selective adsorption of MB over MO in mixed dye solutions, underscoring their potential for efficient dye removal from contaminated solutions.

Another research article reported the preparation of porous CNC/MnO_2/TiO_2 microspheres by air bubble annealing and ionic crosslinking, achieving a highly porous structure with uniformly distributed TiO_2 and MnO_2 nanoparticles. These microspheres demonstrated an impressive MB (31) removal capacity, achieving a staggering 97.5% removal in just five minutes, thanks to the synergistic effect between the two metal oxides. Furthermore, they could be regenerated with UV light and reused for ten cycles without losing efficiency. Their performance was due to improved light absorption, reduced band gap, and the oxidative degradation facilitated by MnO_2, highlighting their potential for treating water contaminated with dyes. The case studies presented so far reflect the applications given to cellulose microspheres for degrading synthetic dyes. To condense the information seen so far, Table 12.1 shows the properties and performance of dye removal using cellulose microspheres.

Table 12.1 summarizes key properties and performance metrics of cellulose microspheres used for dye removal. It highlights the relationship between microsphere characteristics, removal capacity, and reusability, providing insights into their effectiveness and functional mechanisms.

This table presents information on dye removal using cellulose microspheres. It includes the dye type, microsphere type and size, dye name, removal capacity, removal mechanism, functional groups involved, and material reusability.

12.5.5 Cellulose Microspheres—Composite Materials in the Degradation of Synthetic Dyes

Cellulose microspheres, used alone or combined with other materials, are a key player in the innovative solution for degrading synthetic dyes in wastewater. Their adsorption and degradation capacity can be significantly enhanced by combining with metal nanoparticles, activated carbon, or organic compounds (101) (see Figure 12.4). Metal nanoparticles can act as catalysts in reduction or oxidation processes, accelerating the breakdown of toxic dyes into less harmful products (102). Consequently, composite microspheres could revolutionize the efficiency of wastewater treatment, offering a more sustainable and cost-effective alternative to conventional technologies.

Developing hybrid or composite materials based on cellulose microspheres and other advanced materials, such as metal nanoparticles or biological catalysts, represents one of the most promising frontiers in wastewater treatment research (103). These combined materials not only improve the adsorption capacities of dyes but can also accelerate degradation processes through advanced mechanisms such as photocatalysis, providing the latest insights into wastewater treatment (104).

A recent example of the application of metal nanoparticles and cellulose microsphere composites is reported by Yuqing Yu et al. (98). This study, which describes

TABLE 12.1
Properties and performance of dye removal using cellulose microspheres

Dye Type	Cellulose Microsphere Type	Size of Microspheres	Dye Name	Removal Capacity	Mechanism of Removal	Functional Groups	Reusability
Acid (anionic)	Amphoteric cellulose (20)	2.80 ± 0.24 mm	Methyl Orange	345.32 mg/g	Electrostatic attraction	Carboxyl and amino	Eight cycles
	Pd@SA/Na-CMC/GA (94)	NA		Reduced within 15 seconds	Electron transfer	Hydroxy and carboxylate	Six cycles
	TOCNM1 (96)	3 μm		286.5 mg/g	Hydrogen bonding	Carboxyl	NA
	Cellulose/silica (97)	NA		588 mg/g	Electrostatic attraction	Amino	NA
	Ionic Liquid Modified Cellulose (76)	90–110 μm	Acid Orange 7	218.6 mg/g	Electrostatic attraction, hydrogen bonding, and π-π interactions	Imidazole-based groups	Five cycles
	PdNPs/cellulose (37)	3.1 ± 1.4 nm to 3.8 ± 1.1 nm	Congo red	99% in nine minutes	Electron transfer	Hydroxyl	Five cycles
	Cellulose acetate	270 μm		95.6%	Physisorption, electrostatic interactions, and hydrogen bonding	Hydroxyl and acetate	Five cycles
Reactive	Cellulose acetate (CA) (88)	75–200 μm	Reactive blue 19	91%	Chemisorption	APTES-CA	NA
	Chitosan/cellulose (29)	50 μm (aprox)	Reactive black 5	214.36 mg/g	Electrostatic attraction, hydrogen bonding, and π-π interactions	Amino and hydroxyl	Five cycles

(Continued)

TABLE 12.1 (*Continued*)
Properties and performance of dye removal using cellulose microspheres

Dye Type	Cellulose Microsphere Type	Size of Microspheres	Dye Name	Removal Capacity	Mechanism of Removal	Functional Groups	Reusability
Disperse (non-ionic)	Cellulose triacetate (92)	10–16 μm	Red S-R	97%	Hydrogen bonding	Hydroxyl and acetyl	NA
Basic (cationic)	Pd@SA/Na-CMC/GA (94)	NA	Methylene Blue/ Rhodamine B	MB—immediately decolorized RhB—reduced within 35 s	Electron transfer	Hydroxy and carboxylate	Six cycles
	CNC/MnO_2/SA (95)	0.01 mm	Methylene Blue	Decolorization capacity 114.5 mg/g/95.4% within 10 min	Electrostatic interactions and oxidized and degraded	Carboxyl	Ten cycles
	TOCNM1 (96)	3 μm		412.1 mg/g	Electrostatic interaction and hydrogen bonding	Carboxyl	NA
	CNC/MnO_2/TiO_2 (31)	4 mm		97.0%/310.2 mg/g	Synergistic charge transfer and enhanced photocatalytic activation	Carboxyl	Ten regeneration cycles
	Pd NPs/CMs (98)	753.67 ± 107.70 μm		Decolorization capacity 99.8%	Electron transfer	Hydroxyl	Five cycles
	Tannin-immobilized cellulose (99)	408 ± 15 μm		57.15 mg/g	Electrostatic attraction	Hydroxyl	NA
	Cellulose/Fe_3O_4 hydrogel (100)	Ranged from 39 to 62 μm	Crystal violet	60 mg/g	Electrostatic attraction	Sulfate, carboxyl, and phenolic	NA

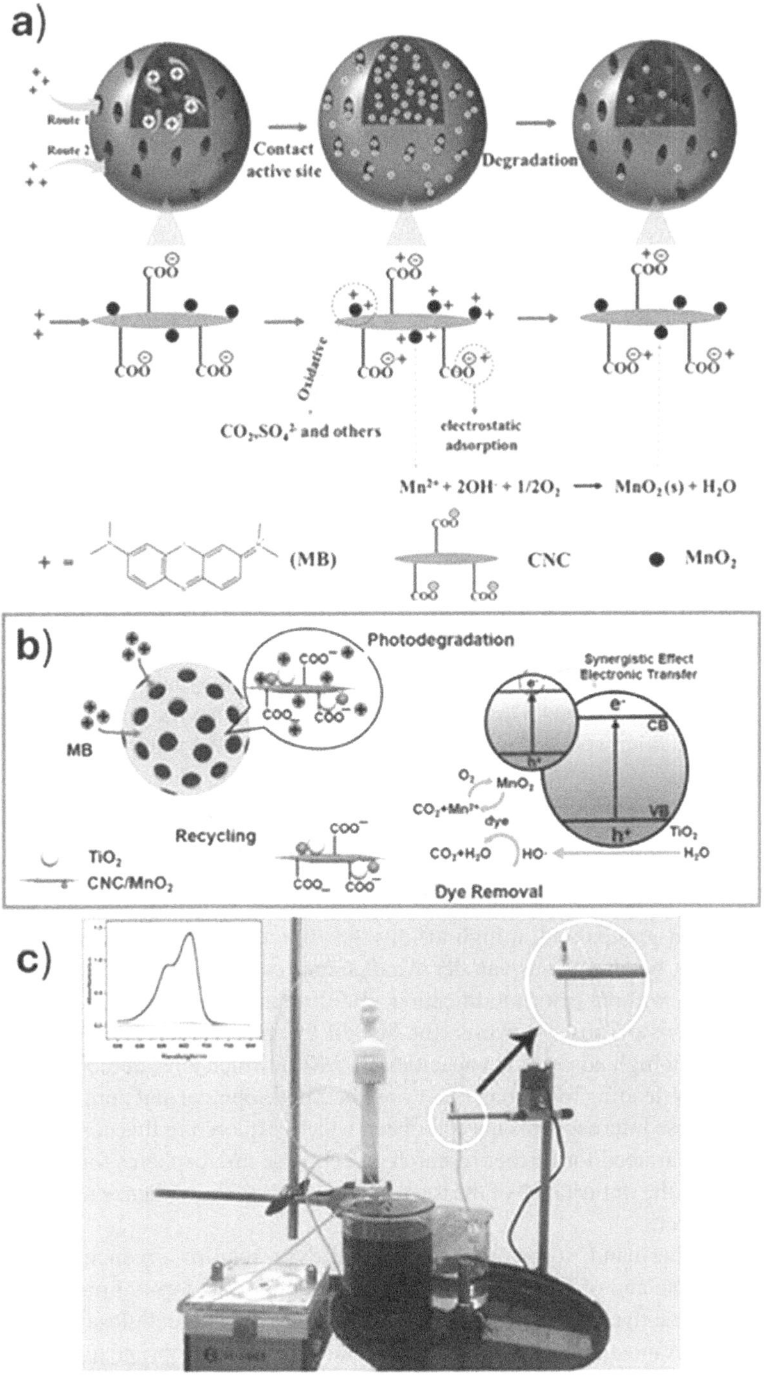

FIGURE 12.4 (a) Pictorial diagram of porous microspheres ($CNC/MnO_2/SA$)-based MB decolorization mechanism (95), (b) efficient dye removal (31), and (c) decolorization setup digital photograph with MB (8 mg/L) and $NaBH_4$ (10 mg/L) in the beaker (94).

using PdNPs for the catalytic decolorization of MB up to 99.8%, involved chemical interactions such as electrostatic, hydrogen bonding, and π–π interactions. These interactions are significant as they provide insights into the potential of cellulose-based composites in environmental remediation. The destruction of the C=N and S–S chromophores, which describes the proposed decolorization mechanism, was found to be influenced by the hydrogen bonding interaction between the hydroxyl (–OH) groups of cellulose and the dimethylamino ($-N(CH_3)_2$) group of MB. However, the π–π stacking interaction between the glucose unit in the cellulose structure and the benzene ring in the MB molecule played a particularly enlightening role in the decolorization mechanism. This report also demonstrated that the composite material could be reused for up to five cycles with a catalytic rate exceeding 90%.

12.6 BENEFITS AND LIMITATIONS

Cellulose microspheres offer several advantages as a material for synthetic dye removal. Their low cost, availability as a biodegradable material, and ability to be modified to improve their performance make them an attractive option compared with more expensive and complex technologies. However, their use in real-world conditions faces certain challenges. The effectiveness of cellulose microspheres for dye removal can be affected by contaminant concentration, pH, and water temperature. Furthermore, the scalability of processes using cellulose microspheres must be carefully evaluated, as controlled laboratory conditions may not be easily replicated in large-scale applications. Despite these challenges, research continues to advance, and cellulose microspheres remain promising for treating water contaminated by synthetic dyes.

There are still significant opportunities in literature, as not all dye categories have been studied using cellulose microspheres. For example, studies on removing direct dyes using cellulose microspheres are limited. Several factors can explain this lack of information. First, in its natural form, cellulose is a hydrophilic material lacking functional groups with a high affinity for direct dyes, which typically interact through weak bonds such as van der Waals forces or electrostatic interactions. This suggests that, without prior modification, cellulose microspheres may not be highly efficient for this purpose. However, the field of dye removal has been dominated by materials with high adsorption capacities (105,106), which have demonstrated superior efficiency, leading to a greater focus on their development and optimization. As a result, cellulose microspheres have not been widely explored in this context. Yet, this presents a clear need for further research on cellulose microspheres for dye removal, underscoring the importance of the topic and motivating the audience to delve deeper into the subject.

On the other hand, some dyes, such as vat dyes, require a reduction process to become soluble and often generate residual products in the form of particles or precipitates. These dyes are difficult to remove or degrade with cellulose microspheres since, for their most effective removal, it would be more appropriate to use filtration, coagulation–flocculation, or biodegradation techniques (107,108) instead of adsorption processes with materials such as cellulose microspheres. Vat dyes present challenges similar to direct dyes in their removal by conventional adsorbents. However, in this case, the main reason is the lack of studies on the chemical properties of these dyes, such as their low solubility in water and their low affinity for cellulose. This underscores the need for further research in this area. Furthermore, there are potentially more effective techniques to treat these contaminants, such as electrocoagulation (109), which makes studies on the use of cellulose microspheres in their removal practically nonexistent.

Similarly, other types of dyes, such as metalliferous and sulfur-based dyes, can be found in aggregates or precipitates, making conventional adsorption methods not the optimal strategy for their removal. The scarcity of studies on the adsorption of these dyes with cellulose microspheres can be explained by the limited interaction between cellulose and these substances due to their chemical properties, which affect their interaction with the adsorbent materials. Furthermore, fluorescent, photochromic, and thermochromic dyes have been scarcely explored in the scientific literature regarding their removal with cellulose microspheres, probably because they have very specific chemical structures that necessitate using more advanced technologies for their removal. This underscores the urgent need for further research and the development of advanced technologies in the field of dye removal.

12.7 CONCLUSIONS AND FUTURE PERSPECTIVES

The increasing contamination of water by synthetic dyes, originating from industries as diverse as textiles, food, and cosmetics, represents a significant threat to aquatic ecosystems and human health. These compounds, characterized by their toxicity and persistence, not only affect water quality but also threaten biodiversity and can trigger adverse effects in the food chain. Although international regulations governing permissible levels of these contaminants exist, their enforcement is insufficient in many regions, underscoring the urgent need for more effective solutions.

In this context, cellulose microspheres, either in their pure form or combined with other materials to form innovative composites, are emerging as a promising alternative for removing and degrading synthetic dyes. These materials stand out for their high adsorption capacity and biodegradability, offering an environmentally friendly and efficient solution compared with traditional wastewater treatment methods. Furthermore, integrating cellulose microsphere-based technologies could constitute an important step toward sustainability in improving water quality and responsible management of natural resources. However, engaging in further research and optimization of these solutions is essential to assess their long-term impact on the environment rigorously.

Therefore, environmental authorities must strengthen regulations on permissible limits for synthetic dyes in wastewater. This can only be achieved through a robust regulatory framework and ongoing research. Such measures are necessary to protect aquatic ecosystems and public health. Equally important is the need to assess the environmental impact of proposed solutions throughout their life cycle. This is a crucial step to ensure the sustainability of any alternative. For instance, being biodegradable and derived from renewable sources, cellulose microspheres could be a more sustainable alternative to conventional adsorbent materials, such as activated carbon, which require costly and potentially polluting regeneration processes.

Further studies are needed on the permanence and biodegradability of cellulose microspheres after use in aquatic environments. The effective implementation of technologies such as cellulose microspheres is a collective effort that requires close collaboration between the scientific community, regulatory authorities, and the industries involved. Creating incentives for applied research and continuously improving environmental regulations will not only mitigate pollution but also foster the development of innovative and globally accessible solutions.

As a perspective, using cellulose microspheres as a solution for degrading synthetic dyes in wastewater opens up an innovative perspective for treating contaminated water. However, several challenges remain that must be addressed to optimize the performance of these technologies and ensure their viability on a large scale. One of the main challenges is improving the efficiency of microspheres in removing dyes at low concentrations, a crucial aspect for complying with the permissible limits established by environmental regulations. Furthermore, it is necessary to further study their behavior under real-life treatment conditions, considering variables such as pH, temperature, and the presence of other contaminants. As research progresses, the potential for developing hybrid or composite materials, combining cellulose microspheres with nanoparticles or biologically derived materials to enhance their adsorbent and degradation capacities, is emerging. This interdisciplinary approach could generate more efficient solutions adaptable to diverse treatment conditions.

As we peer into the future, it's clear that wastewater treatment is complex and demands a comprehensive approach. This approach must effectively address the presence of synthetic dyes, heavy metals, pharmaceuticals, and other hazardous compounds in water. Integrating cellulose microsphere-based technologies with advanced systems like photocatalysis or electrochemical processes could be the key to more comprehensive and efficient treatment, thereby strengthening sustainability in water resource management.

Likewise, public policies and regulations regarding water pollution must evolve in line with scientific advances. Developing stricter regulations on eliminating synthetic dyes and promoting applied research is essential to address the growing water pollution problem. In this sense, innovative solutions must be supported by a regulatory framework that facilitates their implementation in various industries. Furthermore, the application of cellulose microspheres not only contributes to the degradation of dyes but also presents an economically viable and environmentally sustainable option in line with the principles of the circular economy. Cellulose, obtained from renewable sources such as biomass, could be a way to transform agricultural or industrial waste into a useful material for environmental remediation, significantly reducing both the impact of pollutants and the carbon footprint of the process. The future of remediation of waters contaminated with synthetic dyes is favorably influenced by emerging technologies based on cellulose microspheres.

ACKNOWLEDGMENT

The authors acknowledge the financial support provided by DGAPA-UNAM (Dirección General de Asuntos del Personal Académico) under Project No. PAPIIT-IA202226.

REFERENCES

1. Slama H Ben, Chenari Bouket A, Pourhassan Z, Alenezi FN, Silini A, Cherif-Silini H, et al. Diversity of synthetic dyes from textile industries, discharge impacts and treatment methods. *Appl Sci*. 2021 Jul 6;11(14):6255.
2. Mehta M, Sharma M, Pathania K, Jena PK, Bhushan I. Degradation of synthetic dyes using nanoparticles: A mini-review. *Environ Sci Pollut Res*. 2021 Sep 4;28(36): 49434–49446.
3. Wang Q, Chen J, Zhao L, Wang S, Li P, Du K. A sphere-on-sphere structure: Cu nanoparticles grown on cellulose microspheres for the removal of environmental contaminants. *Ind Crops Prod*. 2024 Jul;213:118401.
4. Yadav S, Yadav A, Bagotia N, Sharma AK, Kumar S. Adsorptive potential of modified plant-based adsorbents for sequestration of dyes and heavy metals from wastewater - A review. *J Water Process Eng*. 2021 Aug;42:102148.
5. Sharma J, Sharma S, Soni V. Classification and impact of synthetic textile dyes on Aquatic Flora: A review. *Reg Stud Mar Sci*. 2021 Jun;45:101802.
6. Lin J, Ye W, Xie M, Seo DH, Luo J, Wan Y, et al. Environmental impacts and remediation of dye-containing wastewater. *Nat Rev Earth Environ*. 2023 Oct 26;4(11):785–803.
7. Hemashenpagam N, Selvajeyanthi S. Textile dyes and their effect on human beings. In: Ahmad A, Jawaid M, Mohamad Ibrahim MN, et al., editors. *Nanohybrid Materials for Treatment of Textiles Dyes*. Springer, Singapore; 2023. pp. 41–60.
8. Al-Tohamy R, Ali SS, Li F, Okasha KM, Mahmoud YAG, Elsamahy T, et al. A critical review on the treatment of dye-containing wastewater: Ecotoxicological and health concerns of textile dyes and possible remediation approaches for environmental safety. *Ecotoxicol Environ Saf*. 2022 Feb;231:113160.
9. Mishra D. Food colors and associated oxidative stress in chemical carcinogenesis. In: Chakraborti S, Ray BK, Roychoudhury S, editors. *Handbook of Oxidative Stress in Cancer: Mechanistic Aspects*. Springer, Singapore; 2022. pp. 773–786.
10. Ardila-Leal LD, Poutou-Piñales RA, Pedroza-Rodríguez AM, Quevedo-Hidalgo BE. A brief history of colour, the environmental impact of synthetic dyes and removal by using laccases. *Molecules*. 2021 Jun 22;26(13):3813.
11. U.S. Food and Drug Administration F. https://www.fda.gov/industry/color-additives/compliance-enforcement-color-additives. 2018. Compliance & Enforcement of Color Additives.
12. Food and Agriculture Organization and World Health Organization F. https://www.fao.org/fao-who-codexalimentarius/thematic-areas/contaminants/tr/. 2025. Codex Alimentarius, International Food Standards.
13. International Organization for Standardization (ISO). ISO/TC 256 Pigments, dyestuffs and extenders. 2025.
14. U.S. Environmental Protection Agency (EPA). https://www.epa.gov/regulatory-information-sector/textile-manufacturing-sector-naics-313. 2025. Textile Manufacturing Sector (NAICS 313).
15. Mota IGC, Neves RAM Das, Nascimento SSDC, Maciel BLL, Morais AHDA, Passos TS. Artificial dyes: Health risks and the need for revision of international regulations. *Food Rev Int*. 2023 Apr 3;39(3):1578–1593.
16. Hamidon TS, Adnan R, Haafiz MKM, Hussin MH. Cellulose-based beads for the adsorptive removal of wastewater effluents: A review. *Environ Chem Lett*. 2022 Jun 24;20(3):1965–2017.
17. Lee J, Patel R. Wastewater treatment by polymeric microspheres: A review. *Polymers (Basel)*. 2022 May 5;14(9):1890.
18. Persano F, Malitesta C, Mazzotta E. Cellulose-based hydrogels for wastewater treatment: A focus on metal ions removal. *Polymers (Basel)*. 2024 May 5;16(9):1292.

19. Peng L, Li W, Du J, Zhang M, Zhao L. Efficient removal of p-nitrophenol from water by imidazolium ionic liquids functionalized cellulose microsphere. *Int J Biol Macromol.* 2024 Jul;273:133117.
20. Zheng K, Gong WL, Wu M, Liu L, Qiu X, Shan Y, et al. Amphoteric cellulose microspheres for the efficient remediation of anionic and cationic dyeing wastewater. *Sep Purif Technol.* 2023 Mar;309:123035.
21. Yang Y, Li X, Wan C, Zhang Z, Cao W, Wang G, et al. A comprehensive review of cellulose nanomaterials for adsorption of wastewater pollutants: Focus on dye and heavy metal Cr adsorption and oil/water separation. *Collagen and Leather.* 2024 Dec 20;6(1):35.
22. Varghese AG, Paul SA, Latha MS. Remediation of heavy metals and dyes from wastewater using cellulose-based adsorbents. *Environ Chem Lett.* 2019 Jun 10;17(2): 867–877.
23. Rol F, Belgacem MN, Gandini A, Bras J. Recent advances in surface-modified cellulose nanofibrils. *Prog Polym Sci.* 2019 Jan;88:241–264.
24. Hokkanen S, Bhatnagar A, Sillanpää M. A review on modification methods to cellulose-based adsorbents to improve adsorption capacity. *Water Res.* 2016 Mar;91:156–173.
25. Yang HR, Li SS, An QD, Zhai SR, Xiao ZY, Zhang LP. Facile transformation of carboxymethyl cellulose beads into hollow composites for dye adsorption. *Int J Biol Macromol.* 2021 Nov;190:919–926.
26. Akter M, Bhattacharjee M, Dhar AK, Rahman FBA, Haque S, Rashid TU, et al. Cellulose-based hydrogels for wastewater treatment: A concise review. *Gels.* 2021 Mar 18;7(1):30.
27. Radoor S, Karayil J, Jayakumar A, Kandel DR, Kim JT, Siengchin S, et al. Recent advances in cellulose- and alginate-based hydrogels for water and wastewater treatment: A review. *Carbohydr Polym.* 2024 Jan;323:121339.
28. Lamaming J, Rahman NSA. Cellulose-based adsorbent for removal of color compounds in wastewater. In: Yaser AZ, Lamaming J, Johnson D, Hilal N, editors. *Nutrients and Colored Compounds in Wastewater: Advanced Treatment and Recovery.* Elsevier, Amsterdam; 2025. pp. 341–358.
29. Qiao L, Wang S, Wang T, Yu S, Guo S, Du K. High-strength and low-swelling chitosan/cellulose microspheres as a high-efficiency adsorbent for dye removal. *Cellulose.* 2021 Sep 30;28(14):9323–9333.
30. Peramune D, Manatunga DC, Dassanayake RS, Premalal V, Liyanage RN, Gunathilake C, et al. Recent advances in biopolymer-based advanced oxidation processes for dye removal applications: A review. *Environ Res.* 2022 Dec;215:114242.
31. Song Y, Jiang M, Zhou L, Yang H, Zhang J. Rapidly regenerated CNC/TiO2/MnO2 porous microspheres for high-efficient dye removal. *Carbohydr Polym.* 2022 Sep;292:119644.
32. Tan SN, Yuen ML, Ramli RA. Photocatalysis of dyes: Operational parameters, mechanisms, and degradation pathway. *Green Analytical Chemistry.* 2025 Mar;12:100230.
33. Li Y, Xiao H, Pan Y, Zhang M, Jin Y. Thermal and pH dual-responsive cellulose microfilament spheres for dye removal in single and binary systems. *J Hazard Mater.* 2019 Sep;377:88–97.
34. Hassanisaadi M, Saberi Riseh R, Rabiei A, Varma RS, Kennedy JF. Nano/micro-cellulose-based materials as remarkable sorbents for the remediation of agricultural resources from chemical pollutants. *Int J Biol Macromol.* 2023;246:125763.
35. Zhang DY, Zhang XQ, Yao XH, Wan Y, Song P, Liu ZY, et al. Microwave-assisted synthesis of PdNPs by cellulose solution to prepare 3D porous microspheres applied on dyes discoloration. *Carbohydr Polym.* 2020 Nov;247:116569.
36. Rane A, Joshi SJ. Biodecolorization and biodegradation of dyes: A review. *Open Biotechnol J.* 2021 Jul 15;15(1):97–108.

37. Singh K, Arora S. Removal of synthetic textile dyes from wastewaters: A critical review on present treatment technologies. *Crit Rev Environ Sci Technol.* 2011 Apr 6;41(9):807–878.
38. Tomar T, Kahandawala N, Kaur J, Thounaojam L, Choudhary I, Bera S. Bioremediation of synthetic dyes from wastewater by using microbial nanocomposites: An emerging field for water pollution management. *Biocatal Agric Biotechnol.* 2023 Aug; 51:102767.
39. Blaga AC, Zaharia C, Suteu D. Polysaccharides as support for microbial biomass-based adsorbents with applications in removal of heavy metals and dyes. *Polymers (Basel).* 2021 Aug 27;13(17):2893.
40. Shi X, Zhang X, Ma L, Xiang C, Li L. TiO2-doped chitosan microspheres supported on cellulose acetate fibers for adsorption and photocatalytic degradation of methyl orange. *Polymers (Basel).* 2019 Aug 2;11(8):1293.
41. Zhu JL, Chen SP, Lin W, Huang HD, Li ZM. Cellulose mineralization with in-situ synthesized amorphous titanium dioxide for enhanced adsorption and auto-accelerating photocatalysis on water pollutant. *Chem Eng J.* 2023 Jan;456:141036.
42. Salama A, Abouzeid RE, Owda ME, Cruz-Maya I, Guarino V. Cellulose–silver composites materials: Preparation and applications. *Biomolecules.* 2021 Nov 12;11(11):1684.
43. Sankar Sana S, Haldhar R, Parameswaranpillai J, Chavali M, Kim SC. Silver nanoparticles-based composite for dye removal: A comprehensive review. *Cleaner Materials.* 2022 Dec;6:100161.
44. Li G, Nandgaonkar AG, Wang Q, Zhang J, Krause WE, Wei Q, et al. Laccase-immobilized bacterial cellulose/TiO2 functionalized composite membranes: Evaluation for photo- and bio-catalytic dye degradation. *J Memb Sci.* 2017 Mar;525:89–98.
45. Costa AF de S, de Amorim JDP, Almeida FCG, de Lima ID, de Paiva SC, Rocha MA V., et al. Dyeing of bacterial cellulose films using plant-based natural dyes. *Int J Biol Macromol.* 2019 Jan;121:580–587.
46. Adesibikan AA, Emmanuel SS, Olawoyin CO, Ndungu P. Cellulosic metallic nanocomposites for photocatalytic degradation of persistent dye pollutants in aquatic bodies: A pragmatic review. *J Organomet Chem.* 2024 Apr;1010:123087.
47. Kausar A, Zohra ST, Ijaz S, Iqbal M, Iqbal J, Bibi I, et al. Cellulose-based materials and their adsorptive removal efficiency for dyes: A review. *Int J Biol Macromol.* 2023 Jan;224:1337–1355.
48. Zafar S, Bukhari DA, Rehman A. Azo dyes degradation by microorganisms – An efficient and sustainable approach. *Saudi J Biol Sci.* 2022 Dec;29(12):103437.
49. Jamal M, Awadasseid A, Su X. Exploring potential bacterial populations for enhanced anthraquinone dyes biodegradation: A critical review. *Biotechnol Lett.* 2022 Sep 24;44(9):1011–1125.
50. Benkhaya S, M'rabet S, Lgaz H, El Bachiri A, El Harfi A. Dyes: Classification, pollution, and environmental effects. In: Muthu SS, Khadir A, editors. *Dye Biodegradation, Mechanisms and Techniques.* Springer, Singapore; 2022. pp. 1–50.
51. Kamino S, Uchiyama M. Xanthene-based functional dyes: Towards new molecules operating in the near-infrared region. *Org Biomol Chem.* 2023;21(12):2458–2471.
52. Christie R, Abel A. Phthalocyanine pigments: General principles. *Phys Sci Rev.* 2021 Nov 26;6(11):671–677.
53. Essawy AA, Alsohaimi IH, Hassan HMA, El Agammy EF, Hussein MF, Hasanin THA, et al. Basic fuchsin dye as the first fluorophore for optical sensing of morpholine in fruits crust and urine samples. *Anal Chem.* 2024 Jan 9;96(1):373–380.
54. Umeh CT, Akinyele AB, Okoye NH, Emmanuel SS, Iwuozor KO, Oyekunle IP, et al. Recent approach in the application of nanoadsorbents for malachite green (MG) dye uptake from contaminated water: A critical review. *Environ Nanotechnol Monit Manag.* 2023 Dec;20:100891.

55. Gopika GS, Prasad PMH, Lekshmi AG, Lekshmypriya S, Sreesaila S, Arunima C, et al. Chemistry of cyanine dyes-A review. *Mater Today Proc.* 2021;46:3102–3108.
56. Srivastava A, Rani RM, Patle DS, Kumar S. Emerging bioremediation technologies for the treatment of textile wastewater containing synthetic dyes: A comprehensive review. *J Chem Technol Biotechnol.* 2022 Jan 17;97(1):26–41.
57. Rosaline DR, Keerthana C, Vasthi SS, Rubini SH, Ratna Monica JH, Manikandan A, et al. Magnetic nanoparticles and nanocomposites for the applications of photocatalytic degradation of organic dyes. In: Khan I, Khan A, Khan MMA, editors. *Magnetic Nanoparticles and Polymer Nanocomposites: Fundamentals and Biological, Environmental and Energy Applications.* Woodhead Publishing; 2024. pp. 459–497.
58. Zahid M, Ahmad H, Drioli E, Rehan ZA, Rashid A, Akram S, et al. Role of polymeric nanocomposite membranes for the removal of textile dyes from wastewater. In: Abd-Elsalam KA, Zahid M, editors. *Aquananotechnology.* Elsevier; 2021. pp. 91–103.
59. Khodabandeloo F, Shahsavarifar S, Nayebi B, Niavol KP, Nayebi B, Varma RS, et al. Applications of nanostructured semiconductor photocatalysts for the decontamination of assorted pollutants from wastewater. *Inorg Chem Commun.* 2023 Nov;157:111357.
60. Aditya DS, Mahadevaprasad KN, Nataraj SK. Polymer-Clay Nanocomposites for dyes abatement. In: Ray SS, editor. Comprehensive Polymer Science (Second Edition). Elsevier, Amsterdam; 2026. pp. 502–530.
61. Kumar A, Dixit U, Singh K, Prakash Gupta S, S. Jamal Beg M. Structure and properties of dyes and pigments. In: Papadakis R, editor. *Dyes and Pigments - Novel Applications and Waste Treatment.* IntechOpen, London; 2021. pp. 1–19.
62. Sekar N. Acid dyes. In: Clark M, editor. *Handbook of Textile and Industrial Dyeing.* Woodhead Publishing, Cambridge; 2011. pp. 486–514.
63. Shokri A. Employing electro-peroxone process for degradation of Acid Red 88 in aqueous environment by Central Composite Design: A new kinetic study and energy consumption. *Chemosphere.* 2022 Jun;296:133817.
64. Tsouko E, Pilafidis S, Kourmentza K, Gomes HI, Sarris G, Koralli P, et al. A sustainable bioprocess to produce bacterial cellulose (BC) using waste streams from wine distilleries and the biodiesel industry: Evaluation of BC for adsorption of phenolic compounds, dyes and metals. *Biotechnol Biofuels Bioprod.* 2024 Mar 12;17(1):40.
65. Hashemi E, Norouzi MM, Sadeghi-Kiakhani M. Magnetic biochar as a revolutionizing approach for diverse dye pollutants elimination: A comprehensive review. *Environ Res.* 2024 Nov;261:119548.
66. Valli Nachiyar C, Rakshi AD, Sandhya S, Britlin Deva Jebasta N, Nellore J. Developments in treatment technologies of dye-containing effluent: A review. *Case Stud Chem Environ Eng.* 2023 Jun;7:100339.
67. Sudhakaran G, Karthikeyan R, Madesh S, Dhaareshwar VN, Pachaiappan R, Rajagopal R, et al. Neurotoxic effects and cognitive decline induced by Aniline Blue dye in an in-vivo zebrafish model. *Toxicology.* 2025 Feb;511:154065.
68. Ahmadi S, Ganjidoust H. Using banana peel waste to synthesize BPAC/ZnO nanocomposite for photocatalytic degradation of Acid Blue 25: Influential parameters, mineralization, biodegradability studies. *J Environ Chem Eng.* 2021 Oct;9(5):106010.
69. Shinohara S, Funabiki K, Kikuchi M, Takebayashi S, Hamaguchi K, Hara S, et al. Real-time imaging of head and neck squamous cell carcinomas using confocal micro-endoscopy and applicable dye: A preliminary study. *Auris Nasus Larynx.* 2020 Aug;47(4):668–675.
70. Ahmed HR, Shekh AOB, Salih MA, Saeed DNH, Ealias AM, George G. Copper oxide-graphene frameworks as a novel catalyst for oxidative adsorption removal of eosin from aqueous solutions. *J Mol Struct.* 2025 Jun;1331:141646.
71. Legentil P, Leroux F, Therias S, Mahiou R, Chadeyron G. Revisiting fluorescein and layered double hydroxide using a synergistic approach: A complete optical study. *J Lumin.* 2019 Nov;215:116634.

72. Tsuboy MS, Angeli JPF, Mantovani MS, Knasmüller S, Umbuzeiro GA, Ribeiro LR. Genotoxic, mutagenic and cytotoxic effects of the commercial dye CI Disperse Blue 291 in the human hepatic cell line HepG2. *Toxicol in Vitro.* 2007 Dec;21(8):1650–1655.
73. Ghalkhani M, Zare N, Karimi F, Karaman C, Alizadeh M, Vasseghian Y. Recent advances in Ponceau dyes monitoring as food colorant substances by electrochemical sensors and developed procedures for their removal from real samples. *Food Chem Toxicol.* 2022 Mar;161:112830.
74. Islam Md T, Islam T, Islam T, Repon Md R. Synthetic dyes for textile colouration: Process, factors and environmental impact. *Text Leath Rev.* 2022 Aug 8;5:327–373.
75. Tang AYL, Lo CKY, Kan C. Textile dyes and human health: A systematic and citation network analysis review. *Color Technol.* 2018 Aug 9;134(4):245–257.
76. Wan Y, Liu ZY, Song P, Zhang XQ, Song JC, Fu YJ, et al. Ionic liquid groups modified 3D porous cellulose microspheres for selective adsorption of AO7 dye. *J Clean Prod.* 2019 Dec;240:118201.
77. Lewis DM. The chemistry of reactive dyes and their application processes. In: Clark M, editor *Handbook of Textile and Industrial Dyeing.* Woodhead Publishing, Cambridge; 2011. pp. 303–364.
78. Vickerstaff T. Reactive dyes for textiles. *J Soc Dyers Colour.* 1957 Jun 22;73(6):237–245.
79. Nambela L, Haule LV, Mgani Q. A review on source, chemistry, green synthesis and application of textile colorants. *J Clean Prod.* 2020 Feb;246:119036.
80. Shang SM. Process control in dyeing of textiles. In: Majumdar A, Das A, Alagirusamy R, Kothari VK, editors. *Process Control in Textile Manufacturing.* Woodhead Publishing, Oxford; 2013. pp. 300–338.
81. Khatri A, Peerzada MH, Mohsin M, White M. A review on developments in dyeing cotton fabrics with reactive dyes for reducing effluent pollution. *J Clean Prod.* 2015 Jan;87:50–57.
82. Lewis DM. Developments in the chemistry of reactive dyes and their application processes. *Color Technol.* 2014 Dec 26;130(6):382–412.
83. Leme DM, Oliveira GAR de, Meireles G, Brito LB, Rodrigues L de B, Palma de Oliveira D. Eco- and genotoxicological assessments of two reactive textile dyes. *J Toxicol Environ Health A.* 2015 Mar 4;78(5):287–300.
84. Patil R, Deshmukh H, Deolikar R, Vyavahare G, Shinde S, Desai N, et al. Microbial diversity, interactions and biodegradation of hazardous textile wastewater using biological consortium technology. In: Shah MP, Rodriguez-Couto S, editors. *Wastewater Treatment Reactors.* Elsevier, Amsterdam; 2021. pp. 91–117.
85. Khalifeh Z, Rajabi HR, Kohvade AG, Shokrollahi A. Determination of Victoria blue in water samples using ultrasound-assisted liquid-phase microextraction based on deep eutectic solvents: A comparative study of spectrophotometric and smartphone imaging techniques. *Results Chem.* 2025 Mar;14:102054.
86. Nilsson R, Nordlinder R, Wass U, Meding B, Belin L. Asthma, rhinitis, and dermatitis in workers exposed to reactive dyes. *Occup Environ Med.* 1993 Jan 1;50(1):65–70.
87. Islam T, Repon Md R, Islam T, Sarwar Z, Rahman MM. Impact of textile dyes on health and ecosystem: A review of structure, causes, and potential solutions. *Environ Sci Pollut Res.* 2022 Dec 2;30(4):9207–9242.
88. Shaheen TI, Radwan EK, El-Wakeel ST. Unary and binary adsorption of anionic dye and toxic metal from wastewater using 3-aminopropyltriethoxysilane functionalized porous cellulose acetate microspheres. *Microporous Mesoporous Mater.* 2022 Jun;338:111996.
89. Bharatbhai Vishani D, Shrivastav A. Enzymatic decolorization and degradation of azo dyes. In: Shah MP, Rodriguez-Couto S, Kapoor RT, editors. *Development in Wastewater Treatment Research and Processes.* Elsevier, Amsterdam; 2022. pp. 419–432.
90. Fernandes FH, Abe FR, Gomes TB, Santos CB dos, Kempinas WDG, Leite BA, et al. Disperse Red 1 azo dye: Consequences of low-dose/low-concentration exposures in mice and zebrafish. *Mutat Res Genet Toxicol Environ Mutagen.* 2024 Nov;900:503830.

91. Wang C, Gu W, Zhang S, Li L, Kong J, Zhi H, et al. Multigenerational effects of disperse blue 79 at environmentally relevant concentrations on zebrafish (Danio rerio) fecundity: An integrated approach. *J Hazard Mater.* 2024 Oct;478:135442.
92. Fan X, Liu ZT, Liu ZW. Preparation and application of cellulose triacetate microspheres. *J Hazard Mater.* 2010 May;177(1–3):452–457.
93. Shabir M, Yasin M, Hussain M, Shafiq I, Akhter P, Nizami AS, et al. A review on recent advances in the treatment of dye-polluted wastewater. *J Ind Eng Chem.* 2022 Aug;112:1–19.
94. Baran NY, Çalışkan M, Özpala A, Baran T. Fabrication of nano-sized Pd catalyst supported on sodium carboxymethyl cellulose/gum Arabic/sodium alginate functionalized microspheres for catalytic reduction of nitro compounds, organic dyes, K3[Fe(CN)6], and chromium(VI) pollutants. *Int J Biol Macromol.* 2024 Mar;262:130134.
95. Li J, Zhou L, Song Y, Yu X, Li X, Liu Y, et al. Green fabrication of porous microspheres containing cellulose nanocrystal/MnO2 nanohybrid for efficient dye removal. *Carbohydr Polym.* 2021 Oct;270:118340.
96. Zhu H, Zhu E, Xie Y, Liu D, Hu Y, Shi Z, et al. Hydrangea-like nanocellulose microspheres with high dye adsorption and drug encapsulation prepared by emulsion method. *Carbohydr Polym.* 2022 Nov;296:119947.
97. Salama A. Novel cellulose/silica microspheres as sustainable and efficient adsorbents for methyl orange removal. *Fibers Polym.* 2023 Apr 2;24(4):1333–1343.
98. Yu Y, Liu S, Pei Y, Luo X. Growing Pd NPs on cellulose microspheres via in-situ reduction for catalytic decolorization of methylene blue. *Int J Biol Macromol.* 2021 Jan;166:1419–1428.
99. Pei Y, Wu X, Xu G, Sun Z, Zheng X, Liu J, et al. Tannin-immobilized cellulose microspheres as effective adsorbents for removing cationic dye (Methylene Blue) from aqueous solution. *J Chem Technol Biotechnol.* 2017 Jun;92(6):1276–1284.
100. Park S, Oh Y, Yun J, Yoo E, Jung D, Oh KK, et al. Cellulose/biopolymer/Fe3O4 hydrogel microbeads for dye and protein adsorption. *Cellulose.* 2020 Mar 9;27(5):2757–2773.
101. Benhadria E houceine, Bahsis L, Ablouh EH, Hanani Z, Bakhouch M, Labjar N, et al. Copper oxide nanoparticles-decorated cellulose acetate: Eco-friendly catalysts for reduction of toxic organic dyes in aqueous media. *Int J Biol Macromol.* 2025 Jan;284:137982.
102. Khan MSJ, Sidek LM, Kamal T, Asiri AM, Khan SB, Basri H, et al. An efficient wastewater treatment through reduction of organic dyes using Ag nanoparticles supported on cellulose gum beads. *Int J Biol Macromol.* 2024 Feb;257:128544.
103. Khan A, Raza ZA, Bhatti HN, Sarwar T. Citrate silver nanoparticles impregnated cellulose as a photocatalytic filter in the degradation of organic dye in the aqueous media. *Int J Biol Macromol.* 2024 Mar;261:129881.
104. K P C, Varghese A, T. P. V, K. R. SD. Multifunctional electrospun membranes incorporated with metal oxide nanoparticles, cellulose acetate, and polyvinylpyrrolidone for wastewater treatment: Oil/water separation, dye adsorption, and dye degradation. *Chem Eng J.* 2024 Nov;499:156049.
105. Al-Abbad EA, Rehman R, Hussain MS. A sustainable and cost-effective approach for efficient removal of Direct Blue-14 azo dye from wastewater using North American Zeolite for developing countries. *Phys Chem Earth Parts A/B/C.* 2025 Jun;138:103888.
106. Brahmi A, Agsous M, Benkhaoula BN, Ziani S, Khireddine H, AitAli S, et al. Fluoro-Hydroxyapatite/Chitosan composites as an eco-friendly adsorbent for Direct Red 23 dye removal: Optimization through Response Surface Methodology. *J Mol Struct.* 2025 Feb;1321:140212.
107. Golob V, Ojstršek A. Removal of vat and disperse dyes from residual pad liquors. *Dyes Pigm.* 2005 Jan;64(1):57–61.
108. Sirianuntapiboon S, Saengow W. Removal of vat dyes from textile wastewater using biosludge. *Water Qual Res J.* 2004 Aug 1;39(3):276–284.
109. Hendaoui K, Trabelsi-Ayadi M, Ayari F. Optimization and mechanisms analysis of indigo dye removal using continuous electrocoagulation. *Chin J Chem Eng.* 2021 Jan;29:242–252.

13 Cellulose Microsphere-Based Strategies for Removing Organic Dyes from Wastewater

Rajni Verma, Kulbir Kaur, Mahapara Abbass, Shalom Akhai, Sachin Kumar Mandotra, Bikram Jit Singh, and Rippin Sehgal

GRAPHICAL ABSTRACT

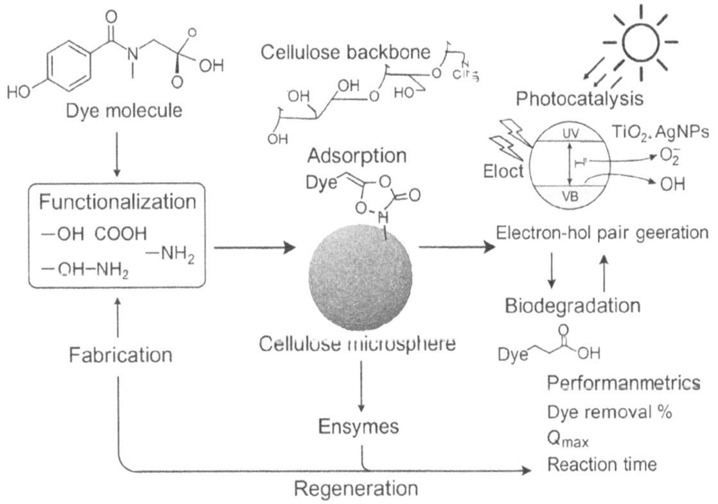

ABBREVIATIONS

AO7: Acid Orange 7 (anionic azo dye)
BC: Bacterial Cellulose
CMs4: Cellulose microsphere modified with 4 methylimidazole (ionic liquid)
CNC: Cellulose Nanocrystal

DOI: 10.1201/9781003596127-15

CNF: Cellulose Nanofibril
CV: Crystal Violet (cationic triphenylmethane dye)
MB: Methylene Blue (cationic thiazine dye)
MG: Malachite Green (cationic triarylmethane dye)
MO: Methyl Orange (anionic azo dye)
MOF: Metal-Organic Framework
NP: Nanoparticle
PDA: Polydopamine
PEI: Polyethylenimine
RhB: Rhodamine B (cationic xanthene dye)
ROS: Reactive Oxygen Species (e.g. •OH radicals)
SA: Sodium Alginate (natural polymer from algae)
TOCNF: TEMPO (oxidant)-Oxidized Cellulose Nanofibril

HIGHLIGHTS

- Cellulose microspheres (including bacterial cellulose, nanocellulose, and modified plant cellulose) are sustainable, high-surface-area materials for removing synthetic dyes from water.
- **Adsorption:** Cellulose microsphere adsorbents achieve high dye uptake (often >200 mg/g) due to abundant hydroxyl and tuned functional groups. Functionalization (e.g. carboxylation, amination, ionic liquids) can push capacities into the hundreds–thousands mg/g range.
- **Catalysis:** Metal-loaded cellulose microspheres act as efficient catalysts for dye degradation. For example, in situ grown Pd nanoparticles on cellulose enable >99% reduction of Methylene Blue within minutes in the presence of $NaBH_4$, with excellent reusability over multiple cycles.
- **Photocatalysis:** Cellulose-based composite microspheres with semiconductors (TiO_2, ZnO, BiOBr, etc.) show rapid photocatalytic dye degradation under UV–visible light. The cellulose support increases catalyst dispersion and can shuttle electrons, improving photodegradation efficiency.
- **Biodegradation:** Enzymes and microbes immobilized on cellulose microspheres decompose dyes under mild conditions. Laccase immobilized on cellulose networks retains high activity (~95% dye removal in three hours) and can be reused for many cycles. Immobilized dye-degrading bacteria achieve >90% decolorization in batch systems.
- **Hybrid Mechanisms:** Many cellulose microspheres exploit synergistic mechanisms – e.g. adsorbing dyes and then photocatalytically oxidizing them, or adsorbing and then bio-degrading – to enhance overall removal and mineralization of pollutants.
- **Diverse Structures:** All forms of cellulose microspheres are covered: from **bacterial cellulose** beads and **nanocellulose**-based microspheres to chemically **modified cellulose** (e.g. carboxymethyl cellulose, cellulose acetate) beads and **composites** with clays, carbon, or MOFs. Each offers unique advantages in porosity, functionality, and stability.

- **Environmental Safety:** Being biodegradable and non-toxic, cellulose microspheres minimize secondary pollution. Spent cellulose adsorbents can often be regenerated or safely disposed; any leached cellulose will naturally decompose, unlike persistent synthetic resins.
- **Recent Innovations:** Cutting-edge developments include magnetic cellulose microspheres for easy recovery, heterojunction photocatalyst–cellulose hybrids for visible-light activity, and microfluidic fabrication of uniform hollow cellulose spheres for controlled delivery of catalytic sites.
- **Outlook:** Cellulose microsphere technologies show high promise for scalable industrial wastewater treatment. Ongoing research is addressing challenges like maintaining performance in real waste streams, improving regeneration efficiency, and integrating these materials into continuous flow systems for practical application.

13.1 INTRODUCTION

Synthetic dyes are widely used in textiles, leather, plastics, cosmetics, and other industries, resulting in large volumes of colored wastewater [1]. It is estimated that over **1 million tons of dyes** are produced globally each year, generating more than **70 billion tons of dye-contaminated wastewater** annually [2–4]. Dye effluents are a major environmental concern because even low concentrations (ppm) impart intense color, blocking sunlight penetration in aquatic systems and disrupting photosynthesis. Moreover, many dyes and their breakdown products are **toxic, carcinogenic,** or **mutagenic** to aquatic life and humans [5,6]. For example, aromatic amines and heavy metal complexes from dyes can cause severe health problems if ingested [7]. Therefore, effective treatment of dye-laden wastewater is critical before discharge into water bodies.

Traditional wastewater treatment methods (flocculation, membrane filtration, chemical oxidation, etc.) often struggle with dyes due to the stable, complex structures of modern synthetic dyes (e.g. azo, anthraquinone, phthalocyanine dyes) that resist degradation [8,9]. **Physical adsorption** and **advanced oxidation** processes are among the most effective approaches [10]. Adsorption using activated carbon or other sorbents can **concentrate dyes** from dilute effluents, and advanced oxidation (e.g. Fenton, photocatalysis) can **mineralize** (chemically break down) the complex dye molecules [11,12]. However, conventional activated carbon, while widely used, has drawbacks: high cost, limited efficiency for certain dyes (especially large or hydrophobic dye molecules), and difficulty in regeneration [13]. Similarly, homogeneous advanced oxidation processes can be costly and generate sludge or require catalysts that are hard to recover [14].

In recent years, biopolymer-based adsorbents have gained attention as sustainable alternatives [15]. Cellulose, the most abundant natural polymer on Earth, stands out as an ideal backbone for developing high-performance dye removal materials [16]. Cellulose is renewable, inexpensive, non-toxic, and biodegradable, yet it has high mechanical strength and a plethora of surface hydroxyl groups that can be chemically modified [17]. By shaping cellulose into microspheres (beads in the micrometer to millimeter scale), researchers can achieve sorbents with high surface area, tunable porosity, and easy handling in water-treatment systems [18]. Unlike cellulose powder or films, which can be hard to separate from water, microspheres are granular and can be readily filtered or used in packed-bed columns [19]. In particular, hollow or porous cellulose microspheres provide a large accessible surface and interior voids that facilitate dye uptake and can host catalysts or microbes [20]. Another key advantage of cellulose microspheres is their versatility. They can be functionalized with various chemical groups to target specific pollutants or doped with catalytic components [21]. This allows a single cellulose microsphere platform to engage in multiple dye degradation mechanisms [22] (Figure 13.1). For instance, a cellulose bead might first adsorb a dye from solution (a physical removal), and subsequently a catalyst on the bead's surface might degrade the dye into smaller, less harmful molecules [23]. The degradation could occur through chemical catalysis (e.g. reduction or oxidation by attached nanoparticles), photocatalysis (if the bead carries a photosensitive semiconductor that generates reactive radicals under light), or biodegradation (if enzymes or microbes are immobilized on the bead to metabolize the dye) [24]. In many cases, these mechanisms work in tandem – adsorption brings the dye into close proximity with the reactive sites on the microsphere, greatly enhancing the overall

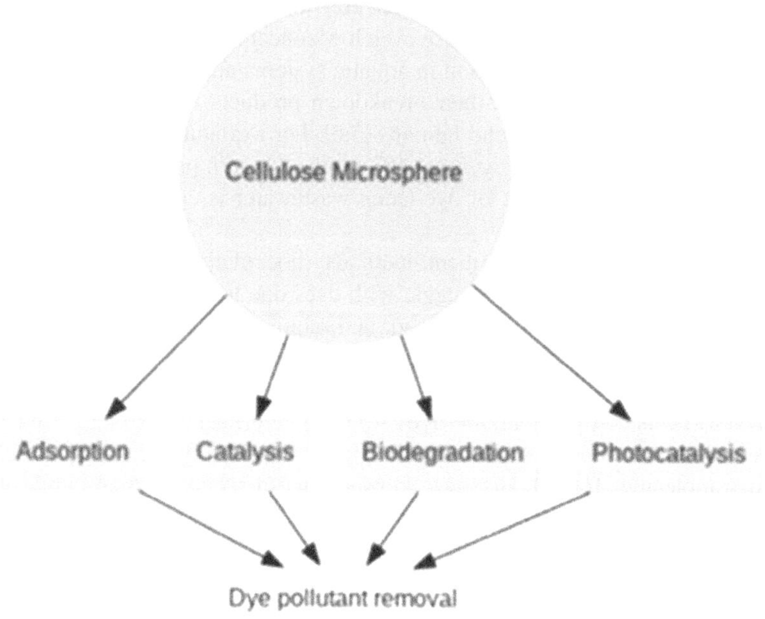

FIGURE 13.1 Schematic representation of a cellulose microsphere enabling multiple mechanisms for dye pollutant removal. Dyes can be captured by adsorption onto the microsphere's surface (physical binding) and subsequently degraded via catalysis (chemical reaction on embedded catalysts), biodegradation (enzymatic or microbial breakdown), or photocatalysis (light-driven oxidation by semiconductor particles). This multi-mechanistic capability allows cellulose microspheres to achieve high removal efficiencies and complete detoxification of dyes.

removal rate [25]. This integrated approach is highly efficient and is a distinct benefit of cellulose-based systems, which often exhibit multi-functionality [26].

In this review, we provide a comprehensive analysis of the degradation of synthetic dye pollutants using cellulose microspheres. We cover all major mechanisms – adsorption, catalysis, photocatalysis, and biodegradation – highlighting how cellulose microspheres are engineered for each purpose. We also discuss all types of cellulose microsphere materials: including bacterial cellulose microspheres produced by microbes, nanocellulose-based microspheres assembled from cellulose nanocrystals/nanofibers, chemically modified cellulose (e.g. carboxymethyl cellulose or cellulose acetate) beads, and various composites (cellulose blended or coated with other polymers or inorganic particles). Recent peer-reviewed studies (primarily from the past ~5 years) are emphasized to illustrate current trends, novel strategies, and performance benchmarks.

TABLE 13.1
Examples of cellulose microsphere-based systems for removal of synthetic dyes from water, illustrating various mechanisms and performance metrics (adsorption capacities Q_{max}, degradation percentages, etc.), as reported in recent literature

Cellulose Microsphere System	Target Dye(s)	Main Mechanism	Performance
CMC–PEI hollow microsphere (carboxymethyl cellulose crosslinked with Al(III), treated with polyethylenimine)	Methylene Blue (MB)	Adsorption (electrostatic & H-bonding)	Q_{max} = 235 mg/g (Langmuir capacity)
Ionic liquid-modified cellulose (CMs4) (cellulose with grafted 4-methylimidazolium groups)	Acid Orange 7 (AO7)	Adsorption (anion exchange)	Q_{max} = 218.6 mg/g; fast uptake (100 mg/g in 10 min)
CNC/MnO_2/alginate porous microsphere (cellulose nanocrystals + MnO_2 in alginate matrix)	MB	Adsorption + oxidation (MnO_2)	95.4% MB removed in 10 min; $Q_e \approx$ 114.5 mg/g at equilibrium
Pd nanoparticles on cellulose (Pd NPs grown in situ on cellulose microsphere)	MB	Catalytic reduction (with $NaBH_4$)	>99.8% MB decolorization; reusable 5+ cycles
Ag on dicarboxylated CNF (silver NPs on oxidized cellulose nanofibrils)	4-Nitrophenol (model) + dyes	Catalytic reduction (electron transfer)	Rapid complete reduction; retained ~72% activity after 10 cycles (for dye)

(Continued)

TABLE 13.1 (*Continued*)
Examples of cellulose microsphere-based systems for removal of synthetic dyes from water, illustrating various mechanisms and performance metrics (adsorption capacities Q_{max}, degradation percentages, etc.), as reported in recent literature

Cellulose Microsphere System	Target Dye(s)	Main Mechanism	Performance
$CNC/MnO_2/TiO_2$ hybrid microsphere (porous microsphere with dual oxides)	MB	Photocatalysis + adsorption	97.0% MB removal under light; Q_{max} = 310.2 mg/g; easily regenerated
Cellulose/GO/TiO_2 hydrogel (cellulose gel with graphene oxide and TiO_2)	MB	Photocatalysis + adsorption	~93% MB degradation under UV; stable over ten cycles
Laccase on BC/TiO_2 composite (enzyme immobilized on bacterial cellulose with TiO_2)	Reactive dyes (e.g. Reactive Yellow)	Enzymatic biodegradation (with photo-assist)	~95% color removal in 3 h; 67% activity retained after ten uses
Bacteria on CA/PEO nanofiber mat (*Bacillus* sp. immobilized on cellulose acetate/polyethylene oxide fibers)	MB	Microbial biodegradation + adsorption	93% MB decolorization in 48 h; ~44% capacity remaining after four cycles
BC/UiO-66 MOF composite(Kombucha-derived BC membrane with in situ grown MOF crystals)	Fluorescein & Bromophenol Blue (anionic dyes)	Adsorption (MOF pores + BC fibers)	>90% of fluorescein adsorbed in 10 min (pH 5); high capacity & rapid kinetics

To give a quick perspective, Table 13.1 summarizes representative examples of cellulose microsphere systems for dye removal, spanning the different mechanisms. As shown, cellulose microspheres have achieved excellent outcomes, such as high adsorption capacities, nearly complete dye decolorization, or fast photocatalytic degradation, often with good reusability [27,28]. These findings underscore the promise of cellulose microspheres as next-generation materials for wastewater remediation. In the subsequent sections, we delve into the details of each mechanism – discussing how the cellulose microspheres are prepared and optimized, the underlying removal mechanisms, kinetic and isotherm aspects, and the key results from recent studies. We also identify remaining challenges (such as stability and scale-up) and propose future research directions to fully realize the potential of cellulose microspheres in environmental cleanup.

13.2 CELLULOSE MICROSPHERES: TYPES AND FABRICATION

Cellulose microspheres can be derived from various sources and prepared by different methods, yielding materials with distinct structures and properties [28,29]. In this section, we outline the main categories of cellulose-based microspheres used in dye pollutant removal and describe how they are fabricated and modified. Understanding the types of cellulose microspheres is important, as their origin and structure influence which degradation mechanisms they are best suited for (adsorption, catalysis, etc.).

13.2.1 Bacterial Cellulose Microspheres

Bacterial cellulose (BC) is a form of nanocellulose produced by certain microbes (e.g. *Komagataeibacter xylinus*, formerly Gluconacetobacter) during fermentation [30]. BC is highly pure (lacks lignin and hemicellulose), has an ultra-fine nanofibrillar network, and features a high crystallinity and water-holding capacity [31]. Typically, BC is obtained as a gelatinous pellicle or membrane [32]. Converting BC into spherical beads requires specialized techniques. One approach is using microfluidics or emulsification: BC precursors or slurry are dripped into a coagulating bath to form spherical gel particles, sometimes followed by freeze-drying to create hollow or porous BC microspheres [33]. A recent study demonstrated microfluidic fabrication of hollow BC microspheres with controlled internal structure, intended for drug delivery and potentially adaptable for pollutant adsorption [34–36]. Another approach is to blend BC with alginate or other biopolymers to improve its formability into beads [37]. For example, BC-incorporated alginate beads have been made to enhance mechanical strength and stability of pure alginate beads [38]. Kombucha biofilms (a form of BC) have also been processed into spherical or granular forms to serve as supports for adsorbents and catalysts [39,40]. BC's nanofiber network imparts a high surface area and porosity to the microspheres, which is beneficial for adsorption [41]. Additionally, BC is easily chemically modified (e.g. oxidized, functionalized) due to its reactive hydroxyls, enabling composite formation with catalysts or grafting of functional moieties [42].

13.2.2 Nanocellulose-Derived Microspheres

Nanocelluloses – including cellulose nanocrystals (CNCs) and cellulose nanofibrils (CNFs) – are typically obtained by breaking down plant cellulose fibers via acid hydrolysis or TEMPO-mediated oxidation [43]. These nanocelluloses can be reassembled into microscale beads through methods like spray-drying, emulsion templating, or gelation [44]. For instance, spray-drying a CNC suspension can yield micro-spherical aggregates of CNCs, which act as adsorbents [45]. However, simple spray-dried CNC beads might be dense and have limited porosity [46]. To address this, researchers use templating techniques: e.g. generating foamed CNC gels via emulsions or gas bubbles, then crosslinking or freeze-drying to lock in a porous microsphere structure [47]. Porous microspheres can be created from CNCs by using a Pickering emulsion stabilized by CNCs, incorporating MnO_2 nanoparticles and sodium alginate as a binder [48]. The result was a honeycomb-like microstructure with ~98% porosity that dramatically improved dye adsorption and catalytic performance [49]. Similarly, TEMPO-oxidized CNFs (TOCNF) have been shaped into beads by dripping into Ca^{2+} solutions (forming ionic crosslinks) followed by further modifications [50]. TOCNF beads exhibit high negative charge (carboxylate groups) and have shown very high adsorption capacities (e.g. a TOCNF bead reached ~185.2 mg dye/g) [51]. Generally, nanocellulose-derived microspheres combine the accessibility of nanoscale building blocks (high functional surface) with the manageability of a granular sorbent [52].

13.2.3 Chemically Modified Cellulose Beads

Plant-derived cellulose (from wood pulp, cotton linters, etc.) can be processed into soluble derivatives or directly into beads by dissolution-regeneration techniques [52]. Common cellulose derivatives used for bead making include carboxymethyl cellulose (CMC) and cellulose acetate (CA) [53]. These can be dissolved in water (CMC) or appropriate solvents (CA in acetone or DMAc) and then dropped into anti-solvent baths to form gelled spheres [54,55]. CMC, for example, can be gelled by ionic crosslinkers (multivalent cations or polycations) [56,57]. Transforming CMC into hollow microspheres by first crosslinking with Al^{3+} (forming CMC–Al gel beads) [58], then introducing polyethyleneimine (PEI), which displaced the Al and created a hollow shell structure [59]. The resulting CMC–PEI hollow beads had abundant amine groups from PEI and were very effective in adsorbing dyes (MB capacity 235 mg/g) [60]. CA is another popular starting material – it can be precipitated into spherical beads, and then hydrolyzed (saponified) to regenerate cellulose or left as acetate. CA beads are hydrophobic relative to pure cellulose, but they can be modified by blending in other components [61]. For example, incorporating graphene oxide into CA beads yielded GO@CA beads with MB capacity ~370 mg/g [62], and adding sulfonated graphene oxide gave SGO@CA beads with capacity ~137mg/g) [63]. The CA matrix provides mechanical stability, while the additive or surface hydrolysis provides adsorption sites [64].

13.2.4 Composite and Hybrid Microspheres

Many advanced cellulose microspheres are in fact composites, combining cellulose (or its derivatives) with other polymers or inorganic materials. Blending with other natural polymers is common – for instance, chitosan/cellulose hybrid microspheres or cellulose/alginate beads – to leverage multiple functional groups. Inorganic additives like clays, metal oxides, and carbon materials have been incorporated to impart catalytic or adsorptive functionalities beyond what cellulose offers alone [65]. One notable class is cellulose–MOF composites, where MOF crystals grow on cellulose substrates [66]. UiO-66 (a Zr-based MOF) can be grown directly on kombucha-derived BC fibers, resulting in a porous hybrid where MOF particles coat the cellulose [67]. This composite (in membrane form, which could be crumbled into particles) showed fast adsorption of anionic dyes due to the MOF's micropores and the cellulose's support [68]. Another example is cellulose-carbon composites: e.g. coating cellulose microspheres with carbon nanotubes or biochar to add π–π interactions for dye adsorption [69]. Moreover, magnetic cellulose microspheres have been prepared by integrating Fe_3O_4 nanoparticles into the bead matrix, enabling magnetic recovery of the sorbent from solution (though often reported with other polymers like alginate or chitosan, the concept extends to cellulose). Composite microspheres are especially useful in photocatalysis and catalysis, where cellulose alone wouldn't degrade dyes but serves as a support for active components. We will encounter many such hybrids (e.g. cellulose-TiO_2, cellulose-metal nanoparticle) in the upcoming sections on specific mechanisms.

In summary, researchers have developed a rich variety of cellulose microspheres – from pure cellulose entities (bacterial or plant-derived) to hybrids with enhanced functionality. The choice of microsphere type is often tailored to the intended mechanism of dye removal. Next, we discuss each degradation mechanism in detail, highlighting how different cellulose microsphere types are utilized and optimized for that mode of action.

13.3 ADSORPTIVE REMOVAL OF DYES BY CELLULOSE MICROSPHERES

13.3.1 Adsorption Mechanism and Factors

Adsorption is frequently the first step in dye removal and is inherently involved in other degradation processes (photocatalysis and biodegradation often benefit from an initial adsorption of the dye onto the catalyst or microbe surface). Cellulose microspheres are highly effective adsorbents for many dyes due to their porous structure and surface chemistry [70]. The primary adsorption mechanisms include: (i) Electrostatic interactions – many dyes are anionic (e.g. sulfonated azo dyes) or cationic (basic dyes); by charging the cellulose surface oppositely, strong attraction can occur [71,72]. Native cellulose has many –OH groups, which at neutral pH can form hydrogen bonds with polar dye groups [73]. If the cellulose is oxidized (introducing $-COO^-$ groups), it gains negative charge and can attract cationic dyes like Methylene Blue (MB) and Crystal Violet (CV) via Coulombic attraction [74]. Conversely, attaching cationic groups (e.g. quaternary amines) to cellulose helps bind anionic dyes like Orange G or Reactive Black [75]. (ii) Hydrophobic interactions and π–π stacking – cellulose itself is somewhat hydrophilic, but chemical modifications or composites (like adding aromatic groups, graphene oxide, etc.) introduce hydrophobic domains that can attract hydrophobic dye molecules or allow π–π stacking with aromatic dye rings [76]. (iii) H-bonding and van der Waals forces – dyes with functional groups ($-NH_2$, $-SO_3H$, $-OH$) can form hydrogen bonds with cellulose's hydroxyls [77]. The overall adsorption capacity is influenced by surface area, pore structure (macropores vs micropores), and the presence of functional groups that provide binding sites.

Cellulose microspheres often exhibit rapid adsorption kinetics for dyes. Their porous nature allows quick diffusion of dye molecules into the interior. Many studies report that adsorption onto these microspheres follows pseudo-second-order (PSO) kinetics, implying chemisorption or strong uptake, and fits the Langmuir isotherm model, indicating monolayer adsorption on a finite number of sites [78]. However, in cases of highly porous or heterogeneous surfaces, dual-mode isotherms or Freundlich behavior (multilayer, heterogeneous adsorption) can be observed [79].

13.3.2 Enhancing Adsorption Capacity via Functionalization

One major advantage of cellulose is the ease of chemical modification to enhance adsorption. Carboxylation of cellulose (via TEMPO oxidation or other acids) introduces carboxylic groups, which not only increase the negative charge for cationic dye uptake but also improve water swellability. TEMPO-oxidized cellulose beads (with high $-COO^-$ content) can reach a maximum MB capacity of 495 mg/g, much higher than unmodified cellulose beads [80]. Sulfation (adding $-SO_3^-$ groups) can similarly boost capacity for basic dyes by increasing anionic sites, though it's less common due to potential cellulose degradation [81]. On the other hand, amino-functionalization is effective for anionic dyes [82]. The CMC–PEI hollow microspheres are a prime example: the introduction of polyethylenimine decorated the interior and exterior of the microsphere with amine groups ($-NH_2$, which can be protonated to $-NH_3^+$ in water). This provided a high density of positive charges to attract anionic dye molecules and also created a unique hollow structure that increased surface area. Consequently, these microspheres could adsorb MB up to 235 mg/g and would likely be even more effective for sulfonated azo dyes (which are anionic) [83]. Ionic liquid (IL) groups have also been grafted onto cellulose to impart ionic binding sites. In one study, seven types of IL-functionalized cellulose microspheres were screened for adsorption of Acid Orange 7 (AO7, an anionic azo dye) [84]. The best performer (denoted CMs4) bore 4-methylimidazolium groups on the cellulose chain and achieved ~218.6 mg/g capacity for AO7, with 100 mg/g uptake attained in only ten minutes. The imidazolium cations likely paired with the sulfonate groups of AO7, pulling the dye out of solution quickly [85]. Furthermore, by adjusting solution pH, the surface charge of these IL-modified beads could be toggled, allowing selective adsorption of either cationic or anionic dyes. This sort of tunable behavior is very attractive for treating complex wastewaters containing mixed dyes.

13.3.3 High-Capacity Porous Structures

Beyond chemistry, the physical structure of the microsphere is crucial. Researchers have learned to maximize the accessible surface area of cellulose microspheres through templating and creating hierarchical porosity. Hollow microspheres (like the CMC–PEI example) expose interior surfaces to solution, effectively doubling surface area compared with solid beads. Mesoporous and macroporous beads allow dyes (which can be large molecules ~1–2 nm in size) to diffuse rapidly into the bulk of the adsorbent [86]. The CNC/MnO_2/alginate microspheres had a foam-like architecture with large, interconnected pores (~98% porosity). This enabled MB to be almost completely taken up within ten minutes (95% removal), a much faster rate than typical commercial activated carbons. The equilibrium capacity reached ~114.5 mg/g, which, while not extremely high in absolute terms, was achieved with very short contact time – highlighting the benefit of a highly porous, low-density structure. In general, by combining cellulose with a porogen (something that creates pores, like gas bubbles, ice crystals in freeze-drying, or sacrificial polymer templates), one can produce microspheres with surface areas comparable to activated carbon but with tailored chemistry.

Some of the highest dye adsorption capacities reported for biopolymer beads approach or even exceed those of activated carbon. A recent review noted that biopolymer-based gel beads (including cellulose-based) have achieved capacities up to 10–1500 mg/g for certain dyes [87]. These exceptionally high values often come from multi-functionalized composites – for example, a cellulose bead intercalated with calcium carbonate and further modified with glutaric anhydride reached ~1186 mg/g for MB [88], and a nanocellulose aerogel modified with polyethyleneimine and graphene oxide adsorbed ~1890 mg/g of a pharmaceutical pollutant [89]. While dye adsorption capacities in the hundreds of mg/g are more common, these numbers highlight that cellulose-based materials can rival or surpass traditional adsorbents when intelligently engineered.

13.3.4 Selectivity and Dye–Dye Separation

In industrial practice, wastewaters often contain mixtures of different dyes. An interesting advantage of "smart" cellulose microspheres is the ability to preferentially bind one class of dye over another by tuning conditions. For instance, the ionic liquid-modified cellulose (CMs4) mentioned above could separate cationic vs. anionic dyes by pH modulation [90]. At low pH, the surface imidazolium is largely protonated and can bind anionic dyes strongly (while cationic dyes remain in solution due to electrostatic repulsion) [91]. At higher pH, those groups might deprotonate or be neutral, reducing affinity for anions and potentially allowing cationic dyes (which can interact via π–π or H-bonding) to be captured [92]. This kind of controllable selectivity is valuable for sequential removal processes or dye recovery. Additionally, some cellulose composites incorporate specific binding motifs – e.g. cyclodextrin for hydrophobic dyes, or metal affinity sites that target metallized dyes [93].

13.3.5 Regeneration and Reusability

For adsorption to be truly practical, adsorbents must be regenerable. Cellulose microspheres generally allow for relatively mild regeneration techniques. Since the binding forces are often electrostatic or H-bonding, simply washing with a high ionic strength solution, changing the pH, or using an organic solvent can desorb the dyes. Many studies report the reusability of cellulose adsorbents for multiple cycles with minimal capacity loss. For example, an ethylenediamine-functionalized cellulose showed <10% drop in capacity over five adsorption–desorption cycles when alternating between acidic (adsorption) and basic (desorption) conditions [94]. The ease of regeneration is partly due to cellulose's chemical stability in a wide pH range (roughly pH 2–12, beyond which it may hydrolyze or swell excessively). Thermal regeneration (as done for activated carbon) is typically not applicable, but the ability to biodegrade spent cellulose adsorbents by composting is an eco-friendly end-of-life option if regeneration fails. Importantly, the mechanical integrity of microspheres helps in regeneration – they can be filtered out, treated, and reintroduced without losing material. If magnetic nanoparticles are embedded, the beads can be magnetically separated and rinsed, enabling near-quantitative recovery of the adsorbent.

In summary, cellulose microspheres are highly effective adsorbents for dye removal, offering high-capacity, fast kinetics, and selective uptake when tailored appropriately. Table 13.1 (above) and numerous studies confirm their performance. Adsorption alone can remove the color and a large fraction of dye mass from wastewater. However, adsorption does not destroy the dye; it concentrates it on the adsorbent. To further degrade the dyes into non-toxic forms, additional catalytic or oxidative steps are often coupled. In the next section, we examine how cellulose microspheres serve as platforms for catalytic degradation of dyes, either after adsorption or simultaneously.

13.4 CATALYTIC DEGRADATION OF DYES USING CELLULOSE MICROSPHERE SUPPORTS

While adsorption entraps dyes, catalytic degradation actually breaks down dye molecules, typically via chemical reactions such as reduction or oxidation. Cellulose by itself is not a catalyst for such transformations, but it is an excellent support matrix for dispersing catalytic nanoparticles or active sites [95]. By hosting catalysts on a recoverable, water-insoluble microsphere, cellulose enables treatment of pollutants while avoiding the introduction of catalyst particles directly into water (solving a separation problem) [96]. Two major catalytic approaches for dye degradation are discussed here: reductive catalysis (often using a reducing agent like $NaBH_4$ to convert dyes to colorless leuco compounds or smaller molecules) and oxidative catalysis (breaking down dyes via reaction with oxidants like H_2O_2 or O_2, often generating radicals) [97,98].

13.4.1 Metal Nanoparticle Catalysts for Reductive Degradation

A common laboratory model for catalytic dye degradation is the reduction of MB or other dyes using sodium borohydride ($NaBH_4$) as a reductant [99]. $NaBH_4$ alone reduces dyes very slowly; however, noble metal nanoparticles (like Pd, Pt, or Au) can catalyze electron transfer from BH_4^- to the dye, rapidly bleaching the solution [100,101]. Cellulose microspheres have been used as green supports to immobilize such metal nanoparticles. For instance, Pd-loaded cellulose microspheres can be prepared by an in situ reduction method – the cellulose microsphere (pre-synthesized from dissolved cellulose) was impregnated with Pd^{2+} and then exposed to a mild reducing agent to nucleate Pd nanoparticles on the surface. The Pd content was about 33% (relative to cellulose) in that study. These Pd/cellulose catalysts demonstrated complete (>99.8%) decolorization of MB within minutes in the presence of $NaBH_4$ [102]. The mechanism involves MB being first adsorbed onto the cellulose/Pd surface, then electrons from BH_4^- (which is also adsorbed or near the catalyst) being transferred to MB via the Pd nanoparticles, reducing MB's chromophore (the –N=– conjugated system) to a leuco form or further degraded products. The reaction can be monitored by UV–vis spectroscopy; the disappearance of MB's blue color is rapid. Importantly, the Pd-loaded microspheres were reused for at least five cycles with minimal loss in efficiency, highlighting the stability imparted by the cellulose support (the Pd NPs remain attached to the cellulose and do not aggregate significantly, thanks to the anchoring effect of cellulose's functional groups).

Similarly, silver nanoparticles (Ag NPs) have been supported on cellulose for catalytic reduction [103]. Ag is a good catalyst for the reduction of nitro compounds and certain dyes, though slightly less active than Pd or Au for some substrates. In one study, dicarboxylated cellulose nanofibrils (DCNF) acted as both the reducing agent and support for in situ formation of Ag NPs [104]. The carboxylate groups on DCNF can bind Ag^+ and also help nucleate Ag^0 under heat. The resulting DCNF-Ag composite efficiently catalyzed the reduction of *p*-nitrophenol (a common test reaction) and also showed activity in degrading dyes like Congo Red [105]. The catalyst retained about 72% of its activity even after ten cycles of dye reduction, indicating good reusability. Another work used cellulose/polypyrrole/graphene composite beads to immobilize guaiacol peroxidase (an enzyme) for dye degradation, but also incidentally generated some conductive silver particles in the matrix for potential catalytic synergy [106].

The advantages of using cellulose supports for metal catalysts are: (i) High dispersion – cellulose's surface (and internal network) allows a high density of nucleation sites, resulting in very fine nanoparticles distributed evenly. In the Pd/cellulose case, SEM-EDS confirmed uniform Pd distribution. (ii) Easy recovery – after the reaction, the catalyst can be simply filtered out as it is a solid bead; in contrast, colloidal metal catalysts would require ultracentrifugation or other steps. (iii) Reduced metal usage – since the substrate (dye) is concentrated on the surface via adsorption, a lower amount of metal can achieve the same effect as a higher loading on a non-adsorbing support. The cellulose can "shuttle" reactants to the catalyst. (iv) Biocompatibility and cost – cellulose is cheaper and more environmentally benign than many synthetic supports (like polymer resins or inorganic supports) [107].

Beyond noble metals, other catalysts like metal sulfides or metal oxides can be attached to cellulose for reductive treatments. For example, zero-valent iron (Fe^0) or sulfides (FeS) on cellulose might reductively cleave azo bonds in dyes via electron transfer [108]. However, noble metals remain most studied for rapid dye decolorization in lab settings.

13.4.2 Catalysts for Oxidative Degradation (Fenton and Related Processes)

Oxidative degradation aims to break dyes into CO_2, H_2O, or at least smaller biodegradable compounds, often via hydroxyl radicals (OH) or other ROS. The Fenton reaction ($Fe^{2+} + H_2O_2 \rightarrow Fe^{3+} + OH + OH^-$) is a classic way to generate radicals for dye destruction. Heterogeneous Fenton catalysts – where iron is immobilized on a solid – are desirable because they avoid adding soluble iron (which needs removal later). Cellulose-based Fenton catalysts have been explored by binding Fe species to cellulose microspheres [109,110]. A noteworthy development is Fe^{3+}-bridged cellulose–alginate gel beads, reported as stable photo-Fenton catalysts. In that approach, alginate (which gelates with Fe^{3+}) was combined with cellulose to form beads crosslinked by Fe^{3+} ions. These Fe (III) centers in the bead can generate OH in the presence of H_2O_2, and under light (photo-Fenton), the Fe^{3+}/Fe^{2+} cycling is accelerated. The Fe-bridged cellulose/alginate beads showed excellent catalytic performance in degrading dyes like Malachite Green, with kinetics following PSO (indicating a strong interaction, possibly due to adsorption + reaction) [111]. Moreover, the structure was stable – the Fe did not leach out significantly because it was bound in the alginate network – and the beads could be reused. This addresses a common issue with Fenton: loss of iron and sludge formation. Here, the biopolymer matrix (cellulose + alginate) effectively traps iron in an active form.

Another approach has been immobilizing iron oxides on cellulose. For example, Fe_3O_4 (magnetite) nanoparticles can be synthesized in situ within cellulose matrixes (taking advantage of cellulose acting as a template and stabilizer). Magnetite can act as a Fenton-like catalyst (especially when doped with Cu or Co to improve activity) and is magnetic for easy recovery [112]. Although much of the literature on Fe_3O_4 for Fenton is with chitosan or synthetic polymer beads, one can infer similar benefits with cellulose. Cellulose acetate/Fe films have been used for photo-Fenton degradation of azo dyes, showing that immobilized Fe(III) in a cellulose matrix remains catalytically active.

13.4.3 Other Catalytic Systems

In addition to metal-based catalysis, cellulose microspheres have been used to support catalytic oxidoreductase enzymes (which overlaps with biodegradation – see next section) and even photocatalysts. There are also reports of palladium on cellulose being used for catalytic oxidation of dyes with oxidants like peroxides or persulfate, although reduction is more common with Pd. Peroxidase enzymes (like horseradish peroxidase, HRP) can degrade dyes in the presence of hydrogen peroxide [113]. Immobilizing HRP on modified cellulose beads has been studied to allow enzyme recovery and reuse. The enzyme catalyzes oxidative polymerization or breakdown of dye molecules. However, enzyme catalysis is covered in Section 6 as part of biodegradation.

13.4.4 Mechanistic Insights

Catalytic degradation often yields complete decolorization, but monitoring intermediate formation is important to confirm breakdown and not just bleaching. In the Pd/cellulose + $NaBH_4$ system, mass spectrometry identified that MB's structure was cleaved – the characteristic conjugated heterocycle was disrupted [114]. In other words, the dye's chromophore was destroyed, which is consistent with true detoxification. For Fenton-based systems, ideally the aromatic rings of dyes are opened to carboxylic acids or CO_2 [115]. Many studies measure TOC (total organic carbon) removal to gauge mineralization. For instance, a Fe-cellulose catalyst might achieve near 100% color removal and high (>80%) TOC reduction for a model dye within a few hours under optimized conditions (with continuous H_2O_2 dosing and optimal pH ~3, which is typical for Fenton) [116]. The presence of cellulose can sometimes consume some radicals (since cellulose is an organic scaffold, it's susceptible to oxidation by OH). Interestingly, in the Fe^{3+}-alginate-cellulose beads, the cellulose may also participate in a redox cycle or at least not excessively scavenge radicals, given the good reusability reported. Some researchers have coated cellulose with inert shells (e.g. silica) to protect it while still allowing the catalyst function, but that adds complexity.

13.4.5 Reusability and Stability

Just as with adsorption, the longevity of a catalyst is crucial. Metal leaching is a concern: however, studies found negligible Pd leaching from cellulose beads after cycling, and similarly Fe leaching from Fe-alginate-cellulose beads was low due to strong binding. The mechanical stability of cellulose in the reaction conditions (which could be alkaline for $NaBH_4$ or acidic for Fenton) must be considered. Cellulose is relatively stable in alkaline $NaBH_4$ solutions (it might even get slightly mercerized but not destroyed in short times), whereas in Fenton's strongly acidic (pH ~3) and oxidative environment, slow cellulose degradation could occur. Nonetheless, over the course of multiple runs in those studies, the beads maintained their form and activity. One strategy to improve stability is crosslinking the cellulose (e.g. with epichlorohydrin or divinyl sulfone) so it doesn't swell or depolymerize easily [117]. Another is to regenerate the catalyst by simple washing or re-depositing metal if needed.

In summary, cellulose microspheres provide a flexible platform for diverse catalytic dye degradation reactions. They allow the use of powerful chemical reactions (like reduction with $NaBH_4$ or oxidation with H_2O_2) in a heterogeneous mode where catalysts are immobilized, making the process more practical. Such catalytic microspheres can be used in batch or continuous flow reactors (e.g. packed in a column where dye-containing water flows through and gets decolorized). The next section will focus on photocatalysis, which is a special subset of catalysis harnessing light energy, and one where cellulose plays an interesting dual role as support and sometimes co-catalyst.

13.5 PHOTOCATALYTIC DEGRADATION OF DYES VIA CELLULOSE MICROSPHERE COMPOSITES

Photocatalysis has emerged as a promising and eco-friendly method for degrading organic pollutants including dyes, using light (especially sunlight or UV) to drive chemical reactions that break down pollutants [118]. Semiconductors like TiO_2, ZnO, and others can absorb photons and generate electron–hole pairs that produce reactive oxygen species (ROS) capable of oxidizing dye molecules to harmless end-products [119]. The combination of photocatalysts with cellulose microspheres merges the advantages of photocatalysis (complete mineralization, no chemical reagents needed besides light and oxygen) with those of a solid support (easy recovery, dispersion of catalyst, mitigation of nanoparticle aggregation).

13.5.1 Cellulose–TiO_2 Photocatalytic Microspheres

TiO_2 is the most widely used photocatalyst due to its strong oxidative power and stability [120]. However, using TiO_2 powder in water treatment has practical issues: the fine particles are difficult to recover after treatment, and they can aggregate, reducing active surface area [121]. Supporting TiO_2 on cellulose microspheres addresses both issues. Cellulose provides a scaffold on which TiO_2 nanoparticles can be either co-synthesized or deposited. For example, cellulose fibers can template the growth of anatase TiO_2 from Ti precursors (like $TiCl_4$ or alkoxides) – the hydroxyl groups on cellulose help bind Ti species, which upon hydrolysis and mild heat form TiO_2 crystallites attached to cellulose [122]. Such in situ crystallization yields a well-dispersed TiO_2 phase intimately mixed with cellulose. In one study, visible-light-active TiO_2 was grown on cellulose by a one-pot hydrolysis at low temperature, creating a composite where TiO_2 particles (5–10 nm) uniformly covered cellulose nanofibers [123].

When applied to dye solutions, cellulose–TiO_2 microspheres show a combination of adsorption and photocatalysis. Initially, the cellulose part might adsorb some dye (especially if surface-modified or if the dye has affinity for cellulose), concentrating it near TiO_2 sites. Under UV or solar illumination, the TiO_2 generates electrons and holes; the holes can directly oxidize dye or more commonly react with water to form OH radicals that attack dye molecules. The electrons typically reduce O_2 to superoxide radicals that also contribute to oxidation. Cellulose/graphene oxide/TiO_2 hydrogel photocatalyst can be made by simply mixing TiO_2, GO, and cellulose solution, followed by gelation [124]. The cellulose hydrogel ensured a porous structure and good dispersion of TiO_2, while graphene oxide improved visible-light absorption and electron transport [125]. The composite achieved ~93% degradation of MB under UV light, and impressively, it maintained performance over ten consecutive cycles with negligible loss [126, 127]. The stability is attributed to the robust network of cellulose keeping the TiO_2 from washing away and possibly preventing catalyst fouling by intermediates (since intermediates can diffuse out or be further oxidized). Another study immobilized TiO_2 on CA fibers and reported efficient photocatalytic degradation of Methyl Orange, benefiting from the large surface area of the fibrous cellulose support [128].

13.5.2 Visible-Light Photocatalysts on Cellulose (Beyond TiO_2)

TiO_2 primarily absorbs UV (λ < 385 nm), which is only ~5% of sunlight [129]. To utilize visible light, other semiconductors or modified TiO_2 are used. Examples include ZnO, BiOCl/BiOBr, CdS, g-C_3N_4, Ag-halide composites, etc. Cellulose microspheres have been combined with these to create visible-light-driven systems. $Bi_4O_5Br_2$/BiOBr heterojunction can be grown on cellulose microspheres (abbreviated as BiOBr/CM). This composite photocatalyst could absorb visible light and showed almost complete removal of Rhodamine B dye under illumination [130]. The cellulose support likely helped by improving the dispersion of the bismuth oxyhalide photocatalyst and possibly adsorbing some dye to bring it in contact with the reactive sites. In another case, a Ag/AgCl/ZnO heterojunction was formed on a polydopamine-coated CMC gel, which significantly enhanced sunlight-driven degradation of dyes (Ag/AgCl extends visible-light absorption via surface plasmon resonance, and the heterojunction reduces electron–hole recombination) [131]. Such multi-component systems take advantage of cellulose's ability to accommodate different materials: the polydopamine on cellulose acted as a glue for Ag/AgCl and also as a photosensitizer itself, and ZnO provided the photocatalytic activity, resulting in a robust composite that responds to sunlight.

13.5.3 Role of Cellulose in Photocatalysis

Interestingly, cellulose is not just a passive support in photocatalysis. Studies have suggested that cellulose (or its carbonized derivatives) can serve as an **electron reservoir or mediator**, improving charge separation in the photocatalyst [132]. As noted in a 2025 review, cellulose's surface can trap photo-excited electrons, shuttling them away from recombination with holes. This prolongs the lifetime of holes that generate ROS, thereby increasing efficiency. In cellulose–TiO_2 composites, the cellulose could accept electrons from TiO_2 conduction band, leaving holes free to produce OH. Those electrons on cellulose might eventually reduce O_2 or other species at a separate site. Essentially, cellulose can act somewhat like a mild co-catalyst, similar to how carbon materials (graphene or activated carbon) are added to photocatalysts to enhance performance. Additionally, cellulose being a soft template means the photocatalyst crystallites often have high surface area and maybe exposed facets that are more active.

However, there is a flip side: under UV, cellulose itself might undergo some photodegradation (particularly in the presence of TiO_2 that generates OH that can attack the cellulose polymer) [133]. To mitigate this, researchers sometimes carbonize the cellulose partially (making a carbonaceous support that is less degradable) or confine the photocatalyst in a way that radicals are mostly confined to near the dye. In the cellulose/GO/TiO_2 hydrogel, the authors did not observe significant cellulose loss over ten cycles, suggesting the cellulose was stable enough or that radicals preferentially attacked the dye (possibly GO also sacrificially took some damage). In aerogels of BC with TiO_2, one can incorporate radical scavenger functional groups to protect cellulose.

13.5.4 Performance Highlights

Photocatalytic cellulose microspheres have shown excellent results across many dye types. Often >90% removal of dyes like MB, Methyl Orange, RhB, and others is achieved within 0.5–2 hours of light exposure [134]. In the CNC/MnO_2/TiO_2 porous microspheres (Table 13.1), a 97% MB removal was achieved, which is notable because it combined adsorption (by the porous matrix and CNC) with dual photocatalysts (MnO_2 and TiO_2). MnO_2 alone has some oxidative capability (even in the dark, it can oxidize certain dyes, as an oxidizing agent), but MnO_2 and TiO_2 together under light can have a synergistic effect, possibly involving MnO_2 scavenging electrons from TiO_2 (preventing recombination) since MnO_2 can be a good electron acceptor [135,136]. The concept of concerted catalysis was invoked to explain the excellent performance – meaning the materials work in concert rather than sequentially. Moreover, that system was rapidly regenerable: after photocatalysis, it could be simply rinsed and reused, and any dye remaining on it would be degraded by the next cycle's OH radicals, cleaning the surface.

Another advantage is handling mixed dyes or real textile effluents. Some studies tested cellulose-based photocatalysts on actual wastewater containing mixtures of dyes and perhaps auxiliary chemicals (salt, surfactants) [137]. Results showed significant decolorization, although rates can be slower than for single dyes due to competitive adsorption and light attenuation. Nonetheless, the broad-spectrum activity of OH radicals means most organic dyes can be attacked. For example, a cellulose-TiO_2 film fully decolorized a simulated textile effluent containing four different dyes under solar light in two hours, and a BiOBr/cellulose catalyst removed not just RhB but also nearly all other organic contaminants in water in a short time [137,138,139].

13.5.5 Immobilization Formats

While we focus on microspheres, it's worth noting that photocatalytic cellulose is also made in other forms (films, membranes, coated fibers). Microspheres, however, are advantageous for packed-bed or fluidized-bed reactors treating flowing water, because they won't cause high pressure drops and can be uniformly irradiated as a suspension. There is ongoing work in developing photocatalytic aerogel beads that float on water (so they can harness sunlight at the air–water interface). For instance, sulfonated graphene oxide impregnated CA beads were designed to float and degrade dyes under sunlight. Floating photocatalyst beads avoid the need for stirring and can be simply scooped out after treatment.

In conclusion, cellulose-supported photocatalysts combine effective dye degradation with material reusability and environmental safety. The cellulose not only helps in handling the catalyst but can improve its performance and adds an adsorption component that makes the overall removal more robust. Photocatalytic microspheres exemplify how green materials (cellulose) and advanced oxidation processes can be married to tackle pollutant dyes in a sustainable way. The next section will explore biodegradation, leveraging biological activity on cellulose microspheres – which is another sustainable route, often complementary to the physical/chemical methods discussed so far.

13.6 BIODEGRADATION OF DYES USING CELLULOSE MICROSPHERES (ENZYMATIC AND MICROBIAL)

Biodegradation involves using biological agents – enzymes or living microorganisms – to break down pollutants [140]. Many dyes, especially complex aromatic dyes, are not easily degraded by naturally occurring microbes in the environment (hence they persist). However, certain specialized bacteria and fungi, as well as isolated enzymes (like laccases and peroxidases), can oxidatively or reductively cleave dye molecules [140]. The challenge is deploying these biocatalysts in wastewater treatment in a controllable, efficient manner. This is where cellulose microspheres come into play: they serve as carriers for immobilizing enzymes or cells, providing a stable environment and overcoming issues like washout of free enzymes or microbes. Additionally, cellulose can adsorb dyes and thus present a higher local concentration of pollutant to the biocatalyst, enhancing degradation rates.

13.6.1 Enzyme-Immobilized Cellulose Microspheres

Laccases and **peroxidases** are the most commonly used enzymes for dye degradation. Laccases (multi-copper oxidases from fungi such as Trametes versicolor) can oxidize a wide range of dyes, particularly phenolic and azo dyes, often requiring mediators for non-phenolic dyes [141]. Peroxidases (like HRP or plant peroxidases) use hydrogen peroxide to oxidize dyes. Immobilizing these enzymes on cellulose can improve their stability (pH, temperature tolerance) and allow reuse [142]. For example, enzymes can be immobilized on a bacterial nanocellulose (BC) sponge via physical adsorption and crosslinking. The BC's nanofibrillar network provided a high surface area for enzyme attachment. They observed that crosslinked laccase on BC retained 69% of its activity after seven reuse cycles, a significant improvement over free laccase, which is typically single use [143]. Similarly, laccase has been immobilized on cellulose beads and membranes functionalized with amine or carboxyl groups (to bind the enzyme via covalent bonding).

One specific study on laccase-immobilized bacterial cellulose/TiO_2 (mentioned in Table 13.1) reported ~95% degradation of a reactive dye in three hours. Here, TiO_2 under UV likely generated H_2O_2 or other ROS that assist laccase in attacking the dye, a synergistic effect combining enzymatic and photocatalytic actions. More straightforwardly, HRP can be immobilized on CA beads (modified with diethylaminoethyl groups) and achieved effective decolorization of several textile dyes, with the immobilized enzyme maintaining activity over multiple batches [144]. Concanavalin A-bound peroxidase on cellulose has also been used to target specific dye classes (direct dyes) [145]. The typical outcome with enzymes is complete or near-complete decolorization, although in some cases, the dyes are polymerized into insoluble products (which could be filtered out).

The benefit of cellulose supports is again easy separation: enzyme-bound beads can be packed in a reactor or stirred in a tank and then recovered. Immobilization often also broadens the enzyme's usable pH range and thermal stability [146]. In wastewater, immobilized enzymes on cellulose could potentially work continuously (columns with enzyme-coated cellulose beads as a bioreactor) [147].

13.6.2 Microbial Cellulose Carriers

Using live microbes (bacteria, fungi, algae) to degrade dyes is essentially bioremediation. However, free microbes in wastewater treatment face dilution and washout, and some dyes at high concentration are toxic to them [148]. Encapsulating or attaching microbes to a support (making a biofilm or bio-bead) can solve these issues. CA nanofibers and cellulose beads have been used to immobilize dye-degrading bacteria [149]. For instance, isolation can be done on a dye-degrading Bacillus strain and immobilized it onto an electrospun cellulose acetate/polyethylene oxide (CA/PEO) nanofibrous mat [150]. The porous mat provided plenty of surface for bacterial cells to attach and form biofilm. In treating MB solution, the bacteria-immobilized membrane achieved ~93% decolorization in 48 hours, compared with much lower removal with free bacteria, and could be reused four times (though activity dropped to ~44% by the 4th cycle) [151,152]. The combination of adsorption (the CA/PEO fiber itself adsorbed some MB) and biodegradation by Bacillus was key – adsorption likely concentrated the dye and also mitigated toxicity by holding some dye on the fiber, while bacteria gradually metabolized the dye.

Another example is using BC as a matrix for immobilizing fungi or bacteria. BC can be produced in situ with bacteria trapped in it (as BC is formed, bacteria get encased in the fibrils). Alginate-cellulose beads loaded with fungal cells have been investigated to treat dyes, leveraging the robust enzymes fungi produce (like laccases and peroxidases). It is found that by entrapping a consortium of dye-degrading bacteria in CA nanofiber mats, they could achieve >90% removal of mixed dyes, benefiting from the cooperative metabolism of the consortia [153–155].

13.6.3 Synergy of Adsorption and Biodegradation

Cellulose microspheres inherently adsorb dyes, which as noted can significantly enhance biodegradation. This is because many dye-degrading enzymes (like laccase) operate better at high substrate concentrations (up to a point), and microbes often need the pollutant to be in their vicinity. Adsorption serves to localize the dye on the support, effectively pre-concentrating it. Then the enzyme or microbe, also located on the same support, can readily attack the dye. This synergy was explicitly observed in the Bacillus on the CA/PEO system: the authors noted that the membrane's adsorption of MB was an "essential part" of the process and when combined with biodegradation led to effective removal. Similarly, a laccase immobilized on a porous support can first immobilize the dye (perhaps via a mediator if needed) and then oxidize it, ensuring even low concentrations of dye are funneled to the enzyme.

Another synergy is between photocatalysis and biodegradation – e.g. the laccase on TiO_2/cellulose system essentially used photocatalysis to generate H_2O_2, which the enzyme then uses to oxidize dye (or the photocatalysis partially oxidizes the dye to a form that enzyme can finish degrading) [154]. Such hybrid advanced oxidation – biotreatment strategies are increasingly researched because they combine the speed of chemical oxidation with the specificity and complete mineralization potential of biological processes.

13.6.4 Limitations and Improvements

While promising, immobilized biocatalysts have some limitations. Enzymes can be expensive and may lose activity over time (e.g. due to conformational changes or fouling by reaction products) [156]. Efforts to improve enzyme immobilization include covalently bonding them to cellulose (to prevent leaching) and using stabilizing agents (like crosslinkers or adding polyols to prevent denaturation). For microbes, a limitation is that not all dyes are easily metabolized; some require cooperative action of multiple enzymes or species. Also, environmental factors like pH, temperature, and presence of other contaminants can affect microbial activity [156]. Cellulose, being an organic substrate, might also be slowly consumed by microbes (especially fungi that produce cellulases), which could limit the long-term stability of the carrier. To counter that, sometimes the cellulose is chemically modified to be less biodegradable or combined with non-biodegradable components (like a synthetic polymer coating) to make a semi-artificial bio-carrier.

Nonetheless, the biodegradation route offers unparalleled sustainability – using naturally occurring processes to detoxify water with minimal chemical inputs. There have been demonstrations of reactor systems like airlift bioreactors containing immobilized laccase or bacteria on cellulose supports to continuously treat dye effluents. These show that scaling up is feasible.

In summary, cellulose microspheres and related structures provide an effective support for deploying biocatalysts in dye removal. They allow reusable enzyme catalysts and stable biofilms to thrive, often in conjunction with the passive adsorption capacity of cellulose. The result is significant dye decolorization and in many cases degradation to less complex molecules. Complete mineralization may sometimes require further oxidation steps, but even partial biotransformation can reduce toxicity (e.g. azo dyes can be cleaved into aromatic amines, which might then be further degraded by other bacteria). The holistic approach to wastewater treatment might involve a sequence: adsorption + catalytic/photocatalytic + biodegradation, and cellulose-based systems are unique in that they can integrate several of these functions in one matrix [157].

13.7 INTEGRATED AND SYNERGISTIC APPROACHES USING CELLULOSE MICROSPHERES

As evident from the prior sections, the lines between adsorption, catalysis, photocatalysis, and biodegradation often blur in practical systems. Modern cellulose microsphere designs tend to be multifunctional, exploiting two or more mechanisms to achieve better performance than any single mechanism alone [158]. In this section, we discuss integrated approaches and synergies, as well as compare the merits of each mechanism when implemented with cellulose microspheres.

13.7.1 Adsorption-Photocatalysis Synergy

One of the most powerful combinations is adsorption + photocatalysis. Many cellulose-based photocatalysts are intentionally made to have strong adsorption affinity for target dyes, sometimes termed "adsorptive photocatalysts" [159]. The sequence of events – often called the "adsorb & degrade" approach – is that the pollutant is first sequestered from bulk solution onto the material and then degraded in situ. This concentrates the pollutant at the reactive sites and also reduces interference from other substances in solution (since the catalyst primarily "sees" the adsorbed pollutant). We saw this in cellulose/GO/TiO_2 hydrogels (high MB uptake then 93% degradation) and CNC/MnO_2/TiO_2 microspheres (huge MB uptake and 97% removal). In both cases, had the materials not adsorbed MB well, the photocatalysis alone might have been less efficient due to slower diffusion of MB to catalyst and more competition from other species (water matrix components). The synergy is also reflected in kinetics: often the rate of dye removal is faster than either pure adsorption or pure photolysis. Another benefit is self-cleaning: even if some dye (or its oxidized intermediate) remains adsorbed, continued illumination can mineralize it, freeing up the adsorption sites again. This means the composite regenerates itself under the reaction conditions, which is ideal [160].

13.7.2 Adsorption-Catalysis (Sequential Batch or Single Pot)

A cellulose microsphere could be used first as an adsorbent to load up with dye, then transferred to another vessel where a reaction (e.g. Fenton oxidation) is performed to destroy the dye on the bead, then reused. This two-step approach ensures no dye is released untreated. However, some systems do this in one pot: for example, Pd/cellulose in $NaBH_4$ is effectively adsorbing MB and simultaneously catalyzing its reduction. The mechanistic description indicates MB was adsorbed prior to reduction on Pd. Similarly, in Pd or Ag catalytic systems, if the dye didn't adsorb, the reaction would rely solely on random collision in solution, which is slower. Thus, adsorption assists catalysis by localizing reactants.

13.7.3 Adsorption-Biodegradation

This synergy has been explicitly highlighted in bio-immobilization studies. The air-lift bioreactor example and combining a physical adsorption process with biological degradation can handle higher loads and lead to more thorough cleanup. The idea is that adsorption provides immediate removal of acute color/toxicity (by binding the dye), and the microbes or enzymes then slowly degrade the bound dye [160]. As degradation proceeds, those adsorption sites get freed and can capture more dye – a regenerating adsorbent concept. It's analogous to a catalytic surface in chemistry but here it's biocatalytic. One challenge can be if the dye binds too tightly and the enzyme can't access it, but since many dyes are not permanently immobilized as they are in dynamic equilibrium on the adsorbent, the enzyme can still work.

13.7.4 Photocatalysis–Biodegradation (Sequential)

In some proposed water-treatment trains, an advanced oxidation process like photocatalysis is used first to break large dye molecules into smaller, more biodegradable fragments, and then a biological treatment (like activated sludge or immobilized microbes) finishes the mineralization. Cellulose microspheres could theoretically serve in both stages – e.g. a TiO_2-cellulose photocatalyst followed by a laccase-cellulose enzyme system. There have been concept papers on integrating photocatalysis and biological filters for recalcitrant pollutants. The polydopamine-CMC/Ag/AgCl/ZnO system is an example where a photosystem was augmented by a bio-inspired polymer (PDA), hinting at combining biological components with photocatalysis.

13.7.5 Comparison of Mechanisms

Each mechanism has its pros and cons:

- **Adsorption** alone is simple and can reach very low residual concentrations of dye (depending on isotherm), but it doesn't eliminate the pollutant; it transfers it. It also can be overwhelmed by high dye concentrations unless capacity is extremely high [161].

- **Catalysis (Chemical)** can be very fast (like reduction with $NaBH_4$) and can handle high concentrations, but usually needs chemical reagents ($NaBH_4$ or H_2O_2) and the catalyst must be precious metal or similar. Also, catalytic reduction of dyes might not mineralize them – e.g. reduction of azo dyes makes aromatic amines, which might still need further treatment [162].
- **Photocatalysis** can fully oxidize dyes to CO_2 and H_2O given enough time and light, and only needs light and oxygen (plus maybe mild oxidant). It can be run continuously. However, it requires light energy (costly if not sunlight) and photocatalyst activity can be inhibited by water matrix constituents (scavengers for radicals or turbidity blocking light) [163].
- **Biodegradation** is cost-effective and eco-friendly, potentially mineralizing pollutants without harsh chemicals, and can handle continuous inflows [164]. But it is slower, often taking hours to days, and sensitive to conditions (pH, toxic spikes, etc.). Some dyes are inherently very resistant to biodegradation.

By integrating these, one can compensate for weaknesses. For instance, adsorption can quickly reduce toxicity and prepare the water for a slower bioprocess. Photocatalysis can take out what bioprocess leaves or vice versa. Cellulose microspheres are attractive in this integrated picture because the same material can serve multiple roles. A single type of microsphere can act as an adsorbent and catalyst, or adsorbent and microbial support. This reduces the need to move water between different treatment units with different media.

13.7.6 Case Study

Complete Treatment: Imagine an integrated system: a reactor packed with a magnetically recoverable cellulose microsphere that has a TiO_2/Fe_3O_4/cellulose composite. As dye wastewater flows through, the cellulose part adsorbs dyes, TiO_2 photocatalytically degrades them under UV lamps, and the presence of Fe_3O_4 allows assisting photo-Fenton reactions plus easy capture of the beads by magnet if needed. Such a system might achieve higher degradation efficiency than a purely photocatalytic reactor with TiO_2 slurry, and be easier to handle. After treatment, any remaining beads can be pulled out magnetically and replaced or regenerated. There are indeed studies trending in this direction – coupling photocatalysis and Fenton on one support.

13.7.7 Performance Metrics

When evaluating integrated systems, metrics like overall decolorization percentage, mineralization (TOC removal), treatment time, and stability over cycles are key. Cellulose microsphere systems have shown decolorization often in the 90%–100% range for model dyes [165]. Achieving high TOC removal may require the photocatalytic or oxidative components to run longer. As for reusability, many reported systems sustain >5–10 cycles with only minor efficiency drops. This is promising for real applications since frequent replacement of the material would hurt the economics.

In comparing with traditional materials: cellulose microspheres often match or exceed the performance of synthetic resin beads or inorganic supports in dye removal, while being superior in sustainability (renewable and degradable) and often cost [166]. A direct comparison in one paper showed a cellulose-based adsorbent had an order of magnitude higher capacity for a cationic dye than a polystyrene resin under the same conditions.

In conclusion of this section, the holistic design of dye treatment systems is moving toward multifunctional materials. Cellulose microspheres, with their inherent multi-mechanistic capabilities, are at the forefront of this innovation. The synergy between adsorption, catalysis, and biodegradation on cellulose platforms can achieve what each alone might not: rapid, thorough, and economical removal of stubborn synthetic dyes from wastewater.

13.8 CHALLENGES AND FUTURE OUTLOOK

Despite the significant progress and encouraging results reviewed above, there remain several challenges to be addressed before cellulose microsphere-based technologies can be widely implemented in industrial wastewater treatment. In this section, we discuss these challenges and outline future research directions and prospects.

13.8.1 Performance in Real Wastewater

Much of the research to date has been conducted on simulated dye solutions (e.g. one dye in clean water). Real textile or dye manufacturing effluents are complex mixtures containing salts (like NaCl, Na_2SO_4), auxiliaries (surfactants, sizing agents), other organics, and sometimes heavy metals [167]. High ionic strength can compete with electrostatic adsorption (e.g. Na^+ can screen negative charges on cellulose, reducing uptake of cationic dyes) [168]. Organics or surfactants might foul the surface of microspheres or occupy active sites. Furthermore, real waste often has mixtures of dyes. While we saw that some advanced materials can separate or selectively adsorb dyes, in a mix, the capacity for a target dye might be effectively reduced by competition. Future studies should focus on testing cellulose microsphere systems on actual wastewater samples to identify any loss of performance and modify the materials accordingly (e.g. adding antifouling coatings or broad-spectrum adsorptive groups). Encouragingly, some composite microspheres (like cellulose–MOF) have shown efficacy in mixed dye scenarios, but more data is needed on long-term operation in realistic conditions.

13.8.2 Regeneration and Longevity

While many lab studies demonstrate regeneration for a few cycles, in a practical scenario, an adsorbent or catalyst might need to last dozens or hundreds of cycles or have a feasible regeneration method. Thermal regeneration is generally not suitable for cellulose (it will char), so chemical or photochemical regeneration is needed. One can envision, for adsorption-focused systems, a column with two sets of beads: one adsorbing while the other is being regenerated by washing or by advanced oxidation (maybe even using the photocatalytic method to "self-clean") [169]. The mechanical stability of cellulose beads over time in water (especially under flow or agitation) is a concern – pure cellulose might swell or fragment. Crosslinking the cellulose or using a mixed polymer matrix can reinforce it. For example, incorporating a small amount of synthetic polymer (like polyacrylamide or PVC) into the bead can increase its durability. Alternatively, **coatings** like thin layers of silica or layer-by-layer polyelectrolytes could harden the surface without impeding function. Future research could explore such stabilized cellulose microspheres designed for longevity.

13.8.3 Prevention of Microbial Growth and Biofouling

Ironically, while we have considered intentional microbial use, if one is using a purely adsorption or photocatalytic cellulose microsphere in an open system, one might experience biofouling – i.e. other environmental bacteria might colonize the cellulose (since it's an organic substrate). This could clog the microspheres or alter performance. In some cases, that might be benign or even help (forming a biodegradation biofilm), but in others, it could lower adsorption capacity or produce slime that hinders flow. To mitigate unwanted biofouling, one could add biocidal components (like a bit of quaternary ammonium or silver nanoparticles) into the microspheres or operate the system in a way that discourages growth (e.g. periodic drying or UV exposure). This needs to be balanced with not leaching toxic biocides into the treated water.

13.8.4 Safety and Environmental Impact

Cellulose itself is safe, but when combined with other things (like metal nanoparticles or photocatalysts), there is potential release of those components over time [170]. For example, repeated use of Ag-NP loaded cellulose might release some Ag^+ or Ag nanoparticles, which are pollutants of concern (toxic to aquatic life) [171]. It's essential to assess leaching: ensure strong binding of catalysts to cellulose or capture any leached material. If eventually the spent microspheres are disposed, their components should ideally be benign. Using biodegradable catalysts (like enzyme systems or even biochar instead of metal NPs) is an interesting direction. Another consideration is the by-products of dye degradation. Advanced oxidation can produce small organic acids or other intermediates. Usually, photocatalysis and Fenton aim for full mineralization, but if partial, one must ensure no highly toxic by-product is formed (for instance, some dyes can form aromatic amines upon reduction, which might be more toxic). Thus, in future research, comprehensive toxicity assays of treated water should complement the chemical analysis. We want to confirm that the treatment not only removes color but also detoxifies the water.

13.8.5 Scaling Up Production of Cellulose Microspheres

Producing grams in the lab is one thing; producing hundreds of kilograms for a large treatment plant is another. Methods like spray-drying, phase inversion, or microfluidic droplet generation would need to be scaled. Spray-drying can already produce cellulose micro-particles at scale (it's used in making excipients), so adapting it to functionalized cellulose might be viable. Emulsion crosslinking is also scalable, though one has to remove solvents and crosslinkers carefully. **Cost** is a factor: if an extra process (like grafting an IL) is used, is the performance gain worth the cost of those chemicals? Life-cycle and cost analyses will be important to convince industry. BC production can be done in large fermenters, but cost of feed and yield may limit it currently to high-value applications. However, as a waste product (like from kombucha or vinegar industries), BC might be more accessible in the future.

13.8.6 Integration into Existing Systems

Ideally, new materials should retrofit into existing water-treatment processes. For example, could cellulose microspheres replace activated carbon in an adsorber unit? Possibly yes, if they can be used in similar column setups. Could they augment biological treatment by adding them to bioreactors? They might improve dye removal in aerobic or anaerobic digesters by adsorbing dyes that normally inhibit microbes. If magnetic, they could be fluidized and then magnetically settled. Research could explore how to implement these materials in continuous flow systems, perhaps using techniques from ion-exchange resin columns or moving bed biofilm reactors (in which carriers move in water).

13.8.7 Novel Functionalizations and Targets

Looking ahead, researchers are starting to design cellulose microspheres for specific classes of dyes. For instance, tailoring a microsphere for reactive azo dyes vs. basic dyes vs. disperse dyes (which are very hydrophobic) – each may need a different approach (high charge density vs. π-rich surface vs. surfactant-like functionality, respectively). There is also a trend to incorporate sensing capabilities – e.g. a photocatalytic cellulose bead that also has a color indicator to tell when it's saturated or exhausted. Moreover, the scope can expand beyond dyes: many of these systems also work for pharmaceuticals, phenolic pollutants, etc. A future outlook is that a single type of cellulose microsphere could simultaneously remove dyes, heavy metals, and even pathogens (if given an antimicrobial coating), offering a holistic solution for wastewater cleanup.

13.8.8 Circular Economy Aspects

The sustainability of cellulose microspheres is a big selling point. They can be made from waste biomass or bacterial fermentation (like the pineapple peel-derived BC for aerogels). After use, if their capacity or activity decays, they could be **biodegraded or composted**, unlike synthetic resins that become plastic waste [172]. In some cases, if the captured dye is not toxic, the dye-laden cellulose could even be reused as a colorant for some purpose (probably niche, like coloring recycled paper). Or if incinerated, it's mainly biomass so it's carbon-neutral (aside from any captured fossil dye carbon). This aligns with circular economy and green chemistry principles.

Future Prospects:

Research is likely to continue on hybrid materials, for example:

- Cellulose–MOF–photocatalyst tri-composites for broadened light absorption.
- Enzyme–photocatalyst conjugates on cellulose for simultaneous oxidative and enzymatic action (one addresses what the other can't).
- Electroactive cellulose beads (if cellulose is made conductive, perhaps by polymerizing aniline or pyrrole on it) that could be used in an electro-Fenton process, where applying a voltage regenerates Fe^{2+} and produces $H_2O8.8$. Research is likely to continue innovating hybrid materials that push the boundaries of performance. For example, cellulose based beads integrated with metal-organic frameworks (MOFs) and photocatalysts could broaden light absorption and increase adsorption capacity simultaneously. Enzyme–photocatalyst conjugates on cellulose (e.g. laccase co-immobilized with TiO_2) might allow simultaneous biological and photocatalytic degradation, handling a wider range of dye structures. Another promising avenue is developing electrically conductive cellulose composites (by incorporating conductive polymers or carbon) that could be used in electrochemical advanced oxidation (e.g. electro-Fenton processes), where an applied voltage continuously regenerates reactive species on the microsphere for pollutant degradation. As these examples show, the versatility of cellulose chemistry means researchers can impart new functionalities to microspheres beyond what is currently realized.

Looking forward, cellulose microspheres are poised to make the leap from lab scale to field applications. Their alignment with green engineering principles – using renewable, biodegradable materials to remediate pollutants – is a compelling advantage in a world increasingly focused on sustainability [173–176]. With continued development to overcome present challenges (ensuring stability, reusability, and efficacy in complex waters), cellulose microsphere technologies could become mainstream in treating dye-containing effluents. The demand for more effective and eco-friendly treatment methods is rising as regulations on industrial wastewater tighten. By leveraging the recent scientific advances and addressing the outlined challenges, future systems might use cellulose microsphere-based reactors to achieve near-zero discharge of hazardous dyes. In summary, the outlook is highly encouraging: cellulose microspheres represent a convergence of high performance and sustainability in water remediation, and ongoing innovations will further enhance their practicality and impact.

13.9 CONCLUSIONS

Synthetic dyes in wastewater pose a serious environmental threat, and conventional treatment methods often fall short in eliminating these persistent pollutants. Cellulose microsphere-based technologies have emerged as a robust solution, offering a multifaceted approach to dye remediation. In this review, we present a detailed examination of how cellulose microspheres can degrade or remove synthetic dyes via adsorption, catalysis, photocatalysis, and biodegradation. We surveyed all major types of cellulose microspheres – including those derived from BC, plant nanocellulose, and chemically modified cellulose – and discussed their preparation, properties, and performance.

Several key conclusions can be drawn:

- **Efficiency:** Cellulose microspheres demonstrate high efficiency in dye removal, achieving rapid decolorization and high uptake capacities. Through surface functionalization and structural optimization (e.g. creating hollow/porous structures), their performance can meet or exceed traditional adsorbents and catalysts.
- **Multi-Mechanism Capability:** Unlike many single-purpose treatments, cellulose microspheres can integrate multiple removal mechanisms. They can act as sorbents to concentrate dyes and as platforms for chemical/photocatalytic breakdown or as carriers for dye-degrading enzymes/bugs. This integration often leads to synergistic effects, yielding more complete and faster dye elimination.
- **Reusability and Green Nature:** Cellulose microspheres are generally reusable over multiple cycles, and at the end of life, they are biodegradable, avoiding secondary pollution. Their fabrication can utilize renewable resources or even waste biomass, aligning with sustainable and circular economy goals.
- **Recent Trends:** The field is rapidly evolving, with recent trends, including the use of advanced composites (like cellulose–MOF hybrids, semiconductor heterojunctions on cellulose, and magnetic cellulose beads) and the move toward visible-light-responsive and enzyme-coupled systems for greater effectiveness under solar irradiation and mild conditions.

While challenges remain (notably ensuring consistent performance in industrial conditions and improving the mechanical robustness of the beads), ongoing research is actively addressing these issues. The development of smarter functionalization strategies and scalable production methods will further enhance the applicability of cellulose microspheres.

In conclusion, cellulose microspheres offer a high-impact, environmentally conscious approach to combat dye pollution in wastewaters. They exemplify how a naturally derived material can be engineered at the nano- and microscale to solve a modern environmental problem. The literature reviewed provides strong evidence that with appropriate design, cellulose microspheres can achieve outstanding removal of a wide spectrum of synthetic dyes, turning polluted water clear and safe. As we advance, the continued interdisciplinary efforts – spanning cellulose chemistry, materials science, environmental engineering, and microbiology – will undoubtedly expand the capabilities and adoption of these systems. The knowledge compiled in this review serves as both a foundation and inspiration for future innovations, ultimately aiding the transition toward more sustainable water-treatment technologies for the protection of environmental and public health.

REFERENCES

1. Ardila-Leal, L. D., Poutou-Piñales, R. A., Pedroza-Rodríguez, A. M., & Quevedo-Hidalgo, B. E. (2021). A brief history of colour, the environmental impact of synthetic dyes and removal by using laccases. *Molecules*, *26*(13), 3813.

2. Saska, J., Li, Z., Otsuki, A. L., Wei, J., Fettinger, J. C., & Mascal, M. (2019). Butenolide derivatives of biobased furans: sustainable synthetic dyes. *Angewandte Chemie International Edition, 58*(48), 17293–17296.
3. Ayele, A., Getachew, D., Kamaraj, M., & Suresh, A. (2021). Phycoremediation of synthetic dyes: an effective and eco-friendly algal technology for the dye abatement. *Journal of Chemistry, 2021*(1), 9923643.
4. Kathing, C., & Saini, G. (2022). A review of various treatment methods for the removal of dyes from textile effluent. *Recent Progress in Materials, 4*(4), 1–15.
5. Dutta, P., Rabbi, M., Sufian, M., & Mahjebin, S. (2022). Effects of textile dyeing effluent on the environment and its treatment: a review. *Engineering and Applied Science Letters (EASL), 5*, 1–17.
6. Yildirim, O. A., Bahadir, M., & Pehlivan, E. (2022). Detrimental effects of commonly used textile dyes on the aquatic environment and human health–a review. *Fresenius Environmental Bulletin, 31*, 9329–9345.
7. Ismail, M., Akhtar, K., Khan, M. I., Kamal, T., Khan, M. A., M Asiri, A., … & Khan, S. B. (2019). Pollution, toxicity and carcinogenicity of organic dyes and their catalytic bio-remediation. *Current Pharmaceutical Design, 25*(34), 3645–3663.
8. Zinicovscaia, I. (2016). Conventional methods of wastewater treatment. In I. Zinicovscaia & L. Cepoi (Eds.), *Cyanobacteria for Bioremediation of Wastewaters* (pp. 17–25). Springer International Publishing, Cham.
9. Ogola, H. J. O., Ashida, H., Ishikawa, T., & Sawa, Y. (2015). Explorations and applications of enzyme-linked bioremediation of synthetic dyes. In N. Shiomi (Ed.), *Advances in Bioremediation of Wastewater and Polluted Soil* (pp. 111–144). IntechOpen, London.
10. Ruan, W., Hu, J., Qi, J., Hou, Y., Zhou, C., & Wei, X. (2019). Removal of dyes from wastewater by nanomaterials: a review. *Advanced Materials Letters, 10*(1), 9–20.
11. Zare, K., Gupta, V. K., Moradi, O., Makhlouf, A. S. H., Sillanpää, M., Nadagouda, M. N., … & Kazemi, M. (2015). A comparative study on the basis of adsorption capacity between CNTs and activated carbon as adsorbents for removal of noxious synthetic dyes: a review. *Journal of Nanostructure in Chemistry, 5*, 227–236.
12. Bhat, A. P., & Gogate, P. R. (2021). Degradation of nitrogen-containing hazardous compounds using advanced oxidation processes: a review on aliphatic and aromatic amines, dyes, and pesticides. *Journal of Hazardous Materials, 403*, 123657.
13. da Silva Santos, D. H., Xiao, Y., Chaukura, N., Hill, J. M., Selvasembian, R., Zanta, C. L. S., & Meili, L. (2022). Regeneration of dye-saturated activated carbon through advanced oxidative processes: A review. *Heliyon, 8*(8), e10205.
14. Peramune, D., Manatunga, D. C., Dassanayake, R. S., Premalal, V., Liyanage, R. N., Gunathilake, C., & Abidi, N. (2022). Recent advances in biopolymer-based advanced oxidation processes for dye removal applications: a review. *Environmental Research, 215*, 114242.
15. Dassanayake, R. S., Acharya, S., & Abidi, N. (2021). Recent advances in biopolymer-based dye removal technologies. *Molecules, 26*(15), 4697.
16. Zhang, Z., Lu, Y., Gao, S., & Wu, S. (2025). Sustainable and efficient wastewater treatment using cellulose-based hydrogels: a review of heavy metal, dye, and micropollutant removal applications. *Separations, 12*(3), 72.
17. Gupta, P. K., Raghunath, S. S., Prasanna, D. V., Venkat, P., Shree, V., Chithananthan, C., Choudhary, S., Surender, K., & Geetha, K. (2019). An update on overview of cellulose, its structure and applications. In A. R. Pascual & M. E. Eugenio Martín (Eds.), *Cellulose*. IntechOpen, London.
18. Aoudi, B., Boluk, Y., & El Din, M. G. (2022). Recent advances and future perspective on nanocellulose-based materials in diverse water treatment applications. *Science of the Total Environment, 843*, 156903.

19. Patel, H. (2019). Fixed-bed column adsorption study: a comprehensive review. *Applied Water Science*, *9*(3), 45.
20. Naderi, N., Ganjali, F., Eivazzadeh-Keihan, R., Maleki, A., & Sillanpää, M. (2024). Applications of hollow nanostructures in water treatment considering organic, inorganic, and bacterial pollutants. *Journal of Environmental Management*, *356*, 120670.
21. Bi, L., Chen, Y. P., Wang, C., Su, J., & Pan, G. (2020). Microalgae-derived cellulose/inorganic nanocomposite rattle-type microspheres as an advanced sensor for pollutant detection. *Chemical Engineering Journal*, *395*, 125073.
22. Yang, Y., Liao, H., Tong, Z., & Wang, C. (2015). Porous Ag/polymer composite microspheres for adsorption and catalytic degradation of organic dyes in aqueous solutions. *Composites Science and Technology*, *107*, 137–144.
23. Elgarahy, A. M., Eloffy, M. G., Guibal, E., Alghamdi, H. M., & Elwakeel, K. Z. (2023). Use of biopolymers in wastewater treatment: A brief review of current trends and prospects. *Chinese Journal of Chemical Engineering*, *64*, 292–320.
24. Kumari, H., Sonia, S., Ranga, R., Chahal, S., Devi, S., ... & Parmar, R. (2023). A review on photocatalysis used for wastewater treatment: dye degradation. *Water, Air, & Soil Pollution*, *234*(6), 349.
25. Tak, S., Grewal, S., Shreya, Phogat, P., Manisha, Jha, R., & Singh, S. (2024). Mechanistic insights and emerging trends in photocatalytic dye degradation for wastewater treatment. *Chemical Engineering & Technology*, *47*(11), e202400142.
26. Tao, Y., Du, J., Cheng, Y., Lu, J., Min, D., & Wang, H. (2023). Advances in application of cellulose—MOF composites in aquatic environmental treatment: remediation and regeneration. *International Journal of Molecular Sciences*, *24*(9), 7744.
27. Malhotra, M., Sudhaik, A., Raizada, P., Ahamad, T., Nguyen, V. H., Van Le, Q., ... & Singh, P. (2023). An overview on cellulose-supported photocatalytic materials for the efficient removal of toxic dyes. *Industrial Crops and Products*, *202*, 117000.
28. Dhadde, G. S., Mali, H. S., Raut, I. D., Nitalikar, M. M., & Bhutkar, M. A. (2021). A review on microspheres: types, method of preparation, characterization and application. *Asian Journal of Pharmacy and Technology*, *11*(2), 149–155.
29. Abdelhamid, H. N., & Mathew, A. P. (2021). Cellulose-zeolitic imidazolate frameworks (CelloZIFs) for multifunctional environmental remediation: adsorption and catalytic degradation. *Chemical Engineering Journal*, *426*, 131733.
30. Blanco Parte, F. G., Santoso, S. P., Chou, C. C., Verma, V., Wang, H. T., Ismadji, S., & Cheng, K. C. (2020). Current progress on the production, modification, and applications of bacterial cellulose. *Critical Reviews in Biotechnology*, *40*(3), 397–414.
31. Zeng, M. (2014). *Bacterial cellulose: fabrication, characterization and biocompatibility studies*. Universitat Autònoma de Barcelona.
32. Hoshi, T., Suzuki, M., Ishikawa, M., Endo, M., & Aoyagi, T. (2019). Encapsulation of micro-and milli-sized particles with a hollow-type spherical bacterial cellulose gel via particle-preloaded droplet cultivation. *International Journal of Molecular Sciences*, *20*(19), 4919.
33. Abbasi Moud, A. (2022). Cellulose through the lens of microfluidics: a review. *Applied Biosciences*, *1*(1), 1–37.
34. Zhang, Z., Lu, Y., Gao, S., & Wu, S. (2025). Sustainable and efficient wastewater treatment using cellulose-based hydrogels: a Review of heavy metal, dye, and micropollutant removal applications. *Separations*, *12*(3), 72.
35. Gorgieva, S. (2020). Bacterial cellulose as a versatile platform for research and development of biomedical materials. *Processes*, *8*(5), 624.
36. Long, F., Guo, Y., Zhang, Z., Wang, J., Ren, Y., Cheng, Y., & Xu, G. (2023). Recent progress of droplet microfluidic emulsification based synthesis of functional microparticles. *Global Challenges*, *7*(9), 2300063.

37. Fareez, I. M., Lim, S. M., Zulkefli, N. A. A., Mishra, R. K., & Ramasamy, K. (2018). Cellulose derivatives enhanced stability of alginate-based beads loaded with Lactobacillus plantarum LAB12 against low pH, high temperature and prolonged storage. *Probiotics and Antimicrobial Proteins, 10*, 543–557.
38. Zhang, S., He, H., Guan, S., Cai, B., Li, Q., & Rong, S. (2020). Bacterial cellulose-alginate composite beads as yarrowia lipolytica cell carriers for lactone production. *Molecules, 25*(4), 928.
39. Girard, V. D., Chaussé, J., & Vermette, P. (2024). Bacterial cellulose: a comprehensive review. *Journal of Applied Polymer Science, 141*(15), e55163.
40. Kołodziejczyk, A. M., Silarski, M., Kaczmarek, M., Harasymczuk, M., Dziedzic-Kocurek, K., & Uhl, T. (2025). Shielding properties of the kombucha-derived bacterial cellulose. *Cellulose, 32*(2), 1017–1033.
41. Mosaffa, E., Ramsheh, N. A., Banerjee, A., & Ghafuri, H. (2024). Bacterial cellulose microfilament biochar-architectured chitosan/polyethyleneimine beads for enhanced tetracycline and metronidazole adsorption. *International Journal of Biological Macromolecules, 273*, 132953.
42. Zhou, L., Ke, K., Yang, M. B., & Yang, W. (2021). Recent progress on chemical modification of cellulose for high mechanical-performance Poly (lactic acid)/Cellulose composite: a review. *Composites Communications, 23*, 100548.
43. Sharma, P., Hicks, S., Ruggiero, A. R., Sharma, S. K., Hsiao, B. S., & Springstead, J. (2025). Extraction and analysis of carboxycellulose nanofibers from virgin plant fibers using updated TEMPO-mediated oxidation. *Cellulose, 32*(2), 887–902.
44. Patel, A. R. (2020). Functional and engineered colloids from edible materials for emerging applications in designing the food of the future. *Advanced Functional Materials, 30*(18), 1806809.
45. Wu, J., & Andrews, M. P. (2024). Mechanical, morphological and comparative properties of microbeads assembled from carboxylated cellulose nanocrystals. *Journal of Materials Chemistry A, 12*(2), 950–960.
46. Abdallah, W., & Kamal, M. R. (2018). Influence of process variables on physical characteristics of spray freeze dried cellulose nanocrystals. *Cellulose, 25*, 5711–5730.
47. Yao, C., Shi, G., Hu, Y., Zhuo, H., Chen, Z., Peng, X., … & Liu, C. (2021). Emulsion templated advanced functional materials from emerging nano building blocks. *Journal of Materials Chemistry A, 9*(46), 25827–25851.
48. Li, J., Zhou, L., Song, Y., Yu, X., Li, X., Liu, Y., … & Zhang, J. (2021). Green fabrication of porous microspheres containing cellulose nanocrystal/MnO2 nanohybrid for efficient dye removal. *Carbohydrate Polymers, 270*, 118340.
49. Zhan, Q., Xuan, S., Su, L., Hou, Y., Jin, P., Zheng, Y., & Wu, Z. (2025). Exploring the wonders of polysaccharides in porous materials. *Materials Horizons, 12*(14), 5053–5074.
50. Abouzeid, R., Taha, M., Khiari, R., & Wu, Q. (2023). TEMPO-oxidized cellulose nanofibers/alginate nanocomposite as a promising nanocomposite material for the adsorption of cationic dyes. *Chemistry Africa, 6*(5), 2331–2342.
51. Al-Sayed, A., Dacrory, S., & Al-Shemy, M. T. (2022). Fabrication of sodium alginate/graphene oxide/nanocrystalline cellulose scaffold for methylene blue adsorption: Kinetics and thermodynamics study. *Separation and Purification Technology, 299*, 120825.
52. Ivbanikaro, A. E., Okonkwo, J. O., Sadiku, E. R., & Maepa, C. E. (2023). Recent development in the formation and surface modification of cellulose-bead nanocomposites as adsorbents for water purification: a comprehensive review. *Journal of Polymer Engineering, 43*(8), 680–714.
53. Abdelhamid, H. N. (2024). Nanocellulose-based materials for water pollutant removal: a review. *International Journal of Molecular Sciences, 25*(15), 8529.

54. Chandel, N., Jain, K., Jain, A., Raj, T., Patel, A. K., Yang, Y. H., & Bhatia, S. K. (2023). The versatile world of cellulose-based materials in healthcare: from production to applications. *Industrial Crops and Products, 201,* 116929.
55. Kendre, P. N., Lokwani, D., Pote, A., Singh, S., Jayeoye, T. J., & Prajapati, B. G. (2023). Potential technologies to develop cellulose beads and microspheres. In M. Shabbir (Ed.), *Regenerated Cellulose and Composites: Morphology-Property Relationship* (pp. 145–176).
56. Parajuli, P. (2021). *Salt and surfactant induced changes in physicochemical properties of cellulose aerogels* (Doctoral dissertation).
57. Garner, J., & Park, K. (2015). Chemically modified natural polysaccharides to form gels. *Polysaccharides, 2,*1555–1582.
58. Li, S. S., Song, Y. L., Yang, H. R., An, Q. D., Xiao, Z. Y., & Zhai, S. R. (2020). Carboxymethyl cellulose-based cryogels for efficient heavy metal capture: aluminum-mediated assembly process and sorption mechanism. *International Journal of Biological Macromolecules, 164,* 3275–3286.
59. Yang, H. R., Li, S. S., An, Q. D., Zhai, S. R., Xiao, Z. Y., & Zhang, L. P. (2021). Facile transformation of carboxymethyl cellulose beads into hollow composites for dye adsorption. *International Journal of Biological Macromolecules, 190,* 919–926.
60. Lee, J., & Patel, R. (2022). Wastewater treatment by polymeric microspheres: a review. *Polymers, 14*(9), 1890.
61. Kang, W., Ma, X., Zhao, H., Ju, J., Zhao, Y., Yan, J., & Cheng, B. (2016). Electrospun cellulose acetate/poly (vinylidene fluoride) nanofibrous membrane for polymer lithium-ion batteries. *Journal of Solid State Electrochemistry, 20,* 2791–2803.
62. Eltaweil, A. S., Abd El-Monaem, E. M., El-Subruiti, G. M., Ali, B. M., Abd El-Latif, M. M., & Omer, A. M. (2023). Graphene oxide incorporated cellulose acetate beads for efficient removal of methylene blue dye; isotherms, kinetic, mechanism and co-existing ions studies. *Journal of Porous Materials, 30*(2), 607–618.
63. Basha, I. K., Abd El-Monaem, E. M., Khalifa, R. E., Omer, A. M., & Eltaweil, A. S. (2022). Sulfonated graphene oxide impregnated cellulose acetate floated beads for adsorption of methylene blue dye: optimization using response surface methodology. *Scientific Reports, 12*(1), 9339.
64. Islam, M. D., Uddin, F. J., Rashid, T. U., & Shahruzzaman, M. (2023). Cellulose acetate-based membrane for wastewater treatment—A state-of-the-art review. *Materials Advances, 4*(18), 4054–4102.
65. Nithya, R., Vinodhini, P. A., Sudha, P. N., & Vinoth, J. (2017). Biocomposites and polymer blends for wastewater treatment. In P. N. Sudha (Ed.), *Industrial Applications of Marine Biopolymers* (pp. 473–499).CRC Press (Taylor & Francis Group), Boca Raton, FL.
66. Bunge, M. A., Ruckart, K. N., Leavesley, S., Peterson, G. W., Nguyen, N., West, K. N., & Glover, T. G. (2015). Modification of fibers with nanostructures using reactive dye chemistry. *Industrial & Engineering Chemistry Research, 54*(15), 3821–3827.
67. Plaza-Joly, P., Gallois, A., Bosc-Rouessac, F., Drobek, M., & Julbe, A. (2024). Synergistic effect of UiO-66 directly grown on kombucha-derived bacterial cellulose for dye removal. *Molecules, 29*(13), 3057.
68. Hashem, T., Ibrahim, A. H., Wöll, C., & Alkordi, M. H. (2019). Grafting zirconium-based metal–organic framework UiO-66-NH2 nanoparticles on cellulose fibers for the removal of Cr (VI) ions and methyl orange from water. *ACS Applied Nano Materials, 2*(9), 5804–5808.
69. Doyo, A. N., Kumar, R., & Barakat, M. A. (2023). Recent advances in cellulose, chitosan, and alginate based biopolymeric composites for adsorption of heavy metals from wastewater. *Journal of the Taiwan Institute of Chemical Engineers, 151,* 105095.

70. Hua, J., Meng, R., Wang, T., Gao, H., Luo, Z., Jin, Y., ... & Yao, J. (2019). Highly porous cellulose microbeads and their adsorption for methylene blue. *Fibers and Polymers, 20*, 794–803.
71. Negi, A. (2024). Cationized cellulose materials: enhancing surface adsorption properties towards synthetic and natural dyes. *Polymers, 17*(1), 36.
72. Lu, J., Zhou, Y., & Zhou, Y. (2023). Recent advance in enhanced adsorption of ionic dyes from aqueous solution: a review. *Critical Reviews in Environmental Science and Technology, 53*(19), 1709–1730.
73. Wan, Y., Liu, Z. Y., Song, P., Zhang, X. Q., Song, J. C., Fu, Y. J., ... & Shi, C. Y. (2019). Ionic liquid groups modified 3D porous cellulose microspheres for selective adsorption of AO7 dye. *Journal of Cleaner Production, 240*, 118201.
74. Omer, A. M., Elgarhy, G. S., El-Subruiti, G. M., Khalifa, R. E., & Eltaweil, A. S. (2020). Fabrication of novel iminodiacetic acid-functionalized carboxymethyl cellulose microbeads for efficient removal of cationic crystal violet dye from aqueous solutions. *International Journal of Biological Macromolecules, 148*, 1072–1083.
75. Ishak, S. A., Murshed, M. F., Md Akil, H., Ismail, N., Md Rasib, S. Z., & Al-Gheethi, A. A. S. (2020). The application of modified natural polymers in toxicant dye compounds wastewater: a review. *Water, 12*(7), 2032.
76. Georgakilas, V., Tiwari, J. N., Kemp, K. C., Perman, J. A., Bourlinos, A. B., Kim, K. S., & Zboril, R. (2016). Noncovalent functionalization of graphene and graphene oxide for energy materials, biosensing, catalytic, and biomedical applications. *Chemical Reviews, 116*(9), 5464–5519.
77. Negi, A. (2024). Cationized cellulose materials: enhancing surface adsorption properties towards synthetic and natural dyes. *Polymers, 17*(1), 36.
78. Hubbe, M. A., Azizian, S., & Douven, S. (2019). Implications of apparent pseudo-second-order adsorption kinetics onto cellulosic materials: A review. *BioResources, 14*(3), 7582–7626.
79. Li, Y., Liu, Z., Wan, X., Xie, L., Chen, H., Qu, G., ... & Zhao, S. (2023). Selective adsorption and separation of methylene blue by facily preparable xanthan gum/amantadine composites. *International Journal of Biological Macromolecules, 241*, 124640.
80. Wang, Q., Liu, S., Chen, H., Liu, J., & Zhu, Q. (2022). TEMPO-oxidized cellulose beads for cationic dye adsorption. *BioResources, 17*(4), 6056–6066.
81. Shi, X., Ji, H., Ji, X., Tian, Z., & Chen, J. (2024). Cellulose modification using sulfamic acid for adsorption of methylene blue and lead ions. *Cellulose, 31*(7), 4381–4393.
82. Lin, C. R., Ivanova, O. S., Petrov, D. A., Sokolov, A. E., Chen, Y. Z., Gerasimova, M. A., ... & Edelman, I. S. (2021). Amino-functionalized Fe3O4@ SiO2 core-shell magnetic nanoparticles for dye adsorption. *Nanomaterials, 11*(9), 2371.
83. Kong, Q., Wang, X., & Lou, T. (2020). Preparation of millimeter-sized chitosan/carboxymethyl cellulose hollow capsule and its dye adsorption properties. *Carbohydrate Polymers, 244*, 116481.
84. Wan, Y., Liu, Z. Y., Song, P., Zhang, X. Q., Song, J. C., Fu, Y. J., ... & Shi, C. Y. (2019). Ionic liquid groups modified 3D porous cellulose microspheres for selective adsorption of AO7 dye. *Journal of Cleaner Production, 240*, 118201.
85. Cheng, N., Hu, Q., Guo, Y., Wang, Y., & Yu, L. (2015). Efficient and seSafapour, S., Mazhar, M., Nikanfard, M., & Liaghat, F. (2022). Recent advancements on the functionalized cyclodextrin-based adsorbents for dye removal from aqueous solutions. *International Journal of Environmental Science and Technology, 19*(6), 5753–5790.
86. Cheng, N., Hu, Q., Guo, Y., Wang, Y., & Yu, L. (2015). Efficient and selective removal of dyes using imidazolium-based supramolecular gels. *ACS Applied Materials & Interfaces, 7*(19), 10258–10265.
87. Bulavová, P., Parmentier, J., & Slovák, V. (2018). Facile synthesis of soft-templated carbon monoliths with hierarchical porosity for fast adsorption from liquid media. *Microporous and Mesoporous Materials, 272*, 155–165.

88. Park, S., Oh, Y., Yun, J., Yoo, E., Jung, D., Oh, K. K., & Lee, S. H. (2020). Cellulose/biopolymer/Fe 3 O 4 hydrogel microbeads for dye and protein adsorption. *Cellulose, 27*, 2757–2773.
89. Li, B., Zhang, Q., Pan, Y., Li, Y., Huang, Z., Li, M., & Xiao, H. (2020). Functionalized porous magnetic cellulose/Fe3O4 beads prepared from ionic liquid for removal of dyes from aqueous solution. *International Journal of Biological Macromolecules, 163*, 309–316.
90. Afroz, S. (2022). Synthesis of Carboxylated Nanocrystalline Cellulose Reinforced Hydrogels for the Removal of Organic Pollutants. Master's Thesis, Department of Chemistry, Bangladesh University of Engineering and Technology (BUET), Dhaka, Bangladesh.
91. Zhang, S. F., Yang, M. X., Qian, L. W., Hou, C., Tang, R. H., Yang, J. F., & Wang, X. C. (2018). Design and preparation of a cellulose-based adsorbent modified by imidazolium ionic liquid functional groups and their studies on anionic dye adsorption. *Cellulose, 25*, 3557–3569.
92. Patra, N., Mandal, B., & Ghosh, S. (2017). Spectroscopic studies on the interaction of dye and surface active ionic liquid. *Industrial & Engineering Chemistry Research, 56*(36), 10044–10052.
93. Hu, Y., Long, S., Fu, H., She, Y., Xu, Z., & Yoon, J. (2021). Revisiting imidazolium receptors for the recognition of anions: highlighted research during 2010–2019. *Chemical Society Reviews, 50*(1), 589–618.
94. Godiya, C. B., Revadekar, C., Kim, J., & Park, B. J. (2022). Amine-bilayer-functionalized cellulose-chitosan composite hydrogel for the efficient uptake of hazardous metal cations and catalysis in polluted water. *Journal of Hazardous Materials, 436*, 129112.
95. Yang, Y., Chen, Z., Wu, X., Zhang, X., & Yuan, G. (2018). Nanoporous cellulose membrane doped with silver for continuous catalytic decolorization of organic dyes. *Cellulose, 25*, 2547–2558.
96. Rivas, B. L., Urbano, B. F., & Sánchez, J. (2018). Water-soluble and insoluble polymers, nanoparticles, nanocomposites and hybrids with ability to remove hazardous inorganic pollutants in water. *Frontiers in Chemistry, 6*, 320.
97. Naseem, K., Farooqi, Z. H., Begum, R., & Irfan, A. (2018). Removal of Congo red dye from aqueous medium by its catalytic reduction using sodium borohydride in the presence of various inorganic nano-catalysts: a review. *Journal of Cleaner Production, 187*, 296–307.
98. Palas, B. (2015). *Azo dye degradation by advanced oxidation methods using metal impregnated organic waste catalyst and comparatively biological oxidation* (Master's thesis, Fen Bilimleri Enstitüsü).
99. Naseem, K., Farooqi, Z. H., Begum, R., & Irfan, A. (2018). Removal of Congo red dye from aqueous medium by its catalytic reduction using sodium borohydride in the presence of various inorganic nano-catalysts: a review. *Journal of Cleaner Production, 187*, 296–307.
100. Saikia, H. (2017). *Synthesis, characterization and composition dependent catalytic activity of pd-based metal nanoparticles* (Doctoral dissertation, Tezpur University).
101. Li, J., Wu, Z., & Chen, S. (2026). Bacterial cellulose in electronic, optoelectronic, photonics, and energy applications. In F. Dourado, M. Gama, F. Hong, R. Banerjee, & M. Das (Eds.), *Advances in Bacterial Cellulose: Towards Sustainable Production and Emerging Applications* (pp. 161–191). Springer Nature Switzerland, Cham.
102. Baran, N. Y., Çalışkan, M., Özpala, A., & Baran, T. (2024). Fabrication of nano-sized Pd catalyst supported on sodium carboxymethyl cellulose/gum Arabic/sodium alginate functionalized microspheres for catalytic reduction of nitro compounds, organic dyes, K3 [Fe (CN) 6], and chromium (VI) pollutants. *International Journal of Biological Macromolecules, 262*, 130134.

103. Albukhari, S. M., Ismail, M., Akhtar, K., & Danish, E. Y. (2019). Catalytic reduction of nitrophenols and dyes using silver nanoparticles@ cellulose polymer paper for the resolution of waste water treatment challenges. *Colloids and Surfaces A: Physicochemical and Engineering Aspects, 577*, 548–561.
104. Bandi, R., Dadigala, R., Han, S. Y., Van Hai, L., Kwon, G. J., & Lee, S. H. (2024). Dicarboxylate cellulose nanofibrils-supported silver nanoparticles as a novel, green, efficient and recyclable catalyst for 4-nitrophenol and dyes reduction. *International Journal of Biological Macromolecules, 280*, 136023.
105. Wu, G., Liu, X., Zhou, P., Wang, L., Hegazy, M., Huang, X., & Huang, Y. (2019). A facile approach for the reduction of 4-nitrophenol and degradation of congo red using gold nanoparticles or laccase decorated hybrid inorganic nanoparticles/polymer-biomacromolecules vesicles. *Materials Science and Engineering: C, 94*, 524–533.
106. Ali, M., Husain, Q., Sultana, S., & Ahmad, M. (2018). Immobilization of peroxidase on polypyrrole-cellulose-graphene oxide nanocomposite via non-covalent interactions for the degradation of Reactive Blue 4 dye. *Chemosphere, 202*, 198–207.
107. Shaghaleh, H., Xu, X., & Wang, S. (2018). Current progress in production of biopolymeric materials based on cellulose, cellulose nanofibers, and cellulose derivatives. *RSC Advances, 8*(2), 825–842.
108. Xu, C., Zhang, B., Wang, Y., Shao, Q., Zhou, W., Fan, D., ... & Tratnyek, P. G. (2016). Effects of sulfidation, magnetization, and oxygenation on azo dye reduction by zerovalent iron. *Environmental Science & Technology, 50*(21), 11879–11887.
109. Zhou, Y., Shen, J., Bai, Y., Li, T., & Xue, G. (2020). Enhanced degradation of Acid Red 73 by using cellulose-based hydrogel coated Fe3O4 nanocomposite as a Fenton-like catalyst. *International Journal of Biological Macromolecules, 152*, 242–249.
110. Doyo, A. N., Kumar, R., & Barakat, M. A. (2023). Recent advances in cellulose, chitosan, and alginate based biopolymeric composites for adsorption of heavy metals from wastewater. *Journal of the Taiwan Institute of Chemical Engineers, 151*, 105095.
111. Sutar, S., Patil, P., & Jadhav, J. (2022). Recent advances in biochar technology for textile dyes wastewater remediation: a review. *Environmental Research, 209*, 112841.
112. Tatarchuk, T., Shyichuk, A., Naushad, M., Danyliuk, N., & Lapchuk, I. (2024). Copper-substituted magnetite as a Fenton-like catalyst boosted with electromagnetic heating. *Journal of Water Process Engineering, 60*, 105170.
113. Asgher, M., Noreen, S., Bilal, M., & Iqbal, H. M. N. (2017). Enhancing catalytic functionality of lignin peroxidase by immobilization on Ca-alginate beads for efficient degradation of textile dyes. *International Journal of Biological Macromolecules*, 105, 1342–1351.
114. Yu, Y., Liu, S., Pei, Y., & Luo, X. (2021). Growing Pd NPs on cellulose microspheres via in-situ reduction for catalytic decolorization of methylene blue. *International Journal of Biological Macromolecules, 166*, 1419–1428.
115. Soltani, F., Navidjouy, N., & Rahimnejad, M. (2022). A review on bio-electro-Fenton systems as environmentally friendly methods for degradation of environmental organic pollutants in wastewater. *RSC Advances, 12*(9), 5184–5213.
116. Yuan, Z., Chen, Y., Qiu, C., Li, M. C., Qi, J., de Hoop, C. F., ... & Huang, X. (2023). Simple ultrasonic integration of shapeable, rebuildable, and multifunctional MIL-53 (Fe)@ cellulose composite for remediation of aqueous contaminants. *International Journal of Biological Macromolecules, 249*, 126118.
117. Zhang, M., Xu, T., Liu, K., Zhu, L., Miao, C., Chen, T., ...& Si, C. (2025). Modulation and mechanisms of cellulose-based hydrogels for flexible sensors. *Sustainable Materials, 5*(1), e255.
118. Lau, G. E., Che Abdullah, C. A., Wan Ahmad, W. A. N., Assaw, S., & Zheng, A. L. T. (2020). Eco-friendly photocatalysts for degradation of dyes. *Catalysts, 10*(10), 1129.

119. Etafo, N. O., Bamidele, M. O., Bamisaye, A., & Alli, Y. A. (2024). Revolutionizing photocatalysis: unveiling efficient alternatives to titanium (IV) oxide and zinc oxide for comprehensive environmental remediation. *Journal of Water Process Engineering*, *62*, 105369.
120. Garusinghe, U. M., Raghuwanshi, V. S., Batchelor, W., & Garnier, G. (2018). Water resistant cellulose–titanium dioxide composites for photocatalysis. *Scientific Reports*, *8*(1), 2306.
121. Dong, H., Zeng, G., Tang, L., Fan, C., Zhang, C., He, X., & He, Y. (2015). An overview on limitations of TiO2-based particles for photocatalytic degradation of organic pollutants and the corresponding countermeasures. *Water Research*, *79*, 128–146.
122. Lim, C. W. (2020). *Synthesis And Characterisation Of Cellulose/Nitrogen Doped Titanium Dioxide (N-Tio2) Composite Using Hydrothermal Method For Sonocatalytic Degradation Of Organic Dye* (Doctoral dissertation, UTAR).
123. Sun, X., Wang, K., Shu, Y., Zou, F., Zhang, B., Sun, G., ... & Wang, X. (2017). One-pot route towards active TiO2 doped hierarchically porous cellulose: highly efficient photocatalysts for methylene blue degradation. *Materials*, *10*(4), 373.
124. Mao, J., Li, S., Huang, J., Meng, K., Chen, G., & Lai, Y. (2018). Recent advances of multifunctional cellulose-based hydrogels. In J. Zhang (Ed.), *Cellulose-Based Superabsorbent Hydrogels* (pp. 37–64). Springer International Publishing, Cham.
125. Chen, Y., Xiang, Z., Wang, D., Kang, J., & Qi, H. (2020). Effective photocatalytic degradation and physical adsorption of methylene blue using cellulose/GO/TiO 2 hydrogels. *RSC Advances*, *10*(40), 23936–23943.
126. Vernekar, M. R., Gokhale, J. S., & Lele, S. S. (2019). Azo Dye Decoloration by Fungi. In *Fungal Bioremediation* (pp. 125–172). CRC Press.
127. Costantino, F., Armirotti, A., Carzino, R., Gavioli, L., Athanassiou, A., & Fragouli, D. (2020). In situ formation of SnO2 nanoparticles on cellulose acetate fibrous membranes for the photocatalytic degradation of organic dyes. *Journal of Photochemistry and Photobiology A: Chemistry*, *398*, 112599.
128. Mikhaylov, R. V., Lavrik, P. E., Kuznetsov, V. N., & Glazkova, N. I. (2022). Spectral dependence of UV light penetration into powder TiO2 anatase. *Journal of Photochemistry and Photobiology A: Chemistry*, *430*, 113954.
129. Gan, L., Geng, A., Song, C., Xu, L., Wang, L., Fang, X., ... & Mei, C. (2020). Simultaneous removal of rhodamine B and Cr (VI) from water using cellulose carbon nanofiber incorporated with bismuth oxybromide: the effect of cellulose pyrolysis temperature on photocatalytic performance. *Environmental Research*, *185*, 109414.
130. Huang, C., Xiao, M., Cui, H., Wang, J., Cai, Y., & Ke, Y. (2023). Carboxymethyl cellulose gels immobilized Ag/AgCl-ZnO nanoparticles for improving sunlight-catalyzed antibacterial performance. *International Journal of Biological Macromolecules*, *252*, 126495.
131. Jiang, Y., Lawan, I., Zhou, W., Zhang, M., Fernando, G. F., Wang, L., & Yuan, Z. (2020). Synthesis, properties and photocatalytic activity of a semiconductor/cellulose composite for dye degradation-a review. *Cellulose*, *27*, 595–609.
132. Garusinghe, U. M., Raghuwanshi, V. S., Batchelor, W., & Garnier, G. (2018). Water resistant cellulose–titanium dioxide composites for photocatalysis. *Scientific Reports*, *8*(1), 2306.
133. Mousavi, S. M., Hashemi, S. A., Ghasemi, Y., Bahrani, S., & Savardashtaki, A. (2021). Cellulose-based nanocomposites as promising materials for photocatalytic degradation of organic pollutants: A review. *International Journal of Biological Macromolecules*, *183*, 103–121.
134. Lekshmi, K. V., Yesodharan, S., & Yesodharan, E. P. (2017). MnO2 and MnO2/TiO2 mediated, persulphate enhanced photocatalysis for the removal of indigo carmine from water. *European Chemical Bulletin*, *6*(5), 177–191.

135. Rashid, T. U., Kabir, S. F., Biswas, M. C., & Bhuiyan, M. R. (2020). Sustainable wastewater treatment via dye–surfactant interaction: a critical review. *Industrial & Engineering Chemistry Research*, *59*(21), 9719–9745.
136. Rajagopal, S., Paramasivam, B., & Muniyasamy, K. (2020). Photocatalytic removal of cationic and anionic dyes in the textile wastewater by H_2O_2 assisted TiO_2 and micro-cellulose composites. *Separation and Purification Technology*, *252*, 117444.
137. Du, M., Du, Y., Feng, Y., Li, Z., Wang, J., Jiang, N., & Liu, Y. (2019). Advanced photocatalytic performance of novel BiOBr/BiOI/cellulose composites for the removal of organic pollutant. *Cellulose*, *26*, 5543–5557.
138. Fouad, F. A., Youssef, D. G., Shahat, F. M., & Abd El-Ghany, M. N. (2023). Role of microorganisms in biodegradation of pollutants. In G. A. M. Ali & A. S. H. Makhlouf (Eds.), *Handbook of Biodegradable Materials* (pp. 221–260). Springer International Publishing, Cham.
139. Rajendran, S., Kalairaj, A., & Senthilvelan, T. (2024). A comprehensive review on enzymatic decolorization of various azo dyes using laccase for the abatement of industrial pollution. *Biomass Conversion and Biorefinery*, *14*, 1–23.
140. Gupta, V., Capalash, N., & Sharma, P. (2015). Laccases and their role in bioremediation of industrial effluents. In R. Chandra (Ed.), *Advances in Biodegradation and Bioremediation of Industrial Waste* (pp. 111–140). CRC Press (Taylor & Francis Group), Boca Raton, FL.
141. Bilal, M., Rasheed, T., Iqbal, H. M., & Yan, Y. (2018). Peroxidases-assisted removal of environmentally-related hazardous pollutants with reference to the reaction mechanisms of industrial dyes. *Science of the Total Environment*, *644*, 1–13.
142. Chen, L., Zou, M., & Hong, F. F. (2015). Evaluation of fungal laccase immobilized on natural nanostructured bacterial cellulose. *Frontiers in Microbiology*, *6*, 1245.
143. Zhumabekova, A. (2023). *Immobilization of Horseradish Peroxidase Enzyme onto Polymeric Microbeads and Its Usage in Dye Decolorization* (Master's thesis, Bursa Uludag University (Turkey)).
144. Satar, R. (2010). Applications of Soluble and Immobilized Plant Peroxidases in the Decolorization and Remediation of Textile Dyes from Polluted Water (Doctoral dissertation). Aligarh Muslim University, Aligarh, Uttar Pradesh, India.
145. Wu, J. C. Y., Hutchings, C. H., Lindsay, M. J., Werner, C. J., & Bundy, B. C. (2015). Enhanced enzyme stability through site-directed covalent immobilization. *Journal of Biotechnology*, *193*, 83–90.
146. Mota, L. S. O., de Oliveira, P. C. O., Peixoto, B. S., & de Moraes, M. C. (2023). Enzyme-coated biochar as a sustainable solution for water and wastewater treatment. *Environmental Science: Water Research & Technology*, *9*(11), 2772–2786.
147. Kusumlata, Ambade, B., Kumar, A., & Gautam, S. (2024). Sustainable solutions: reviewing the future of textile dye contaminant removal with emerging biological treatments. *Limnological Review*, *24*(2), 126–149.
148. Basak, B., Kumar, R., & Jeon, B. H. (2023). Immobilization of microbes and enzymes for textile wastewater treatment. In S. P. Govindwar, M. B. Kurade, B. H. Jeon, & A. Pandey (Eds.), *Current Developments in Bioengineering and Biotechnology: Advances in Eco-friendly and Sustainability Technologies for the Treatment of Textile Wastewater* (pp. 627–648). Elsevier, Amsterdam.
149. Zamel, D., Hassanin, A. H., Ellethy, R., Singer, G., & Abdelmoneim, A. (2019). Novel bacteria-immobilized cellulose acetate/poly (ethylene oxide) nanofibrous membrane for wastewater treatment. *Scientific Reports*, *9*(1), 18994.
150. Saha, P., & Bhaskara Rao, K. V. (2021). Immobilization as a powerful bioremediation tool for abatement of dye pollution: a review. *Environmental Reviews*, *29*(2), 277–299.
151. Zdarta, J., Meyer, A. S., Jesionowski, T., & Pinelo, M. (2018). A general overview of support materials for enzyme immobilization: Characteristics, properties, practical utility, and recent trends. *Catalysts*, *8*(2), 92.

152. Okeke, C. B., Chukwu, K. O., Ndukwe, J. K., Okechukwu, U. S., Onwosi, C. O., & Odibo, F. J. C. (2022). Role of microbial biofilms in dye degradation during textile wastewater treatment. In A. Khadir & S. S. Muthu (Eds.), *Biological Approaches in Dye-Containing Wastewater*: Volume 2 (pp. 57–85). Springer Singapore, Singapore.
153. Hashempour, Y., Mortezazadeh, F., Rezaei, S., Salehipour, M., Gholami-Borujeni, F., Ebrahimnejad, P., & Mogharabi-Manzari, M. (2025). Co-immobilization of laccase and zinc oxide nanoparticles onto bacterial cellulose to achieve synergistic effect of photo and enzymatic catalysis for biodegradation of favipiravir. *International Journal of Biological Macromolecules*, *292*, 139288.
154. Onaizi, S. A. (2018). Enzymatic removal of protein fouling from self-assembled cellulosic nanofilms: experimental and modeling studies. *European Biophysics Journal*, *47*(8), 951–960.
155. Maier, R. M., & Gentry, T. J. (2015). Microorganisms and organic pollutants. In I. L. Pepper, C. P. Gerba, & T. J. Gentry (Eds.), *Environmental Microbiology* (3rd ed., pp. 377–413). Academic Press (Elsevier), San Diego, CA.
156. Abouzeid, R. E., Khiari, R., El-Wakil, N., & Dufresne, A. (2018). Current state and new trends in the use of cellulose nanomaterials for wastewater treatment. *Biomacromolecules*, *20*(2), 573–597.
157. Kabir, S. F., Sikdar, P. P., Haque, B., Bhuiyan, M. R., Ali, A., & Islam, M. N. (2018). Cellulose-based hydrogel materials: chemistry, properties and their prospective applications. *Progress in Biomaterials*, *7*, 153–174.
158. Zhang, Z., Lu, Y., Gao, S., & Wu, S. (2025). Sustainable and efficient wastewater treatment using cellulose-based hydrogels: a review of heavy metal, dye, and micropollutant removal applications. *Separations*, *12*(3), 72.
159. Song, Y., Jiang, M., Zhou, L., Yang, H., & Zhang, J. (2022). Rapidly regenerated CNC/TiO2/MnO2 porous microspheres for high-efficient dye removal. *Carbohydrate Polymers*, *292*, 119644.
160. Vernekar, M. R., Gokhale, J. S., & Lele, S. S. (2019). Azo dye decoloration by fungi. In A. Tomasini & H. Hugo León-Santiesteban (Eds.), *Fungal Bioremediation: Fundamentals and Applications* (pp. 125–172). CRC Press, Boca Raton, FL.
161. Yagub, M. T., Sen, T. K., Afroze, S., & Ang, H. M. (2014). Dye and its removal from aqueous solution by adsorption: a review. *Advances in Colloid and Interface Science*, *209*, 172–184.
162. Al Rubai, H. F., Sultan, M. S., & Abed, G. M. (2024). Enhancing the removal efficiency of Azo dye using sodium borohydride in the presence of inorganic nano catalysts. *Green Analytical Chemistry*, *2*, 100194.
163. Liu, X., Inagaki, S., & Gong, J. (2016). Heterogeneous molecular systems for photocatalytic CO_2 reduction with water oxidation. *Angewandte Chemie International Edition*, *55*(48), 14924–14950.
164. Gupta, R., & Singh, R. P. (2019). Treatment and Recycling of Wastewater from Textile Industry. In R. L. Singh & R. P. Singh (Eds.), *Advances in Biological Treatment of Industrial Waste Water and Their Recycling for a Sustainable Future* (pp. 225–266). Springer Nature Singapore, Singapore.
165. Kurade, M. B., Waghmode, T. R., Xiong, J. Q., Govindwar, S. P., & Jeon, B. H. (2019). Decolorization of textile industry effluent using immobilized consortium cells in upflow fixed bed reactor. *Journal of Cleaner Production*, *213*, 884–891.
166. Ashori, A., Chiani, E., Shokrollahzadeh, S., Madadi, M., Sun, F., & Zhang, X. (2024). Cellulose-Based aerogels for sustainable dye removal: advances and prospects. *Journal of Polymers and the Environment*, *32*(12), 6149–6181.
167. Yaseen, D. A., & Scholz, M. (2019). Textile dye wastewater characteristics and constituents of synthetic effluents: a critical review. *International Journal of Environmental Science and Technology*, *16*, 1193–1226.

168. Aguado, R. J., Mazega, A., Tarrés, Q., & Delgado-Aguilar, M. (2023). The role of electrostatic interactions of anionic and cationic cellulose derivatives for industrial applications: a critical review. *Industrial Crops and Products*, *201*, 116898.
169. Wang, T., Wang, W., Hu, C., Zheng, J., Zhu, Z., & Liu, B. (2024). Design of carboxymethyl cellulose/alginate aerogels with anti-fouling and light-driven self-cleaning for enhanced oily wastewater remediation. *Carbohydrate Polymers*, *342*, 122358.
170. Malhotra, M., Sudhaik, A., Raizada, P., Ahamad, T., Nguyen, V. H., Van Le, Q., … & Singh, P. (2023). An overview on cellulose-supported photocatalytic materials for the efficient removal of toxic dyes. *Industrial Crops and Products*, *202*, 117000.
171. Goswami, S., Dutta, D., Pandey, S., Chattopadhyay, P., Dubey, R., & Tiwari, D. (2024). Novel fibrous Ag (NP) decorated clay-polymer composite: implications in water purification contaminated with predominant micro-pollutants and bacteria. *Journal of Environmental Management*, *359*, 121063.
172. Akhai, S. (2023). Biogas plants for sustainable municipal waste management: A brief review. In P. Srivastava, A. K. Bedyal, D. Ramteke, M. Gupta, & J. K. Sandhu (Eds.), *Handbook of Research on Safe Disposal Methods of Municipal Solid Wastes for a Sustainable Environment* (pp. 162–179). IGI Global Scientific Publishing.
173. Akhai, S. (2024). Trends and Environmental Impact of Paper Consumption: A Prognostic Scenario for the Indian Market by 2030—A Case Study. In *Proceedings of the International Conference on Interdisciplinary Approaches in Civil Engineering for Sustainable Development (IACESD 2023), Bengaluru, Karnataka, India* (Vol. 464, pp. 11–18). Lecture Notes in Civil Engineering. Springer Nature Singapore.
174. Yadav, V., Singh, S., Garg, M. P., & Akhai, S. (2024). Effect of natural Sapindus mukorossi treatment process on bio-waste bagasse fibers for biocomposite fabrication and application purposes. *Industrial Crops and Products*, *219*, 119066.
175. Khang, A., & Akhai, S. (2025). E-waste and lithium-ion battery recycling insights for sustainable transportation. In A. Khang (Ed.), *Driving Green Transportation System Through Artificial Intelligence and Automation: Approaches, Technologies and Applications* (pp. 203–230). Springer Nature Switzerland, Cham.
176. Akhai, S., & Khang, A. (2024). Efficient hospital waste treatment and management through IoT and bioelectronics. In A. Khang (Ed.), *Revolutionizing Automated Waste Treatment Systems: IoT and Bioelectronics* (pp. 126–140). IGI Global Scientific Publishing.

14 Pharmaceutical Applications of Cellulose Microsphere

Tahreem Fatima, Shumaila Kiran, Rabia Azmat, Muhammad Sakander Majid, Maryam Jameel, and Muhammad Umar Ijaz

14.1 INTRODUCTION

The inception of microspheres as targeted drug delivery system (DDS) has helped circumvent issues such as non-target distribution, drug instability, poor bioavailability, and uncontrolled release (Carvalho et al., 2021). It is a multi-particulate delivery system rather than a monolithic dosage form, offering protection to active pharmaceutical ingredients against fluctuating physiological conditions such as oxidation, temperature, and pH variations (Lengyel et al., 2019). The physicochemical, functional, and thermal properties of the polymer matrix in this excipient determine the precise release of the desired amount of active drug and ensure the compound's protection both pre- and post-administration. Biopolymers are widely employed as matrix-forming agents due to their biocompatibility, biodegradability, and ability to modulate drug-release profile offering inherent advantages over conventional drug-delivery routes (Opriş et al., 2024).

Conventional drug-delivery methods (powders, tablets, capsules, and ointments) have certain limitations such as need for frequent drug administration due to short drug cycle time, risk of potential degradation due to fluctuated physiological condition (stomach acidity, ambient temperature, enzymatic deterioration) (Adepu & Ramakrishna, 2021). Additionally, stability of some potent pharmaceuticals can be influenced by environmental factors such as light, moisture, pH, oxidation, and temperature. These drugs may possess undesirable organoleptic properties (e.g., unpleasant taste or smell) affecting patient compliance and may cause side effects such as local irritation or injury due to high drug concentrations (Ezike et al., 2023). In order to overcome these impediments, drug-delivery carriers are employed to transport the active pharmaceutical ingredients encapsulated within a non-drug component named as an excipient that releases the drug in a precise area at a specific time and concentration in response to external stimuli (Torchilin, 2014). Target-specific drug delivery and controlled drug release can be achieved with drug carriers. Nevertheless, the most problematic shortcomings of effective drug carriers are low solubility, limited biocompatibility, fast clearance rate from the body, biodegradability, and non-specific toxicity. Thus desired-controlled drug release is investigated using responsive polymeric carriers (Cassano et al., 2018)

 DOI: 10.1201/9781003596127-16

Cellulose offers a safer, targeted, and sustainable excipient option due to its smart formulation potential, natural origin, adaptability, stimuli responsiveness, and proven efficacy in controlled drug delivery (Carvalho et al., 2021). Pharmaceutical DDSs are designed to achieve a prolonged therapeutic effect by delivering the desired amount of drug to a specific site in the body and maintaining a continuous release over an extended period (Adepu & Ramakrishna, 2021).

14.2 FUNDAMENTALS OF CELLULOSE MICROSPHERES

Cellulose is a polysaccharide valued for its high stability, hydrophilicity, biodegradability, and nontoxic nature. Polysaccharides are naturally abundant and cost-effective; when used as excipients, they enable drug-loaded nanostructures to extend drug retention time and enhance the absorption of the intended drug (Opriş et al., 2024). Polysaccharides possess various functional groups on their backbone such as hydrophilic segments providing a modifiable scaffold for drug-delivery applications.

Cellulose is a linear chain polymer made out of glucose subunits that are linked by acetal linkages between carbons C1 and C4 (β-1,4 linkages) of glucopyranose rings. The cellulose chains themselves are unbranched and are packed side by side in a parallel manner. These chains are linked by a H bridge between the H atoms and OH groups of glucose monomers, forming microfibrils that assemble to form large units (microfibril bundles) (McNamara et al., 2015). Fibrils are arranged in a highly organized and crystalline fashion within the plant cell wall, giving it characteristic stability and mechanical properties such as strength, aqueous insolubility, and fibrous texture (Seddiqi et al., 2021).

Cellulose, in particular, contains hydroxyl groups positioned in the ring plane, enabling non-covalent interactions with biological tissues, especially epithelial and mucosal membranes (Belgacem & Gandini, 2011). The highly donor-reactive OH groups in the molecular structure of cellulose impart intrinsic properties, including hydrophilicity, chemical modifiability, adhesion, chirality, and degradability (Hamedi et al., 2025). Cellulose and its derivatives are widely employed as biopolymeric coating systems due to their favorable therapeutic properties. Cellulose can be readily modified through the reactive hydroxyl groups by introducing various functional groups, leading to derivatives such as cellulose ethers, cellulose esters, or through depolymerization processes (e.g., to produce nanocellulose or oligosaccharides) (Belgacem & Gandini, 2011).

Starch and cellulose derivatives differ in their structure, chemical composition, and origin. The cellulose may be extracted from plants as well as from some microorganisms by applying various extraction, purification, and separation techniques. These final cellulose products are specified based on their molar mass, monosaccharide constitution, and total sugar concentration (Belgacem & Gandini, 2011). Cellulose and its derivative polymers like ethers, esters, and oxycellulose are widely used in the preparation of DDS excipients. The cellulose ether is a fundamental component of many nano and micro drug-delivery structures used for both pharmaceutical and reservoir drug systems. Most frequently used cellulose ethers are ethylcellulose (EC), methylcellulose (MC), sodium carboxymethyl cellulose (CMC), hydroxypropyl cellulose (HPC), and hydroxypropyl methylcellulose (HPMC) (Shokri & Adibkia, 2013)

TABLE 14.1
Various formulation methods and release rates based on different physiochemical properties depend on cellulose derivatives

Cellulose Derivative	Formulation Method	Release Kinetics	Therapeutic Function	Researched Molecule (Drug/Agent)	Key Findings	References
Methylcellulose (MC)	Emulsion crosslinking	Sustained release	Ocular drug delivery	Timolol maleate	Improved bioavailability & prolonged release	El-Kamel (2002), Gupta et al. (2009); Ismail et al. (2012)
Ethylcellulose (EC)	Solvent evaporation	Controlled release	Oral sustained release	Ibuprofen, Naproxen sodium	Enhanced gastric retention & reduced dosing frequency	Khan and Zhu (1998), Khan et al. (2010), Sarkar et al. (2011)
Hydroxypropyl methylcellulose (HPMC)	Spray drying	pH-dependent	Colon-targeted delivery	5-Fluorouracil	Improved tumor suppression in colorectal cancer	Ma et al. (2021), Othman et al. (2020)
Carboxymethyl cellulose (CMC)	Lonotropic gelation	Burst + sustained	Wound healing	Silver nanoparticles	Antibacterial & enhanced tissue regeneration	Alavi and Varma (2021), Capanema et al. (2023)
Hydroxyethyl cellulose	Freeze drying	Diffusion-controlled	Topical drug delivery	Naringenin, isoliquiritigenin	Enhanced skin permeation & anti-inflammatory effects	Kong et al. (2016), Marwah et al. (2022)
Sodium carboxymethyl cellulose (Na-CMC)	Matrix tableting or Direct Compression	Sustained release	Protein delivery	Insulin	Enhanced oral bioavailability	Marschütz et al. (2000), Marschütz and Bernkop-Schnürch (2000)
Oxidized cellulose	In-situ gelation	Rapid release	Hemostatic agent	Tranexamic acid	Effective in bleeding control	Sammartino et al. (2012), Zhao et al. (2022)
Cellulose acetate phthalate (CAP)	Coacervation	pH Triggered	Gastro-resistant delivery	Diclofenac	Reduced gastric irritation	Khuroo et al. (2021), Rahman et al. (1970)
Hydroxypropyl cellulose (HPC)	Thermal gelation	Thermo-responsive	Injectable implants	Paclitaxel	Localized cancer therapy	Bagheri et al. (2014)
Hydroxyethyl cellulose	Emulsification	Sustained + targeted	Gene delivery	siRNA	Improved transfection efficiency	Park et al. (2023)

(see Table 14.1). During drug delivery, when their excipient enters the human body, the water from the biological system moves into the excipient and develops a hydrogel barrier around the drug core. This hydrogel facilitates water penetration into the polymer matrix, releasing the drug molecules (Lukova et al., 2023).

Cellulose forms the extracellular matrix of the cell wall. In plant cells, these cellulose chains are arranged into a cable-like paracrystalline forms (micro- and macrofibrils of cellulose) embedded in the polysaccharide and glycoprotein matrix. Ordered cellulose microfibrils produced by some bacteria, notably by the species *Gluconacetobacter* have also been reported. The biofilm is a sessile community of microbes, which are associated with an extracellular matrix made up of nucleic acids, proteins, and polysaccharides. Because of their ability to adhere to the surfaces and resist antimicrobial therapy, biofilm microbes cause many nosocomial infections. Eukaryote-associated cellulose is arranged in a paracrystalline structure of a single glucan chain that forms cellullose micro- and macrofibrils (McNamara et al., 2015).

14.3 DRUG-DELIVERY APPLICATIONS OF CELLULOSE MICROSPHERES

14.3.1 Drug Encapsulation and Release Kinetics

The advancement in drug-release controlling methods has become very popular over the past few years. The incorporation of drug substances into microcarriers formed by polymers has been described as an effective method to develop drug-delivery systems (DDS) that facilitate drug release at a specific rate and prolong. Controlled release of drug-containing microparticles also promotes stability in drugs, particularly in molecules with susceptibility to fast degradation in the human body. These drugs are mostly characterized by a high decrease in the number of uses and improved acceptance by patients (Basinska et al., 2021). The microcarrier drug features are mostly based on the materials and principles behind how they are formulated. The material used in forming the microparticles should have the ability to impart certain features to the final product and ensure drug substance entry into it (Paccione et al., 2022). The active substance should have high stability, drug release should be controllable, and targeting the drug to its target area in the human body should be feasible. The key limitations to develop them include maintaining stability, getting them with a fine particle size, maintaining drug-release rate and site control, and scaling-up (Yadav et al., 2022).

Cellulose and its associated polymers have been commonly utilized for DDS development in oral, nasal, ophthalmic, and dermal applications because of their intrinsic tolerance, nontoxicity, mucoadhesive ability, possesses a high encapsulating ability and low price. Starch and cellulose polymers can be readily formed by the emulsification technique, which makes it a promising candidate for drug microcarriers. Sphere-shaped cellulose microparticles can be manufactured via emulsification (Druel et al., 2020), spray-assisted technique (Guarino et al., 2007; F. Zhang et al., 2016), layer by layer and coacervation techniques (Paulraj et al., 2018; F. J. Wang et al., 2015), etc. (Lukova et al., 2023). These manufactured cellulose products such as CMC (Liu et al., 2017), CA (Simões et al., 2020; M. Zhang et al., 2020), and EC (Yeap et al., 2017), along with pristine plant and bacterial nanocellulose (Yu et al., 2016) and microcrystalline cellulose (MCC) (Druel et al., 2020) have been extensively studies. The microparticles are preferably formed using cellulose polymers due to their favorable solubility characteristics in appropriate aqueous or organic solvents, which results in simpler processability (Carvalho et al., 2021). The emulsion techniques commonly use CA, cellulose acetate butyrate, cellulose acetate phthalate (CAP), CMC, or EC, due to their solubility. Nevertheless, BNC (W. Zhang et al., 2019) and CNFs (Druel et al., 2020), native vegetable cellulose dissolved in NaOH/urea aqueous solutions (M. Zhang et al., 2020), and microcrystalline cellulose dissolved in ionic liquids (Lin & Jana, 2021) have been reported in microplastic production using emulsions.

Polysaccharides tend to possess high molecular weight and can form several inter- and intra-molecular associations because of free hydroxyl groups. Therefore, they also have high potential to increase the medium's viscosity and its gelling. Such starch and cellulose characteristics are vital prerequisites for particles formation when making drug-loaded microstructure and for the mechanism by virtue of which swelling/degradation of the polymer matrix/shell delivers the incorporated active ingredient (Lukova et al., 2023a). Modified release of incorporated drug substances is an attainable and desirable property of starch- and cellulose-based microparticle DDSs. The rate of drug release from microparticles depends primarily on the matrix structure, polymers, and drug substances. The drugs release time is modified to achieve their availability over an extended duration at a constant rate. However, the API release is a complicated procedure due to its initial rapid release from the external region followed by diffusion release and eventually degradation of polymer excipient. Starch and cellulose derivatives can be easily functionalized to achieve long drug retention by performing a process of crosslinking or polyelectrolyte complex formation and matrix structures establishment (Lukova et al., 2023). The DDSs classified on the bases of drug-release control mechanism include dissolution, diffusion, water penetration (osmotic pressure-controlled and swelling-controlled), chemically controlled, and nanoparticle-based systems (Adepu & Ramakrishna, 2021).

In diffusion-controlled release systems, drugs are encapsulated and delivered by diffusion through inert water-insoluble polymeric membranes (reservoir systems) or polymeric matrixes (monolithic systems) . Drug molecules also pass through holes formed during particle formation. The drug is released by diffusion when the outside surface exposed to the solution dissolves first, and drugs diffuse out of the matrix. For instance, HPMC is the most crucial hydrophilic carrier material utilized for oral delivery DDSs (Tundisi et al., 2021). Among its most critical features is high swellability, an aspect that has a remarkable influence on an incorporated drug's release kinetics. When exposed to water or biological fluid, the latter diffuses inside the microcapsule, and polymer chains relax, increasing the excipient volume. Subsequently, incorporated drug diffuses out of the system (Siepmann & Peppas, 2012).

Drug encapsulation is referred to as a sustained or delayed drug-release mechanism due to slow degradation of polymer matrix. There are other release mechanisms and microparticle formulation methods, which are dependent on the physiochemical properties of cellulose derivatives as evident in Table 14.1.

14.3.2 Targeted Drug-Delivery Strategies

14.3.2.1 Functionalization for Site-Specific Drug Release

Stimuli–responsive drug delivery has provided immense possibilities for on-demand drug discharge. They use external (light, temperature) and internal biological (ambient temperature, pH, enzyme concentration) stimuli to customize drug delivery and reduce active drug side effects (Liu et al., 2016). These stimuli have been utilized in creating several passive targeting methods, only utilizing unique ambient conditions existing specific to the targeted organ. Additionally, characteristic properties like polymer–drug compatibility, nontoxicity, facile processability, and bio-affinity make celluloses and its hybrid a suitable candidate for endogenous stimuli–responsive DDS (Gunathilake et al., 2020).

Among all stimuli–response DDS, temperature-responsive DDS provide an excellent benefit over traditional DDS with its phase transition temperature adjustable by its environment, in phase transition and passive targeting DD. So, thermo-responsive DDS are primarily studied in stimulus–responsive strategies for anti-cancer , antibiotic , or anti-inflammatory (Wu et al., 2016) drug delivery. So far, various materials with thermo-sensitivity have been created for temperature-sensitive drug release such as poly (lactic-co-glycolic acid) (PLGA) and poly (*N*-isopropylacrylamide) (poly-NIPAAm) and its derivatives (Kang et al., 2016). These thermo-sensitive hydrogels have been limited to clinical applications because of their limitations such as non-biodegradability, carcinogenic and teratogenic effects (Ullah et al., 2015). Celluloses derived from bacteria are highly pure and exhibit similar characteristics to those obtained from other sources. CNCs have very high mechanical characteristics such as high tensile strength, stiffness, and surface area. Moreover, high reactive OH availability facilitates CNCs' chemical modifications, generating functionalized CNCs to achieve a wide variety of different applications. For instance, PNIPAAm–CNC hybrid hydrogels were prepared by grafting PNIPAAm with cellulose nanocrystal via the free-radical polymerization method (Zubik et al., 2017).

Although microspheres made of polymers are promising thermostable therapy due to their fabricability but their size and poor microsphere distribution, and low drug encapsulation pose serious challenges. Thus, it is still difficult to resolve polymer–drug compatibility, controllable microsphere size, and high drug encapsulation efficiency. Meantime, high interfacial adhesion between carboxylated CNCs and PEG impregnation enhanced drug–polymer compatibility. This strategy achieved regulated on-demand drug release from encapsulation. Such thermo-responsive composite microspheres with controlled microsphere size and high encapsulation rate have great potential to develop on-demand and sophisticated controlled-release DDS (Yang et al., 2023).

Polymers with pH-responsive nature can be defined as polyelectrolytes connected with functional groups that result in acidic or basic characteristics when subjected to variation in environmental pH. The variation in pH affects the ionization of these polyelectrolyte's acidic and basic groups, although complete ionization is restrained due to the influence exerted by nearby ionized groups (Bayer, 2023). The resulting anions or cations create swelling in polymer chains by electrostatic repulsions. This transformation causes structural changes in the polymer, regulating its hydrophilicity, structural conformation, and water solubility in the environment. pH-responsive hyperbranched polymers-functionalized CNCs have been prepared by direct anionic polymerization with surface OH groups on CNCs as an initiator and glycidol as a monomer molecule (Wan et al., 2019). These polymers are potential drug carriers by the oral route of drug delivery where human physiology in different parts of the human body experiences extreme variations in pH. pH-responsive drug delivery has focused on cellulose derivatives for a long time. In this area, CMC, cellulose acetate, and EC are some of the earliest, most successful, and extensively studied polymers. It has been used as an oral drug-delivery agent to deliver drugs like antibiotics (Sood et al., 2017), anti-cancer and antiviral (Joshy et al., 2017) drugs, and drugs for wound healing using hybrid drug-delivery devices. The presence of carboxyl groups gives pH sensitivity to CMC, making it an ideal choice for pH-triggered oral DDSs.

Oral drug-delivery route demands a microsphere excipient that can resist gastric acidity and prevent enzymatic degradation. CAP is a pH-sensitive cellulose derivative that remains intact and stable in the stomach (pH 1–3); however, it dissolves in the ileocolonic region (pH ≥ 6.8); thereby, it is a useful drug carrier in inflammatory bowel disease. Widely used Eudragit L100/S100 coating has a limitation due to variable luminal pH in the colon (pH 6.0 in the ascending colon vs. pH 7.0 in the terminal ileum). The polymer amalgamation and time release mechanism integration with pH sensitivity improves reliability, ensuring drug release even at a pH of 6.8. CAP bypasses the gastric and small intestinal environment to deliver drugs like mesalamine directly to the inflamed colonic mucosa (Béchard et al., 1995).

Magnetic drug delivery via particulate carriers is a new method for desired drug delivery at localized regions like tumor sites. This method allows precise clinically required drug-concentration delivery near target site reservoirs by an external magnetic field. In this regard, magnetically attractive materials like iron, cobalt, nickel, ferrous oxides, and ferrites are utilized because they have a unique property of magnetism, high aqueous phase stability, and suitable biocompatibility features. Cellulose is used as a backbone chain for polymers and also as sites for immobilizing magnetically responsive nanoparticles via H-bonding or electrostatic forces (Lin et al., 2021). It inhibits the clumping of magnetically responsive nanoparticles and forms a stable and water-soluble colloidal dispersion (Aguilera et al., 2019). These nanoparticles are dispersed in celluloses or their polymers by chain surface polymerization (Butun et al., 2011), precipitation, sorption (Tade et al., 2018), or as cellulose covered magnetically responsive nanoparticles. Among cellulose derivatives, CMC is an ideal coating agent for magnetically responsive nanoparticles DDS. These magnetically responsive nanoparticles evade opsonization due to their hydrophilic surface coating, resulting in slower clearance by immune cells. To circumvent this, CMC has been utilized as coating agents for magnetically responsive nanoparticles in earlier research. Additionally, integration with CMC avoids magnetically responsive nanoparticles' aggregations and prevents their oxidization on the surface. A carboxylite group of CMCs perform an additional functionalization by immobilizing drug molecules and pH sensitivity to magnetically responsive nanoparticle DDSs (Aguilera et al., 2019). Hydroxyethyl cellulose magnetically responsive iron oxide nanoparticles were utilized for cancer drug delivery like doxorubicin hydrochloride (Bekaroğlu et al., 2017). Some other commonly used drug-delivery strategies are mentioned in Table 14.2.

TABLE 14.2
Targeted drug-delivery strategies using cellulose-based systems: mechanisms, examples, and applications

Strategy	Mechanism	Example	Application	References
Receptor-mediated	Ligand-receptor binding, endocytosis	HPC + folic acid	Cancer targeting	Metaxa et al. (2014)
Magnetic guidance	External magnetic field directs carriers	Biopolymer-coated iron oxide nanoparticles	Tumor localization	Stanicki et al. (2022), Vangijzegem et al. (2019)
Antibody conjugation	Antibody-antigen interaction	HER2-targeted antibody drug conjugates	Breast cancer therapy	Najjar et al. (2022)
Biochemical	Matrix-mediated clot stabilization	Polymeric hemostat (Oxidized cellulose)	Wound hemostasis	Ghimire et al. (2021)

14.4 NANOCELLULOSE-EMBEDDED MICROSPHERES FOR ADVANCED THERAPY

Nanocellulose-embedded microsphere is an advanced strategy, which combines the metallic and polymeric nanoparticles for synergistic therapeutic effects. These hybrid systems combine the exceptional properties of nanocellulose (NC) and microspheres to deliver the nanoparticles in a controlled manner. Silver (Wang et al., 2021), and iron oxide are commonly integrated with nanoparticles that are combined with polymeric nanoparticles (PLGA, chitosan) within nanocellulose microspheres (NCMs) offering targeted drug delivery, photothermal therapy, magnetic hyperthermia, and imaging guided treatment. Silver nanoparticle-loaded NCM slowly releases the Ag^+ ions on the wound site, ensuring prolonged antibacterial action while minimizing cytotoxicity to neighboring tissues (Jutakridsada et al., 2023). Gold nanoparticles embedded in NC microsphere exhibit photothermal therapy (Singh et al., 2022).

Nanocellulose is a rising material in biomedical diagnostics owing to its low density, biocompatibility, sustainable biopolymer, and eco-friendly properties. Different natural and synthetic cellulose derivatives have been used particularly in bone tissue engineering (BTE) (Singh et al., 2022). These nanocellulose polymer nanocomposites (PNCs) possess mechanical, thermal, compactness attributes. The injectable antibacterial hydrogel is used in wound healing through embedding antibacterial silver nanoclusters (NCs) within chitosan (CH) hydrogel to eradicate bacterial infection and promote hair follicle/tissue regeneration. The injectable hydrogel demonstrated wide-spectrum properties like bactericidal efficacy and hair follicle/tissue regeneration function overcoming traditional disadvantages of skin tissue engineering and wound healings (Wang et al., 2021).

Metal nanoparticles, such as silver nanoparticles (AgNs), have great antibacterial activity; however, silver is a toxic chemical for normal cells and tissues. Therefore, their high mobility and silver ion release threat restrict its biomedical consumption for wound healing. Ag^+ ions were reduced to Ag nanoparticles (AgNPs) via a TA/Fe coating on cellulose acetate/sodium polyacrylate (CA/SP), forming novel litchi-like microspheres that demonstrated excellent biocompatibility. During the bacterial cellulose (BC) application, growth and reproduction of *Staphylococcus aureus* and *Pseudomonas aeruginosa* were entirely suppressed. The hemostatic function was significantly enhanced owing to the material's strong coagulation activity, reduced bleeding time, and lower hemostatic volume in the femoral artery. The TA/Fe-modified microspheres serve as effective carriers for AgNs, enabling efficient bacteriostasis, rapid hemostasis, and accelerated wound healing (Wang et al., 2023).

14.5 CELLULOSE MICROSPHERES IN REGENERATIVE MEDICINE AND TISSUE ENGINEERING

The nanocellulosic pellicle is a critical biomaterial used in 3D bioprinting, playing a pivotal role in tissue engineering. It contains a highly entangled gel network that must be partially modified to enhance its processability and extrusion efficiency. Nanocellulose-interacting polymers and nanoparticles are reinforced to regulate their mechanical and biological features for direct 3D bioprinting. For instance, SCOBY (symbiotic culture of bacteria and yeast-derived nanocellulose-chitosan bioink mixed with kaolin nanoparticles and screw-driven handheld extrusion bioprinter has shown a promising platform for layer-by-layer regeneration of intricate tissues with homogeneous distribution of cells and particles with high cell viability (Bhattacharyya et al., 2024).

Naturally occurring cellulose-based biomaterials are widely used in regenerative medicine and bone engineering. The conventional 3D network structure of cellulose allows its interaction with biological substitutes in host tissues without inducing toxicity and adverse immune reactions. Cellulose-based microspheres demonstrate hemocompatibility and aid in repairing damaged bone tissues (Ji et al., 2022). The COL/BC/BMP-2 porous microspheres are produced through Malaprade and Schiff base reactions using BC microspheres. These microsphere-based experiments showed improved adhesion, proliferation, and osteogenic differentiation in mouse cells. In another case, BC are utilized to form a porous silk fibroin (SF)/BC/MXene ($Ti_3C_2T_x$) scaffold that facilitates rapid hemostasis and exhibits osteosarcoma ablation performance under near-infrared light irradiation (Yang et al., 2023). CMC microspheres with phenol moieties are used to deliver genetically modified rat bone marrow mesenchymal stem cells (rBMSCs), which upregulate the mRNA and protein expression levels of osteogenesis-related genes (Zheng et al., 2021).

Porous microspheres can be combined with various substances that serve as anisotropic building blocks for osteochondral tissues and have been reported to provide ample internal space for cell adhesion and proliferation while maintaining their injectability. Cellulose derivatives have been reported to promote bone generation, as hybrid of alginate/gelatin/cellulose nanocrystals (CNCs) provide a suitable environment for cell proliferation, adhesion, nutrient exchange, and matrix mineralization in bone tissues, thereby expanding the cellulose application in BTE (Ji et al., 2022). The versatile therapeutic applications of CMC are summarized in Figure 14.1.

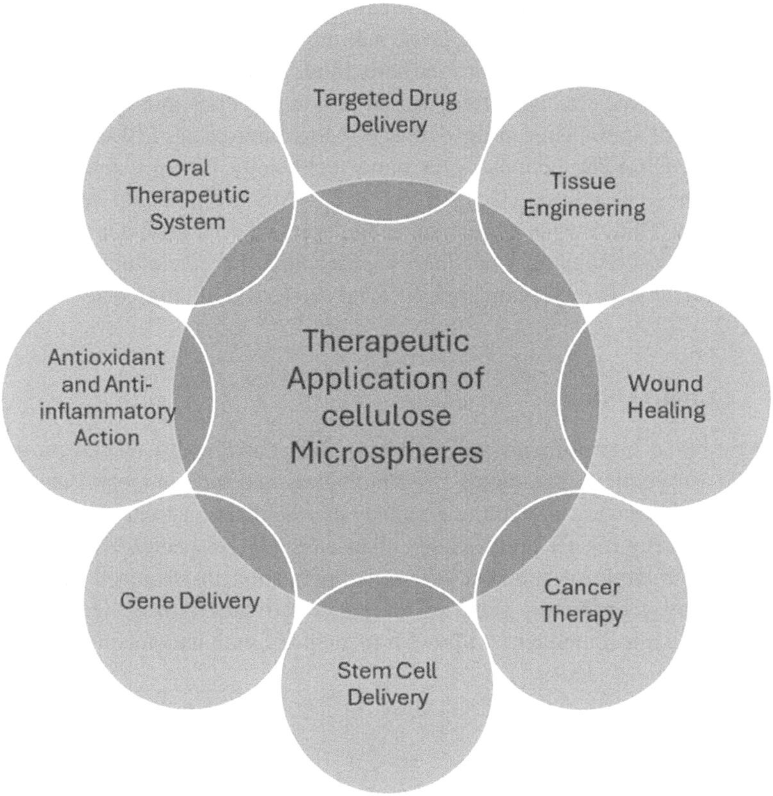

FIGURE 14.1 Therapeutic applications of cellulose microspheres: a multifunctional platform for biomedical advancements.

14.6 CHALLENGES AND FUTURE DIRECTIONS

The cellulose microsphere has remarkably revolutionized the DDS, but its commercialization is still a big challenge due to limited research and resources. Some CM materials have been studied for their biocompatibility and organ toxicity, but their long-term effects and other cellulose polymers have not been studied. There is a research gap in BC compatibility and possible intrinsic toxicity (Sampath Udeni Gunathilake et al., 2020). The short-term cell-based studies are non-reliant for "bench to bedside". The clinical application of CM requires not only *in vivo* and *in vitro* animal testing, but its preparation is also subject to stringent safety regulations. The cellulose derivatives are used in biomedical applications, but their pharmaceutical applications are still in process due to its unregulated production source and characterization. Natural polymers are also prone to impurities, like critical challenge in microcellulose development: the lack of standardized manufacturing quality framework (Lukova et al., 2023a).

Controlled drug delivery can be achieved by combining microsphere development with intelligent biomaterial technology to sense and auto-adjust API release.

For instance, an intelligent microsphere can be engineered to release the drug dose on sensing the patient's biomarker level, making the therapeutic dose delivery more efficient, long-term biocompatible, and safe. Intelligent hydro gels have been devised despite facing some size and strength issues, which means intelligent microsphere is the future of therapeutic drug delivery (Mogoşanu et al., 2016). Incorporation of advanced CRISPR technology in nano technology is a possible prospect for tissue-specific gene editing (Deng et al., 2020). Another potential advancement is quantum dots (QD) that integrate nanotechnology and drug assay (Jain et al., 2022). QD has quantum confinement and fluorescence ability that can influence nano-probes and drug carriers. The QD can label the drug carriers facilitating drug tracing (Jain et al., 2022).

14.7 CONCLUSION

Cellulose-based microspheres are used to deliver liable drugs, which protects the active pharmaceutical ingredients from biological and environmental degradation. The chemical structure of cellulose enables its modification to produce derivatives with tailored drug-release profiles and enhanced loading capacity. The nontoxic and biocompatible nature of cellulose offers its application in advanced therapies like cancer therapy, gene therapy, stem cell therapy, and tissue engineering. Their physiological and chemical uniqueness allows it to integrate with nanotechnology to bridge gap in personalized medicine.

REFERENCES

Adepu, S., & Ramakrishna, S. (2021). Controlled drug delivery systems: current status and future directions. *Molecules*, *26*(19), 5905.

Aguilera, G., Berry, C. C., West, R. M., Gonzalez-Monterrubio, E., Angulo-Molina, A., Arias-Carrión, Ó., & Méndez-Rojas, M. Á. (2019). Carboxymethyl cellulose-coated magnetic nanoparticles transport across a human lung microvascular endothelial cell model of the blood–brain barrier. *Nanoscale Advances, 1*(2), 671–685.

Alavi, M., & Varma, R. S. (2021). Antibacterial and wound healing activities of silver nanoparticles embedded in cellulose compared to other polysaccharides and protein polymers. *Cellulose*, *28*(13), 8295–8311.

Bagheri, M., Shateri, S., Niknejad, H., & Entezami, A. A. (2014). Thermosensitive biotinylated hydroxypropyl cellulose-based polymer micelles as a nano-carrier for cancer-targeted drug delivery. *Journal of Polymer Research*, *21*(10), 567.

Basinska, T., Gadzinowski, M., Mickiewicz, D., & Slomkowski, S. (2021). Functionalized particles designed for targeted delivery. *Polymers*, *13*(12), 2022.

Bayer, I. S. (2023). Controlled drug release from nanoengineered polysaccharides. *Pharmaceutics*, *15*(5), 1364.

Béchard, S. R., Levy, L., & Clas, S.-D. (1995). Thermal, mechanical and functional properties of cellulose acetate phthalate (CAP) coatings obtained from neutralized aqueous solutions. *International Journal of Pharmaceutics*, *114*(2), 205–213.

Bekaroğlu, M. G., İşçi, Y., & İşçi, S. (2017). Colloidal properties and *in vitro* evaluation of hydroxyethyl cellulose-coated iron oxide particles for targeted drug delivery. *Materials Science and Engineering: C, 78*, 847–853.

Belgacem, M. N., & Gandini, A. (2011). Production, chemistry and properties of cellulose-based materials. In D. Plackett (Ed.), *Biopolymers–New Materials for Sustainable Films and Coatings*, (pp. 151–178).

Bhattacharyya, A., Heo, J., Priyajanani, J., Kim, S. H., Khatun, M. R., Nagarajan, R., & Noh, I. (2024). Simultaneous processing of both handheld biomixing and biowriting of kombucha cultured pre-crosslinked nanocellulose bioink for regeneration of irregular and multi-layered tissue defects. *International Journal of Biological Macromolecules, 282*, 136966.

Butun, S., Ince, F. G., Erdugan, H., & Sahiner, N. (2011). One-step fabrication of biocompatible carboxymethyl cellulose polymeric particles for drug delivery systems. *Carbohydrate Polymers, 86*(2), 636–643.

Capanema, N. S. V, Mansur, A. A. P., Carvalho, S. M., Martins, T., Gonçalves, M. S., Andrade, R. S., Dorneles, E. M. S., Lima, L. C. D., de Alvarenga, É. L. F. C., & da Fonseca, E. V. B. (2023). Nanosilver-functionalized hybrid hydrogels of carboxymethyl cellulose/Poly (Vinyl Alcohol) with antibacterial activity for prevention and therapy of infections of diabetic chronic wounds. *Polymers, 15*(23), 4542.

Carvalho, J. P. F., Silva, A. C. Q., Silvestre, A. J. D., Freire, C. S. R., & Vilela, C. (2021). Spherical cellulose micro and nanoparticles: a review of recent developments and applications. *Nanomaterials, 11*(10), 2744.

Cassano, R., Mellace, S., & Trombino, S. (2018). Responsive polymer-biomacromolecule conjugates for drug delivery. In N. Y. Abu-Thabit & A. S. H. Makhlouf (Eds.), *Stimuli Responsive Polymeric Nanocarriers for Drug Delivery Applications, Volume 1* (pp. 433–452). Elsevier.

Deng, H., Huang, W. and Zhang, Z., 2019. Nanotechnology based CRISPR/Cas9 system delivery for genome editing: Progress and prospect. *Nano Research, 12*(10), pp. 2437–2450.

Druel, L., Kenkel, A., Baudron, V., Buwalda, S., & Budtova, T. (2020). Cellulose aerogel microparticles via emulsion-coagulation technique. *Biomacromolecules, 21*(5), 1824–1831.

Druel, L., Kenkel, A., Baudron, V., Buwalda, S., & Budtova, T. (2020b). Cellulose aerogel microparticles via emulsion-coagulation technique. *Biomacromolecules, 21*(5), 1824–1831.

El-Kamel, A. H. (2002). In vitro and in vivo evaluation of Pluronic F127-based ocular delivery system for timolol maleate. *International Journal of Pharmaceutics, 241*(1), 47–55.

Ezike, T. C., Okpala, U. S., Onoja, U. L., Nwike, C. P., Ezeako, E. C., Okpara, O. J., Okoroafor, C. C., Eze, S. C., Kalu, O. L., & Odoh, E. C. (2023). Advances in drug delivery systems, challenges and future directions. *Heliyon, 9*(6), e17488.

Ghimire, S., Sarkar, P., Rigby, K., Maan, A., Mukherjee, S., Crawford, K. E., & Mukhopadhyay, K. (2021). Polymeric materials for hemostatic wound healing. *Pharmaceutics, 13*(12), 2127.

Guarino, M., Rubino, B., & Ballabio, G. (2007). The role of epithelial-mesenchymal transition in cancer pathology. *Pathology, 39*(3), 305–318.

Gunathilake, T. M. S. U., Ching, Y. C., Chuah, C. H., Abd Rahman, N., & Liou, N. S. (2020). Recent advances in celluloses and their hybrids for stimuli-responsive drug delivery. *International Journal of Biological Macromolecules, 158*, 670–688.

Gupta, H., Aqil, M., Khar, R. K., Ali, A., Bhatnagar, A., Mittal, G., & Jain, S. (2009). Development and characterization of 99m Tc timolol maleate for evaluating efficacy of in situ ocular drug delivery system. *American Association of Pharmaceutical Scientists Pharmaceutical Science and Technology, 10*, 540–546.

Hamedi, M. M., Sandberg, M., Olsson, R. T., Pedersen, J., Benselfelt, T., & Wohlert, J. (2025). Wood and cellulose: the most sustainable advanced materials for past, present, and future civilizations. *Advanced Materials, 37*, 2415787.

Ismail, S., Mohammed, F. A., & Ismail, M. A. (2012). Formulation and evaluation of some selected timolol maleate ophthalmic preparations. *Bulletin of Pharmaceutical Sciences Assiut University, 35*(2), 181–197.

Jain, N. S., Somanna, P., & Patil, A. B. (2022). Application of quantum dots in drug delivery. *Nanoscience & Nanotechnology-Asia, 12*(1), 16–31.

Ji, X., Shao, H., Li, X., Ullah, M. W., Luo, G., Xu, Z., Ma, L., He, X., Lei, Z., & Li, Q. (2022). Injectable immunomodulation-based porous chitosan microspheres/HPCH hydrogel composites as a controlled drug delivery system for osteochondral regeneration. *Biomaterials, 285*, 121530.

Joshy, K. S., Snigdha, S., George, A., Kalarikkal, N., Pothen, L. A., & Thomas, S. (2017). Core–shell nanoparticles of carboxymethyl cellulose and Compritol–PEG for antiretroviral drug delivery. *Cellulose, 24*(11), 4759–4771.

Jutakridsada, P., Suwannaruang, T., Kasemsiri, P., Weerapreeyakul, N., Knijnenburg, J. T. N., Theerakulpisut, S., Kamwilaisak, K., & Chindaprasirt, P. (2023). Controllability, antiproliferative activity, Ag+ release, and flow behavior of silver nanoparticles deposited onto cellulose nanocrystals. *International Journal of Biological Macromolecules, 225*, 899–910.

Khan, G. M., & Zhu, J. B. (1998). Ibuprofen release kinetics from controlled-release tablets granulated with aqueous polymeric dispersion of ethylcellulose II: Influence of several parameters and coexcipients. *Journal of Controlled Release, 56*(1–3), 127–134.

Khan, I. U., Ranjha, N. M., & Mehmood, H. Q. (2010). Development of ethylcellulose-polyethylene glycol and ethylcellulose-polyvinyl pyrrolidone blend oral microspheres of ibuprofen. *Journal of Drug Delivery Science and Technology, 20*(6), 439–444.

Khuroo, T., Mohamed, E. M., Dharani, S., Afrooz, H., Barakh Ali, S. F., Cook, P., Khan, M. A., & Rahman, Z. (2021). Coating characterization by hyperspectroscopy and predictive dissolution models of tablets coated with blends of cellulose acetate and cellulose acetate phthalate. *American Association of Pharmaceutical Scientists Pharmaceutical Science and Technology, 22*(3), 122.

Kong, B. J., Kim, A., & Park, S. N. (2016). Properties and in vitro drug release of hyaluronic acid-hydroxyethyl cellulose hydrogels for transdermal delivery of isoliquiritigenin. *Carbohydrate Polymers, 147*, 473–481.

Lengyel, M., Kállai-Szabó, N., Antal, V., Laki, A. J., & Antal, I. (2019). Microparticles, microspheres, and microcapsules for advanced drug delivery. *Scientia Pharmaceutica, 87*(3), 20.

Lin, W.-H., & Jana, S. C. (2021). Analysis of porous structures of cellulose aerogel monoliths and microparticles. *Microporous and Mesoporous Materials, 310*, 110625.

Liu, J., Wei, T., Zhao, J., Huang, Y., Deng, H., Kumar, A., Wang, C., Liang, Z., Ma, X., & Liang, X. J. (2016). Multifunctional aptamer-based nanoparticles for targeted drug delivery to circumvent cancer resistance. *Biomaterials, 91*, 44–56.

Liu, Y., Nambu, N. O., & Taya, M. (2017). Cell-laden microgel prepared using a biocompatible aqueous two-phase strategy. *Biomedical Microdevices, 19*, 1–7.

Lukova, P., Katsarov, P., & Pilicheva, B. (2023). Application of starch, cellulose, and their derivatives in the development of microparticle drug-delivery systems. *Polymers, 15*(17), 3615.

Ma, Y., Thurecht, K. J., & Coombes, A. G. A. (2021). Development of enteric-coated, biphasic chitosan/HPMC microcapsules for colon-targeted delivery of anticancer drug-loaded nanoparticles. *International Journal of Pharmaceutics, 607*, 121026.

Marschütz, M. K., & Bernkop-Schnürch, A. (2000). Oral peptide drug delivery: polymer–inhibitor conjugates protecting insulin from enzymatic degradation in vitro. *Biomaterials, 21*(14), 1499–1507.

Marschütz, M. K., Caliceti, P., & Bernkop-Schnürch, A. (2000). Design and in vivo evaluation of an oral delivery system for insulin. *Pharmaceutical Research, 17*, 1468–1474.

Marwah, M., Badhan, R. K. S., & Lowry, D. (2022). Development of a novel polymer-based carrier for deformable liposomes for the controlled dermal delivery of naringenin. *Journal of Liposome Research, 32*(2), 181–194.

McNamara, J. T., Morgan, J. L., & Zimmer, J. (2015). A molecular description of cellulose biosynthesis. *Annual review of biochemistry, 84*(1), 895–921.

Metaxa, A.-F., Efthimiadou, E. K., Boukos, N., Fragogeorgi, E. A., Loudos, G., & Kordas, G. (2014). Hollow microspheres based on–folic acid modified–hydroxypropyl cellulose and synthetic multi-responsive bio-copolymer for targeted cancer therapy: controlled release of daunorubicin, in vitro and in vivo studies. *Journal of Colloid and Interface Science, 435*, 171–181.

Mogoşanu, G. D., Grumezescu, A. M., Bejenaru, L. E., & Bejenaru, C. (2016). Natural and synthetic polymers for drug delivery and targeting. In A. M. Grumezescu (Ed.), *Nanobiomaterials in Drug Delivery* (pp. 229–284). William Andrew Publishing.

Najjar, M. K., Manore, S. G., Regua, A. T., & Lo, H.-W. (2022). Antibody-drug conjugates for the treatment of HER2-positive breast cancer. *Genes*, *13*(11), 2065.

Opriş, O., Mormile, C., Lung, I., Stegarescu, A., Soran, M.-L., & Soran, A. (2024). An overview of biopolymers for drug delivery applications. *Applied Sciences*, *14*(4), 1383.

Othman, M. H., Zayed, G. M., Ali, U. F., & Abdellatif, A. A. H. (2020). Colon-specific tablets containing 5-fluorouracil microsponges for colon cancer targeting. *Drug Development and Industrial Pharmacy*, *46*(12), 2081–2088.

Paccione, N., Rahmani, M., Barcia, E., & Negro, S. (2022). Antiparkinsonian agents in investigational polymeric micro-and nano-systems. *Pharmaceutics*, *15*(1), 13.

Park, J., Kim, S., & Kim, T. (2023). Polyethylenimine-conjugated hydroxyethyl cellulose for doxorubicin/Bcl-2 siRNA Co-delivery systems. *Pharmaceutics*, *15*(2), 708.

Paulraj, T., Riazanova, A. V, & Svagan, A. J. (2018). Bioinspired capsules based on nanocellulose, xyloglucan and pectin–The influence of capsule wall composition on permeability properties. *Acta Biomaterialia*, *69*, 196–205.

Rahman, M. M., Islam, M. S., Sharmin, N., Chowdhury, J. A., & Jalil, R. (1970). Preparation and evaluation of cellulose acetate phthalate and Ethyl cellulose based microcapsules of diclofenac sodium using emulsification and solvent-evaporation method. *Dhaka University Journal of Pharmaceutical Sciences*, *9*(1), 39–46.

Sammartino, G., Marenzi, G., Miro, A., Ungaro, F., Nappi, A., Sammartino, J. C., Quaglia, F., & Mortellaro, C. (2012). Local delivery of the hemostatic agent tranexamic acid in chronically anticoagulated patients. *Journal of Craniofacial Surgery*, *23*(6), e648–e652.

Sarkar, K., Sadat, S. M. A., Islam, M. S., & Jalil, R. (2011). Study of ethyl cellulose based sustained release microspheres of naproxen sodium. *Dhaka University Journal of Pharmaceutical Sciences*, *10*(2), 123–129.

Seddiqi, H., Oliaei, E., Honarkar, H., Jin, J., Geonzon, L. C., Bacabac, R. G., & Klein-Nulend, J. (2021). Cellulose and its derivatives: towards biomedical applications. *Cellulose*, *28*(4), 1893–1931.

Shokri, J., & Adibkia, K. (2013). Application of cellulose and cellulose derivatives in pharmaceutical industries. In T. G. M. van de Ven & L. Godbout (Eds.), *Cellulose—Medical, Pharmaceutical and Electronic Applications*. IntechOpen. https://doi.org/10.5772/55178

Siepmann, J., & Peppas, N. A. (2012). Modeling of drug release from delivery systems based on hydroxypropyl methylcellulose (HPMC). *Advanced Drug Delivery Reviews*, *64*, 163–174.

Simões, M. G., Coimbra, P., Carreira, A. S., Figueiredo, M. M., Gil, M. H., & Simões, P. N. (2020). Eugenol-loaded microspheres incorporated into textile substrates. *Cellulose*, *27*, 4109–4121.

Singh, A., Kumari, K., & Kundu, P. P. (2022). Nanocellulose biocomposites for bone tissue engineering. In A. Barhoum (Ed.), *Handbook of Nanocelluloses: Classification, Properties, Fabrication, and Emerging Applications* (pp. 597–647). Springer.

Sood, S., Gupta, V. K., Agarwal, S., Dev, K., & Pathania, D. (2017). Controlled release of antibiotic amoxicillin drug using carboxymethyl cellulose-*cl*-poly(lactic acid-*co*-itaconic acid) hydrogel. *International Journal of Biological Macromolecules*, *101*, 612–620.

Stanicki, D., Vangijzegem, T., Ternad, I., & Laurent, S. (2022). An update on the applications and characteristics of magnetic iron oxide nanoparticles for drug delivery. *Expert Opinion on Drug Delivery*, *19*(3), 321–335.

Tade, R. S., More, M. P., Chatap, V. K., Patil, P. O., & Deshmukh, P. K. (2018). Fabrication and *in vitro* drug release characteristics of magnetic nanocellulose fiber composites for efficient delivery of nystatin. *Materials Research Express*, *5*(11), 116102.

Torchilin, V. P. (2014). Multifunctional, stimuli-sensitive nanoparticulate systems for drug delivery. *Nature Reviews Drug Discovery*, *13*(11), 813–827.

Tundisi, L. L., Yang, R., Borelli, L. P. P., Alves, T., Mehta, M., Chaud, M. V., Mazzola, P. G., & Kohane, D. S. (2021). Enhancement of the Mechanical and Drug-Releasing Properties of Poloxamer 407 Hydrogels with Casein. *Pharmaceutical Research*, *38*(3), 515–522.

Ullah, F., Othman, M. B. H., Javed, F., Ahmad, Z., & Akil, H. M. (2015). Classification, processing and application of hydrogels: A review. *Materials Science and Engineering: C*, *57*, 414–433.

Vangijzegem, T., Stanicki, D., & Laurent, S. (2019). Magnetic iron oxide nanoparticles for drug delivery: applications and characteristics. *Expert Opinion on Drug Delivery*, *16*(1), 69–78.

Wan, J., Xie, J., Kong, X., Liu, Z., Liu, K., Shi, F., Pei, A., Chen, H., Chen, W., Chen, J., & Zhang, X. (2019). Ultrathin, flexible, solid polymer composite electrolyte enabled with aligned nanoporous host for lithium batteries. *Nature Nanotechnology*, *14*(7), 705–711.

Wang, F. J., Lu, F. S., Cui, M., & Shao, Z. Q. (2015). Biocompatible microcapsule of carboxymethyl cellulose/chitosan as drug carrier. *Advanced Materials Research*, *1118*, 227–236.

Wang, N., Guang, S., Xu, H., Dong, H., Zhang, S., & Wang, X. (2023). Litchi-like silver nanoparticles immobilized in microspheres for effective bacteriostasis, rapid hemostasis, and wound healing. *Materials Today Chemistry*, *29*, 101462.

Wang, X., Wang, Z., Fang, S., Hou, Y., Du, X., Xie, Y., Xue, Q., Zhou, X., & Yuan, X. (2021). Injectable Ag nanoclusters-based hydrogel for wound healing via eliminating bacterial infection and promoting tissue regeneration. *Chemical Engineering Journal*, *420*, 127589.

Wu, W., Li, J., Liu, W., & Deng, Y. (2016). Temperature-sensitive, fluorescent poly (N-isopropyl-acrylamide)-grafted cellulose nanocrystals for drug release. *BioResources*, *11*(3), 7026–7035.

Yadav, J., Das, S., Karthikeyan, D., Chug, R., Jyoti, A., Srivastava, V.K., Jain, A., Kumar, S., Sharma, V., & Kaushik, S. (2022). Identification of protein drug targets of biofilm formation and quorum sensing in multidrug resistant Enterococcus faecalis. *Current Protein and Peptide Science*, *23*(4), 248–263.

Yang, K., Liu, B., Bai, Y., Chen, S., Chen, C., & Fan, J.-B. (2026). Emerging nanomedicine strategies for sepsis: Immunomodulation and beyond. *Research*, *9*, 1254.

Yang, Z., Zhang, Y., Chen, Y., Yang, Z., Cui, T., Luo, H., Liu, Y., & Wan, Y. (2023). Ti3C2T x MXene-incorporated macroporous bacterial cellulose scaffolds with physical and chemical crosslinking for enhanced mechanical and biological performance. *Cellulose*, *30*(11), 6773–6785.

Yeap, E. W. Q., Ng, D. Z. L., Prhashanna, A., Somasundar, A., Acevedo, A. J., Xu, Q., Salahioglu, F., Garland, M. V, & Khan, S. A. (2017). Bottom-up structural design of crystalline drug-excipient composite microparticles via microfluidic droplet-based processing. *Crystal Growth & Design*, *17*(6), 3030–3039.

Yu, J., Huang, T., Lim, Z. H., Luo, R., Pasula, R. R., Liao, L., Lim, S., & Chen, C. (2016). Production of hollow bacterial cellulose microspheres using microfluidics to form an injectable porous scaffold for wound healing. *Advanced Healthcare Materials*, *5*(23), 2983–2992.

Zhang, F., Wu, W., Zhang, X., Meng, X., Tong, G., & Deng, Y. (2016). Temperature-sensitive poly-NIPAm modified cellulose nanofibril cryogel microspheres for controlled drug release. *Cellulose*, *23*, 415–425.

Zhang, M., Guo, W., Ren, M., & Ren, X. (2020). Fabrication of porous cellulose microspheres with controllable structures by microfluidic and flash freezing method. *Materials Letters*, *262*, 127193.

Zhang, W., Wang, X., Wang, J., & Zhang, L. (2019). Drugs adsorption and release behavior of collagen/bacterial cellulose porous microspheres. *International Journal of Biological Macromolecules*, *140*, 196–205.

Zhao, X., Sun, Y., Meng, Z., Yang, Z., Fan, S., Ye, T., Yang, L., Li, T., Gu, R., & Wu, Z. (2022). Preparation and characterization of tranexamic acid modified porous starch and its application as a hemostatic agent. *International Journal of Biological Macromolecules, 200*, 273–284.

Zheng, S., Chen, H., Zhang, T., Yao, Y., Chen, Y., Zhang, S., & Bai, B. (2021). Gene-modified BMSCs encapsulated with carboxymethyl cellulose facilitate osteogenesis in vitro and in vivo. *Journal of Biomaterials Applications, 35*(7), 814–822.

Zubik, K., Singhsa, P., Wang, Y., Manuspiya, H., & Narain, R. (2017). Thermo-responsive poly(N-isopropylacrylamide)-cellulose nanocrystals hybrid hydrogels for wound dressing. *Polymers, 9*(4), 119.

Section 3

Cellulose Beads

15 Separation and Environmental Applications of Cellulose Beads

Amara Tahir, Muhammad Umar Ijaz, Qamar Haneef, Muhammad Zaid Salar, Naila Ghafoor, and Shumaila Kiran

15.1 INTRODUCTION

Cellulose accounts for approximately 1.5 × 10^{12} metric tons annually of the total biomass utilized for ecofriendly and biodegradable products, making it the most abundant as well as a naturally available polymer (Rasheed et al., 2020). Cellulose is one of the most ubiquitous biopolymers on Earth that comes from a wide variety of species, primarily plants, bacteria, and algae as shown in Table 15.1 (El Allaoui et al., 2023). Moreover, marine lifeforms such as tunicates biosynthesize cellulose

TABLE 15.1
Physical properties of bacterial and plant cellulose

Properties	Plant Cellulose	Bacterial Cellulose	Reference
Modulus of elasticity	5–130 GPa	60–115 GPa	Mahmud et al. (2021); Ybañez and Camacho (2021)
Tensile strength	750–1080 MPa	200–300 MPa	Wang et al. (2018); Zhang et al. (2020)
Pore size	1–100 nm	10–300 nm	Gao et al. (2011); Larsson et al. (2013)
Thermal stability	Initial disintegration = 299°C; maximum disintegration = 328°C; final decomposition = 345°C	Transition = 191°C	Gibson (2012); Cheng et al. (2017); Moniri et al. (2017)
Crystallinity	Low, 40%–60% for native PC	High, 84%–89%	Esa et al. (2014); Huang et al. (2014); Cheng et al. (2017)
Malleability	Low	High	Dhyani and Bhaskar (2019); Gao et al. (2011)

DOI: 10.1201/9781003596127-18

and produce a variant version of cellulose known as "tunicin". There are a number of ways through which cellulose can be produced *in vitro* through enzyme synthesis and inducible microbial strains like algae, fungi, and certain pathogen-free bacteria like *Acetobacter*, *Agrobacterium*, and *Rhizobium* (Ullah et al., 2016). A variety of sources for cellulose extraction include coniferous wood, sugarcane bagasse, date palm, cotton, straw, hemp, jute, sisal fibers, kenaf, and ramie. Such sources are used further to create multiple forms including membranes, beads, films, fibrous hydrogels, aerogels, sponges, and cellulose derivatives, showing the wide diversity of applications of cellulose (Chakhtouna et al., 2022). It is an important structural component of cell walls of plants and has widespread applications in manufacturing companies, including textile, pulp, and paper industries (Ullah et al., 2016).

Cellulose possesses significant properties, including the biodegradability, non-toxicity, resiliency, endurance, biological compatibility, and availability. Cellulose is a linear homopolymer characterized by the chemical formula $(C_6H_{10}O_5)_n$, consisting of *D*-glucopyranose units linked through β-(1 $\rightarrow$ 4) glycosidic bonds. Usually, there are three separate hydroxyl (OH) groups in carbon 2, 3, and 6 in a unit of cellulose. Hydrophilic behavior of cellulose is attributed to the functional groups (OH) and it is more reactive in the main OH group on C-6, whereas, the OH molecules on C-2 and C-3 are less reactive (Thakur et al., 2021). Reactive OH groups enable the modification of cellulose by introducing multiple functions to its surface or by combining it with different inorganic and organic substances. Cellulose modification aims to enhance existing properties or impart new characteristics, contingent upon the specific application (Idrissi et al., 2022). Functional materials derived from cellulose present a viable option for application across diverse fields, such as the treatment of wastewater, catalytic activity, desalinization, therapeutic uses, packaging of food, agriculture, and air quality improvement (He et al., 2021).

Cellulose beads exhibit an interlinked three-dimensional (3D) network that exemplifies advanced porous polymeric materials. These materials address the requirements of various domains, including protein immobilization, medical applications, and agricultural systems. Beads of cellulose can be prepared after obtaining cellulose by treating the biomass source with alkaline solution and then delignifying (degumming) the obtained fibers via several methods described in the published literature. Disintegration of cellulose or its derivatives in various solvents and then regenerating is a widely recognized method (Azimi et al., 2024). Physical properties of bacterial and plant cellulose are given in Table 15.1.

15.2 SYNTHESIS OF CELLULOSE BEADS

The first spherical cellulose beads came into existence in 1951 when scientists dropped a viscose solution into an aqueous coagulation bath (O'Neill et al., 1951). Numerous production techniques have been established since the first discovery to create cellulose beads that measure between 10 μm and 3 mm in diameter. The entire procedure starts by dissolving cellulose and then proceeds to shape the solution into spheres before solidifying the spheres through a sol–gel transition. The properties of the spheres can be adjusted through pre- and post-treatment processes. The spherical particle formation process employs two main methods, which include dropping and dispersion. The two techniques exist because of their technical differences, yet they also define a rough range for bead sizes where dropping produces beads larger than 250–500 μm and dispersion creates smaller beads (Mazzitelli et al., 2008).

15.2.1 Dropping Procedures

The process of dropping creates spherical droplets from cellulose solutions before they solidify in a non-solvent coagulation bath. A nozzle serves as the primary tool to push the solution through this method. Vibrating nozzles or air jets serve as methods to create smaller droplets during the process. The typical bead diameter produced by dropping ranges between 0.5 and 3 mm (Braun et al., 2010). The spherical shape depends on three main factors: solution viscosity, ejection speed, and falling height. Manufacturing at large volumes requires the use of automated systems. Spinning drop atomization employs rotating cylindrical vessels, which control particle size precisely at sizes as small as 500 μm. The process of spinning disk atomization uses a thin polymer solution film on a rotating disk to create droplets through centrifugal force effects. Jet cutting functions by using a rotating knife, which transforms high-velocity solution streams into spherical particles, and this process works underwater through underwater pelletizing (Maggioris et al., 2000).

15.2.2 Dispersion Procedures

The process of dispersion requires mixing cellulose solution with an immiscible solvent to form an emulsion, which surfactants stabilize during rapid rotational movement. The emulsions contain droplets that solidify into beads (Liu et al., 2011). The diameter of beads from dispersion typically ranges from ten to several hundred micrometers, making them roughly ten times smaller than those produced by dropping methods. The process of dispersion requires no specialized tools to produce uniform outcomes, which makes it perfect for business-level manufacturing. Cellulose beads with different size ranges from 30 to 250 μm exist throughout the commercial market today (Rosenberg et al., 2010).

15.3 FUNCTIONALIZATION OF CELLULOSE BEADS

The surface of the cellulose beads has several accessible hydroxyl functional groups, providing multiple active sites for chemical alteration. Through the chemical change, the covalent bonds are established between the modifying agent as well as the reactive OH groups of the cellulose and produce new material characteristics that vary from the original substrate. These adjustments aim to enhance the versatility of the beads made of cellulose and improve their compatibility with different materials. An adhesive was often placed within the cellulose skeleton and the important structural elements, or a two-step process was used to secure the substituents. Each anhydroglucose unit of cellulose has three alcoholic hydroxyls on C2, C3, and C6 that are the subject of numerous cellulose modification processes such as oxidation, etherification, and grafting of polymer to cellulose (Fiorica et al., 2021).

15.3.1 Etherification

It is predicted that approximately 650,000 tons of cellulose ethers are produced worldwide. This considerable amount of production can be attributed to the fact that they are more soluble in water and are non-toxic thus making them suitable for various applications. Cellulose ether production occurs in an alkaline environment (15%–50% aqueous NaOH) under two phases. The initial stage involves mercerization of cellulose beads to activate hydroxyl groups and reduce crystallinity. The second stage entails functionalizing the reactive hydroxyl groups by means of a nucleophilic attack using etherifying reagents, like alkyl epoxide or halide, such as vinyl or halogen compounds for alkylating, and oxirane for hydroxyalkylation (Naserifar et al., 2022). A prominent approach is the transformation of cellulose beads in alkaline environment utilizing epichlorohydrin. In this method, cyclic ring scission between the hydroxyl and the epoxide groups of the polysaccharide leads to the generation of a halohydrin alternative that eliminates HCl and forms an oxirane group. Industrial ethers are typically produced through established organic reactions that involve the nucleophilic reaction of cellulose under an alkaline environment, utilizing any of the following methods: (a) the formation of Williamson ethers with aryl or alkyl halides, (b) ring-opening reactions employing epoxides, and (c) Michael addition-type reactions involving activated double bonds (Heinze et al., 2018).

Synthesis of hydrophobic cellulose beads can be achieved by synthesizing trimethylsilyl ethers through the reaction of hexamethyldisilazane or trimethylsilyl chloride with cellulose for treatment of oily water and size exclusion chromatography. Xiong et al. (2023) altered the hydrophilic properties of cellulose beads, rendering them water-repellent through reactions with trimethylsilyl chloride or hexamethyl disilazane. Carboxymethylation techniques are mostly carried out for modifying the chemical makeup of cellulose beads. Carboxymethyl cellulose (CMC) is manufactured through the dispersion of cellulose in a solution of sodium hydroxide and ethanol. Sodium chloroacetic acid is subsequently introduced to the mixture (Zennifer et al., 2021). CMC is an anionic cellulose derivative characterized mostly by the presence of water solubility, sensitivity to pH, less harmful, low cost, and ability to form gels. They exhibit remarkable properties that account for their extensive application in wastewater treatment, food, pharmaceuticals, personal care and cosmetics, paper, and various other sectors. Combining CMC with other polymers, including chitosan, starch, and alginate, can be done via the application of cross-linking agents in producing cellulose-based beads. The synthesized beads may serve as matrixes for drug encapsulation. Carboxyl groups (COO−) on CMC chains facilitate electrostatic interactions with multivalent metal cations, including Fe^{3+} and Al^{3+}, leading to the formation of stable ionotropic hydrogel beads. Additionally, hydroxyl groups interact with metallic ions, enhancing the stability and insolubility of polymer aggregates in water (Morozkina et al., 2022).

15.3.2 Oxidation

Cellulose beads undergo activation through oxidation with sodium periodate ($NaIO_4$), which selectively cleaves the bond between the C2 as well as C3 carbons in the AGU repeat unit of cellulose. Consequently, two carbonyl groups are generated, which, due to their reactivity, can be coupled with various amines, including diamines and hydrazines, or sulfonated using $NaHSO_3$ and Na_2SO_3, for protein immobilization. Aldehyde moieties generated through periodate-regulated cellulose oxidation facilitate the binding of enzyme lysine through imine formation (Califano et al., 2021).

According to Chang et al. (2023), oxidized cellulose/chitosan microbeads present a viable option for dye sorption in their research. Cellulose beads undergo oxidation by 2,2,6,6-tetramethylpiperidine-1-oxyl radical (TEMPO), which selectively converts the primary C6–OH groups of cellulose to COOH in an aqueous medium, resulting in the formation of 6-deoxy-6-carboxy-cellulose. Wang et al. (2022) produced TEMPO-oxidized regenerated cellulose beads through a straightforward bead drop method. The synthesized anionic cellulose beads demonstrated effective adsorption of methylene blue, achieving an adsorption capacity of approximately 495 mg/g, thereby confirming their efficacy in dye removal. Oxidized cellulose bead served as a drug-delivery system, showcasing the significant potential of these beads in regulating drug release (Xie et al., 2021).

15.3.3 Polymer Grafting

The widespread applications of cellulose beads are due to their biocompatibility, biodegradability, and abundant availability. The modification of beads through grafting methods introduces new properties to the beads that extend across various scientific fields such as chromatography, analytical chemistry, environmental applications, and biomedical research as well as drug-delivery systems. Cellulose beads are capable of storing biomolecules through the grafting process, which allows better control of their interactions in medical and diagnostic applications. Yu et al. (2020) grafted poly(acrylic acid) (PAA) onto cellulose beads to eliminate low-density lipoprotein (LDL) from human plasma. The grafting process makes cellulose beads more effective at separation tasks while improving the precision of complex mixture identification and quantification. Moreover, the stationary chromatographic phases achieve higher adsorption capabilities through the use of cellulose-g-glycidyl methacrylate (GMA) beads (Wang et al., 2017).

Cellulose beads employed in water treatment efficiently remove toxins and pollutants from water sources. Zhao et al. (2021) grafted macro-porous cellulose beads with poly(glycidyl methacrylate) and iminodiacetic acid modifications (MCBs-pGMA-IDA) to extract Cu(II), Mn(II), and Ni(II) heavy metal ions from water. Moreover, poly(hydroxamic acid)-modified MCBs were developed by Liu et al. (2021), which showed efficient results in removing heavy metals from wastewater. The grafting method applied to cellulose beads enhances their drug carrier functions by providing superior drug storage capacity and regulated release patterns and improved target specificity. The modifications strengthen both the durability and stability of drug-encapsulated beads and simultaneously reduce drug degradation as well as toxic reactions (Jiao et al., 2018). Table 15.2 shows the functionalization methods of cellulose beads.

15.4 SEPARATION APPLICATIONS

Aquatic contaminants must be properly eliminated to enhance water quality, safeguard aquatic life, and mitigate ecological imbalances. Given its dynamic nature,

TABLE 15.2
Functionalization methods of cellulose beads

Method	Key Reagents	Products	Applications	Reference
Etherification	NaOH, alkyl halides, epoxide, vinyl compounds	Carboxymethyl cellulose, trimethylsilyl ether	Oil–water treatment, wastewater treatment, drug encapsulation, food, pharma, cosmetics	Naserifar et al. (2022); Xiong et al. (2023); Zennifer et al. (2021)
Oxidation	$NaIO_4$, TEMPO, $NaHSO_3$, Na_2S_3	Aldehyde cellulose, carboxy cellulose	Dye removal, enzyme immobilization, drug binding, drug delivery	Chang et al. (2018); Califano et al. (2021); Xie et al. (2021)
Polymer grafting	Acrylic acid, GMA, IDA	Cellulose-g-PAA, Cellulose-G-GMA, MCBs-PHA	LDL removal from plasma, chromatographic stationary phase, and heavy metal removal	Wang et al. (2017); Yu et al. (2020); Zhao et al. (2021); Liu et al. (2021); Jiao et al. (2018)

there is a pressing need to advance more sustainable and efficient water treatment technologies in response to increasing human activities (Hokkanen et al., 2016). Among the prevalent techniques for wastewater treatment aimed at the removal of persistent toxins, such as adsorption, filtration, coagulation, chlorine disinfection, biological treatment, and chemical methods (including oxidation, ozonation, and photochemical degradation), adsorption is favored due to its cost-effectiveness, high efficiency, and simplicity of implementation (Chuein et al., 2021).

Adsorption is recognized as an environmentally friendly and adaptable method for wastewater purification, especially through the use of biosorbents that are non-toxic and biodegradable, such as cellulose-based materials (Varghese et al., 2019). Nanocellulose exhibits significant versatility as it is capable of being formed into various structures, including films (Shimizu et al., 2019), hydrogels (Liu et al., 2021a), spheres (Ram and Chauhan, 2018), aerogels (Sun et al., 2021), and core–shell structures (Beyki et al., 2016). Nonetheless, many of these materials exhibit insufficient mechanical integrity in relation to current hydrogels that significantly limit their practical applications. Nanocelluloses demonstrate reinforcing properties that address this challenge. Each year, tons of cellulosic biomass are generated, which makes it the most abundantly produced homopolysaccharide. The rationale for utilizing nanocellulose lies in its abundance, non-toxicity, environmental friendliness, and high efficiency as an adsorbent. Nanocelluloses are widely utilized as reinforcing agents in polymeric matrixes (Köse et al., 2020).

The water treatment method using adsorption treatment technology exhibits multiple advantages as it selects contaminants with high precision and operates with low expense and simple procedures. Moreover, it does not experience negative pollutant effects and provides high regeneration capacity through different available adsorbents. The adsorption process can remove different types of inorganic and organic contaminants while preventing the formation of harmful intermediate substances and unwanted products. The effectiveness of contaminant elimination depends strongly on the correct choice of adsorbent, which relies on both the adsorbent material and adsorbate characteristics (Lobato-Peralta et al., 2022). Several natural substances, including clay minerals, activated carbon, and zeolite, have been explored for their ability to serve as effective pollution adsorbents. Activated carbon functions as a popular adsorbent, which provides a large surface area together with extensive pore structure and highly reactive surfaces that make it suitable for multiple applications (Rathi et al., 2021). Industrial AC faces production cost constraints, low selectivity, and restricted raw material sources. The restricted use of natural adsorbents is due to their costly manufacturing process and poor toxic compound adsorption capabilities (Solangi et al., 2021).

Recent research has focused on the production of adsorbents from sustainable and cost-effective precursors, particularly agricultural waste. Agricultural residues, such as lignocellulosic materials, exhibit advantages over alternative adsorbents due to their potential for use without any chemical change or with very little processing, which substantially lowers the price of production (Tapia-Orozco et al., 2023). Their low ash composition and high carbon content render them valuable precursors for activated carbon production and as constituents in blended materials for the adsorption of inorganic as well as organic contaminants (Ahmed et al., 2016).

The structural modification of cellulose and other powdered biomaterials enhances their compatibility and adsorption capacity for pollutants; however, the soft texture of these adsorbents in aqueous environments can obstruct columns and markedly diminish the flow of the eluent utilized to extract adsorbate from the adsorbent surface during adsorption/desorption investigations. Cellulose beads and hydrogels represent a sophisticated variant of cellulose that has garnered notable attention recently, owing to its unique high porosity, functional surface, and multiple potential uses in filtration, catalysis, energy storage, and drug or cell administration among others. Additionally, its extensive pores provide a substantial surface area for binding to target molecules, facilitating considerable adsorption (Meng et al., 2019).

The physical properties of adsorbent materials determine their success in eliminating wastewater contaminants. The physical structure of cellulose allows transformation into films, membranes, and fiber designs as well as sponges, powders, and spherical particles, such as cellulose beads and hydrogels (Syazwani et al., 2018). The most complex adsorbents among all forms exist in the form of polysaccharide-based microspheres, which include aerogel, hydrogel, and beads. Industries can utilize well-designed porous cellulose beads as candidates because of their renewable nature and biodegradability along with their mechanical and chemical resistance as well as specific functionalities (Ding et al., 2021). The particle size of cellulose beads ranges from micro to millimeter scales and they can be either porous or nonporous. Porous carbon blacks consist of materials that have a low density and high surface area. The materials go by different names, such as microspheres, pellets, beaded cellulose, and pearl cellulose. The CBs show better performance in flow processes than films or irregularly shaped particles because they decrease backpressure and agglomeration during column experiments (Zimmermann et al., 2016).

Furthermore, the separation in aqueous media is challenging when cellulose derivatives are utilized in powdered form as absorbents or ion exchangers. Consequently, cellulose beads can address the aforementioned limitation, while the porous surface enhances the availability of cellulose hydroxyl groups. The extensive surface-area-to-volume ratio of beads enhances the incorporation of chemical moieties, thus increasing the binding efficiency of contaminants to the adsorbent (Tripathy et al., 2021). Functionalized porous beads are garnering significant attention due to their high porosity, customizable surfaces, and various potential applications in filtration, purification, catalysis, energy storage, and drug or cell delivery (Ding et al., 2021).

15.5 ENVIRONMENTAL APPLICATIONS BEYOND SEPARATION

Cellulose beads have long been used in separation technologies, but because of their flexible mechanical properties, environment friendly biodegradability, and potentially adjustable surface chemistry, they are also becoming the preferred material in many environmental applications. Cellulose in bead form has profound applications in chromatography, water treatment, drug delivery, and protein immobilization (Gericke et al., 2013; Coombs OBrine et al., 2017).

15.5.1 Use as Carriers for Bioremediation Agents (Microbes, Enzymes)

Cellulose beads are widely acknowledged as efficient carriers for bioremediation agents, including enzymes and microorganisms, owing to their biocompatibility, mechanical strength, and sustainability. A significant instance pertains to the covalent immobilization of glucose oxidase (GOx) derived from *Aspergillus niger* onto cellulose hydrogel beads (Califano et al., 2021). The beads perform enzymatic functions that generate hydrogen peroxide at levels necessary for bacterial elimination. This leads to the growth retardation of multiple harmful microorganisms such as *Pseudomonas aeruginosa*, *Escherichia coli*, and methicillin-resistant *Staphylococcus aureus* (Homaei et al., 2013; Kim et al., 2017). The manufacturing approach for enzyme activity needs optimization to produce superior durability, mechanical strength, and extended service life. Enzyme-binding functions at its peak performance level in environments that sustain moderately acidic conditions (Liu et al., 2016). Pathogen-fighting cellulose beads demonstrate superior environmental benefits compared with conventional antibiotics because they stop multidrug-resistant bacteria from emerging (Tavakolian et al., 2018).

Bacterial cellulose beads represent the most suitable option for bacterial immobilization procedures. The research conducted by Żywicka et al. (2016) found that *Saccharomyces cerevisiae* cells attached to bacterial cellulose beads generated greater alcohol output than free-floating cells. White rot fungus achieved superior stability and performance in pigment removal during wastewater treatment after being immobilized on cellulose beads (Latif et al., 2023).

15.5.2 Biodegradable Scaffolds for Environmental Sensors

Technological developments in recent times have simplified the creation process for cellulose-based hydrogels and aerogels and foams. The materials show promising properties in functioning as bio-supports together with electrodes and active sensing layers. The scaffolds may be engineered for high sensitivity detection of environmental elements, including pollution, temperature, and humidity, without losing selectivity (Fu et al., 2021). The integration of carbon dots into cellulose hydrogel beads creates a major breakthrough for detecting and capturing mercury and other heavy metals from water environments. The beads showed improved fluorescence-sensing thresholds that dropped to 8.8×10^{-8} M, which is well below WHO mercury water standards. The cellulose matrix is biodegradable, reducing the likelihood of recurring pollution associated with these sensors, thereby rendering them effective and environmentally sustainable (Li et al., 2024).

15.5.3 Soil Remediation and Slow-Release Fertilizers

Cellulose beads are also used as slow-release fertilizer carriers, thereby paving the way to sustainable agricultural practices and soil remediation. Traditional fertilizers not only lose their nutrients quickly but also cause environmental contamination and lower crop yields. However, the application of fertilizers along with cellulose bead carriers address these problems and lower environmental risks (Gomez-Maldonado et al., 2024).

For example, nitrogen fertilizers, formulated as slow carbon spheres derived from cellulose, have been engineered to exhibit enhanced thermal stability and a rich array of functional groups. The nitrogen release rates of these fertilizers are much lower, which aids in sustaining nutrient availability over extended periods and diminishes the danger of groundwater pollution (Zhang et al., 2023). Similarly, urea coated with ethyl cellulose as the inner layer, along with cellulose superabsorbent, which absorbs biochemical inhibitors, demonstrates stable release behavior. It effectively controls nitrate content at lower levels and possesses water-retention, slow-release functions, and inhibits the conversion of ammonium nitrogen to nitrogen nitrate (Zhang et al., 2021). Cellulose-based slow-release fertilizers enhance nutrient delivery while minimizing environmental concerns, aiding in soil remediation and promoting sustainable agricultural productivity over time (Firmanda et al., 2022).

15.5.4 Application in Oil Spill Recovery

Cellulose-based beads and related materials are becoming increasingly prevalent in oil spill recovery because of their large surface area, changeable hydrophobicity, and mechanical stability. Cellulose-derived sorbents, recognized for their hydrophobic properties and selective absorption of oils and organic solvents, are favored for this application (Hammouda et al., 2021). Cellulose-based sorbents possess various synthesis methods, product characterization, hydrophobization techniques, performance, and potential large-scale applications (Budtova, 2019).

Chemical modification of cellulose beads provides oleophilic and hydrophobic properties, facilitating effective oil–water separation during spills. Their high porosity allows for significant oil uptake. Additionally, cellulose-based sorbents can be reused and regenerated multiple times, thereby reducing operational costs and waste (Feng et al., 2015). A study suggests that cellulose beads made with triethoxy(octyl)silane (TEOS) are effective in the oil–water separation (El Allaoui et al., 2023).

15.6 REGENERATION, REUSABILITY, AND DEGENERATION OF CELLULOSE BEADS

15.6.1 Regeneration of Cellulose Beads

Cellulose beads have gathered significant interest due to their prevalence, biodegradable nature, sustainability, and low cost. Chemical modification of cellulose beads may transform it into prospective biomedical engineering materials (Trygg et al., 2013), and cellulose beads are presently being utilized as the stationary phase in chromatography systems to facilitate separation owing to their spherical structure that can be easily packed in a column and withstand extremely high flow rate. Moreover, the cellulose beads have undergone chemical modification with a variety of anionic groups to be applied as the absorbent in the process of cleaning water i.e., decontamination of heavy metals in water resources (Wang et al., 2008). Additionally, cellulose beads' drug loading capacity and release kinetics have been investigated for use in biomedical and pharmaceutical preferences (Voon et al., 2016).

A regenerated cellulose can be processed into films/membranes, hydrogels, fibers, aerogels, beads, bioplastics, etc., which have shown prospective applications in biomedicine, energy storage, energy storage, and packaging. Notably, the ability of cellulosic materials to biodegrade in oceans and soil would reduce environmental contamination (Tu et al., 2021). Despite various cellulosic materials, the regenerated cellulose beads (RCBs) that are recognized for their huge surface area and simple to modify, have emerged as a key focus of research, biomaterials, purification, and catalysis (Gong et al., 2023; Kim et al., 2024). The makeup and characteristics of RCB are critical to their uses, particularly in purification and adsorption (Hamidon et al., 2022).

During regeneration of cellulose beads, the cellulose is thoroughly dissolved in a suitable solvent before being shaped into spherical beads via a dropping or dispersion method. The beads are then regenerated in a coagulation wash containing an anti-solvent (Lindh et al., 2014). The most commonly used solvents for cellulose to dissolve include ionic liquids made from 1-butyl-3-methylimidazolium cations, melts of salt hydrates, and DMSO/ammonium fluorides (Bahsaine et al., 2024).

15.6.2 Reusability of Cellulose Beads

Biosorbents maintain their reusability, which makes them essential for industrial wastewater treatment systems because they remain affordable and sustainable. Research demonstrates that the Hg^{2+} adsorption ability of altered porous microcrystalline cellulose beads remained at 79.4% through ten recycling cycles when regenerated with 1M HNO_3. Hence, altering porous architecture of cellulose beads improves the performance as well as increases its recyclability (Li et al., 2019).

Cellulose-based beads can be easily recycled because they biodegrade naturally, which makes them suitable for industrial wastewater treatment. The financial benefits of industry stem from the affordable production costs of cellulose-based beads (Hamidon et al., 2022). The elevated concentration of carboxyl groups on cellulose beads allowed them to maintain their adsorption performance through multiple adsorption–desorption cycles thus proving their reusability as adsorbent materials (Zhang et al., 2024).

The advantage of cellulose-based materials is that they allow metal surface ions to be removed through acidic washing (Liu et al., 2014). The recovery procedure extracts valuable metals from the cellulose beads during the process. The regeneration methods use severe environmental conditions that involve acidic substances as well as environmentally damaging chemical agents. The efficiency of regeneration processes determines the preservation of cellulose-based material resources that needs additional investigation to enhance these processes (Barbosa et al., 2020).

Although the regeneration and reusability of cellulose beads are important but there are several limitations (Lau et al., 2023). The efficient removal of adsorbed materials such as dyes and heavy metals, from cellulose beads is difficult dye to weak interaction between adsorbed materials as well the surface area of cellulose beads. Therefore, it is indispensable to modify cellulose beads to eliminate these limitations (Ghazali et al., 2025).

15.6.3 Biodegradation in Natural Environments

Biopolymers such as cellulose are naturally biodegradable, but the pace of biodegradation varies greatly depending on the nature of the biopolymer and the environment. Cellulose frequently undergoes chemical modification that is necessary for expanding cellulose's uses, though it can interfere with its natural biodegradability (Erdal and Hakkarainen, 2022). Cellulose beads can be biodegraded by numerous microorganisms and enzymes (Béguin and Aubert, 1994). It can also be degraded in wastewater, compost, sewage sludge, soil burial as well as in water bodies, including sea water and lakes (Miyaji et al., 2023).

Enzymatic degradation of cellulose bead occurs by cellulases that is produced by cellulolytic fungi and bacteria (Joshi et al., 2021). These microorganisms can be separated between cellobiohydrolyses that act specifically on the terminal ends of cellulose and endoglucanases that are able to hydrolyze β-1–4-glycosidic bonds found in amorphous cellulose. Additionally, lytic polysaccharide monooxygenases may aid in the oxidative breakdown of glycosidic linkages. When cellulose beads are fully biodegraded, it leads to the production of water, carbon dioxide, and menthane in anaerobic degradation. While in aerobic degradation, it produces water and carbon dioxide (Erdal and Hakkarainen, 2022).

15.6.4 Life Cycle and End-of-Life Consideration

Foroughi et al. (2021) demonstrated that because of growing demand in the past few years for cellulose products, one of the needed areas of this research is to assess environmental impacts resulting from cellulose extraction, cellulose product manufacturing processes, the use of cellulose products, and their end-of-life removal. For this purpose, life-cycle assessment is a valuable technique for measuring the overall environmental impact of all cellulose phases from extraction to manufacture to use to disposal, or from cradle to grave, as shown in the figure. On the basis of these findings, efforts can be implemented to reduce environmental effects and produce environmentally friendly cellulosic compounds. Figure 15.1 shows the life-cycle assessment of cellulose beads.

Two main obstacles exist when evaluating the life cycle of cellulose products. The first problem is the lack of data availability, particularly during the usage and disposal of cellulose beads. Furthermore, the majority of investigations were carried out on a lab scale and the accuracy of the data remains questionable. While today's economic and climate change obstacles have garnered considerable concern from human cultures, lowering energy-consuming procedures and developing low-carbon materials has become extremely crucial (Foroughi et al., 2021).

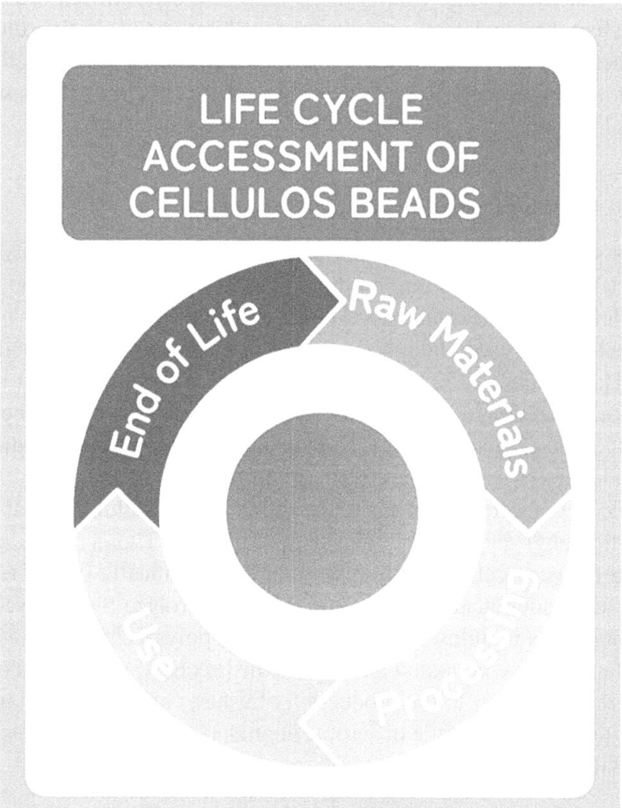

FIGURE 15.1 Life-cycle assessment of cellulose beads.

15.7 COMPARISON WITH SYNTHETIC ADSORBENTS AND SEPARATION MATERIALS

The sustainability and biodegradability of cellulose-based superabsorbent polymers (SAPs) made from lignocellulosic biomass are two key benefits over synthetic adsorbents. Polyacrylates and other conventional petroleum-based SAPs are non-biodegradable and hazardous to the environment because of their extended persistence and propensity to create microplastics. Conversely, bio-based SAPs derived from cellulose-grafted copolymers or modified cellulose nanocrystals are non-toxic, biodegradable, and renewable alternatives (Niu et al., 2025; Omidian et al., 2024).

Although cellulose SAPs' water absorption capacity is somewhat inferior to that of their synthetic equivalents, chemical modifications such as carboxymethylation and graft polymerization can greatly increase it (Xiao et al., 2023). For instance, lignin–cellulose hybrid exhibit biodegradability and better swelling capacity and structural integrity, as compared with lignin or cellulose alone (Zhang et al., 2022). Moreover, the selectivity in separation applications (i.e., heavy metal or dye removal from wastewater) is improved via the application of metal-ion-sensitive groups (Al-Gethami et al., 2024).

The benefits of cellulose SAPs remain despite their limitations that include unpredictable batch variations from different biomass sources together with challenges in scaling production and inconsistent results across different applications. The need for extra processing steps and cross-linking agents in cellulose derivatives might reduce their environmental benefits compared with synthetic polymers when not properly managed (Kasbaji et al., 2023). The integration of cellulose with natural polysaccharides or minimal synthetic elements shows potential for creating hybrid systems that optimize cost-effectiveness and biodegradability alongside performance characteristics. Hence, researchers are studying advanced cross-linking techniques, which include enzymatic catalysis and green solvent processing, to develop viable alternatives based on cellulose SAPs.

15.8 INTEGRATION INTO CIRCULAR ECONOMY MODELS

Sustainable materials engineering has evolved to include SAPs produced from cellulose as essential components for circular economy models. Lignocellulosic biomass with high cellulose content exists in industrial cellulose waste as well as forest leftovers and agricultural waste, which can be obtained to close resource loops and minimize environmental impact. High-performance SAPs made from waste enable recycling of materials, which minimizes the need for fossil fuel-based synthetic materials (Zhang et al., 2022). Recent studies have demonstrated the biodegradability as well as reusability of SAPs that exhibit multiple functions in soil conditioning, agricultural water retention, and hygiene products (Oladosu et al., 2022). The absorbed water retention of cellulose-based hydrogels created from recycled pulp remained above 90% through multiple usage cycles that demonstrates their potential for multiple reuses. Post-use SAPs serve as compostable materials that return nutrients to soils while preventing synthetic pollutant release thus making them suitable for closed-loop systems (Omidian et al., 2024).

The implementation of circular systems faces obstacles because of inefficient processing methods together with unstandardized biomass pre-treatment approaches. The structural differences in input biomass lead to polymer structure differences, which create problems for achieving functional uniformity in scaled operations (Kasbaji et al., 2023). The chemical modification and drying processes needed to transform raw cellulose into cross-linked SAP structures require substantial energy consumption so they might eliminate environmental advantages unless these procedures are enhanced using sustainable chemistry or renewable power sources (Wang et al., 2024). The future achievement of circularity depends on implementing life-cycle assessment (LCA) techniques in the design process of SAPs. Research teams focus on biocatalytic and enzymatic modification approaches to develop SAPs that generate fewer carbon emissions and can be produced with reduced temperature and pH requirements to support industrial symbiosis initiatives (Dodangeh et al., 2024).

15.9 USE OF AGRICULTURAL AND INDUSTRIAL CELLULOSE WASTE FOR BEAD SYNTHESIS

The process of making bead-form SAPs from agricultural and industrial cellulose waste serves as a sustainable material valorization within a circular bioeconomy. The starting materials for SAP production consist of lignocellulosic biomass that includes rice husks, wheat straw, sugarcane bagasse, and forestry pulp waste (Hachimi Alaoui et al., 2023; Xiao et al., 2023). The cellulose extracted from the wastes becomes CMC or hydroxyethyl cellulose (HEC) or graft copolymers through chemical transformation and then cross-linking produces superabsorbent hydrogel beads. These bead-shaped materials perform effectively as controlled-release fertilizers and pollutant sorbents and water-retention agents because of their high surface-area-to-volume ratio and porosity (Bekchanov et al., 2024). Moreover, the wastes from the different industries (paper mill or textile), which creates significant environmental pollution, is now recycled to form SAP beads (Chen et al., 2019).

Bead synthesis from waste sources has several advantages, including low-cost raw materials, waste reduction, and reduced landfill pressures. Furthermore, these bio-based SAP beads may dissolve in soil conditions, contributing nutrients rather than leaving lasting residues (Kumar et al., 2024). However, there are still issues in maintaining consistent quality control, particularly with regard to lignin and hemicellulose content fluctuation, which impacts bead shape and swelling capacity.

Innovative synthesis techniques, such as ionotropic gelation and green cross-linkers (e.g., citric acid, genipin), have demonstrated promising potential in removing hazardous initiators normally employed in SAP synthesis. Furthermore, enhanced bead functionalization using nanoparticles or enzymes is being investigated for environmental cleanup and medicinal delivery systems (Niu et al., 2025).

15.10 ECONOMIC AND ENVIRONMENTAL FEASIBILITY ANALYSIS

The economic and environmental viability of cellulose-based SAPs generated from lignocellulosic biomass is important to their commercialization and widespread acceptance. With increasing pressure to phase out non-degradable petroleum-based

SAPs, cellulose-based replacements have emerged as cost-competitive and ecologically advantageous—particularly when manufactured from waste biomass sources (Zhang et al., 2022; Dodangeh et al., 2024).

Economically, cellulose SAPs have low raw material costs, especially when made from industrial or agricultural byproducts like bagasse, sawdust, or paper mill sludge, which are either freely accessible or reasonably priced because they are considered waste, greatly lowering feedstock costs. Although it requires an initial investment, integrating enzymatic hydrolysis or green solvent extraction has shown long-term cost reductions by reducing the use of chemicals and energy. Additionally, there are significant financial benefits from lower irrigation and waste disposal expenses when SAPs are used in industries like sanitation (as biodegradable diaper materials) or agricultural (as water retention) (Li et al., 2025).

According to LCA research, cellulose-based SAPs have much lower carbon emissions and toxicity levels than their synthetic counterparts (Al-Gethami et al., 2024). Additionally, they reduce long-term ecological impact since they biodegrade in soil in a matter of weeks to months. However, these environmental advantages can be outweighed if harsh chemical treatments or cross-linkers derived from fossil fuels are used during manufacture (Fakhri et al., 2023). From a prospective standpoint, it is very feasible to combine the manufacturing of SAPs with industrial symbiosis, such as paper mills supplying cellulose waste and employing SAPs to treat wastewater. Government subsidies and carbon credits for bio-based products may increase their economic appeal even further.

15.11 CONCLUSION, CURRENT CHALLENGES, AND FUTURE OUTLOOK

15.11.1 Limitations in Scale-Up and Industrial Deployment

The beads made out of cellulose are multipurpose agents, primarily when they are chemically altered so as to promote their functionality. Because of their spherical shape and capacity to endure rapid flow rates, they make excellent filling substances for chromatographic columns. Therefore, cellulose beads could be employed as a stationary-phase medium in size differentiation chromatography as well as a selective adsorbent to biological entities, such as endotoxins, viruses, and proteins (Oliveira and Glasser, 1996). Additionally, cellulose beads have been utilized as a separation medium, to immobilize enzymes and to release active pharmaceutical substances in controlled manners (Fischer et al., 2008). Although these things are performed on laboratory sales and its production on large scale is difficult for both research and business (Wang et al., 2020).

Accomplishing the industrial manufacture and marketing of cellulose beads will face several serious challenges, including the need for a relevant market pull, low-cost raw materials, sustainable production, and greater expense of manufacturing. There is also a lack of global standardization with the supply line along with a deficiency of quick and inexpensive assessment in production techniques. Additionally, the inadequate coordination among organizers, administrations, and research. Thus, proper research and adequate funding for innovation with correct emphasis will be required for industrial-level manufacturing of cellulose beads (Clauser et al., 2022).

15.11.2 Stability and Solubility of Cellulose Beads

Cellulose is an outstanding and potential raw material that can trigger a shift in the use of petroleum-based plastics to more sustainable future because it has excellent chemical stability, good mechanical property, and is easy to functionalize (Mystek, 2022). Cellulose and its resultant products possess great thermal as well as mechanical stability and a three-dimensional open network configuration of high aspect ratio that can be incorporated with other products to bring about a great variety of composite material (Liu et al., 2022). Cellulose has extensive usage in different sectors, including paper textile industries (Lindman et al., 2017). Despite its frequent utilization, the low dissolution of raw cellulose restricts its usage (Yan et al., 2018). This is because the cellulose beads are hydrophobic and unable to dissolve in water and other organic solvents due to the crystalline structure and the presence of hydrogen bonds among cellulose molecules but not with the solvents (O'Brien et al., 2021).

Recycled cellulose beads are widely employed as absorbent substances due to their hydrophilic nature and porous makeup that provide large surface areas for absorption (Zhou et al., 2004). The existence of pores often influences the dispersion of absorbates during the process of absorption and is demonstrated as a major limiting factor of absorption rate (Hamidon et al., 2022). If cellulose beads have abnormal pores, it will affect the absorption capacity and will enhance the obstruction of mass transfer that will severely limit its use. Thus, appropriate arrangement and structure of pores is essential for the proper working of cellulose beads (Wei et al., 2023).

15.11.3 Regulatory Hurdles and Safety Concerns

Due to its numerous applications, which range from medical transportation to environmental remediation, cellulose-based beads have garnered considerable attention. However, safety issues and regulatory barriers continue to be major obstacles to their widespread commercialization. The changes necessary to produce functional beads, such as cross-linking, adding nanoparticles, or grafting stimuli-responsive groups, may result in cytotoxicity or environmental hazards, even though cellulose is naturally benign and biodegradable. Before approving a product for sale, regulatory agencies such as the FDA and EMA require thorough in vitro and in vivo testing to confirm its safety, biological compatibility, and degradation behavior. Furthermore, regulatory compliance depends on maintaining consistency and repeatability in physicochemical parameters from batch to batch (Chang et al., 2023). Furthermore, emerging biomaterials frequently have inadequate regulatory frameworks, which hinder development and raise expenses for developers (Xiao et al., 2023; Rumon, 2025).

15.11.4 Future Research Directions

Research in the future should focus on developing intelligent cellular beads that adapt to environmental factors such as sunlight and biomolecules, as well as pH levels. Specific drug release capabilities of these advanced beads enable improved therapeutic outcomes together with minimal side effects. The release of medications from pH-responsive beads can target infected tissues along with tumor microenvironments because they have reduced pH levels. The development of temperature and enzyme-responsive systems shows potential for future applications in both targeted therapeutic interventions and wound recovery (Abbasi Moud, 2022). Beads that integrate responsive polymers with nanostructures expand their functional capabilities to serve environmental monitoring applications alongside intelligent packaging solutions and biological detection systems (Srivastava and Choudhury, 2022). For future projects to succeed in meeting safety standards, they must address scalability challenges while evaluating production methods' environmental impact and conducting extensive long-term biocompatibility testing (Gong et al., 2024; Mukherjee, 2022).

REFERENCES

Abbasi Moud, A., 2022. Cellulose through the lens of microfluidics: a review. *Applied Biosciences*, *1*(1), pp.1–37.

Ahmed, M.J., 2016. Application of agricultural based activated carbons by microwave and conventional activations for basic dye adsorption: review. *Biochemical Pharmacology*, *4*, pp.89–99, https://doi.org/10.1016/j.jece.2015.10.027.

Al-Gethami, W., Qamar, M.A., Shariq, M., Alaghaz, A.-N.M.A., Farhan, A., Areshi, A.A. and Alnasir, M.H., 2024. Emerging environmentally friendly bio-based nanocomposites for the efficient removal of dyes and micropollutants from wastewater by adsorption: a comprehensive review. *RSC Advances, 14*, pp.2804–2834.

Azimi, B., Sepahvand, S., Ismaeilimoghadam, S., Kargarzadeh, H., Ashori, A., Jonoobi, M. and Danti, S., 2024. Application of cellulose-based materials as water purification filters; a state-of-the-art review. *Journal of Polymers and the Environment*, *32*(1), pp.345–366.

Bahsaine, K., Benzeid, H., El Allaoui, B., Zari, N., El Mahdi, M., el kacem Qaiss, A. and Bouhfid, R., 2024. Porous polyvinyl fluoride coated cellulose beads for efficient removal of Cd (II) from phosphoric acid. *International Journal of Biological Macromolecules*, *254*, p.127867.

Barbosa, R.F., Souza, A.G., Maltez, H.F. and Rosa, D.S., 2020. Chromium removal from contaminated wastewaters using biodegradable membranes containing cellulose nanostructures. *Chemical Engineering Journal*, *395*, p.125055.

Béguin, P. and Aubert, J.P., 1994. The biological degradation of cellulose. *FEMS Microbiology Reviews*, *13*(1), pp.25–58.

Bekchanov, D., Mukhamediev, M., Yarmanov, S., Lieberzeit, P. and Mujahid, A., 2024. Functionalizing natural polymers to develop green adsorbents for wastewater treatment applications. *Carbohydrate Polymers*, *323*, p.121397.

Beyki, M.H., Bayat, M. and Shemirani, F., 2016. Fabrication of core-shell structured magnetic nanocellulose base polymeric ionic liquid for effective biosorption of Congo red dye. *Bioresource Technology*, *218*, pp.326–334.

Braun, M., Guentherberg, N., Lutz, M., Magin, A., Siemer, M., Swaminathan, V. N., Linner, B., Ruslim, F.and Fernandez, R. G. A. U.S. Patent 20,100,331,222 A1, 2010.

Budtova, T., 2019. Cellulose II aerogels: a review. *Cellulose*, *26*, pp.81–121.

Califano, D., Patenall, B.L., Kadowaki, M.A., Mattia, D., Scott, J.L. and Edler, K.J., 2021. Enzyme-functionalized cellulose beads as a promising antimicrobial material. *Biomacromolecules*, *22*(2), pp.754–762.

Chakhtouna, H., Benzeid, H., Zari, N., Qaiss, A. el k. and Bouhfid, R., 2022. Recent advances in Eco-friendly composites derived from lignocellulosic biomass for wastewater treatment. *Biomass Conversion and Biorefinery*. https://doi.org/10.1007/s13399-022-03159-9.

Chang, S., Weng, Z., Zhang, C., Jiang, S. and Duan, G., 2023. Cellulose-based intelligent responsive materials: a review. *Polymers*, *15*(19), p.3905.

Chen, Y.W., Binti Hassan, S.H., Yahya, M. and Lee, H.V., 2019. *Novel superabsorbent cellulose-based hydrogels: present status, synthesis, characterization, and application prospects. Cellulose-based superabsorbent hydrogels*. Springer. pp.155–195.

Cheng, D., An, X., Zhang, J., Tian, X., He, Z., Wen, Y. and Ni, Y., 2017. Facile preparation of regenerated cellulose film from cotton linter using organic electrolyte solution (OES). *Cellulose*, *24*(4), pp.1631–1639.

Chuein, L., Nor N.I., Hamidon, T.S., Nur, F., Nor, S.A., Syazrin, S.S. and Hussin, M.H., 2021. Kinetics and equilibrium studies of methylene blue dye adsorption on oil palm frond adsorbent. *Desalination and Water Treatment, 216*, pp.358–371.

Clauser, N.M., Felissia, F.F., Area, M.C. and Vallejos, M.E., 2022. Technological and economic barriers of industrial-scale production of nanocellulose. In *Green nanomaterials for industrial applications*. Shanker, U., Hussain, C.M., and Rani, M., Eds. (pp.21–39). Elsevier.

Coombs OBrien, J., Torrente-Murciano, L., Mattia, D. and Scott, J.L., 2017. Continuous production of cellulose microbeads via membrane emulsification. *ACS Sustainable Chemistry & Engineering*, *5*(7), pp.5931–5939.

Dhyani, V. and Bhaskar, T., 2019. Pyrolysis of biomass. In *Biofuels: alternative feedstocks and conversion processes for the production of liquid and gaseous biofuels*. Pandey, A., Larroche, C., Dussap, C., Gnansounou, E., Khanal, S.K., and Ricke, S., Eds. (pp. 217-244). Academic Press.

Ding, Y., Song, C., Gong, W., Liu, L., Wu, M., Li, L. and Yao, J., 2021. Robust, sustainable, hierarchical multi-porous cellulose beads via pre-crosslinking strategy for efficient dye adsorption. *Cellulose, 28*(11), pp.7227–7241.

Dodangeh, F., Nabipour, H., Rohani, S. and Xu, C., 2024. Applications, challenges and prospects of superabsorbent polymers based on cellulose derived from lignocellulosic biomass. *Bioresource Technology*, *408*, p.131204.

El Allaoui, B., Benzeid, H., Zari, N., el kacem Qaiss, A. and Bouhfid, R., 2023. Functional cellulose-based beads for drug delivery: preparation, functionalization, and applications. *Journal of Drug Delivery Science and Technology*, *88*, p.104899.

Erdal, N.B. and Hakkarainen, M., 2022. Degradation of cellulose derivatives in laboratory, man-made, and natural environments. *Biomacromolecules*, *23*(7), pp.2713–2729.

Esa, F., Tasirin, S.M. and Abd Rahman, N., 2014. Overview of bacterial cellulose production and application. *Agriculture and Agricultural Science Procedia*, *2*, pp.113–119.

Fakhri, V., Jafari, A., Vahed, F.L., Su, C.-H. and Pirouzfar, V., 2023. Polysaccharides as eco-friendly bio-adsorbents for wastewater remediation: current state and future perspective. *Journal of Water Process Engineering*, *54*, p.103980.

Feng, J., Nguyen, S.T., Fan, Z. and Duong, H.M., 2015. Advanced fabrication and oil absorption properties of super-hydrophobic recycled cellulose aerogels. *Chemical Engineering Journal*, *270*, pp.168–175.

Fiorica, C., Biscari, G., Palumbo, F.S., Pitarresi, G., Martorana, A. and Giammona, G., 2021. Physicochemical and rheological characterization of different low molecular weight gellan gum products and derived ionotropic crosslinked hydrogels. *Gels*, *7*(2), p.62.

Firmanda, A., Fahma, F., Syamsu, K., Suryanegara, L. and Wood, K., 2022. Controlled/slow-release fertilizer based on cellulose composite and its impact on sustainable agriculture. *Biofuels, Bioproducts and Biorefining*, *16*(6), pp.1909–1930.

Fischer, S., Thümmler, K., Volkert, B., Hettrich, K., Schmidt, I. and Fischer, K., 2008, January. Properties and applications of cellulose acetate. In *Macromolecular symposia* (Vol. 262, No. 1, pp.89–96). WILEY-VCH Verlag. https://doi.org/10.1002/masy.200850210

Foroughi, F., Rezvani Ghomi, E., Morshedi Dehaghi, F., Borayek, R. and Ramakrishna, S., 2021. A review on the life cycle assessment of cellulose: from properties to the potential of making it a low carbon material. *Materials*, *14*(4), p.714.

Fu, Q., Cui, C., Meng, L., Hao, S., Dai, R. and Yang, J., 2021. Emerging cellulose-derived materials: a promising platform for the design of flexible wearable sensors toward health and environment monitoring. *Materials Chemistry Frontiers*, *5*(5), pp.2051–2091.

Gao, W.H., Chen, K.F., Yang, R.D., Yang, F. and Han, W.J., 2011. Properties of bacterial cellulose and its influence on the physical properties of paper. *BioResources*, *6*(1), pp.144–153.

Gericke, M., Trygg, J. and Fardim, P., 2013. Functional cellulose beads: preparation, characterization, and applications. *Chemical Reviews*, *113*(7), pp.4812–4836.

Ghazali, N.A., Salleh, K.M., Jafri, N.F., Lau, K.S., Chia, C.H. and Zakaria, S., 2025. High performance adsorption of regenerated cellulose/TEMPO-cellulose for water purification. *International Journal of Biological Macromolecules*, *319*, p.145456.

Gibson, L.J., 2012. The hierarchical structure and mechanics of plant materials. *Journal of the Royal Society Interface*, *9*(76), p.2749.

Gomez-Maldonado, D., Phillips, S.G., Vaidya, S.R., Bartley, P.C., White, J.C., Fairbrother, D.H. and Peresin, M.S., 2024. Modifying soluble NPK release with hydrophobized nanocellulose-based hydrogels for sustainable enhanced efficiency fertilizers. *Environmental Science: Nano*, *11*(2), pp.529–545.

Gong, J., Hou, L., Ching, Y.C., Ching, K.Y., Dai Hai, N. and Chuah, C.H., 2024. A review of recent advances of cellulose-based intelligent-responsive hydrogels as vehicles for controllable drug delivery system. *International Journal of Biological Macromolecules*, *264*, p.130525.

Gong, Y., Yuan, J., Pei, Y., Liu, S. and Luo, X., 2023. One-step quaternization and macromolecular reconstruction to prepare micro-/nano-porous cellulose beads from homogeneous solution for low-concentration amoxicillin removal. *Carbohydrate Polymers*, *315*, p.120985.

Hachimi Alaoui, C., Réthoré, G., Weiss, P. and Fatimi, A., 2023. Sustainable biomass lignin-based hydrogels: a review on properties, formulation, and biomedical applications. *International Journal of Molecular Sciences*, *24*, p.13493.

Hamidon, T.S., Adnan, R., Haafiz, M.M. and Hussin, M.H., 2022. Cellulose-based beads for the adsorptive removal of wastewater effluents: a review. *Environmental Chemistry Letters*, *20*(3), pp.1965–2017.

Hammouda, S., Chen, Z., An, C. and Lee, K., 2021. Recent advances in developing cellulosic sorbent materials for oil spill cleanup: a state-of-the-art review. *Journal of Cleaner Production*, *311*, pp.1–19.

He, B., Chang, P., Zhu, X. and Zhang, S., 2021. Anemone-inspired enzymatic film for cellulose heterogeneous catalysis. *Carbohydrate Polymers*, *260*, p.117795, https://doi.org/10.1016/J.CARBPOL.2021.117795.

Heinze, T., Seoud, O.A. el, and Koschella. A., 2018. Principles of cellulose derivatization. In *Cellulose Derivatives. Springer Series on Polymer and Composite Materials*. Heinze, T., El Seoud, O.A., Koschella, A., Eds. (pp.259–292). Springer.

Hokkanen, S., Bhatnagar, A. and Sillanpää, M., 2016. A review on modification methods to cellulose-based adsorbents to improve adsorption capacity. *Water Research*, *91*, pp.156–173.

Homaei, A.A., Sariri, R., Vianello, F. and Stevanato, R., 2013. Enzyme immobilization: an update. *Journal of Chemical Biology*, *6*, pp.185–205.

Huang, Y., Zhu, C., Yang, J., Nie, Y., Chen, C. and Sun, D., 2014. Recent advances in bacterial cellulose. *Cellulose*, *21*(1), pp.1–30.

Idrissi, A., Gharrak, A. el, Achagri, G., Essamlali, Y., Amadine, O., Akil, A., Sair, S. and Zahouily, M., 2022. Synthesis of urea-containing sodium alginate-g-poly(acrylic acid-coacrylamide) superabsorbent-fertilizer hydrogel reinforced with carboxylated cellulose nanocrystals for efficient water and nitrogen utilization. *Journal of Environmental and Chemical Engineering*, *10*, p.108282, https://doi.org/10.1016/j.jece.2022.108282.

Jiao, Z., Zhang, B., Li, C., Kuang, W., Zhang, J., Xiong, Y., Tan, S., Cai, X. and Huang, L., 2018. Carboxymethyl cellulose-grafted graphene oxide for efficient antitumor drug delivery. *Nanotechnology Reviews*, *7*(4), pp.291–301.

Joshi, S., Choudhary, M. and Srivastava, N., 2021. Cellulase production using different microbial sources. In *Current status and future scope of microbial cellulases.* Tuli, D.K. and Kuila, A., Eds. (pp.1–17). Elsevier.

Kasbaji, M., Mennani, M. Oubenali, M., Benhamou, A. Ait , Boussetta, A., Ablouh, E.-H., Mbarki, M., Grimi, N., Achaby, M. El. and Moubarik, A., 2023. Bio-based functionalized adsorptive polymers for sustainable water decontamination: a systematic review of challenges and real-world implementation. *Environmental Pollution, 335*, p.122349.

Kim, H.J., Jin, J.N., Kan, E., Kim, K.J. and Lee, S.H., 2017. Bacterial cellulose-chitosan composite hydrogel beads for enzyme immobilization. *Biotechnology and Bioprocess Engineering, 22*, pp.89–94.

Kim, J., Kim, Y., Jung, S., Yun, H., Won, S., Yeo, H., Choi, I.G. and Kwak, H.W., 2024. Green lignocellulosic superadsorbent for superior Pd (II) removal and cascade catalytic conversion. *Separation and Purification Technology, 332*, p.125732.

Köse, K., Mavlan, M. and Youngblood, J.P., 2020. Applications and impact of nanocellulose based adsorbents. *Cellulose, 27*(6), pp.2967–2990.

Kumar, S., Kumar, A., Thakur, A. and Kumari, P., 2024. *Valorizing cellulosic biomass waste into valuable nano-biosorbents. Integrated waste management: a sustainable approach from waste to wealth.* Springer. pp.239–268.

Larsson, P.T., Svensson, A. and Wågberg, L., 2013. A new, robust method for measuring average fibre wall pore sizes in cellulose I rich plant fibre walls. *Cellulose, 20*(2), pp.623–631.

Latif, W., Ciniglia, C., Iovinella, M., Shafiq, M. and Papa, S., 2023. Role of white rot fungi in industrial wastewater treatment: a review. *Applied Sciences, 13*(14), pp.8318.

Lau,K.S.,Azmi,N.A.S.,Chin,S.X.,Zakaria,S.andChia,C.H.,2023.Chitosan-bead-encapsulated polystyrene sulfonate for adsorption of methylene blue and regeneration studies: batch and continuous approaches. *Polymers, 15*(5), p.1269.

Li, B., Pan, Y., Zhang, Q., Huang, Z., Liu, J. and Xiao, H., 2019. Porous cellulose beads reconstituted from ionic liquid for adsorption of heavy metal ions from aqueous solutions. *Cellulose, 26*, pp.9163–9178.

Li, M., Zhang, P., Mao, J., Li, J., Zhang, Y., Xu, B., Zhou, J., Cao, Q. and Xiao, H., 2024. Construction of cellulose-based hybrid hydrogel beads containing carbon dots and their high performance in the adsorption and detection of mercury ions in water. *Journal of Environmental Management, 359*, p.121076.

Li, S, Tasnady, D. and Yoldaş, F., 2025. From Waste to Superabsorbent Polymers. In *Superabsorbent Polymers Synthesis, Properties and Applications*. Wang, A. and Zhu, Y., Eds., (vol. 42, ch. 4, pp. 109–147). Royal Society of Chemistry, London.

Lindh, J., Carlsson, D.O., Strømme, M. and Mihranyan, A., 2014. Convenient one-pot formation of 2, 3-dialdehyde cellulose beads via periodate oxidation of cellulose in water. *Biomacromolecules, 15*(5), pp.1928–1932.

Lindman, B., Medronho, B., Alves, L., Costa, C., Edlund, H. and Norgren, M., 2017. The relevance of structural features of cellulose and its interactions to dissolution, regeneration, gelation and plasticization phenomena. *Physical Chemistry Chemical Physics, 19*(35), pp.23704–23718.

Liu, Q., Xia, N., Wan, W., Gao, Y. and Zhu, S., 2021a. Selective capture of toxic anionic dyes of a novel prepared DMDAAC-grafted chitosan/genipin/cellulose hydrogel beads with antibacterial activity. *International Journal of Biological Macromolecules, 189*, pp.722–733.

Liu, P., Sehaqui, H., Tingaut, P., Wichser, A., Oksman, K. and Mathew, A.P., 2014. Cellulose and chitin nanomaterials for capturing silver ions (Ag+) from water via surface adsorption. *Cellulose, 21*, pp.449–461.

Liu, W., Liu, K., Du, H., Zheng, T., Zhang, N., Xu, T., Pang, B., Zhang, X., Si, C. and Zhang, K., 2022. Cellulose nanopaper: fabrication, functionalization, and applications. *Nano-Micro Letters, 14*(1),104.

Liu, W., Budtova, T. and Navard, P., 2011. Influence of ZnO on the properties of dilute and semi-dilute cellulose-NaOH-water solutions. *Cellulose, 18*, pp.911–920.

Liu, Y. and Chen, J.Y., 2016. Enzyme immobilization on cellulose matrixes. *Journal of Bioactive and Compatible Polymers, 31*(6), pp.553–567.

Liu, Y., Qiao, L., Wang, A., Li, Y., Zhao, L. and Du, K., 2021. Tentacle-type poly(hydroxamic acid)-modified macroporous cellulose beads: synthesis, characterization, and application for heavy metal ions adsorption. *Journal of Chromatography A, 1645,* p.104.

Lobato-Peralta, D.R., Duque-Brito, E., Ayala-Cortés, A., Arias, D.M., Longoria, A., Cuentas-Gallegos, A. K., Sebastian, P.J. and Okoye, P.U., 2022. Advances in activated carbon modification, surface heteroatom configuration, reactor strategies, and regeneration methods for enhanced wastewater treatment. *Journal of Environmental and Chemical Engineering, 9*, pp.1–23.

Maggioris, D., Goulas, A., Alexopoulos, A H., Chatzi, E.G. and Kiparissides, C., 2000. Prediction of particle size distribution in suspension polymerization reactors: effect of turbulence nonhomogeneity. *Chemical Engineering Science, 55*, p.4611.

Mahmud, S., Hasan, K.F., Jahid, M.A., Mohiuddin, K., Zhang, R. and Zhu, J., 2021. Comprehensive review on plant fiber-reinforced polymeric biocomposites. *Journal of Materials Science, 56*(12), pp.7231–7264.

Mazzitelli, S., Tosi, A., Balestra, C., Nastruzzi, C., Luca, G., Mancuso, F., Calafiore, R. and Calvitti, M., 2008. Production and characterization of alginate microcapsules produced by a vibrational encapsulation device. *Journal of Biomaterials Applications, 23*, p.123.

Meng, R., Liu, L., Jin, Y., Luo, Z., Gao, H. and Yao, J., 2019. Recyclable carboxylated cellulose beads with tunable pore structure and size for highly efficient dye removal. *Cellulose, 26*, pp.8963–8969.

Miyaji, K., Zhou, C., Maeda, A., Kobayashi, K., Kusumi, R. and Wada, M., 2023. Biodegradation of various forms of cellulose and chitin in natural waters with different salinity. *Polymer Degradation and Stability, 215*, p.110423.

Moniri, M., Boroumand Moghaddam, A., Azizi, S., Abdul Rahim, R., Bin Ariff, A., Zuhainis Saad, W., Navaderi, M. and Mohamad, R., 2017. Production and status of bacterial cellulose in biomedical engineering. *Nanomaterials, 7*(9), p.257.

Morozkina, S., Strekalovskaya, U., Vanina, A., Snetkov, P., Krasichkov, A., Polyakova, V., Uspenskay M., 2022. The fabrication of alginate–carboxymethyl cellulose-based composites and drug release profiles. *Polymers, 14*.p.3604.

Mukherjee, I., 2022. Recent Development of Polysaccharide-Derived Hydrogel. properties, Stimuli-Responsiveness and Bioapplications. *EPRA International Journal of Research and Development, 7*(9), pp.19–31.

Mystek, K., 2022. *Use of Cellulose for the Preparation of Capsules and Beads with Molecularly Tailored Properties* (Doctoral dissertation, KTH Royal Institute of Technology).

Naserifar, P.F. Kuijpers, S., Wojno, R., Kádár, D.and Bernin, M., 2022. Hasani In situ monitoring of cellulose etherification in solution: probing the impact of solvent composition on the synthesis of 3-allyloxy-2-hydroxypropyl-cellulose in aqueous hydroxide systems. *Polymer Chemistry, 13*, pp.4111–4123.

Niu, Q., Xie, J., Li, J., An, Z., Xiao, H., Zhang, X., Su, Z. and Wang, Z., 2025. Superabsorbent polymers: innovations in ecology, environmental, and diverse applications. *Materials* (Basel). *18*, p.823.

O'Neill, J. J., Jr., Reichardt, E. P. U.S. Patent 2,543,928, 1951.

O'Brien, C.T., Virtanen, T., Donets, S., Jennings, J., Guskova, O., Morrell, A.H., Rymaruk, M., Ruusuvirta, L., Salmela, J., Setala, H. and Sommer, J.U., 2021. Control of the aqueous solubility of cellulose by hydroxyl group substitution and its effect on processing. *Polymer, 223*, p.123681.

Oladosu, Y., Rafii, M.Y., Arolu, F., Chukwu, S.C., Salisu, M.A., Fagbohun, I.K., Muftaudeen, T.K., Swaray, S. and Haliru, B.S., 2022. Superabsorbent polymer hydrogels for sustainable agriculture: a review. *Horticulturae*, *8*, p.605.

Oliveira, W.D. and Glasser, W.G., 1996. Hydrogels from polysaccharides. I. Cellulose beads for chromatographic support. *Journal of Applied Polymer Science*, *60*(1), pp.63–73.

Omidian, H., Akhzarmehr, A. and Chowdhury, S.D., 2024. Advancements in cellulose-based superabsorbent hydrogels: sustainable solutions across industries. *Gels*, *10*, p.174.

Ram, B. and Chauhan, G.S., 2018. New spherical nanocellulose and thiol-based adsorbent for rapid and selective removal of mercuric ions. *Chemical Engineering Journal*, *331*, pp.587–596.

Rasheed, M., Jawaid, M., Karim, Z., Abdullah, L.C., 2020. Morphological, physiochemical and thermal properties of microcrystalline cellulose (MCC) extracted from bamboo fiber. *Molecules*, *25*, https://doi.org/10.3390/molecules25122824.

Rathi, B.S., Kumar, P.S. and Show, P., 2021. A review on effective removal of emerging contaminants from aquatic systems: current trends and scope for further research. *Journal of Hazardous Materials*, *409*, p.124413.

Rosenberg, P., Budtova, T., Rom, M. and Fardim, P., 2010. Effect of Enzymatic treatment on solubility of cellulose in 7.6%NaOH-Water and ionic liquid. In *cellulose solvents: for analysis, shaping and chemical modification*. Liebert, T., Heinze, T., and Edgar, K. J., Eds., (pp.213–226). American Chemical Society, Washington, DC.

Rumon, M.M.H., 2025. Advances in cellulose-based hydrogels: tunable swelling dynamics and their versatile real-time applications. *RSC Advances*, *15*(15), pp.11688–11729.

Shimizu, M., Kusumi, R., Saito, T. and Isogai, A., 2019. Thermal and electrical properties of nanocellulose films with different interfibrillar structures of alkyl ammonium carboxylates. *Cellulose*, *26*(3), pp.1657–1665.

Solangi, N.H., Kumar, J., Mazari, S.A., Ahmed, S., Fatima, N. and Mubarak, N.M., 2021. Development of fruit waste derived bio-adsorbents for wastewater treatment: a review. *Journal of Hazardous Materials, 416*, pp.125848.

Srivastava, N. and Choudhury, A.R., 2022. Stimuli-responsive polysaccharide-based smart hydrogels and their emerging applications. *Industrial & Engineering Chemistry Research*, *62*(2), pp.841–866.

Sun, R., Zhu, J., Wu, H., Wang, S., Li, W. and Sun, Q., 2021. Modulating layer-by-layer assembled sodium alginate-chitosan film properties through incorporation of cellulose nanocrystals with different surface charge densities. *International Journal of Biological Macromolecules*, *180*, pp.510–522.

Syazwani, N., Rahman, A., Firdaus, M. and Baharin, Y., 2018. Utilisation of natural cellulose fibres in wastewater treatment. *Cellulose 25*, pp.4887–4903.

Tapia-orozco, N., Ibarra-cabrera, R., Tecante, A., Gimeno, M., Parra, R. and Garcia-arrazola, R., 2023. Removal strategies for endocrine disrupting chemicals using cellulose-based materials as adsorbents: a review. *Biochemical Pharmacology 4*, pp.3122–3142.

Tavakolian, M., Okshevsky, M., van de Ven, T.G. and Tufenkji, N., 2018. Developing antibacterial nanocrystalline cellulose using natural antibacterial agents. *ACS Applied Materials & Interfaces*, *10*(40), pp.33827–33838.

Thakur, V., Guleria, A., Kumar, S., Sharma, S. and Singh, K., 2021. Recent advances in nanocellulose processing, functionalization and applications: a review. *Materials Advances*, *2*, pp.1872–1895, https://doi.org/10.1039/d1ma00049g.

Tripathy, S., Sahu, S., Kishore, R., Bihari, R. and Kar, P.K., 2021. Groundwater for sustainable development efficient removal of Cr (VI) by polyaniline modified biochar from date (phoenix dactylifera) seed. *Groundwater for Sustainable Development 15*, p.100653.

Trygg, J., Fardim, P., Gericke, M., Mäkilä, E. and Salonen, J., 2013. Physicochemical design of the morphology and ultrastructure of cellulose beads. *Carbohydrate Polymers*, *93*(1), pp.291–299.

Tu, H., Zhu, M., Duan, B. and Zhang, L., 2021. Recent progress in high-strength and robust regenerated cellulose materials. *Advanced Materials*, *33*(28), p.2000682.

Ullah, H., Wahid, F., Santos, H.A. and Khan, T., 2016. Advances in biomedical and pharmaceutical applications of functional bacterial cellulose-based nanocomposites. *Carbohydrate Polymers*, *150*, pp.330–352, https://doi.org/10.1016/j.carbpol.2016.05.029.

Varghese, A.G., Paul, S.A. and Latha, M.S., 2019. Remediation of heavy metals and dyes from wastewater using cellulose-based adsorbents. *Environmental Chemistry Letters*, *17*(2), pp.867–877.

Voon, L.K., Pang, S.C. and Chin, S.F., 2016. Highly porous cellulose beads of controllable sizes derived from regenerated cellulose of printed paper wastes. *Materials Letters*, *164*, pp.264–266.

Wang, H., Li, B. and Shi, B., 2008. Preparation and surface acid-base properties of porous cellulose. *BioResources*, *3*(1), pp.3–12.

Wang, L., Gardner, D.J., Wang, J., Yang, Y., Tekinalp, H.L., Tajvidi, M., Li, K., Zhao, X., Neivandt, D.J., Han, Y. and Ozcan, S., 2020. Towards industrial-scale production of cellulose nanocomposites using melt processing: a critical review on structure-processing-property relationships. *Composites Part B: Engineering*, *201*, p.108297.

Wang, R., Feng, Y., Li, D., Li, K. and Yan, Y., 2024. Towards the sustainable production of biomass-derived materials with smart functionality: a tutorial review. *Green Chemistry*, *26*, pp.9075–9103.

Wang, D.M. and Sun, Y., 2017. Fabrication of superporous cellulose beads with grafted anion-exchange polymer chains for protein chromatography. *Biochemical Engineering Journal*, *37*, pp.332–337.

Wang, S., Li, T., Chen, C., Kong, W., Zhu, S., Dai, J., Diaz, A.J., Hitz, E., Solares, S.D., Li, T. and Hu, L., 2018. Transparent, anisotropic biofilm with aligned bacterial cellulose nanofibers. *Advanced Functional Materials*, *28*(24), p.1707491.

Wang, S. Liu, H. Chen, J. Liu, Q. and Zhu, L., 2022. TEMPO-oxidized cellulose beads for cationic dye adsorption. *Bioresources*, *17*, p.6056.

Wei, J., Wang, B., Yuan, H., Kang, Z., Gao, H. and Nie, Y., 2023. Effects of glucose and coagulant on the structure and properties of regenerated cellulose fibers. *Biomacromolecules*, *24*(4), pp.1810–1818.

Xiao, W., Sun, R, Hu, S., Meng, C., Xie, B., Yi, M. and Wu, Y., 2023. Recent advances and future perspective on lignocellulose-based materials as adsorbents in diverse water treatment applications. *International Journal of Biological Macromolecules*, *253*, p.126984.

Xiao, R., Zhou, G., Wen, Y., Ye, J., Li, X. and Wang, X., 2023. Recent advances on stimuli-responsive biopolymer-based nanocomposites for drug delivery. *Composites Part B: Engineering*, *266*, p.111018.

Xie, F. Peter, de W., Fardim, P. and van den Mooter, G., 2021. Tempo-oxidized cellulose beads as potential Ph-responsive carriers for site-specific drug delivery in the gastrointestinal tract. *Molecules*, *26*, p.1030.

Xiong, L. and Zhang, Y., 2023. Wang Polymer fractionation using chromatographic column packed with novel regenerated cellulose beads modified with silane *Journal of Chromatography A*, *1063*, pp.71–77.

Yan, G., Zhang, X., Li, M., Zhao, X., Zeng, X., Sun, Y., Tang, X., Lei, T. and Lin, L., 2018. Stability of soluble dialdehyde cellulose and the formation of hollow microspheres: optimization and characterization. *ACS Sustainable Chemistry & Engineering*, *7*(2), pp.2151–2159.

Ybañez, M.G. and Camacho, D.H., 2021. Designing hydrophobic bacterial cellulose film composites assisted by sound waves. *RSC Advances*, *11*(52), pp.32873–32883.

Yu, H., Fu, G., Zhao, J., Liu, L. and He, B., 2020. Synthesis and in vitro sorption properties of PAA-grafted cellulose beads for selective binding of LDL. *Artificial Cells, Blood Substitutes Biotechnology*, *34*, pp.501–513.

Zennifer, A., Senthilvelan, P., Sethuraman, S. and Sundaramurthi, D., 2021. Key advances of carboxymethyl cellulose in tissue engineering & 3D bioprinting applications. *Carbohydrate Polymers*, *256*, p.117561.

Zhang, Z., Cai, S., Li, Y., Wang, Z., Long, Y., Yu, T. and Shen, Y., 2020. High performances of plant fiber reinforced composites—A new insight from hierarchical microstructures. *Composites science and technology*, *194*, p.108151.

Zhang, H., Liang, H., Xing, L., Ding, W., Geng, Z. and Xu, C., 2023. Cellulose-based slow-release nitrogen fertilizers: synthesis, properties, and effects on pakchoi growth. *International Journal of Biological Macromolecules*, *244*, pp.1–10.

Zhang, M. and Yang, J., 2021. Preparation and characterization of multifunctional slow release fertilizer coated with cellulose derivatives. *International Journal of Polymeric Materials and Polymeric Biomaterials*, *70*(11), pp.774–781.

Zhang, W., Zhang, P., Wang, H., Li, J. and Dai, S.Y., 2022. Design of biomass-based renewable materials for environmental remediation. *Trends in Biotechnology*, *40*(12), pp.1519–1534.

Zhang, Y., Kobayashi, K., Kusumi, R., Kim, U.J. and Wada, M., 2024. Adsorption properties of cellulose beads prepared by emulsion-gelation method and subsequent oxidation. *Colloids and Surfaces A: Physicochemical and Engineering Aspects*, *682*, p.132859.

Zhao, X., Xiao, Z., Qiao, Z. and Zhou, J., 2024. Insights into the assembly process and properties of regenerated cellulose beads prepared in alkali/urea aqueous solutions. *Carbohydrate Polymers*, *338*, p.122184.

Zhao, L., Liang, C., Li, S. and Du, K., 2021. Study of tentacle-like cationic macroporous cellulose spherical adsorbent for heavy metals. *Journal of Cleaner Production*, *303*, p.127114.

Zhou, D., Zhang, L., Zhou, J. and Guo, S., 2004. Cellulose/chitin beads for adsorption of heavy metals in aqueous solution. *Water Research*, *38*(11), pp.2643–2650.

Zimmermann, M.V.G., Borsoi, C., Lavoratti, A., Zanini, M., Zattera, A.J. and Santana, R.M.C., 2016. Drying techniques applied to cellulose nanofibers. *Journal of Reinforced Plastics and Composites 35*, pp.682–697.

Żywicka, A., Peitler, D., Rakoczy, R., Junka, A.F. and Fijałkowski, K., 2016. Wet and dry forms of bacterial cellulose synthetized by different strains of *Gluconacetobacter xylinus* as carriers for yeast immobilization. *Applied Biochemistry and Biotechnology*, *180*, pp.805–816.

16 Cellulose Beads' Role in Drug-Delivery Systems

Muntaha Mahin, Mir Abdul Gaffar, and Md Rashedul Islam

16.1 INTRODUCTION

The introduction of biocompatible and biodegradable carriers has made great advancements in the area of drug-delivery systems. Researchers are currently heavily focusing on natural polysaccharide polymers because of their non-toxic characteristics, cost-effectiveness, biocompatibility, and biodegradability. With an expected yearly worldwide production rate of roughly 1.5 trillion tons, cellulose is the most important organic carbon source among polysaccharides (Xie et al., 2023). Cellulose has been transformed into micro and nano-sized particles in recent years highlighting the increased potential of using carriers based on cellulose for medicinal applications (Carvalho et al., 2021). Among these advances, cellulose beads (spherical microparticles or spheres of millimeter size) derived from regenerated cellulose have attracted great interest in drug-delivery studies.

A key characteristic of cellulose beads is their porous nature and substantial surface area. This mesoporous structure is generated from the regeneration process and influences drug loading and release rates greatly (Zeng et al., 2020). While comparing with synthetic polymers, cellulose shows more chemical stability within the body and generally doesn't induce substantial immunological or toxic responses (Xie et al., 2023). Different approaches, including ionic crosslinking, chemical grafting, and the integration of nanoparticles, which significantly impact their drug-loading capacity and release rate control, help to advance and improve cellulose beads (Alvarez-Lorenzo et al., 2013; Purohit et al., 2023). The stability of cellulose beads in physiological conditions enhances their versatility for various delivery methods, including oral, topical, and injectable applications (Klojdová et al., 2023). However, a significant challenge remains in producing cellulose bead formulations as small variations in their manufacturing process can heavily affect their mechanical properties, drug encapsulation efficiency, and release dynamics. Therefore, establishing sustainable processes to achieve high drug diffusion efficiency now remains the primary focus of research (El Allaoui, Benzeid et al., 2023).

This chapter comprehensively explores cellulose beads in drug delivery, addressing their changes, drug-loading methods, and release strategies. It also discusses new developments and challenges related to enhancing cellulose-based carriers for improved therapeutic effects.

DOI: 10.1201/9781003596127-19

16.2 CHARACTERISTICS AND DRUG-CARRIER INTERACTIONS OF CELLULOSE BEADS

16.2.1 Structural and Chemical Properties

Cellulose beads are biocompatible microspheres with sizes ranging from micrometers to millimeters. They have a highly porous structure with a high surface area and lower density. Furthermore, the presence of additional hydroxyl (–OH) reactive groups provides multiple reactive sites to integrate multi-functionality (El Allaoui, Benzeid et al., 2023). Among the different structures of cellulose, Cellulose I shows high mechanical characteristics. With intrinsic stiffness and strength, it is the preferred choice for biomedical applications when properly structured (Benítez & Walther, 2017). Crystalline cellulose I possesses elastic modulus up to 155 GPa along the main cellulose axis (Muthoka et al., 2020).

Cellulose is insoluble in most of the organic solvents and water as they have a tightly packed structure by forming hydrogen bonds with each other (O'Brien et al., 2021). Hydrophobic and electrostatic interactions within attached fibers reinforce the stability further (Lindman et al., 2017). Advanced ionic liquid solvent systems such as 1-butyl-3-methylimidazolium acetate (BmimAc), 1-butyl-3-methylimidazolium formate (BmimF), 1-butyl-3-methylimidazolium chloride (BmimCl), *N,N*-dimethylacetamide/lithium chloride (DMAc/LiCl), NaOH/urea, LiOH/urea, and NaOH/thiourea are used to dissolve cellulose (F. Hu et al., 2021; Takada & Kadokawa, 2022; Tu et al., 2022). For example, *N*-methyl morpholine-*N*-oxide monohydrate (NMMO-H_2O) is a non-toxic solvent system that is effective in dissolving cellulose. A recovery rate of approximately 90% makes it a favorable option for cellulose processing (Carrillo-Varela et al., 2022).

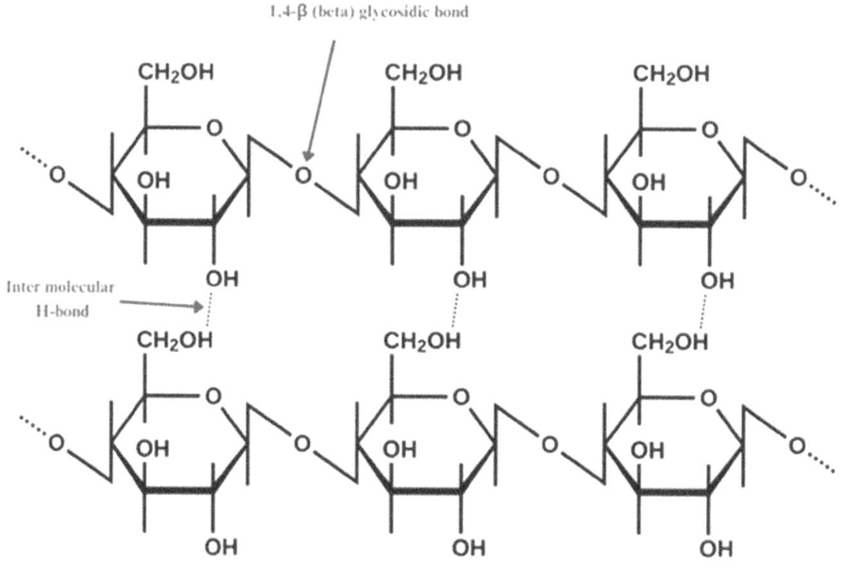

FIGURE 16.1 Intermolecular hydrogen bonding of cellulose.

The existence of hydroxyl groups (–OH) in cellulose structure attracts and retains water molecules making it highly hygroscopic (Figure 16.1). This allows cellulose to swell when exposed to water thus paving the way for aqueous manufacturing of cellulosic-based composites. The inclusion of hydrophilic polymers can lead to more swelling due to moisture adsorption from the air or water (Benítez et al., 2013) (Tables 16.1–16.8).

TABLE 16.1
Impact of water absorption on composite materials' properties

Material Used	Exposure Conditions	Key Findings	Effect on Mechanical Properties	Reference
Hemp fiber-reinforced unsaturated polyester composites	Water absorption at 25°C and 100°C over varying durations	More water uptake in hemp fiber-reinforced samples submerged in boiling water due to swelling of hemp fiber, leading to micro-cracking. Increased cellulose content resulted in higher moisture absorption	Mechanical properties reduced with increased moisture uptake, except for 3-mm thick hemp fiber-reinforced samples, where ultimate tensile stress was higher in aged samples than dry samples	Dhakal et al. (2007)
Flax/polylactic acid biocomposites	Immersion in seawater	Increased Young's modulus in composites exposed to seawater, with the modulus increased up to 32 MPa in the 0.5% range	Improvement in mechanical properties due to combined fiber stiffness and water desorption during tensile testing	Le Duigou et al. (2014)
Flax fibers	Exposure to 90% RH	Exposure to humid conditions (90% RH) increased tensile strength	Increase in tensile strength under humid conditions	Stamboulis et al. (2001)
Bamboo/vinyl ester composites	Water absorption at varying RH levels (60%–90%) at 20°C	Rapid initial water absorption, succeeded decrease until equilibrium was attained. Sharp increase in water content at 80%–90% RH due to capillary spaces in bamboo fibers	Interfacial shear strength decreased with increasing RH, especially beyond 80% RH, due to compromised bonding between polymer and fiber components	H. Chen et al. (2009)

TABLE 16.2
Various types of solvent systems for cellulose dissolution

Solvent System	Key Details	Conditions	Drawbacks	Stabilization/ Pretreatment	References
NMMO-Based dissolution	Cellulose consists of α-D-glucopyranose units (AGUs) linked by β (1→4) glucoside bonds. NMMO breaks intermolecular bonds via hydrogen bonding with cellulose AGUs. Oxidation of hydroxyl groups can degrade cellulose	NMMO monohydrate (13% water); concentrations up to 23%; 80°C–120°C	Thermal instability, especially with additives like magnetic particles or polysaccharide derivatives	Alkaline and radical scavengers such as propyl gallate and sodium hydroxide; microwave heating (210 W, 2450 MHz); ultrasonic pretreatment (37 kHz, 320 W)	Acharya et al. (2021); Baek and Park (2021); El-Wakil et al. (2024); L. Li et al. (2018); Rosenau et al. (2021)
DMAc/ LiCl-based dissolution	Chloride anions (Cl^-) produce hydrogen bonds with cellulose hydroxyl groups; DMAc solvates lithium cations (Li^+); disrupts cellulose's hydrogen bonding network	DMAc with LiCl (5–13 wt%, optimally 8 wt%)	Highly hygroscopic; cellulose degradation above 85°C due to keteniminium intermediates	Solvent exchange (water → methanol → DMAc/ LiCl); freeze-drying	Acharya et al. (2021); Y. Hu et al. (2019); H. Li et al. (2020); Sayyed et al. (2019); Wahlström et al. (2020)
NaOH-based dissolution	NaOH breaks hydrogen bonds in cellulose; additives like urea, thiourea, PEG, and ZnO improve dissolution	NaOH (7%–10%); low temperatures (< 10°C, improves below 0°C)	Limited solubility (5%); only effective for low-DP (< 250) and low-crystallinity cellulose	Chemical, mechanical, and enzymatic pretreatments; additives (urea, thiourea, PEG, ZnO)	Dormanns et al. (2016); Han and Seo (2021); K. Li et al. (2020); Lu et al. (2023); Tu et al. (2022)
Ionic liquids (ILs)–based dissolution	Anions in ILs make hydrogen bonds with hydroxyl groups pf cellulose; breaks intra- and intermolecular hydrogen bonds	ILs like imidazolium and pyridinium-based cations with anions (e.g., chloride)	Highly hygroscopic. Water absorption hinders dissolution	Dehydration of cellulose to avoid moisture. Thermal heating (oil bath, microwave)	Acharya et al. (2021); Bahadur and Phadagi (2019); Sayyed et al. (2019); Taokaew and Kriangkrai (2022)

TABLE 16.3
Nanoparticles in cellulose-based composites for environmental applications

Nanoparticles Used	Matrix Material	Application	Key Findings	Release Mechanism	Performance	Reference
Maghemite (γ-Fe_2O_3)	Cellulose and activated carbon (MCB-AC)	Water pollution remediation (adsorption of organic dyes)	High adsorption capacity for methyl orange and methylene blue; magnetic properties for easy recovery	Pseudo-second-order kinetics	Efficient dye removal and full regeneration, magnetic separation	Luo and Zhang (2009)
Titanium dioxide (TiO_2)	Regenerated cellulose (RC)	Photocatalytic degradation of pollutants	TNW better recyclability and photocatalytic activity than TiO_2 nanoparticles	Fickian diffusion mechanism	Enhanced photocatalytic degradation, good stability, and recyclability	Evyan et al. (2021)
Magnetite (Fe_3O_4)	Cellulose	Heavy metal ion (Cu^{2+}, Pb^{2+}, Zn^{2+}) removal	High adsorption efficiency for heavy metals; pH-dependent behavior	Electrostatic attraction and physical-chemical adsorption mechanisms	Fast adsorption, high capacity, good reusability, sensitive magnetic response	Luo et al. (2016)
Silver nanoparticles (AgNFs)	Cellulose, chitosan, alginate, collagen, etc.	Antibacterial wound dressings	Enhanced antibacterial activity and biocompatibility; stabilization of AgNPs in biopolymers	N/A	Improved antibacterial effects, stabilized AgNPs, promoted wound healing	Alavi and Varma (2021)
Zinc oxide (ZnO), titanium dioxide (TiO_2), silver oxide (Ag_2O)	Cellulose nanocrystals (CNC)	Antimicrobial films	CNC films with metal oxide nanoparticles showed strong antimicrobial properties	N/A	Best antibacterial activity with CNC-Ag_2O films; tunable optical and thermal properties	Lizundia et al. (2018)

TABLE 16.4
Factors affecting drug loading and release in cellulose beads based on hydrophobicity, hydrophilicity, and log P

Factor	Effect on Drug Loading & Release	Examples	Reference
Hydrophobicity of cellulose	More hydrophobic cellulose enables higher loading of lipophilic drugs	Carboxymethyl cellulose (CMC) acetate butyrate (42.9% loading) vs. hydrophilic equivalent (1.8% loading)	Zuppolini et al. (2022)
Drug log P (polarity)	Lipophilic drugs (log P >3) integrate into hydrophobic sections or crystallize in highly hydrophilic beads; hydrophilic drugs (low log P) easily infiltrate but may leach out	Indomethacin (log P ≈ 4.5) remains amorphous; fenofibrate (log P ≈ 5.2) crystallizes	Xie et al. (2021)
Drug–matrix interaction	Hydrogen bonding and pore confinement stabilize certain drugs; high molecular mobility in hydrophobic drugs may cause phase separation	Indomethacin stabilized in oxidized cellulose beads; fenofibrate shows phase separation and reduced loading	Xie et al. (2021)
Stabilization strategies	Surfactants or hydrophobic cellulose derivatives enhance loading; ionic/hydrogen bonding prevents hydrophilic drug leaching	Nanocellulose-stabilized Pickering emulsions encapsulate hydrophobic drugs	H. Yan et al. (2019)
Cellulose bead hydrophilicity	Sufficient hydrophilicity aids swelling and release; hydrophobic modifications slow water absorption, prolonging release	Highly oxidized beads release drugs faster; hydrophobic cellulose extends drug release	Zuppolini et al. (2022)

16.2.2 Process of Cellulose-Based Beads

The creation of cellulose-based beads comes from different cellulose derivatives such as cellulose acetate butyrate, cellulose acetate, and carboxymethyl cellulose (CMC). The produced beads have a diameter ranging from 10 μm to 1–3 mm. The preparation process mainly includes three steps (Carvalho et al., 2021):

TABLE 16.5
Examples of pH-responsive cellulose-based beads for drug release

Hydrogel Type	Drug Used	Method of Synthesis	Key Findings	Drug-Release Mechanism	Reference
Sodium alginate/CMC hydrogel beads	Metformin hydrochloride (MH)	Ionotropic gelation with Fe^{3+} crosslinking	Limited swelling at pH 1.2; increased swelling and drug release at pH 7.4	Fickian diffusion and anomalous transport	Swamy and Yun (2015)
Bacterial cellulose microparticles	Cloxacillin and cefuroxime	Mechanical grinding and molding	Spherical microparticles (350–400 μm); higher drug loading in RBC; rapid drug release (> 85% in 30 min); strong antibacterial activity	Rapid diffusion	Ullah et al. (2019)
Chitosan-coated CMC beads	Amoxicillin trihydrate	Chitosan coating on CMC beads	Drug release influenced by crosslinking, drug–polymer ratio, pH, chitosan coating; encapsulation efficiency 57.61%; maximum drug release 77.40%	Fickian diffusion and anomalous transport	Bulut (2022)
Collagen-bacterial cellulose porous microspheres	Bovine serum albumin (BSA)	Porous microsphere synthesis	Strong BSA–polymer interaction; adsorption follows Langmuir model; first-order drug-release kinetics	First-order kinetics	W. Zhang et al. (2019)
Cellulose-derived microspheres (CCMs)	Insulin	Green hydrolysis (citric/HCl acid)	pH-responsive release: 48.87% in AGF & 85.12% in AIF	Fickian diffusion (Korsmeyer–Peppas model)	Gong et al. (2021)
Phosphorylated cellulose microspheres (CMPs)	Ciprofloxacin (CIP)	Esterification with phosphoric acid	Max adsorption capacity: 139.36 mg/g; cumulative release: 96.28%; maintains drug integrity	First-order kinetics. Fickian diffusion ($n = 0.46$)	S. Wu et al. (2021)

TABLE 16.6
Thermosensitive polymers for drug-delivery applications (K. Zhang et al., 2021)

Polymer	Encapsulated Drug	Medical Application	Reference
Pluronic F-127	Brimonidine tartrate, brinzolamide, cyclosporine	Ophthalmic drug delivery (eye drops)	Y. C. Kim et al. (2020)
PCGA-PEG-PCGA	Liraglutide	Treatment of type 2 diabetes	Y. Chen et al. (2016)
Elastin-like thermogel	Human C-peptide	Management of diabetic complications	Lee et al. (2020)
PEG-PSMCU	Lysozyme	Protein delivery via in-situ gel formation	Phan et al. (2017)
ND-based gelatin	VEGF	Growth factor-based therapies	Pacelli et al. (2017)
Polyurethane thermogel	Anti-VEGF	Treatment of proliferative vascular diseases	Xue et al. (2019)
PLGA-PEG-PLGA	ALA	Transdermal drug administration	Cao et al. (2021)
Alginate-g-PNIPAM	DOX	Cancer treatment	M. Liu et al. (2017)
PHB-based polyester	DOX, PTX	Hepatocellular carcinoma therapy	Y.-L. Wu et al. (2016)
PDLLA-PEG-PDLLA	BP	Photothermal therapy	Shao et al. (2018)

1. Dissolving the cellulosic material.
2. Forming the solution into spherical particles.
3. Solidifying the particles into beads.

Although various solvents can dissolve cellulose, many of them have drawbacks. Issues such as toxicity, volatility, high cost, difficult recovery, and limited solubility are a common occurrence. An ideal solvent for cellulose should be eco-friendly, cheap, recyclable, and chemically and thermally stable. The resulting cellulose solution also should have low viscosity for easier processing. Only specific solvent systems can dissolve cellulose, due to the inter and intra-hydrogen bonding in cellulose chains:

1. NMMO (*N*-methylmorpholine *N*-oxide)
2. DMAc/LiCl (dimethylacetamide/lithium chloride)
3. Alkali/urea solutions
4. Ionic liquids (ILs) (Alexandridis et al., 2018).

After dissolution, cellulose is shaped into spherical particles. The changing of shape is done by various methods such as the dropping method, by a vibratory nozzle, electrostatically, emulsification, or by jet cutter. In the dropping method, the solution is

TABLE 16.7
Applications of cellulose beads in different drug-delivery systems

Drug-Delivery System	Application Area	Drug Type	Mechanism Action	Key Properties/Advantages	Reference
Oral drug delivery	pH-Responsive therapeutic delivery	Diclofenac	The hollow cellulose beads release diclofenac based on environmental pH changes, with diffusion and swelling-controlled-release mechanisms, transforming from hydrophobic to hydrophilic	1. Ensure efficient delivery across the gastrointestinal tract 2. Improved bioavailability	Mohan et al. (2022)
	Oral colorectal cancer treatment	5-Fluorouracil	The system involves loading the drug into LDH(Zn/Al), encapsulating it with CMC for stability and controlled release, and employing a pH-responsive mechanism that optimizes drug release in the colon	1. Maintains minimal drug loss in the acidic stomach environment and targeted release in the colon	Pooresmaeil et al. (2019)
	pH-Sensitive release in treating ocular disorders, popliteal artery disease, and inflammatory conditions (asthma and arthritis)	Dexamethasone	The system uses magnetized nano CMC embedded in alginate-chitosan hydrogel beads, coated with chitosan, to enable pH-sensitive, targeted release of dexamethasone, with enhanced release in acidic cancerous environments and restricted release in the gastrointestinal tract, utilizing magnetic properties for directed delivery	1. Excellent pH-specific release 2. Prevent premature drug release in the stomach 3. Localized and controlled drug delivery in the intestine	Karzar Jeddi and Mahkam (2019)
	Gastrointestinal therapies, cancer treatment, peptic ulcer treatment, probiotic delivery, chronic therapy	Indomethacin	The pH-responsive alginate-CMC hydrogel beads consist of a core-shell structure with an alginate core and a carboxymethyl cellulose shell to protect drugs in acidic conditions (pH 1.2) by minimizing release and enable controlled drug release in neutral conditions (pH 7.4)	1. Increased mechanical stability 2. Sustained drug release in stimulated intestinal fluid	M. Yan et al. (2021)

(Continued)

TABLE 16.7 (*Continued*)
Applications of cellulose beads in different drug-delivery systems

Drug-Delivery System	Application Area	Drug Type	Mechanism Action	Key Properties/Advantages	Reference
	Tuberculosis treatment	Rifampicin	The oral drug-delivery system for rifampicin utilizes alginate-based beads, enhanced with cellulose and C16TMABr-modified cellulose, control drug release, and protect the drug from harsh gastric conditions, addressing issues of poor solubility and short half-life to reduce side effects in tuberculosis treatment	1. Improved drug encapsulation efficiency (EE), 2. Enhanced stability in gastric fluids, and provided controlled release in the intestinal environment	Arianto and Sunardi (2020)
	Gastrointestinal tract	Naproxen (NPX)	The Cu-CMC/NPX/GQD nanocomposite hydrogel beads use copper acetate as a physical crosslinker to create a pH-sensitive, controlled drug-release system that protects drugs like Naproxen in stomach acid and releases them gradually in the intestinal tract, confirmed by low toxicity against Caco-2 cells	1. pH-sensitive protection, 2. Low toxicity, 3. Uniform bead shapes	Javanbakht et al. (2018)
Topical drug delivery	Wound healing	Tetracycline hydrochloride	The composite beads encapsulate tetracycline hydrochloride efficiently and release it in a pH-sensitive, controlled manner according to zero-order, first-order, and Higuchi models under acidic conditions	1. Biocompatible and non-cytotoxic, 2. High EE, 3. Enhanced drug loading, 4. pH-sensitive release and water absorption-resistance	M. Wu et al. (2022)
	Wound healing	5-Fluorouracil	The magnetic β-cyclodextrin/cellulose hydrogel beads utilize Fe3O4 nanoparticles for EMF-responsive rapid swelling-deswelling, enabling stepwise, remotely controlled drug release	1. High loading capacity 2. Excellent biocompatibility	Lin et al. (2019)

(*Continued*)

TABLE 16.7 (*Continued*)
Applications of cellulose beads in different drug-delivery systems

Drug-Delivery System	Application Area	Drug Type	Mechanism Action	Key Properties/Advantages	Reference
Injectable formulations	Localized cancer treatment	Doxorubicin	The CNC-reinforced PLA composite beads releases doxorubicin in a pH-dependent, diffusion and swelling-controlled manner and target triple-negative breast cancer cells by inducing apoptosis locally	1. Provide sustained drug release, 2. Reduce the frequency of administration and minimizing side effects	Hazra et al. (2021)
	Targeted gastrointestinal delivery	Indomethacin and fenofibrate	With release rates flexible by changing the oxidation degree and influenced by drug and solvent properties, the drug-delivery system using oxidized cellulose beads introduces carboxyl groups by means of pH-responsive, controlled release through TEMPO-mediated oxidation, so enabling drug retention at low pH and release at higher pH	1. Adjustable release rate via oxidation levels, 2. Consistent drug levels with reduced dosing frequency, 3. Biocompatible and low cost, 4. Enhanced reswelling and hydrophilicity from TEMPO-mediated oxidation	Xie et al. (2021)
Transdermal drug delivery	Transdermal therapeutic applications and treatment of diseases, such as diabetes, chronic pain, and cancer	Dopamine	The cellulose microspheres, both magnetic and non-magnetic, to encapsulate dopamine, enabling controlled and enhanced drug release (~46% increase) under external magnetic field	1. Biocompatible and biodegradable, 2. Enhanced drug release under magnetic fields, 3. Controlled drug-release kinetics, 4. High EE	Shah et al. (2020)

TABLE 16.8
Key challenges in cellulose-based beads for drug delivery

Challenge Category	Key Issues	Impact	Potential Solutions
Manufacturing & scalability	Difficulty in large-scale production	Variability in bead size and drug loading	Process optimization, automation
	High production costs	Limited industrial feasibility	Cost-effective synthesis methods
	Reproducibility of production methods	Inconsistent batch quality	Standardized protocols, advanced monitoring
Optimization of drug release	Uncontrolled drug-release kinetics	Reduced therapeutic effectiveness	Surface modification, cross-linking control
	Limited control over degradation rate	Inconsistent drug-release profiles	Polymer blending, structural modifications
	Drug–polymer interaction issues	Altered release mechanisms, reduced efficacy	Computational modeling, formulation adjustments
Material quality & consistency	Variability in cellulose properties	Altered mechanical strength and drug release	Strict quality control, source standardization
	Impurities in raw materials	Potential toxicity, reduced efficacy	Purification techniques, refined processing
	Compatibility with different drug types	Limited applicability to various treatments	Functionalization, hybrid materials
Drug stability	Sensitivity to moisture, oxidation, and microbes	Reduced efficacy and shelf life	Protective coatings, optimized storage conditions
	pH and temperature instability	Drug degradation, loss of therapeutic action	Stabilizers, buffer systems
	Long-term storage challenges	Shortened shelf life, compromised quality	Freeze-drying, advanced packaging
Environmental sensitivity	Biodegradability and disposal concerns	Potential environmental accumulation	Eco-friendly formulations, regulatory compliance
	Performance variation under different conditions	Unpredictable drug release	Adaptive formulations, environmental testing
Regulatory & safety considerations	Compliance with drug regulatory standards	Delays in approval, high testing costs	Early regulatory engagement, extensive testing
	Cytotoxicity and biocompatibility issues	Potential adverse effects in patients	In-vitro and in-vivo testing, material screening

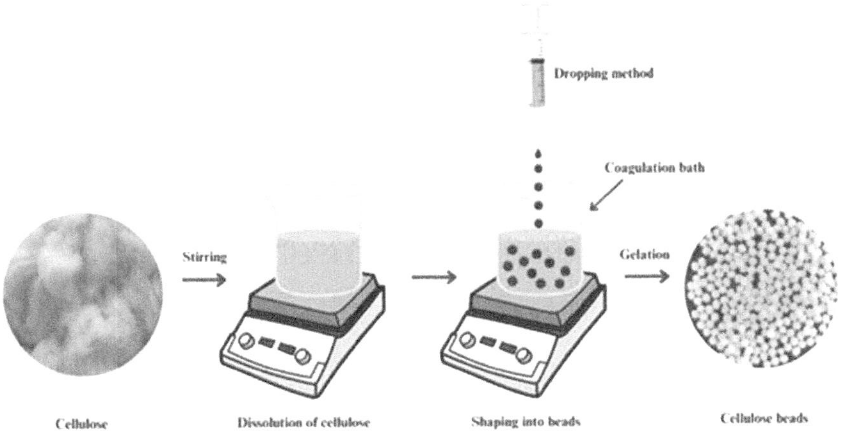

FIGURE 16.2 Preparation method of cellulose beads. Adapted from S. Zhang et al. (2020). Copyright 2025, MDPI.

dispensed into a coagulation bath via nozzle or syringe, with droplet size controlled by factors such as nozzle size, ejection speed, and bath distance (Ganesan et al., 2018; Y. Tan et al., 2020). Monodispersed droplets can be formed by applying vibrations to the fluid through a vibratory nozzle (Auriemma et al., 2020; Ganesan et al., 2018). A high electrical field induces droplet formation in the electrostatic process (C. Li et al., 2016). Moreover, cellulose solution can be cut by the high pressure of jets into droplets, which, upon solidifying, transform into beads (Druel et al., 2018; Schroeter et al., 2021). Droplets can be also produced by emulsification where two immiscible liquids are mixed with surfactants to create stable emulsions (Figure 16.2). The process is then followed by gelation to stabilize the droplets (Druel et al., 2018; S. Zhang et al., 2020).

The next crucial step in producing cellulose-based hydrogel beads is gelation. Among various types of gelation techniques, ionotropic gelation, crosslinking, pH, and temperature-induced gelation, non-solvent-induced phase separation (NIPS) is extensively used for drug delivery. In ionotropic gelation, polyelectrolytes (e.g., cellulose derivatives) interact with oppositely charged ions (e.g., Ca^{2+}) to form hydrogels (Gadziński et al., 2023). This process can occur through external gelation, where the polymer solution is dropped upon a cross-linking bath, internal gelation, where insoluble calcium salts are pre-mixed with the polymer solution before extrusion into an acidic bath, or reverse gelation, where a gelling agent is added dropwise to the polymer solution (Ahmed et al., 2013; Channab et al., 2023; Martins et al., 2015). Strong networks can be formed via chemical crosslinkers, though having environmental concerns (Ganesan et al., 2018). In pH-induced gelation, polymer solutions solidify over contact with acidic or alkaline cross-linking solutions, while temperature-induced gelation relies on temperature changes to induce polymer chain orientation and gelation (Fiorica et al., 2021; Sharma et al., 2019). NIPS process has a polymer extrusion onto a non-solvent bath, which leads to phase separation and gel formation (Gurikov & Smirnova, 2018). Cellulose bead production can also be done mechanically making it an eco-friendly alternative. The process begins with fiber bleaching and is then followed by mechanical stirring of cellulose suspension. Finally, the beads undergo a recovery process resulting in a ribbon-like structure with a smooth surface (El Allaoui, Chakhtouna et al., 2023).

16.3 FUNCTIONALIZATION TECHNIQUES OF CELLULOSE BEADS IN DRUG DELIVERY

16.3.1 Chemical Grafting

Cellulose contains multiple reactive sites for the possibility of chemical modifications. Several hydroxyl functional groups (three per repeating glycosidic unit) are present in these areas making covalent bonding with functional groups possible. The primary reactive sites are found at the C6, C2, and C3 positions. Multi-functionalization processes commonly include etherification, esterification, oxidation, and polymer grafting, which changes the inherent properties of cellulose. Generally, a binder is used to connect the cellulose matrix with functional elements. Modified cellulose beads improve stability, control release behavior, and enhance drug adsorption, thereby raising their efficacy (El Allaoui, Benzeid et al., 2023; Liyanage et al., 2021).

16.3.1.1 Etherification

Etherification of cellulose generally occurs in an alkaline medium (15%–50% NaOH). Cellulose beads go through mercerization under these conditions. Further changes are followed by nucleophilic substitution with etherifying agents like alkyl halides or epoxides (El Allaoui, Benzeid et al., 2023; Naserifar et al., 2022). For example, cellulose reacts with sodium chloroacetate to produce CMC by a process called carboxymethylation. The final product obtained has proper hydrophilicity, pH sensitivity, and gel-forming characteristics (Wahid et al., 2021; S. Wang & Yu, 2013). Moreover, CMC beads can undergo crosslinking with alginate, chitosan, or starch for drug encapsulation (Akalin, 2023; Morozkina et al., 2022). Furthermore, the carboxyl groups ($-COO^-$) present in CMC beads can form stable hydrogels by interacting with metal cations (Al^{3+}, Fe^{3+}) (El Allaoui, Benzeid et al., 2023; Zennifer et al., 2021). Etherification with 2-chloroethyl diethylamine can further enhance the performance of cellulose beads (Gericke et al., 2013).

16.3.1.2 Esterification

Esterification changes the free hydroxyl groups of cellulose into ester linkages. The reaction occurs mainly with carboxylic acids or their derivatives, especially acid anhydrides or acyl chlorides (Aziz et al., 2022; Ciolacu et al., 2020). Tosyl chloride acts as an excellent leaving group when cellulose beads go through nucleophilic substitution, making the cellulose beads more reactive for subsequent functionalization (Opriş et al., 2024).

16.3.1.3 Oxidation

Oxidation can attach functional groups like carboxyl ($-COOH$) and aldehyde ($-CHO$) onto cellulose. Oxidizing agents such as sodium periodate ($NaIO_4$), TEMPO (2,2,6,6-tetramethylpiperidine-1-oxyl), and potassium permanganate ($KMnO_4$) are used generally. These agents can selectively cleave β-1,4-glycosidic linkages in cellulose. $NaIO_4$ targets the C2–C3 bond inside the anhydrous glucose units of cellulose and produces active carbonyl groups. These groups can subsequently link with different amines or undergo sulfonation for protein immobilization. TEMPO oxidation can transform the main hydroxyl groups at the C6 position into carboxyl groups giving 6-deoxy-6-carboxy-cellulose. These structural changes make them proficient in the adsorption of metals, polycations, and dyes as they alter the anionic properties of the beads (El Allaoui, Benzeid et al., 2023; Gericke et al., 2013).

16.3.1.4 Polymer Grafting

Polymer grafting onto cellulose beads is the covalent bonding connection of synthetic polymer chains to generate graft copolymers. Usually by using radical initiators or irradiation, this is accomplished using a "grafting-from" technique by polymerization via free-radical or controlled radical starting directly on the cellulose surface. The development of the chain is predominantly facilitated by Ce^{4+} ions, which interact with cellulose to create a chelate, later decomposing to provide free oxygen radicals to begin the polymerization of acrylic acid monomers (Gericke et al., 2013). The resultant copolymers are far more soluble, biocompatible, and drug encapsulation capable when functional polymer chains covalently connect to the cellulose backbone. Polymers, for example, poly (acrylic acid), polyacrylamide, and poly(*N*-isopropylacrylamide) are grafted onto cellulose substrates in general. Functional qualities are improved as the bead's surface gets interwoven with polymer chains (Purohit et al., 2023). For example, grafting poly (acrylic acid) onto cellulose produces site-specific drug-delivery beads. This modified hydrogel remains stable in gastric acid but expands and releases drugs in the intestines' higher pH. Likewise, the grafting of thermo-responsive poly(*N*-isopropylacrylamide) (PNIPAM) onto cellulose generates drug-release characteristics sensitive to temperature. At 25°C, cellulose nanocarriers fitted with PNIPAM brushes show minimal drug release, whereas at body temperature (37°C), release rate increases (W. Wu et al., 2016) (Figure 16.3).

16.3.2 Biopolymer Coating

Cellulosic beads can obtain improved physicochemical stability, controlled release, and bioavailability of encapsulated medicines by biopolymer coating (Zeng et al., 2020). To increase drug encapsulation efficiency, stability, and controlled-release kinetics, several functionalizing techniques—including cross-linking mechanisms,

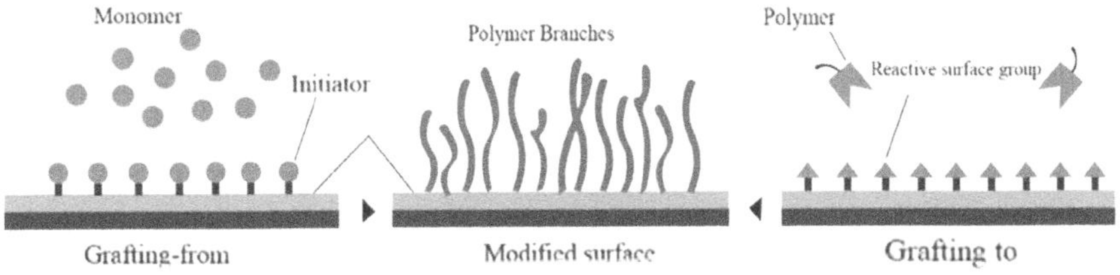

FIGURE 16.3 Polymer grafting-from and grafting-to approach. Reprinted with permission from Gaffar et al. (2025). Copyright 2025, Springer Nature.

surface modifications, and particular bonding interactions between cellulose and biopolymers are applied (Nathanael & Oh, 2020). For example, chitosan, alginate, gelatin, and polylactic acid (PLA) can be employed as coating materials. Every biopolymer has special qualities (biodegradability, biocompatibility, sustainability, accessibility) that can improve the efficiency of medication administration (Opriş et al., 2024). Different methods are used to functionalize the surface of the cellulose beads depending on the chosen appropriate polymer for coating. In general, these techniques can be categorized as chemical and physical approaches.

16.3.2.1 Chemical Functionalization

Chemical functionalization: Chemical functionalization is the development of stronger cross-links or covalent bonds between the biopolymer coating and the cellulose surface, or inside the coating itself. The cellulose backbone is rich in hydroxyl groups that can be chemically activated for the attachment of polymers or functional groups. Oxidation with periodate can produce aldehyde groups on the cellulose chain resulting in dialdehyde cellulose (DAC). The process is then followed by the formation of Sciff-base linkages with amine-containing polymers that can be reduced to stable C–N bonds. Usually cross-linking agents, including glutaraldehyde, genipin, or carbodiimide coupling reagents, are used to connect biopolymer chains to the bead. The stability of the coating is enhanced and some functional groups can be attached to the surface by use of the grafting and cross-linking techniques. Covalently linked coating is less prone to rapid erosion and results in better-controlled drug release. However, the drug remains unaffected by chemical changes if pre-loaded and may require additional reagents (Opriş et al., 2024).

16.3.2.2 Physical Functionalization

Physical functionalization: Physical functionalization techniques incorporate the coating of cellulose beads via the adsorption or deposition of polymers, without the formation of new covalent bonds. These techniques are appealing because they exclude dangerous chemicals. Layer-by-layer assembly is a commonly used physical coating method whereby polyelectrolytes with opposing charges are progressively adsorbed (Mahmoud et al., 2021; Opriş et al., 2024).

16.3.3 Ionic Crosslinking

Cellulose beads can be functionalized with anionic and cationic groups to create ion exchangers with varying strength (Gericke et al., 2013). This type of physical crosslinking does not require covalent bonding rather involves electrostatic interactions between oppositely charged species, such as multivalent metal cations and negatively charged functional groups on the polymer backbone. It is widely used due to its simplicity, cost-effectiveness, and efficiency in forming stable polymer networks (Sun et al., 2021). An established approach of ionic crosslinking, sometimes referred to as polyelectrolyte complexation, is ionotropic gelation. This is the interaction of an anionic polymer matrix with a metal cation without requiring strong chemical treatments under high temperatures (Kurozawa & Hubinger, 2017). There are two basic types of ionotropic gelation:

1. **External ionotropic gelation** – The metal cation is externally injected to enable slow crosslinking of the polymer network.
2. **Internal ionotropic gelation (emulsification)** – The metal cation is embedded in the matrix of polymer at first in an inert state and then later re-activated to induce crosslinking.

The effectiveness of the crosslinking is highly dependent on the type and charge of the metal cations involved. Multivalent metal cations (e.g., Ca^{2+}, Fe^{3+}, Al^{3+}, Zr^{4+}) show stronger electrostatic attraction with functional anionic groups (e.g., carboxylate ($-COO^-$) or sulfate ($-SO_4^{2-}$) in comparison with monovalent cations (e.g., Na^+, K^+, Li^+, Cs^+) (Phan-Xuan et al., 2016). For example, Benhalima et al. (2017) developed a highly stable, crosslinked CMC hydrogel beads using Al^{3+} ions as ionic crosslinkers. The stabilized network was formed due to the interaction between multivalent cations (such as Al^{3+}, Zr^{4+}, Ca^{2+}, and Fe^{3+}) with carboxylate ($-COO^-$) groups on CMC, which also contributed to improved water insolubility. Anionic biopolymers such as sodium alginate can be used as a polymeric matrix where environmental applications such as wastewater treatment are needed due to their biocompatibility, better biodegradability, high adsorption capacity, toxic-free, and economical (Chang et al., 2015).

16.3.4 Nanoparticle Embedding

Incorporating nanoparticles into the porous structure of cellulose beads can significantly enhance their physiochemical properties. The porous network of cellulose beads makes nanoparticle loading suitable. Nanoparticles, such as metal oxides, noble metals, or carbon-based nanomaterials, can be embedded using in-situ synthesis, impregnation, or electrostatic adsorption. Metal precursors and reducing agents can be produced inside the cellulose matrix to create nanoparticles via in-situ synthesis. The beads can be also impregnated in nanoparticle dispersions to allow diffusion and attachment. Electrostatic adsorption relies on charge interactions between functionalized cellulose and nanoparticles. Further stabilizing the implanted nanoparticles are post-functionalization treatments, comprising chemical changes or crosslinking. Because of their increased reactivity, adsorption capacity, and selectivity, these functionalized cellulose beads have wide use in environmental monitoring, drug delivery, catalysis, and air-quality sensing.

16.4 MECHANISM OF DRUG LOADING AND RELEASE IN CELLULOSE BEADS

16.4.1 Drug-Loading Mechanisms

16.4.1.1 Encapsulation

Encapsulation is the physical entrapment of drug molecules within the matrix of cellulose beads providing a protective enclosure (Klojdová et al., 2023). The main process generally starts with the formation of cellulose beads by methods such as solvent evaporation, gelation, or crosslinking. Encapsulation efficiency can be improved by cross-linking agents and structural modifications. For example, CMC can encapsulate

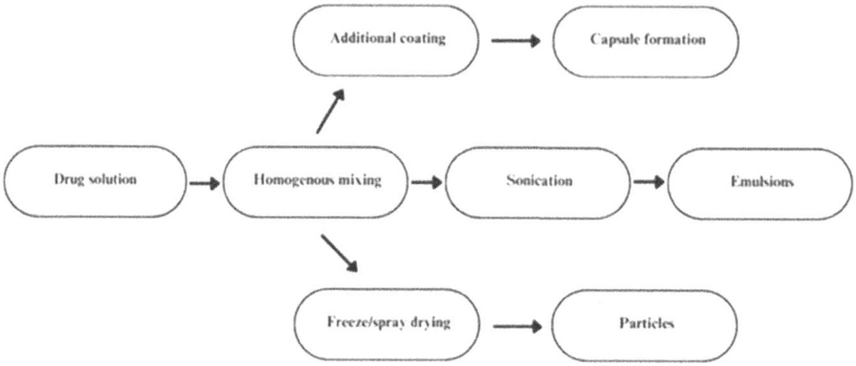

FIGURE 16.4 Different approaches for drug encapsulation. Adapted from Klojdová et al. (2023). Copyright 2025, MDPI.

pharmaceuticals like 5-fluorouracil, with an encapsulation efficiency of 87%. The crosslinking with iron ions can improve the overall drug-release profile (Pooresmaeil et al., 2019). The solubility and chemical properties of both the drugs and the cellulose matrix are also important in drug encapsulation. Cellulose with high porosity can encapsulate both hydrophilic and hydrophobic drugs through a solvent-releasing technique obtaining a loading capacity above 80% (Omura et al., 2017). To encapsulate sensitive pharmaceuticals and dietary supplements spray-drying and freeze-drying techniques are employed. These processes provide minimal thermal stress and transform liquids into powders in a single process (Klojdová et al., 2023). Hollow spherical cellulose beads with an internal cavity for drug encapsulation can be produced from cellulose acetate through base-catalyzed deacetylation. Depending upon preparation conditions, these beads can have high encapsulation efficiency (EE) varying from 20% to 85% (Mohan et al., 2022) (Figure 16.4).

16.4.1.2 Adsorption

Adsorption regulates the contact between drug molecules and the cellulose surface making it a fundamental step in drug loading. Drug molecules get attached to the surface or pore walls of the beads by hydrogen bonding, van der Waals forces, and electrostatic attractions. Polar drugs establish hydrogen bonds with the hydroxyl groups of cellulose but nonpolar drugs may react through π–π or hydrophobic interactions. The rate of adsorption is affected by the time of contact between the drug, the adsorbent, drug concentration, and the accessibility of binding sites (Zhou et al., 2004). This behavior is understood by the linearized version of the pseudo-second-order model:

$$\frac{t}{q_t} = \frac{1}{K_2 q_e^2} + \frac{t}{q_e} \tag{1}$$

where q_t is the adsorption capacity at time (mg/g), q_e is the equilibrium adsorption capacity (mg/g), and K_2 indicates the pseudo-second-order rate constant (Peñafiel &

Flores, 2023). In the case of ionic drugs, they can get adsorped onto carboxyl and sulfate groups of functional cellulose. Hydrogen bonding is essential for the adsorption of smaller compounds such as acrinol. The bonding takes place between the hydroxyl groups of cellulose surface and polar groups of the drugs (Okada et al., 1987). The adsorption capacity is affected by the specific surface area of the cellulose beads and the concentration of the drug. For example, the adsorption isotherms of Congo red dye on cellulose-graphene nanoplatelet beads (GNP) conform to the Langmuir model, indicating monolayer adsorption. The results showing the maximum adsorption capacities are 98.1 mg/g for cellulose beads and 139.6 mg/g for cellulose-GNP beads (González-López et al., 2021). The adsorption of drugs onto cellulose beads is also influenced by temperature and pH levels. Generally, adsorption rises with a reduction in temperature referring to an exothermic reaction.

16.4.1.3 Ionic and Covalent Bonding

Ionic and/or covalent bonding can greatly increase drug-loading rate when reacted with charged groups or reactive groups. Ionic bonding commonly occurs in between cationic drugs with anionic groups on cellulose (carboxyl or phosphate groups). In the case of anionic drugs (e.g., sulfonates or acidic drug anions), they get attracted to the positively charged amine-functionalized beads. Such bonding is specifically beneficial for water-soluble pharmaceuticals that ionize in physiological surroundings (Xie et al., 2023). Generally, the capacity to affect drug-bead binding via ionic interactions is generally assessed using binding constants (e.g., equilibrium association constants) or capacity factors in ion-exchange methods.

Covalent bonding forms strong, permanent bindings between a pharmacological drug and functional groups on a cellulose matrix. This type of bonding enhances the drug's stability and prolongs its release profile. The integration of both ionic and covalent bonds can create a structured drug-release mechanism. In this scenario, the initial release is driven by ionic interactions, followed by a sustained release resulting from the degradation of covalent bonds. Overall, introducing ionic or covalent binding sites into cellulose beads using methods such as carboxylation, phosphorylation, or amination can improve loading capacities and customize affinity for particular medicines.

16.4.1.4 Hydrophobic and Hydrophilic Interactions

The interaction between drugs and cellulose beads is fundamentally influenced by the drug's polarity, and the hydrophilic or hydrophobic characteristics of the beads. Hydrophilic interactions mainly consist of hydrogen bonding and electrostatic attraction between the polar functional groups of the cellulose matrix and hydrophilic drug molecules. In contrast, hydrophobic interactions arise when nonpolar drug molecules adhere to hydrophobic areas of modified cellulose matrixes. Such hydrophobic areas may incorporate graphene oxide or amphiphilic derivatives and, as a result, improve the loading of drugs that are poorly soluble in water (Y. Li & Yang, 2015; Rasoulzadeh & Namazi, 2017).

Hydrophobic drugs may encounter difficulties in adsorbing into a hydrophilic matrix. However, they can adsorb into the less polar areas of cellulose or create aggregate within its pores. Cellulose can be functionalized with hydrophobic groups or amphiphilic derivatives to accommodate such drugs. These amphiphilic carriers show effective swelling in water and enhanced drug solubilization. Achieving efficient drug loading within bead polymer requires the following conditions:

1. The incorporation of hydrophobicity into the bead polymer improves its compatibility with hydrophobic drugs, allowing for increased drug-loading capacities.
2. Proper hydrophilicity is needed for effective swelling and drug release in a biological fluid.
3. A high glass-transition temperature (T_g) in the polymer matrix is a must-have for the immobilization of drug molecules in an amorphous state and inhibition of their recrystallization.

16.4.1.5 Inclusion Complexation

Cellulose beads function as a matrix to encapsulate drug molecules through non-covalent bonding. This molecular encapsulation process is called inclusion complexation. It is mostly used where the solubility of drugs is quite low. Most commonly, cyclodextrins use this host–guest inclusion mechanism for drug loading. Cyclodextrins (CD) are cyclic oligosaccharides derived from starch. They have a hydrophobic inner cavity and a hydrophilic outer surface suitable for drug loading within their cavity forming complexes by van der Waals forces, hydrogen bonding, and hydrophobic interactions (Gandhi et al., 2020). Cellulose beads can embody CDs through physical inclusion or chemical attachment. The effectiveness of employing the inclusion complexation technique is dependent upon the following factors:

1. **Drug–Cyclodextrin Interaction:** The drug-release profile is determined with the stability of the inclusion complex identified by the stability constant (K). The optimal range for this constant is between 100 and 10,000 M^{-1} (Gandhi et al., 2020).
2. **Bead Porosity and Surface Area:** Highly porous cellulose beads provide improved efficiency in drug loading and superior diffusion-controlled-release characteristics.
3. **Solubility Enhancement:** The formation of inclusion complexes enhances the water solubility of poorly soluble drugs. Potential solubility increases 10–100 times (Shahriari-Khalaji et al., 2021).
4. **Sustained Drug Release:** The inclusion of complexation in cellulose beads extends the release duration from hours to days, reduces burst release, and enhances therapeutic efficacy. It also improves drug retention at tumor sites for over 48 hours, achieves up to 80% tumor growth inhibition, and demonstrates reduced systemic toxicity relative to free drug formulations (Mittal et al., 2023).

Amphiphilic drugs reside within the hydrophobic cavity segment, while its hydrophilic portions interact with the aqueous environment of the pore. As a result of this interaction, the bead's ability to retain a larger quantity of the drug in a non-crystalline state increases. Generally, the release of drugs from these systems

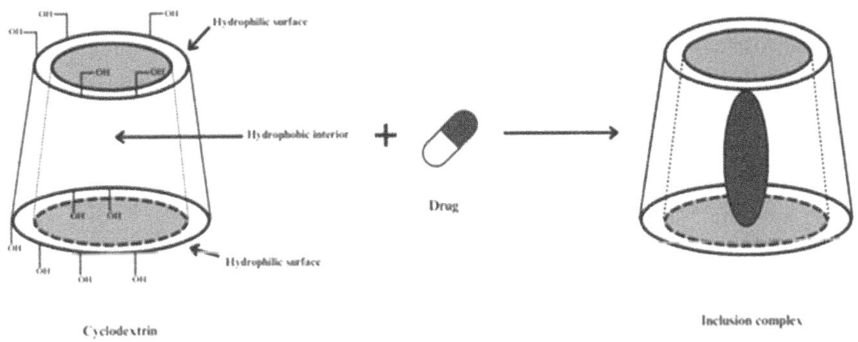

FIGURE 16.5 Visual representation of inclusion complexation technique. Reprinted from Yang et al. (2016). Copyright 2025, Canadian Society for Pharmaceutical Sciences.

relies heavily on the dissociation of the drug–CD complex. Several types of cyclodextrins are available for customization of drug inclusion. For example, α, β, γ, or modified CDs have different cavity sizes and affinities. Among them, γ-CD is optimal for larger drug loading as it possesses a bigger cavity (Cova et al., 2018) (Figure 16.5).

16.4.2 Drug-Release Mechanisms

16.4.2.1 Diffusion-Controlled Release

Drug release happens mainly through diffusion and erosion in cellulose-based beads. Firstly, water is rapidly absorbed by the beads leading to swelling. Increased swelling results in larger pore size and number. The water adsorption rate is faster than the drug-release rate allowing for proper drug diffusion (Orhan et al., 2022). These drug-release mechanisms are generally classified as diffusion-controlled release, erosion-controlled release, and diffusion through water-filled pores. These mechanisms influence the rate and duration of drug delivery, and impact its effectiveness in controlled-release systems.

1. **Diffusion-controlled release:** It is the primary mechanism of drug release in non-degradable polymer-based systems. Here, drug molecules move from the interior of the polymer matrix to the exterior through molecular diffusion. The release rate is dependent on the material's permeability and thickness but independent of concentration gradients (Kamaly et al., 2016).
2. **Diffusion Through Water-Filled Pores:** It mainly occurs in degradable polymer-based drug-delivery systems. In such systems, the drug molecules diffuse through water-filled channels. These porous structures are formed as the polymer absorbs moisture. This process is influenced by convective effects, osmotic pressure, and the random motion of drug molecules. The extent of pore formation and water uptake significantly impacts the drug-release rate (Keraliya et al., 2012).

3. **Erosion-controlled release:** In degradable polymer-based systems, drug release can occur by the breakdown of the matrix, which can be categorized into two types:

    - **Surface Erosion:** This occurs when the polymer degrades externally at a faster rate than water penetrates the matrix. As the outer layers of the polymer dissolve, it leads to a gradual shrinking of the matrix from the exterior inward (Ciolacu et al., 2020).
    - **Bulk Erosion:** When water penetration into the polymer matrix is faster than the rate of degradation, the entire matrix degrades more homogeneously. This results in an unpredictable and inconsistent release profile, making bulk erosion less favorable for controlled drug delivery (Borandeh et al., 2021; Uhrich et al., 1999).

However, drug release generally occurs through a combination of diffusion and erosion in cellulose-based beads. Hydrophilic beads give faster release rates due to rapid pore collapse in an aqueous environment whereas hydrophobic beads maintain a more stable porous structure, leading to slower drug diffusion (Esquivel-Castro et al., 2019) (Figure 16.6).

16.4.2.2 pH-Responsive Release

pH-Responsive cellulose beads have emerged as a promising strategy due to their precise control over effective drug delivery. pH-Responsive beads can undergo solubility, surface activity, and structural modifications in response to environmental pH variations. These polymeric carriers go through changes in conformation and chain behavior when acidic or basic functional groups of such materials either accept or donate protons. In general, pH-sensitive polymers are classified into weak acids, weak bases, degradable biopolymers, and synthetic polymers (Rezaei et al., 2021). These types of polymers are well suited for oral drug delivery due to the presence of a pH gradient in the gastrointestinal (GI) tract. These kinds of systems release the drug by changing their swelling properties or by going through pH-sensitive bond cleavage in response to a change in pH (Rizwan et al., 2017; R. Y. H. Tan et al., 2022). However, some issues occur in regard to drug delivery due to the low

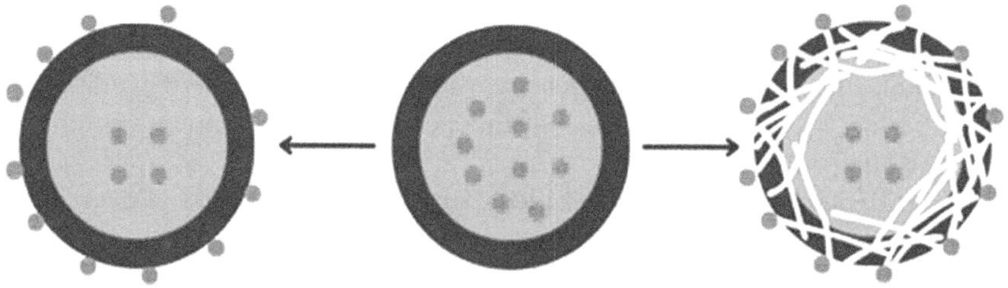

FIGURE 16.6 Diffusion and erosion-controlled drug release. Reprinted with permission from El Allaoui, Benzeid et al. (2023). Copyright 2025, Elsevier.

permeability of drugs through the gastrointestinal mucosa. Moreover, there is a degradation issue caused by gastric acids and digestive enzymes. Bio-based polymers containing carboxyl (–COOH) and sulfonic (–SO_3H) functional groups help mitigate these issues. These carriers prevent excessive swelling under acidic conditions in the stomach (pH 1.2). This premature drug release does not occur as these groups remain protonated in an acidic medium. However, these groups become deprotonated in the neutral-to-slightly alkaline conditions of the intestines (pH 7.4), resulting in increased swelling and higher drug-release rates (Y. Qiu & Park, 2001).

16.4.2.3 Enzyme-Responsive Release

Drugs taken orally are exposed to a high acidic environment of the stomach. This acidic condition may lead to degradation in the upper gastrointestinal (GI) tract before reaching the delivery target for the drug (Park et al., 2010). pH-Dependent drug-delivery systems lack this specific targeting due to the difference in pH between the colon (pH 6.8) and the intestines (pH 7.4). Enzyme-controlled self-regulated systems can properly release drugs in accordance with the physiological needs of the body without the intervention of external factors. These systems function by incorporating a specific enzyme within a polymeric matrix, where the enzyme reacts with a particular substrate (or trigger molecule) to initiate drug release. The rate of drug release is controlled by the concentration of the substrate. This presents itself as a significant advantage over conventional drug-delivery systems as it maintains a fixed release rate without adapting to physiological variations (Goldbart et al., 2002). For example, an enzymatic system could comprise a biodegradable starch matrix combined with hydroxypropyl methylcellulose (HPMC) and an inactive form of α-amylase. In the presence of calcium (Ca^{2+}) ions, the matrix degrades and releases the drug ensuring controlled drug delivery. Another system can be designed for automated insulin release in response to fluctuating blood glucose levels. It includes glucose oxidase (GOX) to react with glucose, catalase to neutralize byproducts and insulin as the therapeutic agent regulating insulin diffusion in a physiologically adaptive manner.

16.4.2.4 Temperature-Responsive Release

Temperature-responsive drug-delivery systems use polymers that go through phase changes and solubility in response to temperature shifts. Cellulose-based beads have emerged as promising carriers due to the possibility of enhanced mechanical stability when combined with thermo-responsive polymers. Such polymers show a sharp phase transition at a certain temperature and can be classified into two types:

1. **Lower Critical Solution Temperature (LCST) Polymers:** LCST polymers are soluble in water at lower temperatures but become insoluble when heated above their transition temperature. Their transition temperatures can be controlled for precise drug release at physiological or pathological temperatures (Karimi et al., 2016).
2. **Upper Critical Solution Temperature (UCST) Polymers:** UCST polymers remain insoluble at lower temperatures and dissolve when heated.

> They are also called dual-responsive drug carriers as their transition can be adjusted by modifying polymer concentration and molecular weight (Boustta et al., 2014).

Such temperature-responsive changes show the potential of smart drug delivery. For example, Pluronic F127 (PF127) has been widely explored among LCST polymers. This polymer is specifically ideal for drug carriers due to its sol-to-gel transition properties. PF127 allows easy drug encapsulation and handling at temperatures below its phase transition (~25°C) where it remains in a sol (liquid) phase. It entraps the drug in its stable gel state (above the transition temperature, ~37°C) enabling sustained release. However, PF127 gives poor mechanical strength and stability. To mitigate such limitations, PF127 has been incorporated with cellulose and its derivatives. For example, Kim et al. (2012) developed a methylcellulose (MC)-PF127 hydrogel for delivering the anti-cancer drug docetaxel (DTX). CMC can be combined with PF127 for the treatment of atopic dermatitis, using cortex moutan as a drug model (W. Wang et al., 2016).

16.4.2.5 Magnetic Field-Triggered Release

These types of cellulose beads are good for site-specific drug delivery. They get induced when exposed to an external magnetic field. Integrating magnetic nanoparticles (MNPs) into the cellulose matrix, such as Fe_3O_4 enables magnetic functionality. These types of matrixes can be synthesized via three methods:

1. **Blending Method:** Fe_3O_4 nanoparticles are mixed into cellulose solution in a coagulation bath before bead formation. Although the method is very simple and scalable, it is prone to nanoparticle aggregation and leakage.
2. **In-situ Method:** Fe^{2+}/Fe^{3+} ions diffuse into pre-formed cellulose beads and are converted into Fe_3O_4 via alkaline treatment. This improves nanoparticle dispersion and retention but may weaken polymer stability.
3. **Grafting method:** Functionalized Fe_3O_4 nanoparticles can be covalently attached to cellulose, which can provide a strong magnetic response upon exposure to a magnetic field (Liao & Huang, 2020).

The orientation of Fe_3O_4 nanoparticles (MNPs) within the hydrogel structure plays an important role in drug diffusion. MNPs align perpendicularly to the drug diffusion path when exposed to a left-to-right magnetic field. This creates a barrier that reduces drug-release rates. Conversely, maximum drug release is ensured under a top-down magnetic field. Optical microscopy has revealed that MNPs transition from a random distribution to an ordered pearl-chain structure under a magnetic field, directly influencing release dynamics. In-vitro studies with vitamin B12 demonstrated that the drug diffusion rate follows the trend: parallel alignment > random (no field) > perpendicular alignment. However, larger molecules face diffusion limitations, which is why this method is most effective for drugs with molecular weights below 100 kDa (T.-Y. Liu et al., 2015).

Another method to regulate drug release is by switching the magnetic field on and off. When the magnetic field is turned on, it creates MNP aggregation. This aggregation compresses the polymer and a reduction in drug release. When the field is turned off, the polymer relaxes. The increased porosity then allows for enhanced drug diffusion (T.-Y. Liu et al., 2006a, 2006b). However, the effect can be reversed by a process called micelle compression. This mechanism is induced when the matrix contains closely packed hydrophobic micelles, which upon compressing via an external field, expel the encapsulated drug (Qin et al., 2009).

Drug release can be also controlled by constraining magnetic moments, a mechanism called the magneto-caloric effect. This effect creates localized heat, which can alter the bead's structure or drug solubility, further modulating release rates (K. Hu et al., 2015).

16.4.2.6 Light-Triggered Release

These types of cellulosic beads are engineered by the addition of photosensitive materials into their polymer matrix. The types of photosensitizer used decide whether the response will be reversible or irreversible (Bustamante-Torres et al., 2021). Increased exposure to light might induce processes such as cleavage, isomerization, or dimerization of the photosensitive groups. Such reactions occurring at photosensitizers may bring out changes like decrosslinking, degradation, expansion, or contraction (Xing et al., 2022). The efficiency of the drug release also depends on the light sources as it is the main element to activate the photosensitive groups within the beads and there is a safety issue of how much the light would penetrate during application. Due to the availability of UV-sensitive photosensitizers, ultraviolet (UV) light has been used extensively for many years. Despite its popularity, UV light is largely confined to in-vitro experiments with limitations in its cytotoxicity and the relatively poor penetration of UV light through tissue restricting its effectiveness in living organisms. In contrast, near-infrared (NIR) proves as a superior alternative to in-vivo applications. It became highly appealing for clinical applications because of its deeper penetration into tissues and enhanced safety profile (Xing et al., 2022).

The light-responsive drug-delivery systems can be broadly classified into three main types:

1. **Photoisomerization-Based Drug Release:** Photoisomerization is the transition of molecular structures from one form to another (generally from a *trans* to a *cis* configuration) when exposed to light. This process allows for the controlled release of drug by altering the pore's structure. The process is often reversible and repeatable since no chemical changes occur (Tomatsu et al., 2011)
2. **Photochemical-Based Drug Release:** Photochemical reactions such as photocleavage involve the embedding of photocleavable linkers (nanoparticles) within the polymer network, which can be disassembled under light. Covalent bonds are formed between the drug and specific sites of the hydrogel via these photocleavable linkers ensuring proper release (Xing et al., 2022).
3. **Photothermal-Based Drug Release:** This process utilizes materials that are capable of converting light energy into heat. The heat energy releases the drug by subsequently disrupting thermally sensitive drug carriers.

The photothermal approach requires two essential components: (i) a photosensitizer that absorbs light and generates heat, and (ii) a temperature-sensitive material that responds to heat by releasing the encapsulated drug (Bordbar-Khiabani & Gasik, 2022; M. Qiu et al., 2018).

16.5 APPLICATIONS OF CELLULOSE BEADS IN DRUG-DELIVERY SYSTEMS

Because of their biocompatibility, biodegradability, and ability to offer regulated and targeted release, cellulose beads are extensively applied in drug administration. They aid in targeted cancer treatment and sustained anti-inflammatory drug delivery by enhancing drug stability and release when used orally. They also serve as carriers for antibacterial agents for wound healing applications. Thus, ensuring prolonged therapeutic effects when used topically. Cellulose beads when used with injectable formulations can reduce injection frequency and improve treatment efficacy. Transdermal delivery systems avoid first-pass metabolism and ensure sustained drug release as they enable drug absorption through the skin in patch-based systems.

16.6 CHALLENGES ASSOCIATED WITH CELLULOSE BEADS FOR DRUG DELIVERY

Developing cellulose-based beads for drug delivery comes with several challenges when ensuring efficient production, maintaining material consistency, and keeping the drug stable under different conditions. Moving to industrial-scale manufacturing adds further layers to issues requiring careful attention to reproducibility, cost management, and quality control. Adjusting factors like drug loading, release rates, and biodegradability requires a great amount of effort. Furthermore, external conditions like humidity, temperature shifts, and chemical degradation can impact the stability and effectiveness of the beads.

16.7 CONCLUSION

Cellulose beads have become a promising component in drug-delivery systems because of its biocompatibility, biodegradability, and ability to provide controlled drug release. Their porous structure enables effective drug encapsulation and chemical modification improves their targeting capabilities and stability in physiological environments. Precise drug delivery in cancer treatment, antimicrobial medicines, chronic illness management, and other therapeutic applications has been possible because of advanced nanotechnology and polymer chemistry. However, issues regarding scalability, reproducibility, and regulatory compliance limit their widespread adoption. Future research should concentrate on optimizing cellulose bead compositions, including smart stimuli–responsive mechanisms, and investigating new production procedures to improve their efficiency and application in contemporary medicine.

Declaration: The authors declare that Generative AI and AI-assisted technologies like ChatGPT 4.0 have been used very sparingly in the writing process to improve the readability and language of the work only. After using this tool, the authors carefully reviewed and revised the content to ensure accuracy and quality. The authors take full responsibility for the final version of this publication.

REFERENCES

Acharya, S., Liyanage, S., Parajuli, P., Rumi, S. S., Shamshina, J. L., & Abidi, N. (2021). Utilization of cellulose to its full potential: A review on cellulose dissolution, regeneration, and applications. *Polymers*, *13*(24), Article 24. https://doi.org/10.3390/polym13244344

Ahmed, M. M., El-Rasoul, S. A., Auda, S. H., & Ibrahim, M. A. (2013). Emulsification/internal gelation as a method for preparation of diclofenac sodium–sodium alginate microparticles. *Saudi Pharmaceutical Journal*, *21*(1), 61–69. https://doi.org/10.1016/j.jsps.2011.08.004

Akalin, G. O. (2023). Alginate/carboxymethylcellulose-based hydrogels as pH-sensitive drug delivery systems: Facile production with enhanced degradation, thermal and mechanical properties. *Iranian Polymer Journal*, *32*(8), 1013–1032. https://doi.org/10.1007/s13726-023-01182-9

Alavi, M., & Varma, R. S. (2021). Antibacterial and wound healing activities of silver nanoparticles embedded in cellulose compared to other polysaccharides and protein polymers. *Cellulose*, *28*(13), 8295–8311. https://doi.org/10.1007/s10570-021-04067-3

Alexandridis, P., Ghasemi, M., Furlani, E. P., & Tsianou, M. (2018). Solvent processing of cellulose for effective bioresource utilization. *Current Opinion in Green and Sustainable Chemistry*, *14*, 40–52. https://doi.org/10.1016/j.cogsc.2018.05.008

Alvarez-Lorenzo, C., Blanco-Fernandez, B., Puga, A. M., & Concheiro, A. (2013). Crosslinked ionic polysaccharides for stimuli-sensitive drug delivery. *Advanced Drug Delivery Reviews*, *65*(9), 1148–1171. https://doi.org/10.1016/j.addr.2013.04.016

Arianto, R. J. D., & Sunardi. (2020). Properties of cellulose and modified cellulose-alginate for rifampicin drug delivery. *IOP Conference Series: Materials Science and Engineering*, *980*(1), 012016. https://doi.org/10.1088/1757-899X/980/1/012016

Auriemma, G., Russo, P., Del Gaudio, P., García-González, C. A., Landín, M., & Aquino, R. P. (2020). Technologies and formulation design of polysaccharide-based hydrogels for drug delivery. *Molecules*, *25*(14), Article 14. https://doi.org/10.3390/molecules25143156

Aziz, T., Farid, A., Haq, F., Kiran, M., Ullah, A., Zhang, K., Li, C., Ghazanfar, S., Sun, H., Ullah, R., Ali, A., Muzammal, M., Shah, M., Akhtar, N., Selim, S., Hagagy, N., Samy, M., & Al Jaouni, S. K. (2022). A review on the modification of cellulose and its applications. *Polymers*, *14*(15), 3206. https://doi.org/10.3390/polym14153206

Baek, S.-Y., & Park, S.-Y. (2021). Highly-porous uniformly-sized amidoxime-functionalized cellulose beads prepared by microfluidics with N-methylmorpholine N oxide. *Cellulose*, *28*(9), 5401–5419. https://doi.org/10.1007/s10570-021-03872-0

Bahadur, I., & Phadagi, R (2019). Ionic liquids as environmental benign solvents for cellulose chemistry: A review. In M. Shahid, S. Maiti, S. A. Khan, & R. V. Adivarekar (Eds.), *Solvents, Ionic Liquids and Solvent Effects*. (pp. 173–183). IntechOpen. https://doi.org/10.5772/intechopen.76433

Benhalima, T., Ferfera-Harrar, H., & Lerari, D. (2017). Optimization of carboxymethyl cellulose hydrogels beads generated by an anionic surfactant micelle templating for cationic dye uptake: Swelling, sorption and reusability studies. *International Journal of Biological Macromolecules*, *105*, 1025–1042. https://doi.org/10.1016/j.ijbiomac.2017.07.135

Benítez, A. J., Torres-Rendon, J., Poutanen, M., & Walther, A. (2013). Humidity and multiscale structure govern mechanical properties and deformation modes in films of native cellulose nanofibrils. *Biomacromolecules*, *14*(12), 4497–4506. https://doi.org/10.1021/bm401451m

Benítez, A. J., & Walther, A. (2017). Cellulose nanofibril nanopapers and bioinspired nanocomposites: A review to understand the mechanical property space. *Journal of Materials Chemistry A*, *5*(31), 16003–16024. https://doi.org/10.1039/C7TA02006F

Borandeh, S., van Bochove, B., Teotia, A., & Seppälä, J. (2021). Polymeric drug delivery systems by additive manufacturing. *Advanced Drug Delivery Reviews*, *173*, 349–373. https://doi.org/10.1016/j.addr.2021.03.022

Bordbar-Khiabani, A., & Gasik, M. (2022). Smart hydrogels for advanced drug delivery systems. *International Journal of Molecular Sciences*, *23*(7), Article 7. https://doi.org/10.3390/ijms23073665

Boustta, M., Colombo, P.-E., Lenglet, S., Poujol, S., & Vert, M. (2014). Versatile UCST-based thermoresponsive hydrogels for loco-regional sustained drug delivery. *Journal of Controlled Release: Official Journal of the Controlled Release Society*, *174*, 1–6. https://doi.org/10.1016/j.jconrel.2013.10.040

Bulut, E. (2022). Design and optimization of pH-responsive chitosan-coated Zn-carboxymethyl cellulose hydrogel bead carriers for amoxicillin trihydrate delivery. *ChemistrySelect*, *7*(24), e202200471. https://doi.org/10.1002/slct.202200471

Bustamante-Torres, M., Romero-Fierro, D., Arcentales-Vera, B., Palomino, K., Magaña, H., & Bucio, E. (2021). Hydrogels classification according to the physical or chemical interactions and as stimuli-sensitive materials. *Gels (Basel, Switzerland)*, *7*(4), 182. https://doi.org/10.3390/gels7040182

Cao, D., Chen, X., Cao, F., Guo, W., Tang, J., Cai, C., Cui, S., Yang, X., Yu, L., Su, Y., & Ding, J. (2021). An intelligent transdermal formulation of ALA-loaded copolymer thermogel with spontaneous asymmetry by using temperature-induced Sol–Gel transition and Gel–Sol (Suspension) transition on different sides. *Advanced Functional Materials*, *31*(22), 2100349. https://doi.org/10.1002/adfm.202100349

Carrillo-Varela, I., Mendonça, R. T., Pereira, M., Reyes-Contreras, P., & Contreras, D. (2022). Methylene blue adsorption onto hydrogels made from different Eucalyptus dissolving pulps. *Cellulose*, *29*(1), 445–468. https://doi.org/10.1007/s10570-021-04301-y

Carvalho, J. P. F., Silva, A. C. Q., Silvestre, A. J. D., Freire, C. S. R., & Vilela, C. (2021). Spherical cellulose micro and nanoparticles: A review of recent developments and applications. *Nanomaterials*, *11*(10), 2744. https://doi.org/10.3390/nano11102744

Chang, J., Woo, H., Ko, M.-S., Lee, J., Lee, S., Yun, S.-T., & Lee, S. (2015). Targeted removal of trichlorophenol in water by oleic acid-coated nanoscale palladium/zero-valent iron alginate beads1. *Journal of Hazardous Materials*, *293*, 30–36. https://doi.org/10.1016/j.jhazmat.2015.03.021

Channab, B.-E., Ouardi, M. E., Marrane, S. E., Layachi, O. A., Idrissi, A. E., Farsad, S., Mazkad, D., BaQais, A., Lasri, M., & Ahsaine, H. A. (2023). Alginate@ZnCO2O4 for efficient peroxymonosulfate activation towards effective rhodamine B degradation: Optimization using response surface methodology. *RSC Advances*, *13*(29), 20150–20163. https://doi.org/10.1039/D3RA02865H

Chen, H., Miao, M., & Ding, X. (2009). Influence of moisture absorption on the interfacial strength of bamboo/vinyl ester composites. *Composites Part A: Applied Science and Manufacturing*, *40*(12), 2013–2019. https://doi.org/10.1016/j.compositesa.2009.09.003

Chen, Y., Li, Y., Shen, W., Li, K., Yu, L., Chen, Q., & Ding, J. (2016). Controlled release of liraglutide using thermogelling polymers in treatment of diabetes. *Scientific Reports*, *6*(1), 31593. https://doi.org/10.1038/srep31593

Ciolacu, D. E., Nicu, R., & Ciolacu, F. (2020). Cellulose-based hydrogels as sustained drug-delivery systems. *Materials*, *13*(22), 5270. https://doi.org/10.3390/ma13225270

Cova, T. F., Murtinho, D., Pais, A. A. C. C., & Valente, A. J. M. (2018). Combining cellulose and cyclodextrins: Fascinating designs for materials and pharmaceutics. *Frontiers in Chemistry*, *6*, 271. https://doi.org/10.3389/fchem.2018.00271

Dhakal, H. N., Zhang, Z. Y., & Richardson, M. O. W. (2007). Effect of water absorption on the mechanical properties of hemp fibre reinforced unsaturated polyester composites. *Composites Science and Technology*, *67*(7), 1674–1683. https://doi.org/10.1016/j.compscitech.2006.06.019

Dormanns, J. W., Schuermann, J., Müssig, J., Duchemin, B. J. C., & Staiger, M. P. (2016). Solvent infusion processing of all-cellulose composite laminates using an aqueous NaOH/urea solvent system. *Composites Part A: Applied Science and Manufacturing*, *82*, 130–140. https://doi.org/10.1016/j.compositesa.2015.12.002

Druel, L., Niemeyer, P., Milow, B., & Budtova, T. (2018). Rheology of cellulose-[DBNH][CO2Et] solutions and shaping into aerogel beads. *Green Chemistry*, *20*(17), 3993–4002. https://doi.org/10.1039/C8GC01189C

El Allaoui, B., Benzeid, H., Zari, N., Qaiss, A. E. K., & Bouhfid, R. (2023). Functional cellulose-based beads for drug delivery: Preparation, functionalization, and applications. *Journal of Drug Delivery Science and Technology*, *88*, 104899. https://doi.org/10.1016/j.jddst.2023.104899

El Allaoui, B., Chakhtouna, H., Zari, N., Benzeid, H., Qaiss, A. el kacem, & Bouhfid, R. (2023). Superhydrophobic alkylsilane functionalized cellulose beads for efficient oil/water separation. *Journal of Water Process Engineering*, *54*, 104015. https://doi.org/10.1016/j.jwpe.2023.104015

El-Wakil, N., Taha, M., Abouzeid, R., & Dufresne, A. (2024). Dissolution and regeneration of cellulose from N-methylmorpholine N-oxide and fabrication of nanofibrillated cellulose. *Biomass Conversion and Biorefinery*, *14*(4), 5399–5410. https://doi.org/10.1007/s13399-022-02841-2

Esquivel-Castro, T. A., Ibarra-Alonso, M. C., Oliva, J., & Martínez-Luévanos, A. (2019). Porous aerogel and core/shell nanoparticles for controlled drug delivery: A review. *Materials Science and Engineering: C*, *96*, 915–940. https://doi.org/10.1016/j.msec.2018.11.067

Evyan, C.-Y. Y., Salleh, K. M., Chong, M. Y., Chia, C. H., & Zakaria, S. (2021). Effect of dimensionality of nanosized TiO embedded in regenerated cellulose beads as a portable catalyst for reusable decomposition system. *Polymers for Advanced Technologies*, *32*(9), 3549–3562. https://doi.org/10.1002/pat.5365

Fiorica, C., Biscari, G., Palumbo, F. S., Pitarresi, G., Martorana, A., & Giammona, G. (2021). Physicochemical and rheological characterization of different low molecular weight gellan gum products and derived ionotropic crosslinked hydrogels. *Gels*, *7*(2), Article 2. https://doi.org/10.3390/gels7020062

Gadziński, P., Froelich, A., Jadach, B., Wojtyłko, M., Tatarek, A., Białek, A., Krysztofiak, J., Gackowski, M., Otto, F., & Osmałek, T. (2023). Ionotropic gelation and chemical crosslinking as methods for fabrication of modified-release gellan gum-based drug delivery systems. *Pharmaceutics*, *15*(1), Article 1. https://doi.org/10.3390/pharmaceutics15010108

Gaffar, M. A., Ahammed, B., Hoque, T., Maksura, S., & Islam, Md. R. (2025). Advanced surface modification techniques for sustainable coloration of textile materials. In M. Shahid, S. Maiti, S. A. Khan, & R. V. Adivarekar (Eds.), *Advancements in Textile Coloration: Techniques, Technologies, and Trends* (pp. 287–315). Springer Nature. https://doi.org/10.1007/978-981-96-5091-0_11

Gandhi, S. R., Quintans, J. D. S. S., Gandhi, G. R., Araújo, A. A. D. S., & Quintans Júnior, L. J. (2020). The use of cyclodextrin inclusion complexes to improve anticancer drug profiles: A systematic review. *Expert Opinion on Drug Delivery*, *17*(8), 1069–1080. https://doi.org/10.1080/17425247.2020.1776261

Ganesan, K., Budtova, T., Ratke, L., Gurikov, P., Baudron, V., Preibisch, I., Niemeyer, P., Smirnova, I., & Milow, B. (2018). Review on the production of polysaccharide aerogel particles. *Materials*, *11*(11), Article 11. https://doi.org/10.3390/ma11112144

Gericke, M., Trygg, J., & Fardim, P. (2013). Functional cellulose beads: Preparation, characterization, and applications. *Chemical Reviews*, *113*(7), 4812–4836. https://doi.org/10.1021/cr300242j

Goldbart, R., Traitel, T., Lapidot, S. A., & Kost, J. (2002). Enzymatically controlled responsive drug delivery systems. *Polymers for Advanced Technologies, 13*(10–12), 1006–1018. https://doi.org/10.1002/pat.275

Gong, Y., Mohd, S., Wu, S., Liu, S., Pei, Y., & Luo, X. (2021). pH-responsive cellulose-based microspheres designed as an effective oral delivery system for insulin. *ACS Omega, 6*(4), 2734–2741. https://doi.org/10.1021/acsomega.0c04946

González-López, M. E., Laureano-Anzaldo, C. M., Pérez-Fonseca, A. A., Gómez, C., & Robledo-Ortíz, J. R. (2021). Congo red adsorption with cellulose-graphene nanoplatelets beads by differential column batch reactor. *Journal of Environmental Chemical Engineering, 9*(2), 105029. https://doi.org/10.1016/j.jece.2021.105029

Gurikov, P., & Smirnova, I. (2018). Non-conventional methods for gelation of alginate. *Gels, 4*(1), Article 1. https://doi.org/10.3390/gels4010014

Han, J., & Seo, Y. (2021). Comparison of cellulose dissolution behavior by alkaline and sulfuric acid solvents and their films' physical properties. *Materials, 14*(18), Article 18. https://doi.org/10.3390/ma14185273

Hazra, R. S., Dutta, D., Mamnoon, B., Nair, G., Knight, A., Mallik, S., Ganai, S., Reindl, K., Jiang, L., & Quadir, M. (2021). Polymeric composite matrix with high biobased content as pharmaceutically relevant molecular encapsulation and release platform. *ACS Applied Materials & Interfaces, 13*(34), 40229–40248. https://doi.org/10.1021/acsami.1c03805

Hu, F., Wang, M., Wang, N., Hu, Y., Gan, M., Liu, D., Xie, Y., & Feng, Q. (2021). All-cellulose composites prepared by partially dissolving cellulose using NaOH/thiourea aqueous solution. *Journal of Applied Polymer Science, 138*(48), 51298. https://doi.org/10.1002/app.51298

Hu, K., Sun, J., Guo, Z., Wang, P., Chen, Q., Ma, M., & Gu, N. (2015). A novel magnetic hydrogel with aligned magnetic colloidal assemblies showing controllable enhancement of magnetothermal effect in the presence of alternating magnetic field. *Advanced Materials (Deerfield Beach, Fla.), 27*(15), 2507–2514. https://doi.org/10.1002/adma.201405757

Hu, Y., Acharya, S., & Abidi, N. (2019). Cellulose porosity improves its dissolution by facilitating solvent diffusion. *International Journal of Biological Macromolecules, 123*, 1289–1296. https://doi.org/10.1016/j.ijbiomac.2018.10.062

Javanbakht, S., Nazari, N., Rakhshaei, R., & Namazi, H. (2018). Cu-crosslinked carboxymethylcellulose/naproxen/graphene quantum dot nanocomposite hydrogel beads for naproxen oral delivery. *Carbohydrate Polymers, 195*, 453–459. https://doi.org/10.1016/j.carbpol.2018.04.103

Kamaly, N., Yameen, B., Wu, J., & Farokhzad, O. C. (2016). Degradable controlled-release polymers and polymeric nanoparticles: Mechanisms of controlling drug release. *Chemical Reviews, 116*(4), 2602–2663. https://doi.org/10.1021/acs.chemrev.5b00346

Karimi, M., Sahandi Zangabad, P., Ghasemi, A., Amiri, M., Bahrami, M., Malekzad, H., Ghahramanzadeh Asl, H., Mahdieh, Z., Bozorgomid, M., Ghasemi, A., Rahmani Taji Boyuk, M. R., & Hamblin, M. R. (2016). Temperature-responsive smart nanocarriers for delivery of therapeutic agents: Applications and recent advances. *ACS Applied Materials & Interfaces, 8*(33), 21107–21133. https://doi.org/10.1021/acsami.6b00371

Karzar Jeddi, M., & Mahkam, M. (2019). Magnetic nano carboxymethyl cellulose-alginate/chitosan hydrogel beads as biodegradable devices for controlled drug delivery. *International Journal of Biological Macromolecules, 135*, 829–838. https://doi.org/10.1016/j.ijbiomac.2019.05.210

Keraliya, R. A., Patel, C., Patel, P., Keraliya, V., Soni, T. G., Patel, R. C., & Patel, M. M. (2012). Osmotic drug delivery system as a part of modified release dosage form. *International Scholarly Research Notices, 2012*(1), 528079. https://doi.org/10.5402/2012/528079

Kim, J. K., Won, Y.-W., Lim, K. S., & Kim, Y.-H. (2012). Low-molecular-weight methyl cellulose-based thermo-reversible Gel/Pluronic micelle combination system for local and sustained docetaxel delivery. *Pharmaceutical Research, 29*(2), 525–534. https://doi.org/10.1007/s11095-011-0581-8

Kim, Y. C., Shin, M. D., Hackett, S. F., Hsueh, H. T., Lima e Silva, R., Date, A., Han, H., Kim, B.-J., Xiao, A., Kim, Y., Ogunnaike, L., Anders, N. M., Hemingway, A., He, P., Jun, A. S., McDonnell, P. J., Eberhart, C., Pitha, I., Zack, D. J., … Ensign, L. M. (2020). Gelling hypotonic polymer solution for extended topical drug delivery to the eye. *Nature Biomedical Engineering*, *4*(11), 1053–1062. https://doi.org/10.1038/s41551-020-00606-8

Klojdová, I., Milota, T., Smetanová, J., & Stathopoulos, C. (2023). Encapsulation: A strategy to deliver therapeutics and bioactive compounds? *Pharmaceuticals*, *16*(3), 362. https://doi.org/10.3390/ph16030362

Kurozawa, L. E., & Hubinger, M. D. (2017). Hydrophilic food compounds encapsulation by ionic gelation. *Current Opinion in Food Science*, *15*, 50–55. https://doi.org/10.1016/j.cofs.2017.06.004

Le Duigou, A., Bourmaud, A., Davies, P., & Baley, C. (2014). Long term immersion in natural seawater of Flax/PLA biocomposite. *Ocean Engineering*, *90*, 140–148. https://doi.org/10.1016/j.oceaneng.2014.07.021

Lee, A.-J., Lee, Y.-J., Jeon, H.-Y., Kim, M., Han, E.-T., Park, W. S., Hong, S.-H., Kim, Y.-M., & Ha, K.-S. (2020). Application of elastin-like biopolymer-conjugated C-peptide hydrogel for systemic long-term delivery against diabetic aortic dysfunction. *Acta Biomaterialia*, *118*, 32–43. https://doi.org/10.1016/j.actbio.2020.09.055

Li, C., He, M., Tong, Z., Li, Y., Sheng, W., Luo, L., Tong, Y., Yu, H., Huselstein, C., & Chen, Y. (2016). Construction of biocompatible regenerated cellulose/SPI composite beads using high-voltage electrostatic technique. *RSC Advances*, *6*(58), 52528–52538. https://doi.org/10.1039/C6RA02364A

Li, H., Kruteva, M., Mystek, K., Dulle, M., Ji, W., Pettersson, T., & Wågberg, L. (2020). Macro- and microstructural evolution during drying of regenerated cellulose beads. *ACS Nano*, *14*(6), 6774–6784. https://doi.org/10.1021/acsnano.0c00171

Li, K., Yang, H., Jiang, L., Liu, X., Lang, P., Deng, B., Li, N., & Xu, W. (2020). Glycerin/NaOH aqueous solution as a green solvent system for dissolution of cellulose. *Polymers*, *12*(8), Article 8. https://doi.org/10.3390/polym12081735

Li, L., Zhu, Y., & Yang, J. (2018). 3D bioprinting of cellulose with controlled porous structures from NMMO. *Materials Letters*, *210*, 136–138. https://doi.org/10.1016/j.matlet.2017.09.015

Li, Y., & Yang, L. (2015). Driving forces for drug loading in drug carriers. *Journal of Microencapsulation*, *32*(3), 255–272. https://doi.org/10.3109/02652048.2015.1010459

Liao, J., & Huang, H. (2020). Review on magnetic natural polymer constructed hydrogels as vehicles for drug delivery. *Biomacromolecules*, *21*(7), 2574–2594. https://doi.org/10.1021/acs.biomac.0c00566

Lindman, B., Medronho, B., Alves, L., Costa, C., Edlund, H., & Norgren, M. (2017). The relevance of structural features of cellulose and its interactions to dissolution, regeneration, gelation and plasticization phenomena. *Physical Chemistry Chemical Physics*, *19*(35), 23704–23718. https://doi.org/10.1039/C7CP02409F

Lin, F., Zheng, J., Guo, W., Zhu, Z., Wang, Z., Dong, B., Lin, C., Huang, B., & Lu, B. (2019). Smart cellulose-derived magnetic hydrogel with rapid swelling and deswelling properties for remotely controlled drug release. *Cellulose*, *26*(11), 6861–6877. https://doi.org/10.1007/s10570-019-02572-0

Liu, M., Song, X., Wen, Y., Zhu, J.-L., & Li, J. (2017). Injectable thermoresponsive hydrogel formed by alginate-g-Poly(N-isopropylacrylamide) That releases doxorubicin-encapsulated micelles as a smart drug delivery system. *ACS Applied Materials & Interfaces*, *9*(41), 35673–35682. https://doi.org/10.1021/acsami.7b12849

Liu, T.-Y., Chan, T.-Y., Wang, K.-S., & Tsou, H.-M. (2015). Influence of magnetic nanoparticle arrangement in ferrogels for tunable biomolecule diffusion. *RSC Advances*, *5*(109), 90098–90102. https://doi.org/10.1039/C5RA17306J

Liu, T.-Y., Hu, S.-H., Liu, K.-H., Liu, D.-M., & Chen, S.-Y. (2006a). Preparation and characterization of smart magnetic hydrogels and its use for drug release. *Journal of Magnetism and Magnetic Materials*, *304*(1), e397–e399. https://doi.org/10.1016/j.jmmm.2006.01.203

Liu, T.-Y., Hu, S.-H., Liu, T.-Y., Liu, D.-M., & Chen, S.-Y. (2006b). Magnetic-sensitive behavior of intelligent ferrogels for controlled release of drug. *Langmuir: The ACS Journal of Surfaces and Colloids*, *22*(14), 5974–5978. https://doi.org/10.1021/la060371e

Liyanage, S., Acharya, S., Parajuli, P., Shamshina, J. L., & Abidi, N. (2021). Production and surface modification of cellulose bioproducts. *Polymers*, *13*(19), 3433. https://doi.org/10.3390/polym13193433

Lizundia, E., Goikuria, U., Vilas, J. L., Cristofaro, F., Bruni, G., Fortunati, E., Armentano, I., Visai, L., & Torre, L. (2018). Metal nanoparticles embedded in cellulose nanocrystal based films: Material properties and post-use analysis. *Biomacromolecules*, *19*(7), 2618–2628. https://doi.org/10.1021/acs.biomac.8b00243

Lu, L., Fan, W., Meng, X., Xue, L., Ge, S., Wang, C., Foong, S. Y., Tan, C. S. Y., Sonne, C., Aghbashlo, M., Tabatabaei, M., & Lam, S. S. (2023). Current recycling strategies and high-value utilization of waste cotton. *Science of the Total Environment*, *856*, 158798. https://doi.org/10.1016/j.scitotenv.2022.158798

Luo, X., Lei, X., Cai, N., Xie, X., Xue, Y., & Yu, F. (2016). Removal of heavy metal ions from water by magnetic cellulose-based beads with embedded chemically modified magnetite nanoparticles and activated carbon. *ACS Sustainable Chemistry & Engineering*, *4*(7), 3960–3969. https://doi.org/10.1021/acssuschemeng.6b00790

Luo, X., & Zhang, L. (2009). High effective adsorption of organic dyes on magnetic cellulose beads entrapping activated carbon. *Journal of Hazardous Materials*, *171*(1), 340–347. https://doi.org/10.1016/j.jhazmat.2009.06.009

Mahmoud, Y. A.-G., El-Naggar, M. E., Abdel-Megeed, A., & El-Newehy, M. (2021). Recent advancements in microbial polysaccharides: Synthesis and applications. *Polymers*, *13*(23), 4136. https://doi.org/10.3390/polym13234136

Martins, E., Renard, D., Davy, J., Marquis, M., & Poncelet, D. (2015). Oil core microcapsules by inverse gelation technique. *Journal of Microencapsulation*, *32*(1), 86–95. https://doi.org/10.3109/02652048.2014.985342

Mittal, V., Kriplani, P., & Guarve, K. (2023). Ethylcellulose- an amazing polymer for anticancer formulations. *Current Applied Polymer Science*, *6*(2), 61–75. https://doi.org/10.2174/0124522716273253231129104511

Mohan, T., Ajdnik, U., Nagaraj, C., Lackner, F., Dobaj Štiglic, A., Palani, T., Amornkitbamrung, L., Gradišnik, L., Maver, U., Kargl, R., & Stana Kleinschek, K. (2022). One-step fabrication of hollow spherical cellulose beads: Application in pH-responsive therapeutic delivery. *ACS Applied Materials & Interfaces*, *14*(3), 3726–3739. https://doi.org/10.1021/acsami.1c19577

Morozkina, S., Strekalovskaya, U., Vanina, A., Snetkov, P., Krasichkov, A., Polyakova, V., & Uspenskaya, M. (2022). The fabrication of alginate–carboxymethyl cellulose-based composites and drug release profiles. *Polymers*, *14*(17), 3604. https://doi.org/10.3390/polym14173604

Muthoka, R. M., Kim, H. C., Kim, J. W., Zhai, L., Panicker, P. S., & Kim, J. (2020). Steered pull simulation to determine nanomechanical properties of cellulose nanofiber. *Materials*, *13*(3), Article 3. https://doi.org/10.3390/ma13030710

Naserifar, S., Kuijpers, P. F., Wojno, S., Kádár, R., Bernin, D., & Hasani, M. (2022). *In situ* monitoring of cellulose etherification in solution: Probing the impact of solvent composition on the synthesis of 3-allyloxy-2-hydroxypropyl-cellulose in aqueous hydroxide systems. *Polymer Chemistry*, *13*(28), 4111–4123. https://doi.org/10.1039/D2PY00231K

Nathanael, A. J., & Oh, T. H. (2020). Biopolymer coatings for biomedical applications. *Polymers*, *12*(12), 3061. https://doi.org/10.3390/polym12123061

O'Brien, C. T., Virtanen, T., Donets, S., Jennings, J., Guskova, O., Morrell, A. H., Rymaruk, M., Ruusuvirta, L., Salmela, J., Setala, H., Sommer, J.-U., Ryan, A. J., & Mykhaylyk, O. O. (2021). Control of the aqueous solubility of cellulose by hydroxyl group substitution and its effect on processing. *Polymer, 223*, 123681. https://doi.org/10.1016/j.polymer.2021.123681

Okada, S., Nakahara, H., & Isaka, H. (1987). Adsorption of drugs on microcrystalline cellulose suspended in aqueous solutions. *Chemical & Pharmaceutical Bulletin, 35*(2), 761–768. https://doi.org/10.1248/cpb.35.761

Omura, T., Imagawa, K., Kono, K., Suzuki, T., & Minami, H. (2017). Encapsulation of either hydrophilic or hydrophobic substances in spongy cellulose particles. *ACS Applied Materials & Interfaces, 9*(1), 944–949. https://doi.org/10.1021/acsami.6b13261

Opriş, O., Mormile, C., Lung, I., Stegarescu, A., Soran, M.-L., & Soran, A. (2024). An overview of biopolymers for drug delivery applications. *Applied Sciences, 14*(4), 1383. https://doi.org/10.3390/app14041383

Orhan, B., Kaygusuz, H., & Erim, F. B. (2022). Sustainable alginate-carboxymethyl cellulose superabsorbents prepared by a novel quasi-cryogelation method. *Journal of Polymer Research, 29*(8), 333. https://doi.org/10.1007/s10965-022-03185-1

Pacelli, S., Acosta, F., Chakravarti, A. R., Samanta, S. G., Whitlow, J., Modaresi, S., Ahmed, R. P. H., Rajasingh, J., & Paul, A. (2017). Nanodiamond-based injectable hydrogel for sustained growth factor release: Preparation, characterization and *in vitro* analysis. *Acta Biomaterialia, 58*, 479–491. https://doi.org/10.1016/j.actbio.2017.05.026

Park, J. H., Saravanakumar, G., Kim, K., & Kwon, I. C. (2010). Targeted delivery of low molecular drugs using chitosan and its derivatives. *Advanced Drug Delivery Reviews, 62*(1), 28–41. https://doi.org/10.1016/j.addr.2009.10.003

Peñafiel, M. E., & Flores, D. (2023). Competitive adsorption of drugs from a multi-component mixture on sugarcane bagasse. *Water, 15*(11), 2127. https://doi.org/10.3390/w15112127

Phan, V. H. G., Thambi, T., Gil, M. S., & Lee, D. S. (2017). Temperature and pH-sensitive injectable hydrogels based on poly (sulfamethazine carbonate urethane) for sustained delivery of cationic proteins. *Polymer, 109*, 38–48. https://doi.org/10.1016/j.polymer.2016.12.039

Phan-Xuan, T., Thuresson, A., Skepö, M., Labrador, A., Bordes, R., & Matic, A. (2016). Aggregation behavior of aqueous cellulose nanocrystals: The effect of inorganic salts. *Cellulose, 23*(6), 3653–3663. https://doi.org/10.1007/s10570-016-1080-1

Pooresmaeil, M., Behzadi Nia, S., & Namazi, H. (2019). Green encapsulation of LDH(Zn/Al)-5-Fu with carboxymethyl cellulose biopolymer; new nanovehicle for oral colorectal cancer treatment. *International Journal of Biological Macromolecules, 139*, 994–1001. https://doi.org/10.1016/j.ijbiomac.2019.08.060

Purohit, P., Bhatt, A., Mittal, R. K., Abdellattif, M. H., & Farghaly, T. A. (2023). Polymer Grafting and its chemical reactions. *Frontiers in Bioengineering and Biotechnology, 10*, 1044927. https://doi.org/10.3389/fbioe.2022.1044927

Qin, J., Asempah, I., Laurent, S., Fornara, A., Muller, R. N., & Muhammed, M. (2009). Injectable superparamagnetic ferrogels for controlled release of hydrophobic drugs. *Advanced Materials, 21*(13), 1354–1357. Doi: 10.1002/adma.200800764

Qiu, M., Wang, D., Liang, W., Liu, L., Zhang, Y., Chen, X., Sang, D. K., Xing, C., Li, Z., Dong, B., Xing, F., Fan, D., Bao, S., Zhang, H., & Cao, Y. (2018). Novel concept of the smart NIR-light–controlled drug release of black phosphorus nanostructure for cancer therapy. *Proceedings of the National Academy of Sciences, 115*(3), 501–506. https://doi.org/10.1073/pnas.1714421115

Qiu, Y., & Park, K. (2001). Environment-sensitive hydrogels for drug delivery. *Advanced Drug Delivery Reviews, 53*(3), 321–339. https://doi.org/10.1016/S0169-409X(01)00203-4

Rasoulzadeh, M., & Namazı, H. (2017). Carboxymethyl cellulose/graphene oxide bionanocomposite hydrogel beads as anticancer drug carrier agent. *Carbohydrate Polymers, 168*, 320–326. https://doi.org/10.1016/j.carbpol.2017.03.014

Rezaei, N., Akbarzadeh, I., Kazemi, S., Montazeri, L., Zarkesh, I., Hossein-Khannazer, N., Hassan, M., & Vosough, M. (2021). Smart materials in regenerative medicine. *Modern Medical Laboratory Journal*, *4*(1), 39–51. https://doi.org/10.30699/mmlj17.4.1.39

Rizwan, M., Yahya, R., Hassan, A., Yar, M., Azzahari, A. D., Selvanathan, V., Sonsudin, F., & Abouloula, C. N. (2017). pH sensitive hydrogels in drug delivery: Brief history, properties, swelling, and release mechanism, material selection and applications. *Polymers*, *9*(4), Article 4. https://doi.org/10.3390/polym9040137

Rosenau, T., Potthast, A., Hettegger, H., Bacher, M., Opietnik, M., Röder, T., & Adorjan, I. (2021). On the role of N-methylmorpholine-N-oxide (NMMO) in the generation of elemental transition metal precipitates in cellulosic materials. *Cellulose*, *28*(16), 10143–10161. https://doi.org/10.1007/s10570-021-04181-2

Sayyed, A. J., Deshmukh, N. A., & Pinjari, D. V. (2019). A critical review of manufacturing processes used in regenerated cellulosic fibres: Viscose, cellulose acetate, cuprammonium, LiCl/DMAc, ionic liquids, and NMMO based lyocell. *Cellulose*, *26*(5), 2913–2940. https://doi.org/10.1007/s10570-019-02318-y

Schroeter, B., Yonkova, V. P., Niemeyer, N. A. M., Jung, I., Preibisch, I., Gurikov, P., & Smirnova, I. (2021). Cellulose aerogel particles: Control of particle and textural properties in jet cutting process. *Cellulose*, *28*(1), 223–239. https://doi.org/10.1007/s10570-020-03555-2

Shah, A., Kuddushi, M., Mondal, K., Jain, M., & Malek, N. (2020). Magnetically driven release of dopamine from magnetic-non-magnetic cellulose beads. *Journal of Molecular Liquids*, *320*, 114290. https://doi.org/10.1016/j.molliq.2020.114290

Shahriari-Khalaji, M., Zarkesh, M., & Nozhat, Z. (2021). Application of bacterial nanocellulose in cancer drug delivery: A review. *Current Pharmaceutical Design*, *27*(34), 3656–3665. https://doi.org/10.2174/1381612827666210412150445

Shao, J., Ruan, C., Xie, H., Li, Z., Wang, H., Chu, P. K., & Yu, X.-F. (2018). Black-phosphorus-incorporated hydrogel as a sprayable and biodegradable photothermal platform for postsurgical treatment of cancer. *Advanced Science*, *5*(5), 1700848. https://doi.org/10.1002/advs.201700848

Sharma, M., Deohra, A., Reddy, K. R., & Sadhu, V. (2019). Biocompatible *in-situ* gelling polymer hydrogels for treating ocular infection. In V. Gurtler, A. S. Ball, & S. Soni (Eds.), *Methods in Microbiology* (Vol. 46, pp. 93–114). Academic Press. https://doi.org/10.1016/bs.mim.2019.01.001

Stamboulis, A., Baillie, C. A., & Peijs, T. (2001). Effects of environmental conditions on mechanical and physical properties of flax fibers. *Composites Part A: Applied Science and Manufacturing*, *32*(8), 1105–1115. https://doi.org/10.1016/S1359-835X(01)00032-X

Sun, R., Zhu, J., Wu, H., Wang, S., Li, W., & Sun, Q. (2021). Modulating layer-by-layer assembled sodium alginate-chitosan film properties through incorporation of cellulose nanocrystals with different surface charge densities. *International Journal of Biological Macromolecules*, *180*, 510–522. https://doi.org/10.1016/j.ijbiomac.2021.03.092

Swamy, B. Y., & Yun, Y.-S. (2015). *In vitro* release of metformin from iron (III) cross-linked alginate–carboxymethyl cellulose hydrogel beads. *International Journal of Biological Macromolecules*, *77*, 114–119. https://doi.org/10.1016/j.ijbiomac.2015.03.019

Takada, A., & Kadokawa, J. (2022). Preparation of cellulosic soft and composite materials using ionic liquid media and ion gels. *Cellulose*, *29*(5), 2745–2754. https://doi.org/10.1007/s10570-021-04215-9

Tan, R. Y. H., Lee, C. S., Pichika, M. R., Cheng, S. F., & Lam, K. Y. (2022). PH responsive polyurethane for the advancement of biomedical and drug delivery. *Polymers*, *14*(9), 1672. https://doi.org/10.3390/polym14091672

Tan, Y., Wu, H., Xie, T., Chen, L., Hu, S., Tian, H., Wang, Y., & Wang, J. (2020). Characterization and antibacterial effect of quaternized chitosan anchored cellulose beads. *International Journal of Biological Macromolecules*, *155*, 1325–1332. https://doi.org/10.1016/j.ijbiomac.2019.11.104

Taokaew, S., & Kriangkrai, W. (2022). Recent progress in processing cellulose using ionic liquids as solvents. *Polysaccharides, 3*(4), Article 4. https://doi.org/10.3390/polysaccharides3040039

Tomatsu, I., Peng, K., & Kros, A. (2011). Photoresponsive hydrogels for biomedical applications. *Advanced Drug Delivery Reviews, 63*(14–15), 1257–1266. https://doi.org/10.1016/j.addr.2011.06.009

Tu, H., Li, X., Liu, Y., Luo, L., Duan, B., & Zhang, R. (2022). Recent progress in regenerated cellulose-based fibers from alkali/urea system via spinning process. *Carbohydrate Polymers, 296*, 119942. https://doi.org/10.1016/j.carbpol.2022.119942

Uhrich, K. E., Cannizzaro, S. M., Langer, R. S., & Shakesheff, K. M. (1999). Polymeric systems for controlled drug release. *Chemical Reviews, 99*(11), 3181–3198. https://doi.org/10.1021/cr940351u

Ullah, H., Badshah, M., Correia, A., Wahid, F., Santos, H. A., & Khan, T. (2019). Functionalized bacterial cellulose microparticles for drug delivery in biomedical applications. *Current Pharmaceutical Design, 25*(34), 3692–3701. https://doi.org/10.2174/1381612825666191011103851

Wahid, F., Huang, L.-H., Zhao, X.-Q., Li, W.-C., Wang, Y.-Y., Jia, S.-R., & Zhong, C. (2021). Bacterial cellulose and its potential for biomedical applications. *Biotechnology Advances, 53*, 107856. https://doi.org/10.1016/j.biotechadv.2021.107856

Wahlström, N., Edlund, U., Pavia, H., Toth, G., Jaworski, A., Pell, A. J., Choong, F. X., Shirani, H., Nilsson, K. P. R., & Richter-Dahlfors, A. (2020). Cellulose from the green macroalgae Ulva lactuca: Isolation, characterization, optotracing, and production of cellulose nanofibrils. *Cellulose, 27*(7), 3707–3725. https://doi.org/10.1007/s10570-020-03029-5

Wang, S., & Yu, Y. (2013). Bioactive bead type cellulosic adsorbent for blood purification. In T. G. M. Van De Ven (Ed.), *Cellulose—Medical, Pharmaceutical and Electronic Applications* (pp. 195–214). InTech. https://doi.org/10.5772/56223

Wang, W., Wat, E., Hui, P. C. L., Chan, B., Ng, F. S. F., Kan, C.-W., Wang, X., Hu, H., Wong, E. C. W., Lau, C. B. S., & Leung, P.-C. (2016). Dual-functional transdermal drug delivery system with controllable drug loading based on thermosensitive poloxamer hydrogel for atopic dermatitis treatment. *Scientific Reports, 6*(1), 24112. https://doi.org/10.1038/srep24112

Wu, M., Deng, W., Zhang, Y., Chen, C., Liu, Z., Fatehi, P., & Li, B. (2022). Facile fabrication of cellulose nanofibrils/Chitosan beads as the potential pH-Sensitive drug carriers. *Polymers, 14*(11), 2286. https://doi.org/10.3390/polym14112286

Wu, S., Gong, Y., Liu, S., Pei, Y., & Luo, X. (2021). Functionalized phosphorylated cellulose microspheres: Design, characterization and ciprofloxacin loading and releasing properties. *Carbohydrate Polymers, 254*, 117421. https://doi.org/10.1016/j.carbpol.2020.117421

Wu, W., Li, J., Liu, W., & Deng, Y. (2016). Temperature-sensitive, fluorescent poly (N-Isopropyl-acrylamide)-grafted cellulose nanocrystals for drug release. *BioResources, 11*(3), 7026–7035. https://doi.org/10.15376/biores.11.3.7026-7035

Wu, Y.-L., Wang, H., Qiu, Y.-K., Liow, S. S., Li, Z., & Loh, X. J. (2016). PHB-based gels as delivery agents of chemotherapeutics for the effective shrinkage of tumors. *Advanced Healthcare Materials, 5*(20), 2679–2685. https://doi.org/10.1002/adhm.201600723

Xie, F., De Wever, P., Fardim, P., & Van Den Mooter, G. (2021). TEMPO-oxidized cellulose beads as potential pH-responsive carriers for site-specific drug delivery in the gastrointestinal tract. *Molecules, 26*(4), 1030. https://doi.org/10.3390/molecules26041030

Xie, F., Slak, J., Fardim, P., & Van Den Mooter, G. (2023). Novel cationic cellulose beads for oral delivery of poorly water-soluble drugs. *International Journal of Pharmaceutics: X, 5*, 100146. https://doi.org/10.1016/j.ijpx.2022.100146

Xing, Y., Zeng, B., & Yang, W. (2022). Light responsive hydrogels for controlled drug delivery. *Frontiers in Bioengineering and Biotechnology, 10*, 1075670. https://doi.org/10.3389/fbioe.2022.1075670

Xue, K., Zhao, X., Zhang, Z., Qiu, B., Tan, Q. S. W., Ong, K. H., Liu, Z., Parikh, B. H., Barathi, V. A., Yu, W., Wang, X., Lingam, G., Hunziker, W., Su, X., & Loh, X. J. (2019). Sustained delivery of anti-VEGFs from thermogel depots inhibits angiogenesis without the need for multiple injections. *Biomaterials Science*, *7*(11), 4603–4614. https://doi.org/10.1039/C9BM01049A

Yan, H., Chen, X., Feng, M., Shi, Z., Zhang, W., Wang, Y., Ke, C., & Lin, Q. (2019). Entrapment of bacterial cellulose nanocrystals stabilized Pickering emulsions droplets in alginate beads for hydrophobic drug delivery. *Colloids and Surfaces B: Biointerfaces*, *177*, 112–120. https://doi.org/10.1016/j.colsurfb.2019.01.057

Yan, M., Chen, T., Zhang, S., Lu, T., & Sun, X. (2021). A core-shell structured alginate hydrogel beads with tunable thickness of carboxymethyl cellulose coating for pH responsive drug delivery. *Journal of Biomaterials Science, Polymer Edition*, *32*(6), 763–778. https://doi.org/10.1080/09205063.2020.1866350

Yang, M., Gong, W., Wang, Y., Shan, L., Li, Y., & Gao, C. (2016). Bioavailability improvement strategies for poorly water-soluble drugs based on the supersaturation mechanism: An update. *Journal of Pharmacy & Pharmaceutical Sciences*, *19*(2), Article 2. https://doi.org/10.18433/J3W904

Zeng, B., Wang, X., & Byrne, N. (2020). Cellulose beads derived from waste textiles for drug delivery. *Polymers*, *12*(7), 1621. https://doi.org/10.3390/polym12071621

Zennifer, A., Senthilvelan, P., Sethuraman, S., & Sundaramurthi, D. (2021). Key advances of carboxymethyl cellulose in tissue engineering & 3D bioprinting applications. *Carbohydrate Polymers*, *256*, 117561. https://doi.org/10.1016/j.carbpol.2020.117561

Zhang, K., Xue, K., & Loh, X. J. (2021). Thermo-responsive hydrogels: From recent progress to biomedical applications. *Gels*, *7*(3), Article 3. https://doi.org/10.3390/gels7030077

Zhang, S., He, H., Guan, S., Cai, B., Li, Q., & Rong, S. (2020). Bacterial cellulose-alginate composite beads as Yarrowia lipolytica cell carriers for lactone production. *Molecules*, *25*(4), Article 4. https://doi.org/10.3390/molecules25040928

Zhang, W., Wang, X., Wang, J., & Zhang, L. (2019). Drugs adsorption and release behavior of collagen/bacterial cellulose porous microspheres. *International Journal of Biological Macromolecules*, *140*, 196–205. https://doi.org/10.1016/j.ijbiomac.2019.08.139

Zhou, D., Zhang, L., Zhou, J., & Guo, S. (2004). Cellulose/chitin beads for adsorption of heavy metals in aqueous solution. *Water Research*, *38*(11), 2643–2650. https://doi.org/10.1016/j.watres.2004.03.026

Zuppolini, S., Salama, A., Cruz-Maya, I., Guarino, V., & Borriello, A. (2022). Cellulose amphiphilic materials: Chemistry, process and applications. *Pharmaceutics*, *14*(2), 386. https://doi.org/10.3390/pharmaceutics14020386

17 Regenerated Cellulose-Based Wound Dressing from Waste-Biomass

Source, Preparation, Characterization, and Biomedical Application

Popat Mohite, Riteshkumar Patil, Rajesh Bogati, Aishwarya Gorad, and Sudarshan Singh

17.1 INTRODUCTION

Wound healing is a complex biological process that requires an optimal environment to promote tissue regeneration, prevent infections, and minimize complications. The increasing prevalence of chronic wounds, burns, diabetic ulcers, and surgical injuries has driven the demand for advanced wound-dressing materials that offer improved biocompatibility, moisture management, and antibacterial properties (1). However, conventional wound dressings, such as synthetic polymers and traditional gauze, often fall short in terms of effectiveness, sustainability, and affordability. These limitations have led to the exploration of natural and biodegradable materials for wound care applications, with regenerated cellulose emerging as a highly promising candidate (2).

Cellulose, the most abundant biopolymer on Earth, is primarily sourced from woody and nonwoody plants. However, its extraction and purification from conventional sources require energy-intensive processes, including mechanical, chemical, or enzymatic treatments, which can generate hazardous waste and contribute to environmental pollution. As a sustainable alternative, waste-biomass has gained significant attention as a renewable and cost-effective source of cellulose. Agricultural residues, food processing waste, forestry by-products, and industrial plant waste provide abundant sources of cellulose that can be regenerated into high-quality wound-dressing materials. Utilizing these waste streams for cellulose production not only reduces environmental impact but also adds value to biomass waste, supporting circular economy initiatives, and sustainable healthcare solutions (3,4).

DOI: 10.1201/9781003596127-20

The process of regenerating cellulose from waste-biomass involves multiple steps, including extraction, purification, and structural modification, to tailor its properties for biomedical applications. Regenerated cellulose can be processed into various forms, such as films, hydrogels, and nanofibers, which mimic the extracellular matrix and facilitate cell adhesion, proliferation, and tissue regeneration. These materials possess remarkable characteristics, such as high moisture retention, excellent mechanical strength, breathability, and non-toxicity, making them ideal for wound-dressing applications. Additionally, regenerated cellulose-based wound dressings can be functionalized with bioactive agents, such as antimicrobial compounds, growth factors, and nanoparticles, to enhance their therapeutic potential and promote faster wound healing (5) (Figure 17.1).

This chapter provides a comprehensive overview of the sources of waste-biomass for cellulose extraction, various preparation and regeneration techniques, and the characterization of regenerated cellulose-based wound dressings. Furthermore, it explores their biomedical applications, emphasizing their advantages over conventional wound care materials. By integrating sustainability with biomedical innovation, regenerated cellulose-based dressings have the potential to revolutionize wound-healing strategies, offering an eco-friendly, cost-effective, and highly efficient solution for addressing global healthcare challenges.

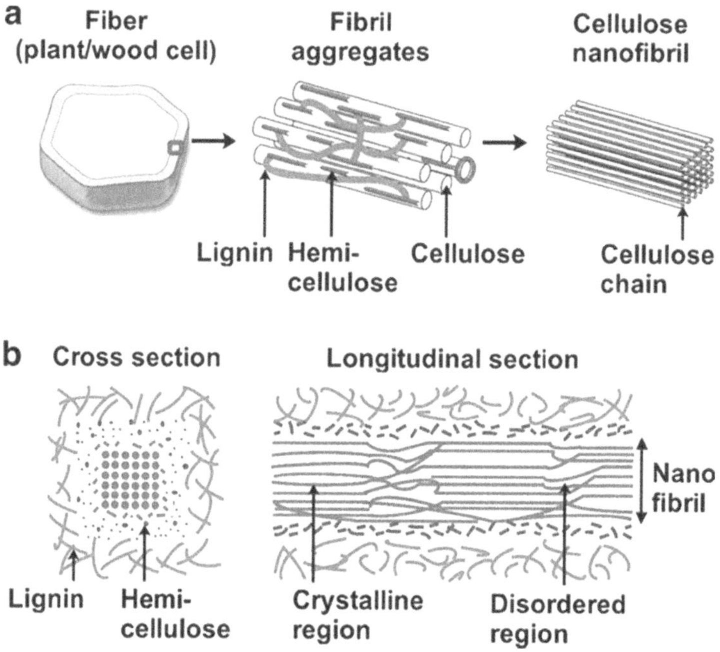

FIGURE 17.1 Schematic illustration of the microstructure of a cellulosic fiber. a Plant/wood fiber, fibril aggregate, and nanofibril (containing cellulose chains) associated with lignin and hemicellulose. b Cross-section and longitudinal section showing the crystalline and disordered regions of a cellulose nanofibril embedded in lignin and hemicellulose matrix (6).

17.2 BACTERIAL AND FUNGAL INVOLVE THE PRODUCTION AND BIODEGRADATION OF CELLULOSE

The field of wound healing, tissue repair, and tissue engineering has gained significant attention in recent years due to multiple factors, including the rising aging population, the prevalence of chronic diseases such as diabetes, and the limitations of current wound care treatments. The global wound-healing market was valued at approximately USD 19.88 billion in 2020 and is projected to grow at a compound annual growth rate of 4.1% in the coming years (7).

Bacterial cellulose (BC) has emerged as a promising biomaterial in regenerative medicine, particularly in wound-healing applications. Due to its unique properties—such as high purity, nanofibrillar structure, excellent biocompatibility, and high water retention—BC has demonstrated superior performance compared with conventional wound dressings. Commercial BC-based membranes, including Bionext®, Membracell®, and Xcell®, have been widely adopted for their ability to accelerate epithelialization, promote tissue regeneration, and enhance wound-healing outcomes (8).

Moreover, BC offers advantages such as reduced pain, prevention of bacterial infections, and potential use in drug-delivery systems. Its non-adherent nature, moisture-retaining properties, and biocompatibility make it an ideal material for wound care, including applications in burns, chronic wounds, and diabetic foot ulcers. With continuous advancements in BC-based biomaterials, novel strategies are being explored to further enhance its effectiveness in biomedical applications (9).

Cellulose, primarily derived from woody and nonwoody plants, is produced at an estimated 1×10^{12} tons per year, making the Plantae kingdom the dominant source. However, plant cellulose (PC) extraction is complex, requiring intensive mechanical, chemical, or biological treatments that generate hazardous waste and contribute to environmental concerns. In contrast, BC is a highly pure form of cellulose that requires minimal purification using eco-friendly methods, making it a sustainable alternative.

Although BC and PC share the same molecular structure, their properties differ significantly. BC exhibits higher crystallinity, a greater degree of polymerization, and retains its fibrous architecture after purification, unlike processed PC. These unique characteristics make BC a superior material for biomedical and industrial applications, particularly in wound healing and tissue engineering (10).

17.3 THE PRODUCTION PROCESS OF FIBERS FROM CELLULOSE-RICH BIOMASS

The process of converting cellulose-abundant biomass into regenerated cellulose fibers involves several essential steps, each aimed at extracting and treating cellulose for various applications. It begins with biomass selection and pretreatment, where suitable sources such as agricultural residues (e.g., corn stalks, wheat straw) or industrial waste (e.g., discarded cotton fabrics) are chosen. These materials contain cellulose, lignin, and hemicellulose, which must be removed to obtain pure

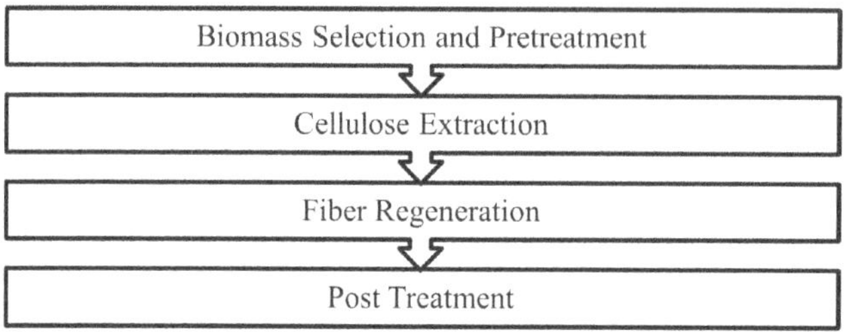

FIGURE 17.2 Production process steps.

cellulose (11). Chemical pretreatment methods, including acid or alkaline treatments, help break down lignin and hemicellulose, making cellulose more accessible. For example, dilute acid treatments effectively degrade hemicellulose, facilitating subsequent extraction. Following this, cellulose extraction is carried out using methods like organosolv pulping, in which organic solvents such as ethanol or methanol dissolve lignin, leaving behind cellulose fibers (12). This process is advantageous due to its efficiency in lignin removal and solvent recovery capabilities. The next step, fiber regeneration, involves dissolving purified cellulose in suitable solvents to form a viscous solution, which is then extruded through spinnerets into a coagulation bath, where it regenerates into continuous fibers. Innovative technologies, such as those developed by Spinnova, enable the production of textile fibers from wood pulp or waste without toxic chemicals, offering an eco-friendly alternative to conventional methods (13). Finally, post-treatment and applications involve washing, bleaching, and drying the fibers to enhance their characteristics for specific uses. These regenerated cellulose fibers are widely utilized in industries ranging from textiles to biocomposites, contributing to sustainable material development. For example, Zeoform, derived from lignocellulosic biomass, serves as an environmentally friendly alternative to conventional plastics. Through these processes, cellulose-rich waste-biomass is transformed into valuable regenerated fibers, promoting sustainability and reducing reliance on non-renewable resources (14) (Figure 17.2).

Natural products involved in the production of cellulose.

17.3.1 *Crotalaria juncea*

Sunn hemp (*Crotalaria juncea* L.), commonly called Indian hemp, is a multipurpose tropical and subtropical legume cultivated in several regions, including India, for its high-quality fiber (15). India is among the top producers of Sunn hemp, following Brazil and Bangladesh (16). This crop is highly adaptable, drought-resistant, and fast-growing, yielding significant amounts of fiber in a short period, and has been used for making ropes, twines, fishing nets, mats, and other utility items. Additionally, it serves as an important green manure crop, contributing to soil health (17). The fibers also find application in producing tissue paper, cigarette paper, and currency notes, owing to their high cellulose content (15) (Figure 17.3).

Sunn hemp fibers primarily consist of cellulose (approximately 78%), along with hemicellulose (10%), lignin (4%), and minor amounts of pectin. This composition

FIGURE 17.3 *Crotalaria juncea* plant (18).

makes it a highly suitable raw material for industrial applications, particularly for producing paper, pulp, and advanced materials (15,19,17). Studies have also demonstrated the feasibility of utilizing Sunn hemp pulp as a sustainable alternative to conventional fibers in paper production (20,21). In recent years, Sunn hemp has been recognized as a potential source for nanocrystalline cellulose (NCC), which is widely researched for its use in coatings, nanocomposites, and other high-performance materials (19). Its cellulose content, superior to many other natural fibers, makes it an excellent alternative for developing sustainable products. With its renewable, biodegradable nature and diverse applications, Sunn hemp stands out as an environmentally friendly and sustainable resource for both traditional and modern industries.

Y Jyothi et al. has demonstrated that *C. juncea* has therapeutic potential, as evidenced by a study on rats with phenylhydrazine-induced hemolytic anemia. The study revealed that treatment with *C. juncea* significantly improved reduced levels of red blood cells, hemoglobin, and packed cell volume in the blood (22).

17.3.2 *Musa acuminata*

Southeast Asia is home to the wild banana species *Musa acuminata*. Several parts of the *Musa* plants have been applied topically or taken orally as therapeutic in traditional medicine. All plant components, including the roots, stem, pseudostems, leaves, fruits, and flowers, have long been used in traditional medicine throughout America, Asia, Oceania, India, and Africa (23). Although *M. acuminata* is widely distributed, Malaysia is known to be the primary source of this fruit, which generates a substantial amount of peel waste (30%–40% of the fruit's weight). Because of its high levels of water, phosphorus, and nitrogen, which encourage microbial growth, this waste presents environmental problems. Banana peels, since they're rich in cellulose, can be processed into useful goods. In the pharmaceutical industry, cellulose is frequently employed as thickening agents and fillers in medication tablets (24). The study's analysis of the chemical makeup of banana pseudo-stem biomass shows that 38% (by dry weight) of it is cellulose (25). In their research, Charu

Lata Mahanta et al. produced cellulose fiber (CF) and nanocellulose (NC) from banana rachis waste from *M. acuminata*, which were then utilized to create bioactive aerogels. High porosity (>90%), superior water absorption (up to 1821.66%), and a progressive release of bioactive chemicals from butterfly pea flower extracts were all demonstrated by aerogels containing CF and NC. These aerogels exhibit the possibility of bioactive delivery systems for applications requiring controlled release (26). The cellulose acetate films made from *M. acuminata* were shown to possess antioxidant and antibacterial qualities by Achuth Jayakrishnanthe et al. Due to these films' potential, they can be used in packaging to improve food safety along with using them for encapsulating bioactive substances (23) (Figures 17.4 and 17.5).

17.3.3 *Actinidia deliciosa*

The genus *Actinidia* and family Actinidiaceae include kiwi fruit. Of the 76 species of the genus *Actinidia*, only two are cultivated for commercial purposes: *Actinidia chinensis*, also known as golden kiwifruit, and *Actinidia deliciosa*, also known as fuzzy kiwifruit. Italy is the world's second-largest producer of kiwi fruit, after China. New Zealand and Italy are the world's two biggest exporters of kiwis (29). Stems and leaves from pruning are the primary plant materials used to obtain cellulose (Figure 17.6).

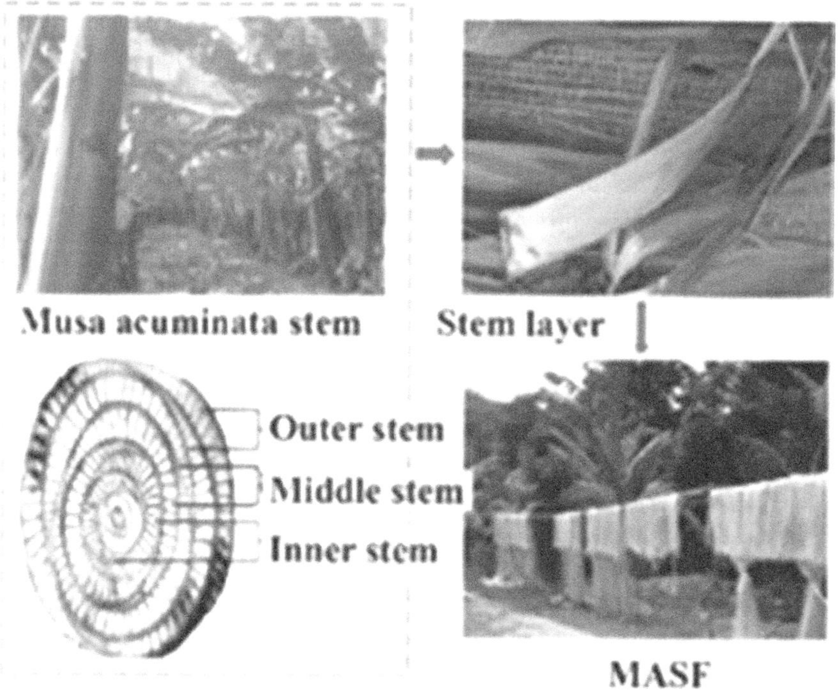

FIGURE 17.4 *Musa acuminata* stem (27).

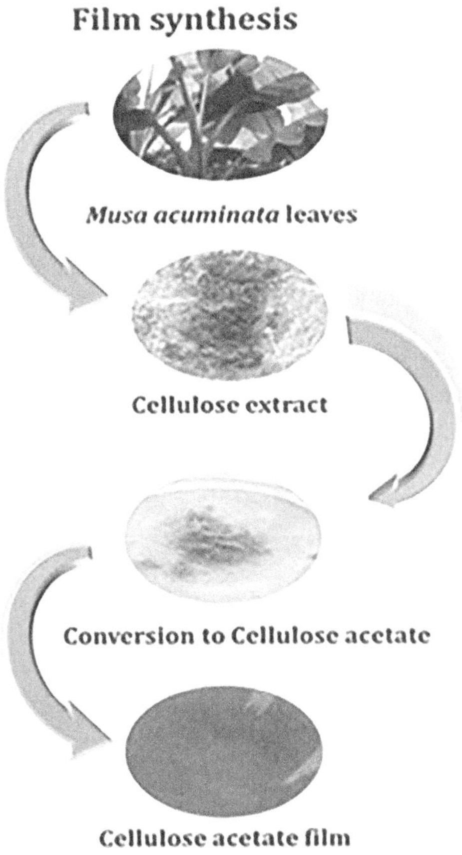

FIGURE 17.5 *M. acuminata* leaves converted to cellulose acetate film (28).

FIGURE 17.6 *Actinidia deliciosa* (30).

These residues are abundant and considered agricultural waste, making them a viable source for cellulose synthesis. Using the standard TAPPI techniques, the composition of kiwi wood was previously determined. This revealed a notable amount of holocellulose (73.50%), cellulose (38.30%), and lignin (25.30%) (31). Elena Fortunati et al. has shown that the wood of kiwi can be used to make cellulose nanocrystals, which can further enhance the mechanical and thermal properties of films made from chitosan (CH) and polyvinyl alcohol (PVA). The incorporation of CNCs improved the tensile strength and Young's modulus, while also reducing moisture absorption, making these films more resistant to environmental conditions and suitable for biodegradable packaging. Additionally, kiwi pruning residues, being a renewable and cellulose-rich resource, provide a sustainable alternative for kraft paper production (31,32). Furthermore, Yuanying Ni et al. showed that the fibers (cellulose and hemicellulose) present in kiwifruits can inhibit the activity of α-amylase, which would result in the lowering of high blood sugar levels and be beneficial, by increasing the wound-healing process in diabetic people (33,34).

FIGURE 17.7 *Azadirachta indica* bark (35).

17.3.4 *Azadirachta indica*

The neem tree, *Azadirachta indica*, is indigenous to India and a member of the Meliaceae family of mahogany. It is extensively grown in India and other areas with comparable environmental circumstances because it flourishes in tropical and semi-tropical temperatures. With its products being used for more than two millennia, the tree has historical significance due to its industrial and medical uses (Figure 17.7).

A. indica bark is specifically used to extract cellulose. The retting method is used to separate the bark fibers, which are found to bind tightly. To make fiber extraction easier, the bark is soaked in water for up to 15 days. After that, the fibers are washed, dried, and prepared for examination and use. *A. indica* bark fibers exhibit an excellent cellulose content (68.42%) (36). With this high amount, the fibers can be used in applications requiring durability and strength. Oilseed cake waste from the food sector can be converted into goods with added value, lowering manufacturing costs and solving environmental problems. Neem oil cake (NOC), the residue that remains after neem seed oil extraction, is an abundant source of cellulose, hemicelluloses, proteins, and lipids, claim Rantheesh J et al. In order to investigate its potential as a biofiller in polymer composites, cellulose was isolated from NOC. Given its high cellulose content, crystallinity, thermal stability, and desired qualities, including low density and rough surface, neem cake cellulose (NCC) can be utilized to make ecologically friendly composites (37). Also, cellulose from the NOC can be used as a biopolymer in modified drug-delivery systems and dressing wounds (38).

17.3.5 *Ceiba pentandra*

The *Ceiba pentandra* tree, also known as the kapok tree, is a member of the Malvaceae family (subfamily Bombacoideae) and is naturally found in seasonally dry tropical forests across regions such as Central America, the Caribbean, and South America (39). It is widely cultivated in Southeast Asia, Sri Lanka, East Asia, and parts of Africa, where its fibers are processed and exported (40). Historically, kapok fibers were a valuable commercial product before World War II, with significant use in items like life jackets and aviation clothing. However, the global trade of kapok fibers saw a decline in the 1960s with the increasing production of synthetic alternatives. Even so, these fibers continue to hold potential for industrial and scientific applications (40) (Figure 17.8).

Kapok fibers are predominantly made up of cellulose, with a cellulose content ranging from 55.69% to 64%. In contrast, the alpha-cellulose content in wood dust derived from *Ceiba* is about 42% (42–45). Kapok fibers have found utility in manufacturing recycled papers and are recognized for their biocompatibility and non-toxic nature, making them suitable for medical and pharmaceutical use. For example, both untreated and chemically crosslinked fibers from *C. pentandra* have shown promise in pharmaceutical applications, particularly when modified with agents like citric acid (46). Additionally, chemically treating kapok fibers with monochloroacetic acid transforms their cellulose content into carboxymethyl cellulose (CMC), which broadens their industrial applicability (39). Combining kapok fibers with other natural

FIGURE 17.8 *Ceiba pentandra* tree (41).

fibers like cotton or rayon enhances the stability of textiles and reduces water and resource consumption during production. This integration contributes to a reduced environmental footprint, making the resulting materials more sustainable (47). Furthermore, kapok fibers are being studied for advanced uses, such as their incorporation into alginate beads containing nanocrystalline cellulose (NCC). This combination has been shown to enhance the mechanical properties of the beads, including their hardness and resistance to compressive forces (43). Microcrystalline cellulose (MCC), derived from kapok fibers, is another innovation with significant potential. MCC is being utilized in dye separation membranes, where its hydrophilic properties make it highly effective (48). These applications emphasize the versatility of kapok fibers and their derivatives, underlining their importance in sustainable industries, pharmaceutical advancements, and environmentally friendly technologies. Seyed Davoud Jazayeri et al. revealed that *C. pentandra* fiber incorporated with silver nanoparticles showed antibacterial properties against both Gram-positive and negative bacteria (49).

17.3.6 *Morus alba* L

Morus alba L, commonly known as white mulberry, is a species belonging to the Moraceae family. Native to Asia, particularly regions with warm climates, this deciduous tree is now widely cultivated in subtropical, tropical, and temperate regions, including the United States, Mexico, Australia, Kyrgyzstan, and Argentina (50,51). The mulberry tree is a fast-growing species, capable of reaching heights between 10

FIGURE 17.9 *Morus alba* (54).

and 20 m (50). Its various parts exhibit significant cellulose content: the stem contains approximately 57.4% cellulose, the leaves contain 37.5% cellulose, and the bark contains about 43.8% cellulose (51,52). Additionally, the cellulose content of *Morus alba* fiber ranges between 60% and 70%. This high cellulose content significantly enhances the mechanical and thermal properties of the fiber, making it highly suitable for reinforcement in composite materials (53) (Figure 17.9).

The mulberry tree has long been valued for its edible fruit, traditionally consumed fresh or used in the production of wine, fruit juice, jam, and canned foods. Mulberry fruit is recognized for its medicinal benefits, including liver and kidney protection, strengthening joints, improving eyesight, and anti-aging properties. It has also been employed in natural medicine for the treatment of conditions such as sore throat, fever, hypertension, and anemia (55).

Various parts of the mulberry plant, including its fruits, bark, leaves, and roots, have been integral to traditional Chinese medicine. They have been used to manage ailments such as diabetes, atherosclerosis, and cancer, and to boost the immune system due to their strong antioxidant properties. The leaves of *Morus alba* are particularly noteworthy for their high fiber content, which makes them nutritious, palatable, and non-toxic (56). Moreover, studies have revealed that CMC derived from mulberry leaves can be used to create edible packaging films. These films improve the appearance of food products, enhance mechanical properties, and provide microbial protection, making them a sustainable option for food packaging. The rich cellulose content and diverse applications of *Morus alba* highlight its significance not only as a source of nutrition but also as a valuable resource for traditional medicine, cellulose-based products, composite reinforcement, and modern industrial innovations (57).

17.3.7 *Ziziphus nummularia*

Ziziphus nummularia, commonly referred to as small ber or wild ber, is a perennial plant indigenous to southeastern Pakistan and India. Known for its nutritional and therapeutic benefits, various parts of the plant such as fruits, leaves, roots, and tubers—have been traditionally used to manage ailments like ulcers, respiratory issues, fevers, wounds, liver problems, and asthma. Additionally, the plant's roots are valued for their cooling properties, often employed to alleviate headaches and digestive imbalances (58) (Figure 17.10).

The fibers of *Z. nummularia* contain around 65.72% cellulose, while its bark has a cellulose content of approximately 52.34% (60,61). The fibers possess notable surface roughness, which enhances their compatibility with polymer matrixes, making them suitable for use in composite material development. This property highlights the plant's potential in advanced material applications (60).

Furthermore, *Z. nummularia* has shown significant promise in nanotechnology. Irfan Ijaz et al. *revealed* that gold nanoparticles synthesized using *Z. nummularia* exhibit notable anticancer and antioxidant activities (62). Similarly, Polymer Capped Silver Nanoparticles produced from the aqueous extract of *Z. nummularia* demonstrate antibacterial activity against Gram-positive and negative bacteria with antioxidant properties as shown by Kokila Parmar et al. (63). These findings emphasize the versatility of *Ziziphus* species in medical, material science, and nanotechnological applications.

FIGURE 17.10 *Ziziphus nummularia* (59).

17.3.8 *ZEA MAYS*

Maize or corn (*Zea mays*) is an important cereal crop of the world and it is native to Mexico, with its cultivation widely spread across countries such as the United States, China, Brazil, India, France, South Africa, and Nigeria (64,65). It is the third most widely grown crop in the world, after rice and wheat, and due to its exceptional yield potential, it is often referred to as the "queen of cereals" (64). *Z. mays* is used for diverse applications as its various parts are rich in cellulose (Figure 17.11).

The tassels contain 41% cellulose, while corn cobs comprise 40% cellulose, and are primarily used in the production of paper. Corn bran contains approximately 28% cellulose, and maize husks, with about 31.3% cellulose, are particularly valuable for research purposes (67–70). Julien Brasa et al. extracted CNCs from maize husks through processes like bleaching, hydrolysis, and washing (70). These CNCs have versatile applications and can be engineered into zero-dimensional nanostructures for drug delivery, one-dimensional fibers for increased strength, two-dimensional films for flexibility, and three-dimensional hydrogels or aerogels for advanced material development (71).

FIGURE 17.11 *Zea mays* (66).

Furthermore, extracts from maize husks have shown anti-inflammatory properties by downregulating iNOS expression and inhibiting NF-kB and AP-1 pathways, in the mouse macrophage cell line (RAW264.7), as shown by Eunsun Jung et al. (72). They have also been combined with other materials to produce biodegradable paper cups. In addition, maize holds potential in the packaging industry as a sustainable alternative to plastic, especially in food packaging (73). Cellulose from various parts of *Z. mays* has also been processed into CMC, which has demonstrated potential for use in pharmaceutical applications (74). These versatile uses underscore the importance of maize not only as a food crop but also as a resource for innovative and sustainable industrial applications.

17.4 SKIN WOUND, TYPES OF WOUNDS, AND HEALING PROCESS

The biggest organ in the human body, the skin acts as the body's main defense against environmental threats such as chemical exposure, physical damage, and microbial invasion. A wound develops when the skin is injured, impairing its ability to protect. Injuries that impact muscles, bones, and internal tissues can vary in depth from minor abrasions.

17.4.1 Types of Wounds

Wounds can be categorized based on their origin, depth, and healing characteristics (75) (Figure 17.12).

Acute wounds occur suddenly and include punctures, which are small, deep holes caused by sharp objects like needles or nails; lacerations, which are irregular skin tears resulting from blunt trauma; incisions, which are clean, straight cuts often made by surgical instruments; and abrasions, which are superficial injuries caused by friction that affect only the epidermis (76). In contrast, chronic wounds are persistent injuries that either take a long time to heal or fail to heal completely. These include venous ulcers, commonly found on the legs due to impaired venous blood flow; diabetic ulcers, which arise from poor circulation and nerve damage in individuals with diabetes; and pressure ulcers, caused by prolonged pressure and frequently occurring in individuals with limited mobility (77,78). Burn injuries are classified into first, second, or third-degree burns based on their depth and severity (79). Additionally, infected wounds occur when bacteria colonize the injury site, leading to inflammation, pus formation, and delayed healing (Tables 17.1–17.3).

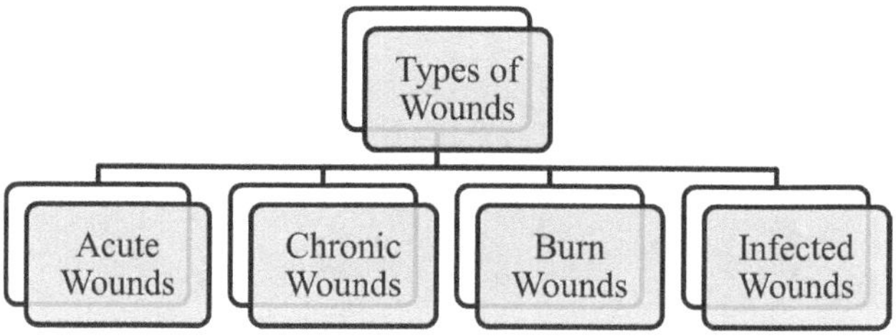

FIGURE 17.12 Types of wounds.

TABLE 17.1
Types of intention (76)

Primary Intention	Secondary Intention	Tertiary Intention
• Skin edges are approximated with sutures, staples, or adhesive	• Large, deep wounds in which edges cannot be approximated	• Healing is promoted by leaving the wound open for a prescribed period before approximating the skin edges for primary closure
• Enables quick closure with minimal scarring	• Wound depth determines degree of new tissue matrix and epidermal surface needed for complete closure	• Used in infected or dehisced surgical wounds
• Used when there is minimal epithelialization and new tissue	• Lenghthier than healing the primary intention • Can be associated with substantial scarring	

TABLE 17.2
Timeline of wound phases

Phases of Healing	Timeline
Homeostasis	Within hours of injury
Inflammation	1–5 days post-injury
Proliferation	4–21 days post-injury
Remodeling	21 days to 1 year post-injury

Wound healing is a complex biological process at the cellular and molecular level that aims to restore tissue integrity (80). It occurs in four overlapping phases, beginning with hemostasis, where blood vessels constrict immediately after injury to prevent blood loss. Platelets aggregate at the site, forming a fibrin clot that acts as a temporary barrier to stabilize the wound (81). The inflammatory phase follows, lasting 24–48 hours post-injury, during which immune cells migrate to the site to clear debris and pathogens, while TGF-β and other cytokines initiate tissue repair. Next, the proliferative phase is marked by fibroblast activity, angiogenesis, and collagen deposition. Endothelial cells form new capillaries to restore oxygen and nutrient supply, fibroblasts synthesize extracellular matrix proteins, and epithelial cells migrate to cover the wound, facilitating the formation of granulation tissue (77). Finally, the remodeling (maturation) phase, lasting weeks to months, strengthens the tissue as collagen type III is replaced by collagen type I, leading to increased tensile strength, wound contraction, and scar formation, ultimately restoring skin integrity (82) (Figure 17.13).

TABLE 17.3
Application of cellulose in wound dressings

Case Study	Description	References
Bacterial cellulose (BC) dressings in burn wound management	A clinical study evaluated the effectiveness of BC dressings (EpiProtect) for treating burn wounds. The results showed improved wound coverage, moisture retention, and microbial protection, leading to reduced pain and faster healing as per Mendes et al., 2021	(98)
Biofill for burn treatment	It is a BC-based skin substitute, was used in burn wound management. It demonstrated high biocompatibility, pain reduction, and a significant decrease in infection rates.	(99)
BC composites with antimicrobial agents	BC dressings were enhanced with tetracycline hydrochloride (TCH) to improve antibacterial efficacy. The composite exhibited strong antimicrobial activity, biocompatibility, and controlled drug release, making it suitable for wound-healing applications.	(92)
Xcell for chronic wound care	Xcell dressings were used for venous ulcers. They maintained moisture levels and aided in autolytic debridement, accelerating wound healing.	(100)

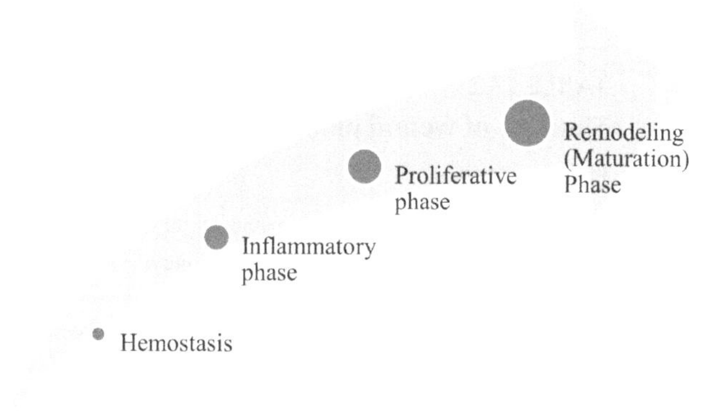

FIGURE 17.13 Phases of wound healing.

17.4.2 Fabrication Techniques Involve with Wound Dressing Including Conventional and Advanced Technology Such as Electrospinning and Additive Manufacturing

The development of wound dressings has evolved significantly, integrating both conventional and advanced fabrication technologies to meet diverse medical requirements. These processes facilitate the production of wound dressings with desired features such as biocompatibility, moisture retention, and antimicrobial properties, which are critical for fostering an environment conducive to healing (83) (Figure 17.14).

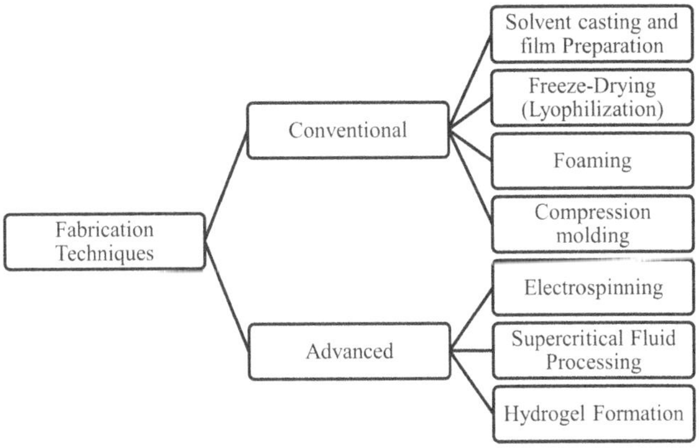

FIGURE 17.14 Fabrication techniques.

17.5 CONVENTIONAL TECHNIQUES FOR MAKING WOUND DRESSINGS INCLUDE

17.5.1 Solvent Casting and Film Preparation

The required solution is poured into thin films, which are then left to dry. It is a common technique for the preparation of dressings made out of cellulose with good moisture absorption and flexibility (84).

17.5.2 Freeze-Drying (Lyophilization)

Porous structures are obtained by first freezing the material and then evaporating the solvent. These wound dressings are lightweight and highly absorbent, making them optimal for exudate management (84,85).

17.5.3 Foaming

Soft, sponge-like dressings that are comfortable and suitable to be used for wounds and exudate absorption are produced (86).

17.5.4 Compression Molding

Blanks made of these materials are heated and compressed in order to obtain the desired shape. This technique is used to obtain flat and uniform wound dressings with stable structures (87).

17.6 ADVANCED MANUFACTURING TECHNIQUES

17.6.1 Electrospinning

Electrospinning produces new nanofiber-based wound dressings using bleeding-edge technology. This process involves the use of a polymer solution, a high-voltage charge, and the formation of ultrafine fibers. These otherworldly fibers mimic extracellular matrixes (ECM) and support a home environment for cell seeding and/or repair. Electrospin cellulose-based wound dressings weigh little, breathe, and can load drugs over a considerable time (88).

17.6.2 Supercritical Fluid Processing

Supercritical fluid processing is a process wherein supercritical fluids such as CO_2 are used to prepare highly porous dressing materials. This technology is advantageous for the controlled release of drug properties in the dressing materials (89).

17.6.3 Hydrogel Formation

Advanced crosslinking methods are used to synthesize hydrogels from cellulose derivatives. These hydrogels are very useful in maintaining a moist and heterogeneous wound-healing environment with reduced scar formation (87).

17.7 APPLICATION OF CELLULOSE-BASED WOUND DRESSING

The cellulose-based wound dressings are well-recognized medicinal apparel due to their biocompatibility, biodegradability, and excellent properties in wound healing. The wound dressings are mainly sourced from natural cellulose, including bacteria cellulose, plant fibers, and regenerated cellulose from waste-biomass. Cellulose is known for the ability of water absorption, which provides a protective barrier and promotes cell proliferation; thus, it is an excellent material in wounds management (90).

Regenerated cellulose wound dressings offer a distinct advantage by being fabricated to cater to specific medical needs. Cellulose modified with bioactive compounds such as antimicrobial agents or growth factors can enhance the wound-healing process and prevent the onset of infections. Moreover, other benefits from these cover materials consist of their ability to replicate the extracellular matrix of human skin thus fostering tissue regeneration, which makes these dressings suitable for the treatment of acute and chronic wounds such as burns, diabetic ulcers, and surgical wounds (91).

The benefits of these cellulose-based dressings extend beyond the medicinal realm; they also serve to promote environmental sustainability. The use of waste-biomass for the production of these materials decreases environmental pollution and propagates the circular economy (92). The manufacturing procedure is also low-cost, thus rendering them available for mega medical applications, particularly in developing sections of the world where advanced wound care products may not be readily available (93).

17.7.1 Moisture Maintenance and Wound Healing

Moisture Maintenance and Wound Healing: Cellulose-based dressings keep moisture levels ideal at the wound site so that it is neither too dry nor unduly drenched with exudates (94).

17.7.2 Antimicrobial Protection

Antimicrobial Protection: Dresses can be infused with antibacterial agents like silver nanoparticles, chitosan, or plant-based anti-infective agents to prevent infection.

17.7.3 Burn and Chronic Wound Treatment

Burn and Chronic Wound Treatment: BC dressings are highly effective against burn injuries due to their water retention property and close adherence to the wound surface (95). BC dresses are also used in treating chronic wounds, such as diabetic ulcers and pressure sores.

17.7.4 Biodegradable and Eco-Friendly

Biodegradable and Eco-Friendly: Naturally derived, these dresses decompose harmlessly without precipitating hazards, thus reducing medical waste (96).

17.7.5 Cheap and Accessible

Cheap and Accessible: Regenerated cellulose from waste-biomass renders the manufacturing of these dresses inexpensive, hence benefiting health systems in resource-limited settings.

Cellulose-based wound dressings are an appropriate answer for the wound care of contemporary times combining efficacy, sustainability, and affordability (97).

17.8 BIOMEDICAL APPLICATIONS OF REGENERATED CELLULOSE

Regenerated cellulose has gained significant attention in biomedical applications due to its biocompatibility, biodegradability, and excellent mechanical properties. Derived from natural cellulose through chemical processing, regenerated cellulose retains its inherent non-toxicity and hydrophilicity, making it suitable for various medical uses. It has been extensively explored for wound dressings, drug-delivery systems, tissue engineering scaffolds, and hemodialysis membranes. Its ability to be functionalized further enhances its versatility, allowing for improved bioactivity and controlled drug release. With advancements in nanotechnology and biomaterial engineering, regenerated cellulose continues to evolve as a promising material for innovative biomedical solutions.

A study by Alexa Scheer et al. explored the integration of cellulose microspheres (CMS) into paper products, demonstrating the successful incorporation of up to 50 wt.% CMS using retention aids. Despite high CMS content, the paper retained its tensile strength while exhibiting increased density and enhanced surface smoothness due to the formation of a CMS layer atop the fiber base, effectively filling voids and correcting surface irregularities. These modified paper sheets hold significant potential for various applications, including medical wound dressings, point-of-care diagnostic test strips, sustainable packaging with antimicrobial or moisture-absorbing properties, and high-efficiency filtration materials for air and water purification. The study highlights the feasibility of CMS-containing paper as a versatile platform for functional and sustainable material innovations in biomedical, environmental, and industrial applications (101).

The study by Wei Tang and colleagues successfully developed a low-surfactant emulsification technique to fabricate physically crosslinked CMs with tunable particle sizes ranging from 150 μm to 1000 μm. By employing different dehydration methods, the mechanical properties of the CMs were effectively regulated, ensuring their durability under high blood flow conditions. Notably, the perfusion column composed of ethanol-treated CMs withstood the pressure induced by a 190 ml/min flow rate, meeting the critical requirement for hemoperfusion devices. Furthermore, the study demonstrated that C18-modified CMs exhibited specific binding properties for bilirubin, following the Langmuir isothermal adsorption model, indicating strong interactions between the hydrophobic C18 chains and bilirubin. This research presents a scalable and efficient strategy for synthesizing high-strength, porous CMS, making them a promising sorbent for hemoperfusion applications, addressing key challenges in blood detoxification therapies (102).

The study by Bikendra Maharjan et al. successfully addressed the mechanical limitations of natural hydrogel scaffolds in bone tissue engineering by incorporating regenerated cellulose (rCL) nanofibers into a chitosan (CS) hydrogel matrix. The rCL/CS composite scaffold exhibited enhanced compressive strength, surpassing that of pure CS scaffolds, while maintaining a unique porous morphology conducive to cellular activities. Biomineralization studies and in vitro cell culture assays demonstrated improved pre-osteoblast (MC3T3-E1) viability, attachment, and proliferation on the composite scaffold. Additionally, alkaline phosphatase and alizarin red (ARS) staining confirmed enhanced osteogenic differentiation, indicating the scaffold's potential to support bone regeneration. These findings suggest that the rCL/CS composite hydrogel scaffold is a promising three-dimensional bio-scaffold for bone tissue engineering, offering a novel fabrication approach with improved mechanical and biological properties for future regenerative applications (103).

Snyder et al. conducted a study demonstrating the effectiveness of cellulose nanofibers (CNF) derived from Eucalyptus pulp, doped with titanium dioxide (TiO_2), gold (Au), and silver (Ag) particles, in removing methylene blue, an organic contaminant, from polluted water. Their findings indicated that the photocatalytic performance of TiO_2-CNF improved when functionalized with Au and Ag particles under simulated sunlight (104). Additionally, research by Stephen et al. highlighted the efficiency of cellulose acetate nanofibers modified with oxolane-2,5-dione in adsorbing heavy metal ions such as lead and cadmium from contaminated water sources. These nanofibers demonstrated excellent reusability while maintaining high adsorption efficiency. Moreover, CNF-based aerogels have been successfully utilized for water purification, further emphasizing their potential in environmental remediation (105).

17.9 CONCLUSIONS

Cellulose, as one of the most abundant biopolymers, plays a crucial role in numerous industrial and biomedical applications. Its production and biodegradation are influenced by bacterial and fungal processes, contributing to sustainable material development and environmental balance. The transformation of cellulose-rich biomass into fibers is a key step in producing functional materials for pharmaceuticals, textiles, and advanced biomedical applications. The natural sources of cellulose explored in this chapter, including *Crotalaria juncea*, *Musa acuminata*, *Acacia concinna*, *Actinidia deliciosa*, *Azadirachta indica*, *Ceiba pentandra*, *Morus alba*, *Ziziphus nummularia*, and *Zea mays*, demonstrate the vast potential of plant-derived cellulose in diverse fields. The cellulose content varies across different plant parts, with bast fibers, stems, leaves, and fruit peels being the primary sources. These natural fibers have been widely used in textile manufacturing, drug-delivery systems, wound dressings, and tissue engineering scaffolds. Studies, including microspheres, nanofibers and hydrogels and biocompatibility, antimicrobial properties, and antioxidant potential, further validate their role in biomedical applications. The exploration of novel sources and the optimization of extraction processes will further enhance the efficiency and applicability of cellulose, paving the way for eco-friendly and high-performance biomaterials.

REFERENCES

1. Han G. Chronic wound healing : A review of current management and treatments. *Adv Ther.* 2017;34(3):599–610.
2. Srivastava GK, Martinez-Rodriguez S, Md Fadilah NI, Looi Qi Hao D, Markey G, Shukla P, Fauzi MB, Panetsos F. Progress in wound-healing products based on natural compounds, stem cells, and microRNA-based biopolymers in the European, USA, and Asian markets: Opportunities, barriers, and regulatory issues. *Polymers.* 2024 May 3;16(9):1280.
3. Abolore RS, Jaiswal S, Jaiswal AK. Green and sustainable pretreatment methods for cellulose extraction from lignocellulosic biomass and its applications : A review. *Carbohydr Polym Technol Appl* [Internet]. 2024;7(November 2023):100396. Available from: https://doi.org/10.1016/j.carpta.2023.100396
4. Magalhães S, Fernandes C, Pedrosa JF, Alves L, Medronho B, Ferreira PJ, Rasteiro MD. Eco-friendly methods for extraction and modification of cellulose: an overview. *Polymers.* 2023 Jul 24;15(14):3138.
5. Husnaini S, Hanisah N, Suhaili N, Hannan F, Mat A, Othaman R. Preparation of cellulose-based hydrogel : A review. *J Mater Res Technol* [Internet]. 2020;10:935–952. Available from: https://doi.org/10.1016/j.jmrt.2020.12.012
6. Seddiqi H, Oliaei E, Honarkar H, Jin J. Cellulose and its derivatives : Towards bio medical applications [Internet]. *Cellulose.* Springer Netherlands. 2021;28:1893–1931. Available from: https://doi.org/10.1007/s10570-020-03674-w
7. Yang C, Yang C, Chen Y, Liu J, Liu Z, Chen H jiuan. Journal of science : Advanced materials and devices the trends in wound management : Sensing, therapeutic treatment, and "theranostics." *J Sci Adv Mater Devices* [Internet]. 2023;8(4):100619. Available from: https://doi.org/10.1016/j.jsamd.2023.100619
8. de Amorim JDP, da Silva Junior CJG, de Medeiros ADM, do Nascimento HA, Sarubbo M, de Medeiros TPM, Costa AF, Sarubbo, LA. Bacterial cellulose as a versatile biomaterial for wound dressing application. *Molecules.* 2022;27(17):5580. https://doi.org/10.3390/molecules27175580
9. Horue, M., Silva, J.M., Berti, I.R., Brandão, L.R., Barud, H.D.S. and Castro, G.R., 2023. Bacterial cellulose-based materials as dressings for wound healing. *Pharmaceutics, 15*(2), p.424.
10. Ullah MW, Manan S, Kiprono SJ. Synthesis, structure, and properties of bacterial cellulose. *Nanocellulose.* 2019;6:81–113.

11. Samrot AV, Ngaakudzwe KT, Rajalakshmi D, Prakash P, Suresh Kumar S, Chandramohan M, Alex Anand D, Lilly Mercy J, Simon Y, Saigeetha S. Waste-derived cellulosic fibers and their applications. *Adv Mater Sci Eng*. 2022;2022(1):7314694.
12. Vaidya AA, Murton KD, Smith DA, Dedual G. A review on organosolv pretreatment of softwood with a focus on enzymatic hydrolysis of cellulose. *Biomass Convers Biorefin* [Internet]. 2022;5427–5442. Available from: https://doi.org/10.1007/s13399-022-02373-9
13. Frazier RM, Lendewig M, Vera RE, Vivas KA, Forfora N, Azuaje I, et al. Textiles from non-wood feedstocks : Challenges and opportunities of current and emerging fiber spinning technologies. *J Bioresour Bioprod*. 2024;9(July):410–432.
14. Dungani R, Karina M, Sulaeman A, Hermawan D, Hadiyane A. Asian journal of plant sciences review article agricultural waste fibers towards sustainability and advanced utilization : A review. *Asian J Plant Sci* [Internet]. 2016;15(1):42–55. Available from: https://doi.org/10.3923/ajps.2016.42.55
15. Mahur BK, Ahuja A, Singh S, Maji PK, Rastogi VK. Different nanocellulose morphologies (cellulose nanofibers, nanocrystals and nanospheres) extracted from Sunn hemp (Crotalaria Juncea). *Int J Biol Macromol*. 2023;253:126657.
16. Choudhary M, Singh D, Jain SK, Sonawane SRS, Singh D, Devnani GL, et al. Kinetics modeling & comparative examine on thermal degradation of alkali treated Crotalaria juncea fiber using model fitting method. *J Indian Chem Soc*. 2023;100(2):100918.
17. Robinson MB, Reynolds OL. Australian Sunn Hemp (Crotalaria juncea L.) Strategic RD & E Plan. 2022;(November 2021):1–30.
18. Permaculture Plants. https://permacultureplants.com/ [Internet]. Available from: https://permacultureplants.com/
19. Ghosh G, Saha S, Bhowmick S. Investigation of the effect of process parameters on fabrication of nanocrystalline cellulose from crotalaria juncea nanocrystalline cellulose from crotalaria juncea. *J Nat Fibers* [Internet]. 2023;20(1). Available from: https://doi.org/10.1080/15440478.2023.2181275
20. Kong AT, Olmedo-velarde A, Borth W, Wang K hui, Hu JS, Melzer MJ. Molecular and biological characterization of a novel tobamovirus infecting sunn hemp (Crotalaria juncea) in Hawaii. *Plant Dis*. 2023;107(10):3106–3112.
21. Yimlamai P, Ardsamang T, Puthson P, Somboon P. Soda pulping of sunn hemp (Crotalaria juncea L.) and its usage in molded pulp packaging. *J Bioresour Bioprod* [Internet]. 2023;8(3):280–291. Available from: https://doi.org/10.1016/j.jobab.2023.04.003
22. Lalhriatpuii JY, Chhetri D, Sohaila S, Rachana G. Evaluation of Haemopocitic activity of crotalaria juncea seed powder suspension in Phenylhydrazine induced haemolytic anaemia in rats. *Ind J Sci Technol*. 2023 Dec 30;16(47):4594–604.
23. Yadav A. Banana (Musa acuminata): Most popular and common Indian plant with multiple pharmacological potentials. *World J Biol Pharm Health Sci*. 2021;7(1):036–44.
24. Zaini NS, Hakeem WA, Kamal N. Preliminary Study of Cellulose Extraction from Banana (Musa Acuminata Colla cv. Mas) Peel. *J Pharm Negat Results*. 2022 Oct 8;13.
25. Sango T, Maxime A, Yona C, Duchatel L, Marin A, Kor M, et al. Industrial crops & products step – wise multi – scale deconstruction of banana pseudo – stem (Musa acuminata) biomass and morpho – mechanical characterization of extracted long fi bres for sustainable applications. *Ind Crop Prod* [Internet]. 2018;122(June):657–668. Available from: https://doi.org/10.1016/j.indcrop.2018.06.050
26. Basumatary B, Mahanta CL. Isolation of cellulose and synthesis of nanocellulose from banana (Musa acuminata) rachis fibre and their utilization and characterization as bioactive aerogels. *Biomass Convers Biorefinery* [Internet]. 2024;14(18):22297–22314. Available from: https://doi.org/10.1007/s13399-023-04260-3
27. Sujita S, Sari NH. The greatness of the characteristics fiber pseudo stem outer layer of Musa acuminata origin lombok Indonesia as reinforcing polyester composite. *Eastern-European J Enterp Technol*. 2022;4(12–118):38–43.

28. [PDF] The greatness of the characteristics fiber pseudo stem outer layer of Musa acuminata origin lombok Indonesia as reinforcing polyester composite | Semantic Scholar [Internet]. Available from: https://www.semanticscholar.org/paper/The-greatness-of-the-characteristics-fiber-pseudo-Sujita-Sari/d7a3e2f436fc0609ff9fff033177ade411d815f
29. Satpal D, Kaur J, Bhadariya V, Sharma K. Actinidia deliciosa (Kiwi fruit): A comprehensive review on the nutritional composition, health benefits, traditional utilization, and commercialization. *J Food Process Preser.* 2021 Jun;45(6):e15588.
30. Actinidia deliciosa | Características, cultivo, propiedades, beneficios | Kiwi [Internet]. Available from: https://www.flores.ninja/actinidia-deliciosa/
31. Engi - FP. The Utilization of Kiwi (Actinidia Deliciosa) Pruning Waste for Kraft Paper Production and the Effect of the Bark on Paper. 2015;58(194).
32. Luzi F, Fortunati E, Giovanale G, Mazzaglia A, Torre L, Mariano G. International Journal of Biological Macromolecules Cellulose nanocrystals from Actinidia deliciosa pruning residues combined with carvacrol in PVA CH films with antioxidant/antimicrobial properties for packaging applications. *Int J Biol Macromol* [Internet]. 2017;104:43–55. Available from: https://doi.org/10.1016/j.ijbiomac.2017.05.176
33. Wang K, Li M, Han Q, Fu R, Ni Y. Food bioscience inhibition of α -amylase activity by insoluble and soluble dietary fibers from kiwifruit (Actinidia deliciosa). *Food Bioscience.* 2021;42(17):101057.
34. Burgess, J. L., Wyant, W. A., Abdo Abujamra, B., Kirsner, R. S., & Jozic, I. (2021). Diabetic Wound-Healing Science. *Medicina, 57*(10), 1072. https://doi.org/10.3390/medicina57101072
35. Dhama A, Ahmad S, Kumar P, Paharia AK. Evaluation of antidiarrhoeal potentioal of Azadirachta indica stem bark extact in rats. *Int J Sci Res Publ.* 2020;10(3):9913.
36. Manimaran P, Senthamaraikannan P, Murugananthan K, Sanjay MR, Senthamaraikannan P, Murugananthan K, et al. Physicochemical properties of new cellulosic fibers from Azadirachta indica plant physicochemical properties of new cellulosic fibers from Azadirachta indica plant. *J Nat Fibers* [Internet]. 2017;00(00):1–10. Available from: https://doi.org/10.1080/15440478.2017.1302388
37. Rantheesh J, Indran S, Raja S, Siengchin S. Isolation and characterization of novel micro cellulose from Azadirachta indica A. Juss agro-industrial residual waste oil cake for futuristic applications. *Biomass Convers Biorefinery* [Internet]. 2023;13(5):4393–4411. Available from: https://doi.org/10.1007/s13399-022-03467-0
38. Tudoroiu EE, Dinu-Pîrvu CE, Albu Kaya MG, Popa L, Anuţa V, Prisada RM, Ghica MV. An overview of cellulose derivatives-based dressings for wound-healing management. *Pharmaceuticals.* 2021 Nov 24;14(12):1215
39. Chan EW, Yeong SW, Wong CW, Soo OY, Phua AC, Ng YK. Ceiba pentandra (L.) Gaertn.: An overview of its botany, uses, reproductive biology, pharmacological properties, and industrial potentials. *J Appl Biol Biotechnol.* 2023;11(1):1–7.
40. Bozaci E. Optimization of the alternative treatment methods for Ceiba pentandra (L.) Gaertn (kapok) fiber using response surface methodology. *J Text Inst* [Internet]. 2019; 110(10):1404–1414. Available from: https://doi.org/10.1080/00405000.2019.1602897
41. Kapok (Ceiba pentandra). Botanical characteristics and uses of the fiber - Sbenny's Blog [Internet]. Available from: https://blog.sbenny.com/gardening/trees/fruit-trees/kapok-ceiba-pentandra-botanical-characteristics-and-uses-of-the-fiber/
42. Baraniak J, Kania-dobrowolska M. Multi-purpose utilization of kapok Ffber and properties of ceiba pentandra tree in various branches of industry multi-purpose utilization of kapok fiber and properties of ceiba. *J Nat Fibers* [Internet]. 2023;20(1). Available from: https://doi.org/10.1080/15440478.2023.2192542
43. Sartika D, Syamsu K, Warsiki E, Fahma F, Arnata IW. Nanocrystalline cellulose from kapok fiber (Ceiba pentandra) and its reinforcement effect on alginate hydrogel bead. *Starch-Stärke.* 2021 Sep;73(9–10):2100033.

44. Rezekinta FA, Kasim A, Syafri E, Ridwan F, Chaniago I, Ichiura H. The effect of using cellulose nanofiber from Kapok (*Ceiba pentandra*, L) as reinforcement on the properties of recycled papers. *J Appl Agri Sci Technol.* 2024;8(4):426–436.
45. Walia YK, Gupta DK. Isolation, Extraction and Purification of Hemicelluloses of Ceiba pentandra and Isolation, Extraction and Purification of Hemicelluloses of Ceiba pentandra and Morus nigra. 2015;(January).
46. Peraza-ku SA, Escobar-morales B, Rodríguez-fuentes N, Cervantes-uc M, Uribe-calderon JA. Ceiba pentandra cellulose crosslinked with citric acid for drug release systems. 2021;504(April).
47. Gómez-Maqueo X, Gamboa-deBuen A. The biology of the genus Ceiba, a potential source for sustainable production of natural fiber. *Plants.* 2022 Feb 15;11(4):521.
48. Abdullah R, Astira D, Zulfiani U, Rahmad A, Rahmawati Z. Ultrafiltration membranes for dye wastewater treatment : Utilizing cellulose acetate and microcrystalline cellulose fillers from Ceiba Pentandra. *Commun Sci Technol.* 2024;9(1):7–15.
49. Khaima W, Wan A, Khalir M, Shameli K, Jazayeri SD. In-situ biofabrication of silver nanoparticles in Ceiba pentandra natural fiber using Entada spiralis extract with their antibacterial and catalytic dye reduction properties. *Nanomaterials.* 2020 Jun 3;10(6):1104.
50. Kadam RA, Dhumal ND. The Mulberry, Morus alba (L.): The medicinal herbal source for human health the mulberry, Morus alba (L.): The medicinal herbal source for human health department of home science, Shardabai Pawar Mahila Mahavidyalaya, Shardanagar Tal. A Sericulture Unit, Malegaon Sheti Farm, Agricultural Development Trust Baramati. 2019;(November).
51. Łochyńska M. Energy and nutritional properties of the white mulberry (Morus alba L.). *J Agri Sci Technol A.* 2015;5(9):709–716.
52. Prithivirajan R, Balasundar P, Shyamkumar R, Al-harbi NS, Ramkumar T, Narayanasamy P. Characterization of cellulosic fibers from Morus alba L. *stem. J Nat Fibers* [Internet]. 2018;00(00):1–9. Available from: https://doi.org/10.1080/15440478.2018.1426079
53. Kumar G, Madda R, Schuster J, Shaik YP. Development of morus alba reinforced Poly-lactic acid with elevated mechanical and thermal properties. *Mater Sci Appl.* 2024;15(7):186–200.
54. Morus alba L. | Plants of the World Online | Kew Science [Internet]. Available from: https://powo.science.kew.org/taxon/urn:lsid:ipni.org:names:30051955-2
55. Saikia B, Buragohain L, Barua, CC, Sarma, J, Tamuli, SM, Kalita, DJ, Barua, AG, Barua, IC, Elancheran, R. Evaluation of anti-amnesic effect of Conyza bonariensis in rats. *Ind J Pharmacol.* Mar–Apr 2022;54(2):102–109. DOI: 10.4103/ijp.ijp_201_19
56. Shehata N, Rdwan H. Potential protective effects of White mulberry (Morus Alba L.) leaves on rat liver injuries induced by carbon tetrachloride. *J Home Econom.* 2017; 27(1):1–24.
57. Akhtar HMS, Ahmed S, Olewnik-Kruszkowska E, Gierszewska M, Brzezinska MS, Dembińska K, et al. CMC use Carboxymethyl cellulose based films enriched with polysaccharides from mulberry leaves (Morus alba L.) as new biodegradable packaging material. *Int J Biol Macromol.* 2023;253:127633.
58. Uddin N, Muhammad N, Ali SS, Ullah R, Bari A, Hussain H, et al. Characterization of the genetic variability within Ziziphus nummularia genotypes by phenotypic traits and SSR markers with special reference to geographic distribution. *Genes.* 2023 Jan 6;14(1):155.
59. Ziziphus nummularia – eFlora of India [Internet]. Available from: https://efloraofindia.com/efi/ziziphus-nummularia/
60. Vijaya GS, Gounder RI, Natarajan SS. Bark use-characterization of natural cellulose fibers from the barks of Ziziphus nummularia as a reinforcement for lightweight composite applications. *J Nat Fibers.* 2022;19(17):15663–15679.

61. SV G, I R, N SS. Suitability assessment of raw-alkalized Ziziphus nummularia bark fibers and its polymeric composites for lightweight applications. *Polym Compos* [Internet]. 2022;43(8):5059–5075. Available from: https://4spepublications.onlinelibrary.wiley.com/doi/abs/10.1002/pc.26782
62. Bukhari A, Ijaz I, Gilani E, Nazir A, Zain H, Saeed R. Green synthesis of metal and metal oxide nanoparticles using different plants' parts for antimicrobial activity and anticancer activity : A review article. *Coatings*. 2021 Nov 9;11(11):1374.
63. Arodiya F, Makvana C, Parmar K. Polymer capped silver nanoparticles from Ziziphus nummularia leaves extract : Potent antibacterial and antioxidant activity. *Biosci Biotechnol Res Asia*. 2021;18(December):691–701.
64. Shah TR, Prasad K, Kumar P. Maize — A potential source of human nutrition and health : A review Maize — A potential source of human nutrition and health : A review. *Cogent Food Agric* [Internet]. 2016;54(1). Available from: https://doi.org/10.1080/23311932.2016.1166995
65. Adiaha, MS. The impact of Maize (Zea mays L.) and it uses for human development: A review. *Int J Sci World*. 2017; 5(1):93–95. https://doi.org/10.14419/ijsw.v5i1.7585
66. Zea mays (Corn, Indian corn, Maize) | North Carolina Extension Gardener Plant Toolbox [Internet]. Available from: https://plants.ces.ncsu.edu/plants/zea-mays/
67. Maepa CE, Jayaramudu J, Okonkwo JO, Ray SS, Sadiku ER. International journal of polymer analysis and extraction and characterization of natural cellulose fibers from maize tassel extraction and characterization of natural cellulose fibers from maize tassel. *GPAC* [Internet]. 2015;20(2):99–109. Available from: https://doi.org/10.1080/1023666X.2014.961118
68. Lau, T., Harbourne, N. and Oruna-Concha, M.J., 2019. Valorisation of sweet corn (Zea mays) cob by extraction of valuable compounds. *International Journal of Food Science and Technology, 54*(4), pp. 1240–1246.
69. Rose DJ, Inglett GE, Liu SX. Utilisation of corn (Zea mays) bran and corn fiber in the production of food components. *J Sci Food Agri*. 2010 Apr 30;90(6):915–924.
70. Smyth M, García A, Rader C, Foster EJ, Bras J. Industrial crops & products extraction and process analysis of high aspect ratio cellulose nanocrystals from corn (Zea mays) agricultural residue. *Ind Crop Prod* [Internet]. 2017;108(June):257–266. Available from: https://doi.org/10.1016/j.indcrop.2017.06.006
71. Grishkewich N, Mohammed N, Tang J, Tam KC. Current opinion in colloid & interface science recent advances in the application of cellulose nanocrystals. *Curr Opin Colloid Interface Sci* [Internet]. 2017;29:32–45. Available from: https://doi.org/10.1016/j.cocis.2017.01.005
72. Roh K baeg, Kim H, Shin S, Kim Y soo, Lee J a, Kim MO, et al. Anti-inflammatory effects of Zea mays L. husk extracts. *BMC Complement Altern Med* [Internet]. 2016; 1–8. Available from: https://doi.org/10.1186/s12906-016-1284-9
73. Torres-vargas O, Pa MC, Gonz ML. Industrial Crops & Products Corn starch based biocomposite films reinforced with cellulosic nanocrystals extracted from corn husks (Zea mays L.): Characterization and application in cherry tomato packaging. 2025;225(January).
74. Excipient P, Rahman S, Mondal IH, Yeasmin MS, Sayeed MA. Conversion of Lignocellulosic Corn Agro-Waste into Cellulose Derivative and Its Potential Application as. 6.
75. Mele E. Electrospinning of natural polymers for advanced wound care: Towards responsive and adaptive dressings. *J Mater Chem B* [Internet]. 2016;4(28):4801–12. Available from: https://doi.org/10.1039/C6TB00804F
76. Tottoli EM, Dorati R, Genta I, Chiesa E, Pisani S, Conti B. Skin wound healing process and new emerging technologies for skin wound care and regeneration. *Pharmaceutics* [Internet]. 2020;12(8). Available from: https://www.mdpi.com/1999-4923/12/8/735

77. Raziyeva K, Kim Y, Zharkinbekov Z, Kassymbek K, Jimi S, Saparov A. Immunology of acute and chronic wound healing. *Biomolecules* [Internet]. 2021;11(5). Available from: https://www.mdpi.com/2218-273X/11/5/700
78. Okur ME, Karantas ID, Şenyiğit Z, Okur NÜ, Siafaka PI. Recent trends on wound management: New therapeutic choices based on polymeric carriers. *Asian J Pharmaceut Sci*. 2020;15(6):661–684. https://doi.org/10.1016/j.ajps.2019.11.008.
79. Rodrigues M, Kosaric N, Bonham CA, Gurtner GC, Rodrigues M, Kosaric N, et al. Wound healing : A cellular perspective. *Physiol Rev*. 2025; 99(1):665–706. Doi: 10.1152/physrev.00067.2017
80. Wilkinson HN, Hardman MJ, Wilkinson HN. Wound healing : Cellular mechanisms and pathological outcomes. *Open Biol*. 2020;10(9):200223. https://doi.org/10.1098/rsob.200223
81. Almadani, Y.H., Vorstenbosch, J., Davison, P.G. and Murphy, A.M., 2021, August. Wound healing: a comprehensive review. In *Seminars in plastic surgery* (Vol. 35, No. 03, pp. 141–144). Thieme Medical Publishers, Inc.
82. Gonzalez AC, Costa TF, Andrade ZD, Medrado AR. Wound healing-A literature review. *Anais brasileiros de dermatologia*. 2016;91:614–620.
83. Jonidi Shariatzadeh F, Currie S, Logsetty S, Spiwak R, Liu S. Enhancing wound healing and minimizing scarring: A comprehensive review of nanofiber technology in wound dressings. *Prog Mater Sci* [Internet]. 2025;147:101350. Available from: https://www.sciencedirect.com/science/article/pii/S0079642524001191
84. Di Muzio L, Simonetti P, Carriero VC, Brandelli C, Trilli J, Sergi C, et al. Solvent casting and UV photocuring for easy and safe fabrication of nanocomposite film dressings. *Molecules* [Internet]. 2022;27(9). Available from: https://www.mdpi.com/1420-3049/27/9/2959
85. Ng SF. Freeze-dried wafers for wound healing. *Ther Dressings Wound Heal Appl* [Internet]. 2020;137–155. Available from: https://onlinelibrary.wiley.com/doi/full/10.1002/9781119433316.ch7
86. Hargis A, Yaghi M, Bermudez NM, Gefen A. Foam dressings for wound healing. *Curr Dermatol Rep* [Internet]. 2024;13(1):28–35. Available from: https://doi.org/10.1007/s13671-024-00422-2
87. Qu J, Zhao X, Liang Y, Zhang T, Ma PX, Guo B. Antibacterial adhesive injectable hydrogels with rapid self-healing, extensibility and compressibility as wound dressing for joints skin wound healing. *Biomaterials* [Internet]. 2018;183:185–199. Available from: https://www.sciencedirect.com/science/article/pii/S0142961218306045
88. Uhljar LÉ, Ambrus R. Electrospinning of potential medical devices (wound dressings, tissue engineering scaffolds, face masks) and their regulatory approach. *Pharmaceutics* [Internet]. 2023;15(2). Available from: https://www.mdpi.com/1999-4923/15/2/417
89. Kravanja KA, Finšgar M, Knez Ž, Knez Marevci M. Supercritical fluid technologies for the incorporation of synthetic and natural active compounds into materials for drug formulation and delivery. *Pharmaceutics* [Internet]. 2022;14(8). Available from: https://www.mdpi.com/1999-4923/14/8/1670
90. Varnaitė-Žuravliova S, Baltušnikaitė-Guzaitienė J. Properties, production, and recycling of regenerated cellulose fibers: Special medical applications. *J Funct Biomater* [Internet]. 2024;15(11). Available from: https://www.mdpi.com/2079-4983/15/11/348
91. Pita-Vilar M, Concheiro A, Alvarez-Lorenzo C, Diaz-Gomez L. Recent advances in 3D printed cellulose-based wound dressings: A review on in vitro and in vivo achievements. *Carbohydr Polym* [Internet]. 2023;321:121298. Available from: https://www.sciencedirect.com/science/article/pii/S0144861723007634
92. Zheng L, Li S, Luo J, Wang X. Latest advances on bacterial cellulose-based antibacterial materials as wound dressings. *Front Bioeng Biotechnol* [Internet]. 2020;8. Available from: https://www.frontiersin.org/journals/bioengineering-and-biotechnology/articles/10.3389/fbioe.2020.593768

93. Xiao J, Li P, Zhang X, Wang X. Study on preparation of regenerated cellulose fiber from biomass based on mixed solvents. *Materials (Basel)* [Internet]. 2024;17(4). Available from: https://www.mdpi.com/1996-1944/17/4/819
94. Ho TTP, Tran HA, Doan VK, Maitz J, Li Z, Wise SG, et al. Natural polymer-based materials for wound healing applications. *Adv NanoBiomed Res* [Internet]. 2024;4(5):2300131. Available from: https://onlinelibrary.wiley.com/doi/full/10.1002/anbr.202300131
95. Pang Q, Jiang Z, Wu K, Hou R, Zhu Y. Nanomaterials-based wound dressing for advanced management of infected wound. *Antibiotics* [Internet]. 2023;12(2). Available from: https://www.mdpi.com/2079-6382/12/2/351
96. Atiwesh G, Mikhael A, Parrish CC, Banoub J, Le TAT. Environmental impact of bioplastic use: A review. *Heliyon* [Internet]. 2021 Sep 1;7(9). Available from: https://doi.org/10.1016/j.heliyon.2021.e07918
97. Zhang L, Wang S, Tan M, Zhou H, Tang Y, Zou Y. [Retracted] Efficacy of oxidized regenerated cellulose/collagen dressing for management of skin wounds: A systematic review and meta-analysis. *Evidence-Based Complement Altern Med* [Internet]. 2021;2021(1):1058671. Available from: https://onlinelibrary.wiley.com/doi/full/10.1155/2021/1058671
98. Portela R, Leal CR, Almeida PL, Sobral RG. Minireview bacterial cellulose : A versatile biopolymer for wound dressing applications. *Microbial Biotechnol.* 2019 Jul;12(4):586–610.
99. He W, Wu J, Xu J, Mosselhy DA, Zheng Y, Yang S. Bacterial cellulose: Functional modification and wound healing applications. *Adv Wound Care.* 2021 Nov;10(11):623–640.
100. He W, Wu J, Xu J, Mosselhy DA, Zheng Y, Yang S. Bacterial cellulose: Functional modification and wound healing applications. *Adv Wound Care* [Internet]. 2021;10(11):623–640. Available from: https://doi.org/10.1089/wound.2020.1219
101. Scheer A, Fischer J, Bakhshi A, Bauer W, Fischer S, Spirk S. Fusion of cellulose microspheres with pulp fibers : Creating an unconventional type of paper. *Carbohydr Polym* [Internet]. 2024;338(April):122207. Available from: https://doi.org/10.1016/j.carbpol.2024.122207
102. Wei Tang Ziting Yuan BSKLYL, Zhong J. Facile and scalable fabrication of regenerated cellulose microspheres with high strength and porosity as a potential matrix for hemoperfusion. *J Macromol Sci Part B* [Internet]. 2024;63(7):588–603. Available from: https://doi.org/10.1080/00222348.2024.2314413
103. Maharjan B, Park J, Kaliannagounder VK, Awasthi GP, Joshi MK, Park CH, et al. Regenerated cellulose nanofiber reinforced chitosan hydrogel scaffolds for bone tissue engineering. *Carbohydr Polym* [Internet]. 2021;251:117023. Available from: https://www.sciencedirect.com/science/article/pii/S0144861720311966
104. Snyder A, Bo Z, Moon R, Rochet JC, Stanciu L. Reusable photocatalytic titanium dioxide–cellulose nanofiber films. *J Colloid Interface Sci* [Internet]. 2013;399:92–98. Available from: https://www.sciencedirect.com/science/article/pii/S0021979713001963
105. Stephen M, Catherine N, Brenda M, Andrew K, Leslie P, Corrine G. Oxolane-2,5-dione modified electrospun cellulose nanofibers for heavy metals adsorption. *J Hazard Mater* [Internet]. 2011;192(2):922–927. Available from: https://www.sciencedirect.com/science/article/pii/S030438941100762X

Index

D

E

F

G

H

I

J

K

L

M